Lecture Notes in Mathematics

Edited by A. Dold and B. Eckmann

T0184798

1064

Probability Measures on Groups VII

Proceedings of a Conference held in Oberwolfach, 24–30 April 1983

Edited by H. Heyer

Springer-Verlag
Berlin Heidelberg New York Tokyo 1984

Editor

Herbert Heyer
Universität Tübingen, Mathematisches Institut
Auf der Morgenstelle 10
7400 Tübingen, Federal Republic of Germany

AMS Subject Classification (1980): 60B15, 60J15, 60H25, 43A05, 60A10, 60B11; 43A05, 43A10, 43A33, 46L50, 47D05

ISBN 3-540-13341-0 Springer-Verlag Berlin Heidelberg New York Tokyo
ISBN 0-387-13341-0 Springer-Verlag New York Heidelberg Berlin Tokyo

Library of Congress Cataloging in Publication Data. Main entry under title: Probability measures on groups, VII. (Lecture notes in mathematics; 1064) 1. Probabilities–Congresses. 2. Groups, Theory of–Congresses. 3. Stochastic processes–Congresses. 4. Measure theory–Congresses. I. Heyer, Herbert. II. Title: Probability measures on groups, 7. III. Title: Probability measures on groups, seven. IV. Series: Lecture notes in mathematics (Springer-Verlag); 1064.
QA3.L28 no. 1064 [QA273.A1] 510 s [519.2] 84-5636
ISBN 0-387-13341-0

Printing and binding: Beltz Offsetdruck, Hemsbach/Bergstr.
2146/3140-543210

PREFACE

This volume reflects the activity of a group of mathematicians who assembled at the Forschungsinstitut in Oberwolfach between April 24th and April 30th 1983 in order to take part in the 7th Conference on "Probability Measures on Groups". 40 participants from 10 different countries contributed to the success of the meeting by presenting their results and by surveying new developments of the theory. The editor is particularly grateful to those colleagues who accepted the invitation to prepare talks of expository nature and to add the reworked manuscripts to the present volume. The 31 research and 4 survey papers included in this volume indicate the wide range of problems discussed within the theory of probability on algebraic-topological-geometrical structures such as groups, semigroups, hypergroups, algebras, Sturm-Liouville systems, Gelfand pairs, general homogeneous spaces and manifolds. The contributions to these Proceedings can be roughly classified under the following topics (i) Probability measures on locally compact groups (Decomposition, infinite divisibility, convolution semigroups, holomorphy, stability) (ii) Random walks on groups and homogeneous spaces (Dichotomy theorems, recurrence, polynomial growth, cogrowth) (iii) Markov processes on hypergroups (Transience, Lévy-Khinchin formulae, central limit theorems) (iv) Noncommutative probability theory (Subadditive ergodic theorems, Gaussian functionals) (v) Random matrices and operators (Laws of large numbers, random Schrödinger operators, characteristic exponents). A little difficult to classify are the papers deepening the existing theory in a very specific direction (f.e. the computation of supports of absolutely continuous Gauss measures on $SL(2,\mathbb{R})$) or those papers extending well known theorems beyond the scope of probability theory (f.e. the study of complex Lévy measures), not to speak of the more fundamental contributions to the harmonic analysis of nonabelian groups or of the remarkable implications to theoretical physics.

In the discussions during the conference numerous problems were posed most of which remained open. The authors of the survey papers tried to make some of them precise, hoping that progress can be achieved prior to the 8th Conference on "Probability Measures on Groups" planned for late 1985.

Tübingen, Januar 1984

C O N T E N T S

Survey articles

The authors Y. Guivarc'h and R. Jajte kindly provided their manuscripts for publication, although they were unable to participate at the conference.

PAPERS GIVEN AT THE CONFERENCE

BUT NOT PUBLISHED IN THIS VOLUME

E. Dettweiler:	Diffusions on Banach spaces
S. Pincus:	Strong law of large numbers for products of random matrices
K. Urbanik:	Generalized convolutions

LIST OF PARTICIPANTS

M. S. Bingham	Hull, England
W. R. Bloom	Perth, Western Australia
Ph. Bougerol	Paris, France
H. Byczkowska	Wrocław, Poland
T. Byczkowski	Wrocław, Poland
H. Carnal	Bern, Schweiz
Y. Derriennic	Brest, France
E. Dettweiler	Tübingen, West-Germany
T. Drisch	Dortmund, West-Germany
L. Elie	Paris, France
B.-J. Falkowski	Neubiberg, West-Germany
Ph. J. Feinsilver	Carbondale, Illinois, USA
G. Forst	København, Denmark
L. Gallardo	Nancy, France
M.-O. Gebuhrer	Strasbourg, France
P. Gerl	Salzburg, Austria
W. Hazod	Dortmund, West-Germany
H. Heyer	Tübingen, West-Germany
A. Janssen	Dortmund, West-Germany
E. Kaniuth	Paderborn, West-Germany
F. Kinzl	Salzburg, Austria
J. Kisyński	Warsawa, Poland
J. Lacroix	Rennes, France
R. Lasser	München, West-Germany
E. Le Page	Rennes, France
G. Letac	Toulouse, France
M. McCrudden	Manchester, England
A. Mukherjea	Tampa, Florida, USA
S. Pincus	Holmdel, New Jersey, USA
A. Raugi	Toulouse, France
H. Rindler	Wien, Austria
I. Z. Ruzsa	Budapest, Hungary
R. Schott	Nancy, France

E. Siebert	Tübingen, West-Germany
G. J. Székely	Budapest, Hungary
K. Urbanik	Wrocław, Poland
W. von Waldenfels	Heidelberg, West-Germany
M. E. Walter	Boulder, Colorado, USA
W. Woess	Salzburg, Austria
Hm. Zeuner	Tübingen, West-Germany

TRANSLATION BOUNDED MEASURES AND THE ORLICZ-PALEY-SIDON THEOREM

Walter R. Bloom
Murdoch University
Perth, Western Australia, 6150
AUSTRALIA

Let G denote a locally compact abelian group with character group Γ. Haar measures λ, θ will be chosen on G, Γ respectively, normalised so that Plancherel's theorem holds. We shall write $M(G)$ for the space of complex Radon measures on G, that is, $M(G)$ is the dual of $C_c(G)$, the space of continuous functions with compact support endowed with its usual inductive limit topology.

A measure $\mu \in M(G)$ is said to be translation (or shift) bounded if $\{|\mu|(x+K) : x \in G\}$ is bounded for each compact $K \subset G$. Translation bounded measures arise naturally in the study of convolution. Measures $\mu, \nu \in M(G)$ are called convolvable if, for every $f \in C_c(G)$, the function $(x,y) \to f(x+y)$ is integrable over $G \times G$ with respect to the product measure $|\mu| \times |\nu|$. It is known (Argabright and de Lamadrid [1], Theorem 1.2) that μ is translation bounded if and only if μ is convolvable with every bounded Radon measure on G.

Translation bounded measures also arise in the study of potential theory. Suppose the family $(\mu_t)_{t>0}$ of positive measures on G with $\mu_t(G) \leq 1$ for all $t > 0$ forms a transient convolution semigroup in the sense of Berg and Forst [2], Definition 13.2, so that the potential kernel

$$\kappa = \int_0^\infty \mu_t \, dt$$

exists as a measure in $M(G)$. In general κ is unbounded. However we always have that κ is translation bounded (Berg and Forst [2], Proposition 13.10).

The space $M_T(G)$ of translation bounded measures first seems to have been studied in detail by Lin [14], in the case where G is the Euclidean space R^n, although it did appear earlier in Choquet and Deny [7] in connection with the solution of convolution equations. Extensions of Lin's results are to be found in Thornett [17], and Robertson and Thornett [15]. In another direction $M_T(G)$ has been studied as the dual of a certain amalgam function space, and the origins of this can be traced back to the work of Wiener [18].

In this paper we present some results on translation bounded measures and use these to prove an extension of the Orlicz-Paley-Sidon theorem.

We take Hewitt and Ross [12] as our standard reference for harmonic analysis on G ; any unexplained notation will be found there. Note that ξ_A denotes the characteristic function of the set A .

1. Amalgams and translation bounded measures

Every locally compact abelian group G can, by the structure theorem (Hewitt and Ross [12], (24.30)), be written as $R^a \times G_0$, where a is a nonnegative integer and G_0 contains a compact open subgroup H . We write $J = Z^a \times G_0/H$, $K = [0,1)^a \times H$ and G as the disjoint union $\bigcup_\alpha K_\alpha$ where, for each $\alpha = (n_1,n_2,\ldots,n_a,y+H) \in J$, $K_\alpha = \alpha' + K$ and $\alpha' = (n_1,n_2,\ldots,n_a,y)$. For each s, $t \in [1,\infty]$ and $f \in L^s_{loc}(G)$ write

$$\|f\|_{s,t} = \left\{\sum_\alpha \left(\int_{K_\alpha} |f|^s d\lambda\right)^{t/s}\right\}^{1/t} ,$$

with the usual modification if $\max\{s,t\} = \infty$, and

$$(L^s,\ell^t) = \{f \in L^s_{loc}(G) : \|f\|_{s,t} < \infty\} .$$

We write (C,ℓ^t) for the subspace of continuous functions in (L^∞,ℓ^t) , and (L^s,c_0) for the subspace of (L^s,ℓ^∞) consisting of those functions f for which

$$\alpha \to \int_{K_\alpha} |f|^s d\lambda$$

tends to zero at infinity. The subspaces (M,ℓ^t) and (M,c_0) of $M(G)$ are defined in a similar way; we write these as $M_t(G)$ and $M_0(G)$ respectively. Note the inclusion $M_s(G) \subset M_t(G)$ for $1 \le s \le t \le \infty$.

Amalgams have been studied in some detail for the real line by Holland [13], and by Stewart [16], Bertrandias, Datry and Dupuis [3], Busby and Smith [6] and Feichtinger [10] for locally compact abelian groups. These authors define amalgams in a variety of ways. We have followed Stewart's definition; see also W.R. Bloom [5].

It is clear that each of the above spaces, with two functions being identified if they agree locally almost everywhere, is a Banach space with the norm $\|\cdot\|_{s,t}$. It is known (Bertrandias, Datry and Dupuis [3], Property (g)(β), p.78) that the dual of (C,ℓ^t) is isomorphic with $M_{t'}$, for each $t \in [1,\infty)$, where t,t' are conjugate indices $(t^{-1}+t'^{-1}=1)$. It is also easy to see that M_∞ consists precisely of the translation bounded measures defined in the introduction. For convenience we present this as a separate result.

Proposition 1 $M_\infty(G) = M_T(G)$.

Proof The inclusion $M_T(G) \subset M_\infty(G)$ is clear since each K_α is compact. In the other direction consider $\mu \in M_\infty(G)$ and compact $K' \subset G$. Then there exists an integer m such that for each $x \in G$ there exists $\alpha_1, \alpha_2, \ldots, \alpha_m \in J$ with

$$x + K' \subset \cup\{K_{\alpha_n} : n = 1, \ldots, m\} .$$

(Note that m can be chosen to be independent of x .) Then

$$|\mu|(x+K') \le \sum_{n=1}^{m} |\mu|(K_{\alpha_n}) \le m\|\mu\|_\infty ,$$

so that $\mu \in M_T(G)$. //

Proposition 2 The following conditions on $\mu \in M(G)$ are equivalent:

(a) $\mu \in M_T(G)$;

(b) $\{|\mu|(\tau_x f) : x \in G\}$ is bounded for each $f \in C_c(G)$;

(c) $\{\mu(\tau_x f) : x \in G\}$ is bounded for each $f \in C_c(G)$;

(d) $\mu * f$ is continuous and bounded for each $f \in C_c(G)$.

In the statement of Proposition 2, $\tau_x f(y) = f(-x+y)$. The equivalence of (a), (b) and (c) is given in Berg and Forst [2], Proposition 1.12 and Exercise 1.17, and that of (a) and (d) in Argabright and de Lamadrid [1], Theorem 1.1, p.5.

In [14], Lin proved that translation boundedness of μ ensured a μ-integrability condition for a certain class of spectral functions. (For a statement of this and related results see Thornett [17], Section 4.) Lin's proof, which was given for R^a only, was based on the property that functions on the real line with compactly supported Fourier transforms have extensions to functions entire in the complex plane. In Theorem 1 below we present an extension of Lin's result for all locally compact abelian groups, our proof requiring only basic properties of amalgams. First we require a preliminary lemma.

We write $L^p_X(G)$ for the space of functions in $L^p(G)$ with compactly supported Fourier transforms (for $p > 2$ the Fourier transform is defined in a distributional sense; see W.R. Bloom [5], p.266, for example).

Lemma Let $p \in [1,\infty]$. Then

(a) $L^p_\chi(G) \subset (C,\ell^p)$;

(b) For every $g \in (C,\ell^p)$ there exists $f \in L^p_\chi(G)$ with $|f| \geq |g|$.

Proof (a). Consider $f \in L^p_\chi(G)$ and write $\Lambda = \mathrm{supp}(\hat{f})$. By Stewart [16], Theorem 3.1, there exists $k \in (L^\infty,\ell^1)$ with $\hat{k} = 1$ on Λ . Then

$$f = k*f \in (L^\infty,\ell^1)*L^p \subset L^{p'}*L^p \cap (L^\infty,\ell^1)*(L^1,\ell^p) \subset C(G) \cap (L^\infty,\ell^p) = (C,\ell^p) .$$

(b). First assume $p < \infty$. For each $\alpha \in J$ write $\psi(\alpha) = \sup\{|g(x)| : x \in K_\alpha\}$. Pick $k \in L^1_\chi(G)$ with $k > \xi_K$ (just use Berg and Forst [2], Proposition 2.4; the Fourier transform of the function constructed there is easily seen to belong to $L^1(G)$) and write $\phi(\alpha) = \sup\{|k(x)| : x \in K_\alpha\}$. Define $f = \sum_\alpha \psi(\alpha)\tau_{\alpha'}k$. We first show that $f \in L^p(G)$. Choose $\beta \in J$ and $x \in K_\beta$. Then

$$f(x) = \sum_\alpha \psi(\alpha)\tau_{\alpha'}k(x) = \sum_\alpha \psi(\alpha)k(x-\alpha') \leq \sum_\alpha \psi(\alpha)\phi(\beta-\alpha)$$

and

$$\begin{aligned}\int_G |f(x)|^p d\lambda &= \sum_\beta \int_{K_\beta} |f(x)|^p d\lambda \\ &\leq \sum_\beta \int_{K_\beta} \Big|\sum_\alpha \psi(\alpha)\phi(\beta-\alpha)\Big|^p d\lambda \\ &\leq \sum_\beta \lambda(K_\beta)\Big|\sum_\alpha \psi(\alpha)\phi(\beta-\alpha)\Big|^p \\ &= \lambda(K)\sum_\beta |\psi*\phi(\beta)|^p .\end{aligned}$$

But $\psi \in \ell^p$ and $\phi \in \ell^1$ so that $\psi*\phi \in \ell^p$. Hence the latter sum is finite and $f \in L^p(G)$.

Now $\mathrm{supp}(\tau_{\alpha'}k)^\wedge = \mathrm{supp}(\hat{k})$ for each α , from which it follows that $\mathrm{supp}(\hat{f}) \subset \mathrm{supp}(\hat{k})$, a compact set; hence $f \in L^p_\chi(G)$. Finally note that for $x \in K_\beta$,

$$f(x) \geq \psi(\beta)\tau_{\beta'}k(x) \geq \psi(\beta)\xi_{K_\beta}(x) = \psi(\beta) \geq |g|(x) ,$$

and this completes the proof for $p < \infty$.

For $p = \infty$ just choose f to be the constant function that takes the value $\|g\|_\infty$. //

Theorem 1 Let $1 \le q \le p \le \infty$, $q \ne \infty$, and write $r' = p/q$. The following conditions on $\mu \in M(G)$ are equivalent:

(a) $\mu \in M_r(G)$

(b) $L^p_\chi(G) \subset L^q(\mu)$

(c) For each compact set $\Lambda \subset \Gamma$ there exists a constant $c = c(p,q,\Lambda)$ such that

$$\|f\|_{L^q(\mu)} \le c\|f\|_p$$

for every $f \in L^p_\chi(G)$ with $\operatorname{supp}(\hat{f}) \subset \Lambda$.

Proof (a) => (b) . Let $f \in L^p_\chi(G)$. From Part (a) of the lemma we see that $f \in (C,\ell^p)$ and hence $f^q \in (C,\ell^{r'})$. It follows from the duality between $(C,\ell^{r'})$ and $M_r(G)$ that $f \in L^q(\mu)$ (in the case $r' = \infty$ we replace (C,ℓ^∞) by C_0).

(b) => (c) . This implication can be shown by a standard argument using the closed graph theorem; the space

$$L^p_\Lambda(G) = \{f \in L^p(G) : \operatorname{supp}(\hat{f}) \subset \Lambda\}$$

is a Banach space with the usual L^p-norm.

(c) => (a) . Choose $g \in (C,\ell^{r'})$. Then $g^{1/q} \in (C,\ell^p)$ and, by Part (b) of the lemma, there exists $f \in L^p_\chi(G)$ with $|f| \ge |g|^{1/q}$. Then

$$\left|\int_G g\, d\mu\right| \le \int_G |g|\, d|\mu| \le \int_G |f|^q d|\mu| \le c\|f\|_p < \infty ,$$

and this shows that $\mu \in (C,\ell^{r'})' = M_r$ (where, if $r' = \infty$, replace (C,ℓ^∞) by C_0). //

2. Fourier multipliers

It is well known that if a sequence $(z(n))$ of complex numbers satisfies

$$\sum_{n=-\infty}^{\infty} |z(n)w(n)| < \infty$$

for every sequence $(w(n))$ belonging to $\ell^2(Z)$ then $(z(n)) \in \ell^2(Z)$; this is a special case of the converse of Hölder's inequality. The so-called Orlicz-Paley-Sidon theorem shows that the same conclusion may be drawn if the sequences $(w(n))$ are restricted to be Fourier transforms of functions in $C(T)$, where T denotes the circle group. Extensions of this result to locally compact abelian groups have been given by Edwards, Hewitt and Ritter [9] and Fournier [11]. For an extension where the $(w(n))$ are further restricted to lie in $\ell^p(Z)$, $p \in [1,2)$, see L.M. Bloom [4].

Such results are part of the multiplier problem; the result of Edwards, Hewitt and Ritter can be stated as saying that the space of Fourier multipliers of $C_c(\Gamma)$ is given by (L^1,ℓ^2), that is, if

$$\int_G |\hat{f}(x)w(x)|\,dx < \infty \qquad (1)$$

for every $f \in C_c(\Gamma)$ then $w \in (L^1,\ell^2)$.

It is immediately apparent that if w satisfies (1) for all $f \in C_c(\Gamma)$ then w is locally integrable. Hence it is natural to extend the problem and ask for which Radon measures μ is it the case that

$$\int_G |\hat{f}(x)|\,d|\mu|(x) < \infty$$

for every $f \in C_c(\Gamma)$?

Robertson and Thornett [15] have characterised the nonnegative translation bounded measures on R^a as those $\mu \in M(R^a)$ for which $\hat{\xi}_\Lambda \in L^2(\mu)$ for each Borel subset Λ of a given (fixed) compact non-degenerate interval I in R^a. Their proof, which relies on the Vitali-Hahn-Saks theorem, actually carries over to locally compact abelian groups, and for all $p \in [1,\infty]$; see Theorem 3 below. We first present a general theorem based on the Nikodým boundedness theorem and use this, together with a result of Fournier, to give a similar characterisation of $M_2(G)$ (Theorem 4).

Theorem 2 Let $1 \le p \le \infty$ and choose any compact set $\Omega \subset \Gamma$ with nonvoid interior. Suppose that $\hat{\xi}_\Lambda \in L^p(\mu)$ for every Borel set $\Lambda \subset \Omega$. Then there exists a constant $c = c(\Omega)$ such that

$$\|\hat{f}\|_{L^p(\mu)} \le c\|f\|_\infty$$

for every $f \in L^\infty(\Omega)$.

Proof Consider the linear operator $T_{K,g} : L^\infty(\Omega) \to L^1(\mu)$ defined by

$$T_{K,g}(f) = \hat{f}\, g\, \xi_K ,$$

where $K \subset G$ is compact and $g \in L^{p'}(\mu)$ with $\|g\|_{L^{p'}(\mu)} \le 1$. The inequalities

$$\int_G |\hat{f}g\xi_K|\, d|\mu| \le \|g\|_{L^{p'}(\mu)} |\mu|(K)^{1/p} \|\hat{f}\|_\infty$$
$$\le \|g\|_{L^{p'}(\mu)} |\mu|(K)^{1/p}\, \theta(\Omega)\|f\|_\infty$$

show that each $T_{K,g}$ is bounded. Also for each Borel set $\Lambda \subset \Omega$,

$$\int_G |\hat{\xi}_\Lambda g \xi_K|\, d|\mu| \le \|\hat{\xi}_\Lambda\|_{L^p(\mu)} \|g\, \xi_K\|_{L^{p'}(\mu)} \le \|\hat{\xi}_\Lambda\|_{L^p(\mu)} .$$

Now apply Diestel and Uhl [8], Corollary 2, p.16 to obtain that

$$\|T_{g,K}(f)\|_{L^1(\mu)} \le c\|f\|_\infty$$

for all $f \in L^\infty(\Omega)$ and for all K,g as above. Making use of Hewitt and Ross [12], (12.13),

$$\|\hat{f}\, \xi_K\|_{L^p(\mu)} \le c\|f\|_\infty$$

for every compact $K \subset G$, from which the result follows. //

Theorem 3 Let $\Omega \subset \Gamma$ be compact with nonvoid interior. Then
$M_T(G) = \{\mu \in M(G) : \hat{\xi}_\Lambda \in L^2(\mu)$ for each Borel set $\Lambda \subset \Omega\}$.

Proof Let $\mu \in M_T(G)$ and choose a Borel subset Λ of Ω. Taking $p = q = 2$ in Theorem 1 we have $\hat{\xi}_\Lambda \in L^2_\chi(G) \subset L^2(\mu)$, and this gives one inclusion.

In the other direction, let $K \subset G$ be compact and, by Fournier [11], Lemma 3.0, choose $k \in L^\infty(\Omega)$ with $\hat{k} \ge \xi_K$. Then, for any $x \in G$,

$$|\mu|(x+K) = \int \xi_{x+K} d|\mu| = \int \tau_x \xi_K^2 d|\mu| \le \int \tau_x \hat{k}^2\, d|\mu| \le c\|k\|_\infty^2 ,$$

the last inequality following from Theorem 2 since $\operatorname{supp}(\tau_x \hat{k})^\wedge = \operatorname{supp}(k) \subset \Omega$, and this shows that $\mu \in M_T(G)$. //

Theorem 4 Let $\Omega \subset \Gamma$ be compact with nonvoid interior. Then
$M_2(G) = \{\mu \in M(G) : \hat{\xi}_\Lambda \in L^1(\mu)$ for each Borel set $\Lambda \subset \Omega\}$.

Proof Let $\mu \in M_2(G)$ and choose a Borel subset Λ of Ω. Taking $p = 2$, $q = 1$ in Theorem 1 we have $\hat{\xi}_\Lambda \in L^2_\chi(G) \subset L^1(\mu)$, and this gives one inclusion.

In the other direction, choose $g \in (C,\ell^2)$. By Fournier [11], Theorem 3.3′, there exists $(f_n) \subset C(\Omega)$ with

$$\sum_{n=1}^\infty \|f_n\|_\infty < \infty \quad \text{and} \quad |g(x)| \le \sum_{n=1}^\infty |\hat{f}_n(x)| , \qquad (2)$$

where the latter inequality holds χ - a.e. . Since

$$\sum_{n=1}^{\infty} |\hat{f}_n(x)| \leq \sum_{n=1}^{\infty} \|f_n\|_1 \leq \theta(\Omega) \sum_{n=1}^{\infty} \|f_n\|_\infty < \infty$$

and each $\hat{f}_n$ is continuous then so is the sum, and we have that the inequality in (2) holds everywhere. Then, using Lebesgue's theorem and Theorem 2 above,

$$\int_G |g| \, d|\mu| \leq \int_G \sum_{n=1}^{\infty} |\hat{f}_n| \, d|\mu| = \sum_{n=1}^{\infty} \int_G |\hat{f}_n| \, d|\mu| \leq \sum_{n=1}^{\infty} c\|f_n\|_\infty < \infty ,$$

so that $(C,\ell^2) \subset L^1(\mu)$.

An application of the closed graph theorem shows that this inclusion is continuous, and hence there is a constant c for which

$$\left|\int_G g \, d\mu\right| \leq \int_G |g| d|\mu| \leq c\|g\|_{\infty,2}$$

for all $g \in (C,\ell^2)$. Thus $\mu \in (C,\ell^2)' = M_2(G)$, as required. //

Both Theorems 3 and 4 can be viewed as a strengthening of Theorem 1, for the cases $p = q = 2$ and $p = 2$, $q = 1$ respectively, inasmuch as the inclusion in Part (b) is weakened. For other values of p, q and r the analogous statements remain undecided.

Our Theorem 4 should be compared with Fournier [11], Theorem 3.3 which, via the argument used above, is easily seen to be equivalent to deducing $\mu \in M_2(G)$ from the inclusion $C(\Omega)^\wedge \subset L^1(\mu)$, where $\Omega \subset \Gamma$ is a fixed compact set with nonvoid interior and $C(\Omega)^\wedge$ denotes the set of Fourier transforms of members of $C(\Omega)$ (compare this with the discussion in the introduction to this section). We have obtained a similar implication, with the "test" set $C(\Omega)$ replaced by the set of simple functions supported in Ω . The key to both results is to establish the inequality in Theorem 2 for all $f \in C(\Omega)$.

REFERENCES

[1] Loren Argabright and Jesús Gil de Lamadrid, *Fourier analysis of unbounded measures on locally compact abelian groups*, Mem. Amer. Math. Soc. No.145, American Mathematical Society, Providence, 1974.

[2] Christian Berg and Gunnar Forst, *Potential theory on locally compact abelian groups*, Ergebnisse der Mathematik und ihrer Grenzgebiete Band 87, Springer-Verlag, Berlin, Heidelberg, New York, 1975.

[3] J. -P. Bertrandias, C. Datry et C. Dupuis, *Unions et intersections d'espaces L^p invariantes par translation ou convolution*, Ann. Inst. Fourier (Grenoble) 28 (1978), 53-84.

[4] Lynette M. Bloom, *The Fourier multiplier problem for spaces of continuous functions with p-summable transforms*, J. Austral. Math. Soc. 17 (1974), 319-331.

[5] Walter R. Bloom, *Strict local inclusion results between spaces of Fourier transforms*, Pacific J. Math. 99 (1982), 265-270.

[6] Robert C. Busby and Harvey A. Smith, *Product-convolution operators and mixed-norm spaces*, Trans. Amer. Math. Soc. 263 (1981), 309-341.

[7] Gustave Choquet et Jacques Deny, *Sur l'équation de convolution* $\mu = \mu \star \sigma$, C.R. Acad. Sci. Paris Sér. A-B 250 (1960), 799-801.

[8] J. Diestel and J.J. Uhl, Jr., *Vector Measures*, Math. Surveys No.15, American Mathematical Society, Providence, 1977.

[9] R.E. Edwards, Edwin Hewitt and Gunter Ritter, *Fourier multipliers for certain spaces of functions with compact supports*, Invent. Math. 40 (1977), 37-57.

[10] Hans G. Feichtinger, *Banach convolution algebras of Wiener's type* (Proc. Conf. "Functions, series, operators", Budapest, August, 1980), Colloq. Math. Soc. J. Bolyai, North Holland, Amsterdam, Oxford, New York, 1982.

[11] John J.F. Fournier, *On the Hausdorff-Young theorem for amalgams*, Monatsh. Math. 95 (1983), 117-135.

[12] Edwin Hewitt and Kenneth A. Ross, *Abstract harmonic analysis, Vol.1*, Die Grundlehren der mathematischen Wissenschaften, Band 115, Springer-Verlag, Berlin, Göttingen, Heidelberg, 1963.

[13] Finbarr Holland, *Harmonic analysis on amalgams of L^p and ℓ^q*, J. London Math. Soc. (2) 10 (1975), 295-305.

[14] V. Ja. Lin, *On equivalent norms in the space of square summable entire functions of exponential type*, Amer. Math. Soc. Transl. (2) 79 (1969), 53-76.

[15] A.P. Robertson and M.L. Thornett, *On translation-bounded measures*, J. Austral. Math. Soc. (Series A).

[16] James Stewart, *Fourier transforms of unbounded measures*, Canad. J. Math. 31 (1979), 1281-1292.

[17] M.L. Thornett, *A class of second-order stationary random measures*, Stochastic Process. Appl. 8 (1979), 323-334.

[18] N. Wiener, *Tauberian theorems*, Ann. of Math. 33 (1932), 1-100.

EXTENSION DU THÉORÈME DE CONTINUITÉ DE PAUL LÉVY AUX GROUPES MOYENNABLES

Philippe BOUGEROL

Le théorème de continuité de Paul Lévy s'énonce ainsi :

"Si les transformées de Fourier d'une suite de probabilités sur $\mathbb{R}^d$ convergent vers une fonction continue en zéro, cette suite converge faiblement vers une probabilité".

Le but de cet exposé est de faire quelques remarques simples concernant l'extension de ce théorème aux groupes localement compacts. Nous montrons en particulier que, convenablement réécrit, ce résultat reste vrai sur un groupe moyennable, généralisant ainsi [1] qui supposait le groupe associé à un couple de Guelfand.

Je tiens à remercier Gérard Letac qui m'a indiqué ce problème et l'article [1].

1.- PRELIMINAIRES

Considérons un groupe G localement compact à base dénombrable, en abrégé L.C.D. Son dual $\hat{G}$ est l'ensemble des classes de représentations unitaires irréductibles. Si T est dans $\hat{G}$, nous noterons encore T un de ses représentants, H_T l'espace de Hilbert sur lequel il opère et H_T^1 les éléments de H_T de norme un (faisant l'abus de notation habituel consistant à confondre une représentation et sa classe).

Si μ est une probabilité sur G, on appelle transformée de Fourier de μ l'ensemble des classes d'opérateurs $T(\mu) = \int T_x d\mu(x)$, lorsque T parcourt $\hat{G}$ (cf. [9]). Nous nous intéressons au problème suivant :

PROBLEME : *Soit* $\{\mu_n, n \geq 0\}$ *une suite de probabilités sur* G *telle que, pour tout* T *de* $\hat{G}$, $T(\mu_n)$ *converge vers un opérateur* D(T). *Suivant les propriétés de* D(T), *que peut-on en déduire sur la convergence faible de la suite* $\{\mu_n,\ n \geq 0\}$?

Rappelons qu'on dit que $\{\mu_n,\ n \geq 0\}$ tend faiblement vers une mesure bornée ν si, pour toute fonction f sur G continue à support compact

$$\lim_{n\to+\infty} \int f\, d\mu_n = \int f\, d\nu$$

et qu'alors ν est une mesure positive de masse inférieure ou égale à un.

Nous allons voir que, sans aucune hypothèse sur D, $\{\mu_n,\ n \geq 0\}$ converge faiblement vers une mesure bornée. Commençons par rappeler la définition du dual réduit $\hat{G}_r$. Si T est un élément de $\hat{G}$ et u dans H_T^1 posons $\varphi_{T,u}(x) = \langle T_x u, u \rangle$, pour tout x de G.

Le dual réduit $\hat{G}_r$ est l'ensemble des T de $\hat{G}$ pour lesquelles il existe un u de H_T^1 tel que $\varphi_{T,u}$ soit limite uniforme sur les compacts d'une suite de fonctions de la forme $f * \tilde{f}$ où f est continue à support compact et $\tilde{f}(x) = f(x^{-1})$.

PROPOSITION 1 : *Soit* $\{\mu_n,\ n \geq 0\}$ *une suite de probabilités sur* G *telle que, pour tout* T *de* $\hat{G}_r$, $T(\mu_n)$ *converge faiblement vers un opérateur* D(T). *Alors* $\{\mu_n,\ n \geq 0\}$ *converge faiblement vers une mesure bornée* ν.

DEMONSTRATION :L'ensemble des fonctions sur G, bornées par 1 en e, limites uniformes sur tout compact de fonctions de la forme $h * \tilde{h}$, h continue à support compact, est un convexe compact dont les points extrémaux sont 0 et les éléments de $\mathcal{A} = \{\varphi_{T,u} : T \in \hat{G}_r,\ u \in H_T^1\}$, (cf. [7], 18.3.5, et [8]). On en déduit en particulier (cf. [7],13.6.8) que si f est continue à support compact, de norme L^2 égale à un, il existe une probabilité σ sur $\mathcal{A}$ telle que

$$(f * \tilde{f})(x) = \int_{\mathcal{A}} \varphi(x)\, d\sigma(\varphi) \qquad \text{pour } x \in G.$$

On a donc, pour tout entier n,

$$\int_G (f * \tilde{f})(x)\, d\mu_n(x) = \int_{\mathcal{A}} \{ \int_G \varphi(x)\, d\mu_n(x)\}\, d\sigma(\varphi).$$

Par hypothèse, si $\varphi = \varphi_{T,u}$, $\int \varphi(x)\, d\mu_n(x) = \langle T(\mu_n)u, u \rangle$ converge vers $\langle D(T)u, u \rangle$ que nous noterons $D(\varphi)$. Toute valeur d'adhérence faible ν de la suite $\{\mu_n,\ n \geq 0\}$ vérifie donc

$$\int_G f * \tilde{f}\, d\nu = \int_{\mathcal{A}} D(\varphi) d\sigma(\varphi) \quad . \tag{1}$$

Par densité des combinaisons linéaires de fonctions de la forme $f * \tilde{f}$, on en déduit que μ_n tend faiblement vers ν.

REMARQUE 1.: Cette démonstration suggère la définition suivante de la continuité de D : D est continue si l'application qui à $\varphi_{T,u}$ dans $\mathcal{A}$ associe $\langle D(T)u,u \rangle$ est continue, $\mathcal{A}$ étant muni de sa topologie usuelle, celle de la convergence compacte. Si G est moyennable et D continu en ce sens en 1, μ_n converge vers une probabilité. (En effet, si G est moyennable, 1 est limite uniforme sur les compacts d'une suite $f_n * \tilde{f}_n$ où f_n est continue à support compact de norme L^2 égale à un, (cf. [7], 18.3.6). L'égalité (1) appliquée à chaque f_n montre qu'il existe une suite σ_n de probabilités sur $\mathcal{A}$ telle que

$$\int f_n * \tilde{f}_n\, d\nu = \int D(\varphi) d\sigma_n(\varphi).$$

Le point 1 étant extrémal, quand n tend vers l'infini, σ_n converge faiblement vers la masse de Dirac en 1. Si D est continue en 1, on peut passer à la limite pour obtenir que $\int 1\, d\nu = D(1) = 1$.) Ceci est une généralisation du théorème de Paul Lévy mais il est en général difficile de vérifier ce type de continuité sur les exemples concrets. C'est pourquoi nous donnons une autre définition de la continuité, plus faible, qui au moins sur les groupes de type I assure le même résultat.

Lorsque le groupe G est de type I (nous en rappelons les principales propriétés au début de la démonstration ci-dessous), on peut préciser le résultat de la proposition précédente.

PROPOSITION 2 : *Soit* G *un groupe L.C.D. de type I et soit* σ *une mesure sur* $\hat{G}$ *associée à une représentation régulière de* G. *Si* $\{\mu_n,\ n \geq 0\}$ *est une suite de probabilités sur* G *telle que, pour* σ *presque tout* T, $T(\mu_n)$ *converge faiblement vers un opérateur* D(T), *alors* μ_n *converge faiblement vers une mesure bornée* ν *vérifiant* $T(\nu) = D(T)$, σ *presque partout.*

DEMONSTRATION : Soit λ la représentation régulière gauche de G dans $L^2(G)$. Puisque G est de type I, séparable, on peut lui associer une mesure σ sur $\hat{G}$, de support $\hat{G}_r$ telle que (cf. [7]) :

$$L^2(G) = \int n(T)H_T d\sigma(T) \ , \qquad n(T) \in \mathbb{N} \cup \{\chi_o\}$$

$$\lambda = \int T \, d\sigma(T) \quad .$$

Par définition, il existe une application $\xi : L^2(G) \to \prod_{T \in \hat{G}} n(T)H_T$, où $n(T)H_T$ est la somme hilbertienne de $n(T)$ copies de H_T, vérifiant:

(a) $\xi(L^2(G))$ est l'ensemble des champs de carré intégrable .

(b) Si f,g sont dans $L^2(G)$, $< f,g > = \int < \xi_T(f), \xi_T(g) > d\sigma(T)$.

(c) Il existe une suite (f_n) dans $L^2(G)$ telle que, pour tout T de $\hat{G}$ $\{\xi_T(f_n), n \geq 0\}$ est total dans $n(T)H_T$.

(d) Pour tous x de G, f et g de $L^2(G)$

$$< \lambda_x f, g > = \int < T_x \xi_T(f), \xi_T(g) > d\sigma(T) \quad .$$

Soient alors f et g deux éléments de $L^2(G)$ et ψ une fonction borélienne bornée sur $\hat{G}$. D'après (a) il existe h dans $L^2(G)$ tel que $\xi_T(h) = \psi(T)\xi_T(g)$ pour tout T de $\hat{G}$. On a

$$\int_G (f * \tilde{h})(x^{-1}) d\mu_n(x) = < \lambda(\mu_n) f, h > =$$

$$= \int_{\hat{G}} < T(\mu_n)\xi_T(f), \psi(T)\xi_T(g) > d\sigma(T) \quad .$$

Puisque $f * \tilde{h}$ tend vers zéro à l'infini, si ν est une valeur d'adhérence faible de la suite $\{\mu_n, n \geq 0\}$, on obtient

$$\int (f * \tilde{h})(x^{-1}) d\nu(x) = \int < D(T)\xi_T(f), \xi_T(g) > \overline{\psi(T)} d\sigma(T) \quad .$$

Ceci entraîne d'une part que μ_n converge vers ν, d'autre part que

$$\int < T(\nu)\xi_T(f), \xi_T(g) > \overline{\psi(T)} d\sigma(T) = \int < D(T)\xi_T(f), \xi_T(g) > \overline{\psi(T)} d\sigma(T) \quad .$$

On en déduit que, σ presque partout,

$$< T(\nu)\xi_T(f), \xi_T(g) > = < D(T)\xi_T(f), \xi_T(g) > \quad .$$

Faisant parcourir à f et g la suite (f_n) du (c), on obtient que

$$T(\nu) = D(T) \qquad \sigma \text{ presque partout.}$$

2.- LE CAS DES GROUPES MOYENNABLES

Le point de $\hat{G}$ où, connaissant la transformée de Fourier d'une mesure bornée on peut lire sa masse totale, est la représentation triviale, notée $\mathbb{1}$, c'est-à-dire la représentation à valeurs dans $\mathbb{C}$ vérifiant $\mathbb{1}_x z = z$, pour x dans G et z dans $\mathbb{C}$. Lorsque G est moyennable $\mathbb{1}$ est dans le support de toute mesure associée à une représentation régulière ([7], 18.3.6) et la proposition 2 va nous permettre de montrer facilement un analogue du théorème de Paul Lévy.

Commençons par définir la continuité de $D(T) = \lim T(\mu_n)$ en $\mathbb{1}$. Afin d'avoir une notion facile à vérifier (voir Remarque 1), notons que D peut être considérée comme une application définie sur $\hat{G}$, à valeurs dans l'ensemble $\mathcal{L}$ des classes d'opérateurs définis sur un Hilbert, unitairement équivalents. Ceci nous amène à poser :

<u>DEFINITION</u> : *Soit* A *une partie de* $\hat{G}$ *contenant* $\mathbb{1}$ *et* D *une application de* A *dans* $\mathcal{L}$ *tel que* $D(\mathbb{1})$ *soit un scalaire. On dit que* $D : A \to \mathcal{L}$ *est continu en* $\mathbb{1}$ *si :*

"Pour tout $\varepsilon > 0$, *il existe un voisinage* V *de* $\mathbb{1}$ *dans* $\hat{G}$ *tel que si* T *est dans* $V \cap A$, *il existe un représentant, noté encore* D(T), *de* D(T) *défini sur un Hilbert* H *et un élément* u *de* H *de norme un vérifiant*
$| \langle D(T)u,u \rangle - D(\mathbb{1}) | < \varepsilon$.

Dans cette définition, nous avons muni $\hat{G}$ de sa topologie usuelle, la topologie de Fell, pour laquelle une base de voisinages de $\mathbb{1}$ est formée des intersections finies de $V(C,\varepsilon)$, C compact de G, $\varepsilon > 0$, où

$$V(C,\varepsilon) = \{T \in \hat{G} : \exists\, u \in H_T^1 \text{ t.q. } | \langle T_x u,u \rangle - 1 | < \varepsilon,\ \forall x \in C\} .$$

Il est clair que la transformée de Fourier d'une mesure bornée est continue en $\mathbb{1}$. Notons aussi que si D(T) est la limite de transformées de Fourier de probabilités biinvariantes sur un couple de Guelfand, cette définition coïncide avec la définition usuelle (dans ce cas D(T) est essentiellement un scalaire).

Sur les groupes moyennables de type I, on a l'analogue du théorème de

Paul Lévy suivant qui généralise [1] (voir la remarque 1 quand le groupe n'est pas de type I):

PROPOSITION 3 : *Soit G un groupe moyennable, L.C.D, de type I, σ une mesure sur $\hat{G}$ associée à une représentation régulière de G. Si $\{\mu_n, n \geq 0\}$ est une suite de probabilités sur G telle que, pour tout T d'un sous-ensemble A de $\hat{G}$ de complémentaire σ négligeable, $T(\mu_n)$ converge faiblement vers un opérateur D(T) et si $D : A \to \mathcal{L}$ est continu en $\mathbb{I}$, la suite $\{\mu_n, n \geq 0\}$ converge faiblement vers une probabilité.*

DEMONSTRATION : Par la proposition 2, la suite $\{\mu_n, n \geq 0\}$ converge faiblement vers une mesure positive ν, de masse inférieure ou égale à un, telle que $T(\nu) = D(T)$, σ presque partout. Pour tout $\varepsilon > 0$, il existe un voisinage V de $\mathbb{I}$ tel que si T est dans $V \cap A$ il existe une réalisation concrète de D(T) à valeurs dans un Hilbert H et un élément u de H de norme un vérifiant

$$| < D(T)u,u > - 1| < \varepsilon \quad .$$

Le support de σ contient $\mathbb{I}$ car G est moyennable et σ charge $V \cap A$, on peut donc choisir un tel T pour lequel $D(T) = T(\nu)$. Alors

$$| < T(\nu)u,u > - 1| < \varepsilon \quad .$$

Comme $| < T(\nu),u,u > | \leq \int | < T_x u,u > | \, d\nu(x) \leq \nu(G)$, on en déduit que pour tout $\varepsilon > 0$, $\nu(G) \geq 1 - \varepsilon$, donc que ν est une probabilité. □

L'intérêt de cet énoncé est qu'il suffit de connaître une partie de $\hat{G}$ portant σ et sa topologie. Voyons comment vérifier les hypothèses de l'énoncé sur deux exemples.

EXEMPLE 1 : Si G est le groupe d'Heisenberg de dimension 3, c'est-à-dire $\mathbb{R}^3$ muni du produit

$$(x_1,y_1,z_1)(x_2,y_2,z_2) = (x_1 + x_2, y_1 + y_2, z_1 + z_2 + x_1 y_2)$$

considérons pour chaque $\lambda > 0$ la représentation T^λ de G à valeurs dans $L^2(\mathbb{R})$ définie par

$$T^\lambda_{(x,y,z)} f(t) = \{\exp i\,\lambda(z + ty)\}\, f(t+x) \quad (f \in L^2(\mathbb{R}), t \in \mathbb{R})$$

On sait ([13], 7.1) que la mesure de Plancherel sur $\widehat{G}$ est portée par ces représentations et que $(\{\Pi\} \cup \{T^\lambda, \lambda < \varepsilon\})_{\varepsilon>0}$ forme une base de voisinage de Π pour la topologie induite. Pour vérifier qu'une suite de probabilités μ_n sur G converge vers une probabilité, il suffit donc de montrer que :

- pour tout $\lambda > 0$, $T^\lambda(\mu_n)$ converge faiblement vers un opérateur $D(\lambda)$ vérifiant : pour tout $\varepsilon > 0$, il existe $\eta > 0$ tel que si $0 < \lambda < \eta$ pour un u_λ de $L^2(\mathbb{R})$ de norme un, $|<D(\lambda)u_\lambda, u_\lambda> - 1| < \varepsilon$.

<u>EXEMPLE 2</u> : Soit G le groupe affine, c'est-à-dire $\mathbb{R}_+^* \times \mathbb{R}$ muni du produit

$$(a_1,b_1)(a_2,b_2) = (a_1a_2, b_1 + a_1b_2) \quad .$$

Les mesures associées à une représentation régulière de G sont portées par $\{\Pi, T^+, T^-\}$ où $T^\pm$ est la représentation à valeurs dans $L^2(\mathbb{R})$ définie par

$$T^\pm_{(a,b)}f(t) = \{\exp \pm ie^t b\}f(t + \log a) \qquad (f \in L^2(\mathbb{R}),\ t \in \mathbb{R}) \quad .$$

La trace de tout voisinage de Π sur cet ensemble contenant T^+ et T^- (cf [13], 7.1), il suffit, pour obtenir la convergence de μ_n vers une probabilité, de véfifier que $T^\pm(\mu_n)$ converge faiblement vers un opérateur $D(\pm)$ et que pour tout $\varepsilon > 0$ il existe $u^\pm$ dans $L^2(\mathbb{R})$ de norme un tel que $|<D(\pm)u^\pm, u^\pm> - 1| < \varepsilon$.

3.- REMARQUES SUR LE CAS GENERAL

Dans des cas particuliers les articles [1] et [10] introduisent les groupes ayant la "propriété de Paul Lévy". De façon générale on peut dire que :

DEFINITION : *On dit qu'un groupe G, L.C.D., a la propriété de Paul Lévy si, lorsque $\{\mu_n, n > 0\}$ est une suite de probabilités sur G vérifiant : "pour tout T de $\hat{G}$, $T(\mu_n)$ converge faiblement vers un opérateur D(T) tel que $D : \hat{G} \to \mathcal{L}$ est continu en $\mathbb{I}$", alors μ_n converge faiblement vers une probabilité.*

Le lemme élémentaire suivant va nous permettre de construire des exemples :

LEMME : *Soit μ une probabilité symétrique sur un groupe G, de support G. Si $\mu_n = \mu^{*2n}$, pour tout T de $\hat{G}$ autre que la représentation triviale, $T(\mu_n)$ converge fortement vers 0.*

DEMONSTRATION : Pour tout T de $\hat{G}$, $T(\mu_1)$ est un opérateur autoadjoint positif, donc admet une décomposition spectrale de la forme $\int_0^1 \lambda\, dE(\lambda)$. On voit que $T(\mu_n) = \int_0^1 \lambda^n dE(\lambda)$ converge fortement vers E(1), projecteur sur $\{v, T(\mu_1)v = v\}$. Mais si $T(\mu_1)v = v$, pour tout g du support de μ_1, donc pour tout g de G, $T_g v = v$. La représentation T étant irréductible non triviale v doit être nul.

Considérons un groupe G ayant la propriété de Kazdhan (cf. [5] et [11]) par définition cela signifie que $\mathbb{I}$ est isolé dans $\hat{G}$ pour la topologie de Fell. Si $\{\mu_n, n \geq 0\}$ est une suite de probabilités sur G du type de celle introduite dans le lemme, pour tout T de $\hat{G}$,

$$\lim T(\mu_n) = \begin{cases} 0 & \text{si } T \neq \mathbb{I} \\ 1 & \text{si } T = \mathbb{I} . \end{cases}$$

Puisque $\mathbb{I}$ est isolé, pour toute définition raisonnable, lim $T(\mu_n)$ est continue sur $\widehat{G}$. Pourtant on sait que si G n'est pas compact, μ_n converge faiblement vers zéro (c est d'ailleurs ici immédiat, il suffit de reprendre la démonstration précédente avec T égal à une représentation régulière). Ces groupes n'ont donc pas la propriété de Paul Lévy. Il en est de même d'un groupe produit dont un facteur non compact a la propriété de Kazdhan. Puisque par exemple $Sl(d,\mathbb{R})$, $d \geq 3$ ([5]) et certains groupes presque périodiques maximaux ([12]) ont la propriété de Kazdhan, on voit que dans les familles de groupes où la propriété de Paul Lévy est introduite dans [1] et [9] il y a des groupes ne l'ayant pas.

En vertu du critère suivant et du résultat de [10], un groupe de Lie simple connexe de centre fini non compact a la propriété de Paul Lévy si et seulement si il n'a pas la propriété de Kazdhan.

<u>CRITERE</u> : *Soit G un groupe L.C.D. tel que tout voisinage de* $\mathbb{1}$ *dans* $\widehat{G}$ *contient une représentatuon T telle que, pour tout u de* H_T, $\varphi(x) = \langle T_x u,u \rangle$ *tend vers 0 quand x tend vers l'infini. Alors G a la propriété de Paul Lévy.*

<u>DEMONSTATION</u> : Soit $\{\mu_n,\ n \geq 0\}$ une suite de probabilités sur G telle que pour tout T de $\widehat{G}$, $T(\mu_n)$ converge faiblement vers un opérateur D(T) continu en $\mathbb{1}$. Par la proposition 1, μ_n converge faiblement vers une mesure positive ν de masse inférieure ou égale à 1. Pour tout $\varepsilon > 0$, il existe un voisinage V de $\mathbb{1}$ tel que si $T \in V$, pour un u de H_T^1

$$|\langle D(T)u,u \rangle - 1| < \varepsilon .$$

Par hypothèse on peut choisir T tel que $\langle T_x u,u \rangle$ tend vers zéro à l'infini. Alors

$$\langle D(T)u,u \rangle = \lim \int \langle T_x u,u \rangle \, d\mu_n(x) = \int \langle T_x u,u \rangle \, d\nu(x)$$

et $\nu(G) \geq 1 - \varepsilon$ pour tout $\varepsilon > 0$, donc ν est une probabilité.

<u>REMARQUE 2</u> : Sur tout groupe de Lie simple G connexe de centre fini, on peut aussi utiliser le critère suivant : soit $\{\mu_n,\ n \geq 0\}$ une suite de probabilités sur G convergeant faiblement vers une mesure ν et S l'ensemble des fonctions sphériques bornées sur G muni de la topologie de la convergence compacte. Si pour tout φ de S, $\mathcal{F}\mu_n(\varphi) = \int \varphi \, d\mu_n$ converge vers une fonction $d(\varphi)$ continue en 1, ν est une probabilité. Ce résultat est une

conséquence du fait que tout voisinage de 1 dans S contient une fonction tendant vers zéro à l'infini [13]. Il reste vrai si G est semi simple.

On peut se poser la question de déterminer quels sont les groupes ayant la propriété de Paul Lévy. Comme le montrent les exemples précédents, cela demande une connaissance assez précise du dual.

4.- UNE EXTENSION DANS LE CADRE DES COUPLES DE GUELFAND

Soit G un groupe L.C.D. possédant un sous-groupe compact K tel que (G,K) soit un couple de Guelfand, c'est-à-dire que le semi groupe $\mathcal{M}^1(G\,;K)$ des probabilités μ vérifiant $\varepsilon_k * \mu * \varepsilon_{k'} = \mu$ pour k et k' dans K, est commutatif.

Rappelons quelques éléments de la théorie des couples de Guelfand (cf.[4] ou [6]). A chaque T de $\hat{G}$ on associe $K_T = \{u \in H_T : T_k u = u,\ \forall\, k \in K\}$ et $\hat{Z} = \{T \in \hat{G} : K_T \neq \{0\}\}$. Pour tout T de $\hat{Z}$, K_T est de dimension un et on peut définir sans ambiguité

$$\varphi_T(x) = \langle T_x u, u \rangle \qquad , \qquad x \in G,\ u \text{ vecteur unitaire de } K_T.$$

L'application qui à T associe φ_T est un homéomorphisme de $\hat{Z}$ sur S^+, l'ensemble des fonctions sphériques de type positif, muni de la topologie de la convergence compacte.

Si μ est dans $\mathcal{M}^1(G\,;K)$ sa transformée de Fourier $T(\mu)$ vérifie

$$T(\mu) = \begin{cases} 0 & \text{si } T \notin \hat{Z} \\ \mathcal{F}\mu(\varphi_T)T(m) & \text{si } T \in \hat{Z} \end{cases}$$

où $\mathcal{F}\mu(\varphi) = \int \varphi\, d\mu$ si φ est dans S^+ et m est la mesure de Haar de K. On appelle encore transformée de Fourier de μ l'application $\mathcal{F}\mu$ de S^+ dans $\mathbb{C}$.

Posons $S_r^+ = \{\varphi_T : T \in \hat{Z} \cap \hat{G}_r\}$. Il existe une unique mesure positive σ sur S^+, de support S_r^+, appelée mesure de Plancherel, telle que si f est une fonction continue intégrable sur G, biinvariante par K, et si $\mathcal{F}f$ est dans $L^1(S^+,d\sigma)$ alors

$$f(e) = \int \mathcal{F}f(\varphi)\,d\sigma(\varphi) \quad .$$

Soit λ la représentation régulière gauche de G dans $L^2(G)$. L'énoncé suivant peut être considéré comme une généralisation du théorème de Paul Lévy. En effet si G est moyennable, $S^+ = S_r^+$ et pour toute mesure positive ν, $\|\lambda(\nu)\| = \nu(G)$ donc dans ce cas la proposition se réduit à ce théorème.

PROPOSITION 4 : Soit (G,K) *un couple de Guelfand. Il existe une unique représentation* T_o *de* $\hat{Z} \cap \hat{G}_r$ *telle que si* $\varphi_o = \varphi_{T_o}$, *pour toute mesure positive bornée* ν *sur* G, *biinvariante par* K, $\|\lambda(\nu)\| = \int \varphi_o d\nu$.

Si $\{\mu_n, n \geq 0\}$ *est une suite d'éléments de* $\mathcal{M}^1(G\,;K)$ *vérifiant :*

a. Pour σ *presque tout* φ *de* S_r^+, $\mathcal{F}\mu_n(\varphi)$ *converge vers un complexe* $d(\varphi)$.

b. $\mathcal{F}\mu_n(\varphi_o)$ *tend vers* $d(\varphi_o)$ *et* d *est continu en* φ_o.

Alors $d(\varphi)$ *est sur* S_r^+ *la transformée de Fourier de la limite faible* ν *de la suite* $\{\mu_n, n \geq 0\}$ *et* $\|\lambda(\nu)\| = \lim\|\lambda(\mu_n)\|$.

Donnons les grandes lignes de la démonstration :

1. On montre que si μ et ν sont deux probabilités symétriques de $\mathcal{M}^1(G\,;K)$

$$\|\lambda(\mu * \nu)\| = \|\lambda(\mu)\|\,\|\lambda(\nu)\| ,$$

pour cela on utilise [2] qui dit que si γ est une probabilité symétrique sur G et V un voisinage compact de l'élément neutre, $\|\lambda(\gamma)\| = \lim\{\gamma^{*2n}(V)\}^{1/2n}$

2. Soit $\mathcal{F}$ l'ensemble des fonctions positives de $L^1(G)$, biinvariante par K, non nulles. A chaque f de $\mathcal{F}$ on associe $A(f) = \{\varphi \in S_r^+ : \|\lambda(f)\| = |\mathcal{F}f(\varphi)|\}$. La formule de Plancherel et le lemme de Riemann-Lebesgue montrent que A(f) est un compact non vide. Si g est une autre fonction de $\mathcal{F}$ on vérifie à l'aide du 1 que $A(f * g)$ est contenu dans A(f) et A(g). La famille des compacts $\{A(f),\ f \in \mathcal{F}\}$ ayant la propriété d'intersection finie, il existe un φ_o de S_r^+ tel que :

$$\forall f \in \mathcal{F} \qquad \|\lambda(f)\| = |\mathcal{F}f(\varphi_o)| .$$

3. Par définition, pour tout φ de S_r^+ il existe une suite f_n d'éléments de $L^2(G)$ de norme un tel que φ soit limite uniforme sur les compacts de $f_n * \tilde{f}_n$. On en déduit que si f est dans $\mathcal{F}$

$$\int |\varphi(x)|\ f(x)dx = \lim \int |(f_n * \tilde{f}_n)(x)|\ f(x)dx$$

$$\leq \overline{\lim} \ < |f_n|, \lambda(f)\ |f_n| > \ \leq \|\lambda(f)\| \ .$$

Ceci montre d'abord que φ_o est positive puis que

$$\varphi_o(x) = \sup_{\varphi \in S_r^+} |\varphi(x)| \ .$$

A partir de ce résultat la démonstration de la proposition est immédiate.

REMARQUE : En fait pour toute probabilité μ sur G (pas nécessairement biinvariante) $\|\lambda(\mu)\| = \|T_o(\mu)\|$. En effet, quitte à remplacer μ par $\mu * \check{\mu}$ on peut supposer μ symétrique. Alors si h est une fonction positive continue à support compact biinvariante $\|\lambda(\mu)\| = \lim \{\mu^{2n}(h * \tilde{h})\}^{1/2n}$. Par la formule de Plancherel sphérique, on en déduit que $\|\lambda(\mu)\| = \lim \{\mathcal{F}\mu^{2n}(\varphi_o)\}^{1/2n}$. Comme $|\mathcal{F}\mu^{2n}(\varphi_o)| \leq \|T_o(\mu)\|^{2n}$ ceci entraîne que $\|\lambda(\mu)\| \leq \|T_o(\mu)\|$. L'inégalité inverse est claire car T_o est dans $\hat{G}_r$.

Dans un contexte différent ces représentations T_o ont été introduites dans [3] sous le nom d'"origine" du dual réduit.

o°o

REFERENCES

[1] BEN MANSOUR, S. : "Le cube comme couple de Guelfand : applications et généralisations", Thèse. Université Paul Sabatier, Toulouse (1981).

[2] BERG, C. ; CHRISTENSEN, J.P.R. : "Sur la norme des opérateurs de convolution", Inventiones ath. 23, pp. 173 178,(1974).

[3] BOUGEROL, P. : "Théorème central limite local sur certains groupes de Lie". Ann. Sc. Ecole Normale Sup., 14, 4° série, pp. 403-432, (1981).

[4] BOUGEROL, P. : "Un mini cours sur les couples de Guelfand", Publ. du Laboratoire de Statistique et Probabilités. Université Paul Sabatier, Toulouse (1982).

[5] DELAROCHE, C. ; KIRILLOV, A. : "Sur les relations entre l'espace dual d'un groupe et la structure de ses sous-groupes fermés", Sém. Bourbaki, Exposé 343, Lecture Notes 180, Springer (1972).

[6] DIEUDONNE, J. : "Eléments d'analyse", t.6, Gauthier-Villars, Paris (1975).

[7] DIXMIER, J. : "Les C^* algèbres et leurs représentations", Gauthier-Villars, Paris (1964).

[8] GODEMENT, R. : "Les fonctions de type positif et la théorie des groupes" Trans. Amer. Math. Soc., 63, pp. 1-84, (1948).

[9] HEYER, H. : "Probability measures on locally compact groups". Ergebnisse der Math.,Berlin Heidelberg New-York, Springer (1977).

[10] HOWE, R. ; MOORE, C. : "Asymptotic properties of unitary representations" J. of Funct. Anal., 32, pp. 72-96, (1979).

[11] KAZDHAN, D. : "Connection of the dual space of a group with the structure of its closed subgroups". Funct. Analysis and Applic., 1, pp. 63-65, (1967).

[12] MARGULIS, G.A. : "Some remarks on invariant means", Monat. Math., 90, pp. 223-235, (1980).

[13] WARNER, G. : "Harmonic analysis on semi simple Lie groups". Grundlehren der Math., 188, Berlin Heidelberg New-York, Springer, (1972).

Philippe BOUGEROL
UER de Mathématiques
Université Paris VII
2, place Jussieu, Paris 75251.

DECOMPOSITION OF CONVOLUTION SEMIGROUPS

OF PROBABILITY MEASURES ON GROUPS

T. Byczkowski and T. Żak

In this note we prove some decomposability properties of semigroups of probability measures on groups. The first section is different from the rest. We prove here a rather general result: any symmetric semigroup of probability measures is continuous. Technique is elementary and is based on that in [2], Lemma 2.3. Results of this type have been known [6], [10] but only for groups having some additional properties.

In the second section we prove a result closely related to 0-1 law for processes with values in a locally compact group. Our basic group there is G^∞ as the space of sample paths of processes, where G is locally compact. We consider a situation when a continuous semigroup $(\mu_t)_{t>0}$ has the property: $\mu_t(H)>0$, for all $t>0$, where H is a Borel subgroup of G^∞. Then the generator of $(\mu_t)_{t>0}$ can be decomposed into a bounded part concentrated on H^c and a generator of a semigroup concentrated on H. This, in particular, yields 0-1 law for Gaussian law. The technique used here is a slight extension of that from [3].

In the third section we obtain an abstract version of Lévy - Khintchine formula on abelian groups analogous to that proved by Tortrat in [12]. Our approach consists here of decomposition of the generator of a continuous semigroup into two parts. Therefore we finally obtain a decomposition of $(\mu_t)_{t>0}$ into two semigroups. A completely satisfactory formula has been obtained for symmetric semigroups.

1. Continuity of semigroups of measures.

In this section we prove that every symmetric semigroup of probability measures is continuous. For standard concepts and terminology concerning weak convergence of measures, convolution, idempotent factors etc., the reader is referred to [1] and [9]. We only recall that by a convolution semigroup of probability measures on a topological group we mean a family $(\mu_t)_{t>0}$ of probability measures such that

$$\mu_t * \mu_s = \mu_{t+s}$$

$(\mu_t)_{t>0}$ is called continuous if the mapping $t \longrightarrow \mu_t$ is continuous with respect to the weak convergence of measures. It is known [11] that it is equivalent to the following condition:

$$\lim_{t\to 0+} \mu_t \text{ exists.}$$

$(\mu_t)_{t>0}$ is called e-continuous if $\lim_{t\to 0} \mu_t = \delta_e$ (the point mass at the identity of G).

Proposition 1. Let $(\mu_t)_{t>0}$ be a semigroup of symmetric Radon probability measures on a Hausdorff group G. Then $(\mu_t)_{t>0}$ is continuous.

Proof. We first need some basic facts concerning idempotent factors. A probability measure λ is called a two-sided idempotent factor of μ if $\mu * \lambda = \lambda * \mu = \mu$. λ will be called briefly an idempotent factor of μ. We first prove the following:

(i) For every probability measure μ there exists the greatest idempotent factor λ of μ. λ has the following property:

if $\mu * \nu = \nu * \mu = \mu$ for a probability then $\operatorname{supp}(\nu) \subseteq \operatorname{supp}(\lambda)$.

Indeed, if λ_1, λ_2 are two idempotent factors of μ then

$$(\lambda_1 * \lambda_2)^{*n} * \mu = \mu * (\lambda_1 * \lambda_2)^{*n} = \mu \quad \text{for } n = 1,2,\ldots$$

It is well known that λ_i are Haar measures of some compact subgroups, say H_i, $i = 1,2$. From Th. 6 in [8] follows that the sequence $(\lambda_1 * \lambda_2)^{*n}$ converges weakly to the Haar measure concentrated on the closed subgroup generated by H_1 and H_2. If we denote this measure by λ then it is clear that λ is an idempotent factor of μ and that $\lambda_i * \lambda = \lambda * \lambda_i = \lambda$, for $i = 1,2$. Therefore, if $\{\lambda_i;\ i \in I\}$ is the family of all idempotent factors of μ then this family is directed by the following partial ordering: $\lambda_i < \lambda_j$ iff $\lambda_i * \lambda_j = \lambda_j * \lambda_i = \lambda_j$. Moreover, this family is conditionally compact. Hence, by Zorn's Lemma there exists a maximal element in this family, and, because of this partial ordering, this element is the greatest idempotent factor of μ. Next, we show

(ii). For every $t_\alpha \longrightarrow 0$ μ_{t_α} is weakly conditionally compact and for every $t > 0$ $\lim_\alpha \mu_{t+t_\alpha} = \mu_t$.

This will end the proof. Indeed, if ε, ε' are two accumulation points of μ_{t_α} and μ_{t_β}, $\mu_{t_{\beta'}}$ are corresponding subnets convergent to ε, ε', respectively, then by the right continuity we obtain that

$$\lim_\beta \mu_{t+t_\beta} = \mu_t = \mu_t * \varepsilon = \varepsilon * \mu_t$$

and

$$\lim_{\beta'} \mu_{t+t_{\beta'}} = \mu_t = \mu_t * \varepsilon' = \varepsilon' * \mu_t .$$

Taking appropriate subnets we obtain

$$\varepsilon = \varepsilon * \varepsilon' = \varepsilon' * \varepsilon = \varepsilon' .$$

To prove (ii), we fix $t > 0$ and let (t_α) be an arbitrary net of positive numbers convergent to 0. Putting $u_\alpha = t_\alpha/2$ we obtain, as in the proof of Lemma 2.3 in [2] that $\{\mu_{u_\alpha}\}$ is weakly conditionally compact and that $\mu_t = \mu_{u_\alpha} * \mu_{t-u_\alpha}$. Now, if ε is an accumulation point of μ_{u_α} we choose a subnet (u_β) of (u_α) such that μ_{u_β} converges to ε and μ_{t-u_β} converges to a probability measure γ. Thus, we have obtained

(1) $$\gamma * \varepsilon = \varepsilon * \gamma = \mu_t .$$

We next choose a subnet (s_β) of (u_β) such that $2s_\beta < u_\beta$. Denote $r_\beta = u_\beta - s_\beta > s_\beta$ and $w_\beta = r_\beta - s_\beta > 0$. Taking appropriate subnets once again we obtain

(2) $$\varepsilon = \varepsilon_0 * \varepsilon = \varepsilon * \varepsilon_0$$

and

(3) $$\varepsilon_0 = \varepsilon * \delta = \delta * \varepsilon ,$$

where ε_0 and δ are accumulation points of μ_{r_β} or μ_{w_β}, respectively. Suppose now that m_F is the Haar measure concentrated on a compact subgroup F of G which is the greatest idempotent factor of ε. Because of (2) we have $\operatorname{supp}(\varepsilon_0) \subseteq F$, so $m_F * \varepsilon_0 = \varepsilon_0 * m_F = m_F$. By (3) we obtain $m_F * \varepsilon * \delta = m_F * \varepsilon_0 = m_F$. (3) again and the familiar formula for the support of convolution yield now

$$\operatorname{supp}(\varepsilon)\, x_0 \subseteq \operatorname{supp}(\varepsilon * \delta) = \operatorname{supp}(\varepsilon_0) = F ,$$

for $x_0 \in \operatorname{supp}\delta$. Hence $m_F * \varepsilon = m_F * x_0^{-1}$ while $m_F * \varepsilon = \varepsilon$ and finally

$$\varepsilon = m_F * x_0^{-1} = x_0 * m_F .$$

Therefore μ_{2u_β} converges to $\varepsilon^2 = m_F$. By (1) we obtain that $\varepsilon^2 * \mu_t = \mu_t * \varepsilon^2 = \mu_t$, which completes the proof of (ii) and ends the proof of Proposition.

2. Decomposition of semigroups of measures on G^∞. Throughout this section G will stand for a locally compact group, satisfying the second countability axiom, unless stated otherwise. By $(G^\infty, \mathcal{B}^\infty)$ we denote the product of countably many copies of G with the product σ-field. $(G^n, \mathcal{B}^n)$ will denote the product of n copies of G.

Let us define the natural projection $\pi_{1,\dots,k}$ from G^∞ into G^k:

$$\pi_{1,\dots,k}(x) = \langle x(1), \dots, x(k) \rangle .$$

Let μ be a probability measure on $(G^\infty, \mathcal{B}^\infty)$. By $\mu_{1\dots k}$ we denote the probability measure on $(G^k, \mathcal{B}^k)$ induced by μ and $\pi_{1,\dots,k}$. Measures $\mu_{1,\dots,k}$ will be referred to as finite-dimensional distributions of μ.

Now, we say that a sequence of probability measures on G^∞ converges weakly to a certain probability measure if all finite-dimensional distributions converge weakly to corresponding finite-dimensional distributions of the limiting measure.

By $C_u(N)$ $[C_0(N)]$ we denote the family of all continuous bounded [vanishing at infinity] cylindrical functions on G^∞, that is functions $f\colon G^\infty \longrightarrow R$ with the property that there exists a finite subset of positive integers N, say $\{1,\dots,k\}$ such that $f(g^1) = f(g^2)$ whenever $g_i^1 = g_i^2$ for $i = 1,\dots,k$, and that f restricted to $G_1 \otimes \dots \otimes G_k$ is a bounded uniformly continuous function on G^k [vanishing at infinity]. $C_u(N)$ $[C_o(N)]$ with the supremum norm is a normed linear space. Every positive linear functional of norm 1 on $C_o(N)$ may be identified with a probability measure on G^∞. Indeed, every such functional defines a consistent family of finite-dimensional distributions, which, in virtue of Kolmogorov's Extension Theorem defines a unique probability measure on $(G^\infty, \mathcal{B}^\infty)$. Applying Banach - Alaoglu Theorem we see that a family (sequence) of probability measures on $(G^\infty, \mathcal{B}^\infty)$ is weakly conditionally compact whenever any corresponding family (sequence) of finite-dimensional distributions is uniformly tight. A sequence $\{\mu_i\}_i$ of probability measures on G^∞ converges weakly to a probability measure μ if $\mu_i(f) \longrightarrow \mu(f)$ for all $f \in C_o(N)$ and all finite-dimensional distributions are uniformly tight.

The main tool used here as well as in the next section is that of probability operators.

We recall that if G is a metric separable group then for a probability measure μ and for any Borel measurable and bounded function $f\colon G \longrightarrow R$ we may define

$$T_\mu f(x) = \int f(xy)\mu(dy).$$

It is easy to see that if μ, ν are probability measures on G and $\mu * \nu$ is their convolution then $T_{\mu*\nu} = T_\mu T_\nu$.

If we restrict considered functions to a subclass $\mathcal{C}$ of left uniformly continuous bounded functions on G then $\mu_n \Rightarrow \mu$ implies $T_{\mu_n} \longrightarrow T_\mu$ strongly. Assume additionally that $\mathcal{C}$ is invariant under the action of all T_μ, where μ is any probability measure on G. Then every e-continuous semigroup $(\mu_t)_{t>0}$ of probability measures on G corresponds to the strongly continuous semigroup of probability operators $(T_t)_{t>0}$ acting on $\mathcal{C}$. In this situation we may apply the whole abstract theory of semigroups on Banach spaces. In particular, any such a semigroup is uniquely determined by its infinitesimal generator A defined on a dense subset $D(A)_{\mathcal{C}}$ of $\mathcal{C}$ by the formula:

$$A\ f = \lim_{t\to 0+} \frac{1}{t}(T_t - I)f$$

In our situation we take as $\mathcal{C}$ either $C_u(N)$ or $C_o(N)$. It is clear that both these subspaces satisfy all the required properties.

Dealing with a fixed semigroup $(\mu_t)_{t>0}$ we will define the space $L_1(\mu)$ for μ given by the formula:

$$\mu(\cdot) = \int e^{-t} \mu_t(\cdot)\, dt \quad .$$

The following proposition is proved in [3].

<u>Proposition 2.</u> Let $(\mu_t)_{t>0}$ be an e-continuous semigroup of probability measures on G^∞. $(\mu_t)_{t>0}$ then acts, as a strongly continuous semigroup on $L_1(\mu)$. If H is a Borel subgroup of G^∞ such that $\mu(H)>0$ then $\mu_t(H)\longrightarrow 1$ as $t\longrightarrow 0$.

Now, suppose that $(\mu_t)_{t>0}$ and H are as in Proposition 2 and, additionally, that H is normal. Let π be the canonical homomorphism of G^∞ into G^∞/H. Endow G^∞/H with the measurable structure induced from G^∞ by π. Let $\lambda_t = \pi(\mu_t)$. We have the following

<u>Corollary.</u> Assume that $(\mu_t)_{t>0}$ and H are as above. Then

$$\lambda_t = \pi(\mu_t) = \exp tc(\gamma - \delta_H), \quad c \geq 0,$$

for a certain probability measure γ on G^∞/H. Hence

$$\lim_{s\to 0+} \frac{1}{s}\left[1 - \mu_s(H)\right] \text{ exists.}$$

For the proof, see [3].

It is not difficult to see that if all μ_t' s are symmetric then $\mu_{t_0}(H)>0$ for a certain t_0 implies that $\mu_t(H)>0$ for all $t>0$, where H is as in Proposition 2. On the other hand, if $\mu(H)>0$ then $\mu_t(H)>0$ for all $t>0$. For proofs, see again [3].

Before the formulation of the main result of this section we state here a theorem which is crucial in our approach.

<u>Trotter Approximation Theorem.</u> Let $T_t^{(n)}$ be a sequence of strongly continuous semigroups of operators on a Banach space X, satysfying the condition

$$\| T_t^{(n)} \| \leq e^{Kt},$$

where K is independent of n and t. Let A_n be the infinitesimal generator of $T_t^{(n)}$. Assume that $\lim A_n x$ exists in the strong sense on a dense linear subspace D. Define

$$Ax = \lim_n A_n x, \quad x \in D.$$

Suppose additionally that for some $\lambda > K$ the range of $\lambda I - A$ is dense in X.

Then the closure of A is the infinitesimal generator of a strongly continuous semigroup T_t such that

$$T_t x = \lim_n T_t^{(n)} x \quad \text{for } x \in X.$$

Theorem 1. Assume that $(\mu_t)_{t>0}$ is an e-continuous semigroup of probability measures on $(G^\infty, \mathcal{B}^\infty)$. Suppose that H is a $\mathcal{B}^\infty$-measurable normal subgroup of G^∞ such that $\mu_t(H) > 0$ for all $t > 0$. Then the generator A of $(\mu_t)_{t>0}$ has a decomposition on $D(A)_{C_u(N)}$

$$A = c(\nu - \delta_e) + A_H$$

where $c \geq 0$, ν is a probability measure concentrated on H^c and A^H is the generator of a semigroup of probability measures concentrated on H.

Proof. Let μ_s^H be the conditional probability of μ_s with respect to H. Since

$$\mu_s = \mu_s|H^c + \mu_s(H)\cdot\mu_s^H$$

and $\mu_s(H) \longrightarrow 1$ as $s \longrightarrow 0$, μ_s^H converges weakly to δ_e, as $s \to 0$, because $\| \mu_s| H^c \| \longrightarrow 0$ (Proposition 2).

Next, we write

$$\frac{1}{s}\left[\mu_s - \delta_e\right] = \frac{1}{s}\left[\mu_s - \mu_s^H\right] + \frac{1}{s}\left[\mu_s^H - \delta_e\right] .$$

Because

$$\frac{1}{s}\left[\mu_s - \mu_s^H\right] = \frac{1}{s}\,\mu_s|H^c - \frac{1}{s}\left[1 - \mu_s(H)\right]\cdot\mu_s^H =$$

$$= \frac{1}{s}\left[1 - \mu_s(H)\right]\cdot\left[\mu_s^{H^c} - \mu_s^H\right]$$

and since, by the corollary, $\frac{1}{s}\left[1 - \mu_s(H)\right] \longrightarrow c \geq 0$, this, together with the fact that $\mu_s^H \longrightarrow \delta_e$, as $s \longrightarrow 0$, implies that $\frac{1}{s}\left[\mu_s - \mu_s^H\right]$ is weakly* compact on $C_o(N)$. Choose a subsequence $s_k \longrightarrow 0$ such that $\mu_{s_k}^{H^c} f \longrightarrow \nu f$ for all $f \in C_o(N)$. Let A be the generator of $(\mu_t)_{t>0}$. ν is a subprobability measure and for all $f \in D(A)_{C_o(N)}$

$$\lim_{k \to \infty} \frac{1}{s_k}\left[\mu_{s_k}^H - \delta_e\right] f$$

exists. Denote this limit by A^H. Then for $f \in D(A)_{C_o(N)}$ we have

$$(4) \qquad A f = c(\nu - \delta_e) f + A^H f .$$

A^H is a sum of a generator and a bounded operator hence it is a generator of some semigroup of measures [4]. Let now $\mathfrak{A}$ be the generator of $(\mu_t)_{t>0}$ acting on $L_1(\mu)$. By a standard reasoning we have

$$\overline{(A)}^{L_1} = \mathfrak{A} .$$

Observe that if $f \in D(A)_{C_o(N)}$ then

$$\frac{1}{s}\left[\mu_s - \delta_e\right]({}_y f) \longrightarrow A\, f(y)$$

uniformly in $y \in G^{\infty}$, hence in $L_1(\mu)$. Also

$$\left[\mu_s^{H^c} - \mu_s^{H}\right]({}_y f) \longrightarrow \left[\nu - \delta_e\right]({}_y f)$$

pointwisely and, since they are all bounded by $2\|f\|_{\infty}$, we have $L_1(\mu)$ convergence as well. Hence

$$A^H f = \lim_k \frac{1}{s_k}\left[\mu_{s_k}^{H} - \delta_e\right] f \qquad \text{in } L_1(\mu)$$

for $f \in D(A)_{C_o(N)}$. Observe that

$$\left\| \exp\left(\frac{t}{s}\left[T_{\mu_s H} - I\right]\right)\right\|_{L_1, L_1} \leqslant \exp\left(\frac{t}{s}\left(\mu_s(H)^{-1}\ e^s - 1\right)\right).$$

and that

$$\lim_{s\to 0+} \frac{1}{s}\left[\mu_s(H)^{-1}\ e^s - 1\right] < 1 + c < \infty .$$

Therefore, and since $(\lambda - A^H)\left(D(A)_{C_o(N)}\right) = C_o(N)$ and $C_o(N)$ is dense in $L_1(\mu)$, by Trotter Approximation Theorem

$$\mathcal{A}^H = \overline{(A^H)}^{L_1}$$

defines a generator of a semigroup T_t^H on $L_1(\mu)$.
We also have

$$T_t^{H,s_k} = \exp\left(\frac{t}{s_k}\left[T_{\mu_{s_k}^H} - I\right]\right) \longrightarrow T_t^H \qquad \text{in } L_1(\mu).$$

Now, T_t^{H,s_k} is concentrated on H, for all $t > 0$, so

$$\int_H T_t^{H,s_k}\, \mathbb{1}_H(y)\mu(dy) = \int_H \mathbb{1}_H(y)\mu(dy) = \mu(H) .$$

If $k \to \infty$, we obtain

$$\int_H T_t^H\, \mathbb{1}_H(y)\mu(dy) = \mu(H) .$$

This, together with $0 \leqslant T_t^H \mathbb{1}_H(y) \leqslant 1$ implies that $T_t^H \mathbb{1}_H(y) = \mathbb{1}_H(y)$. Of course $\mathcal{A}^H \mathbb{1} = 0$ and $\mathcal{A}^H f \geqslant 0$ if $f \geqslant 0$ and $f(e) = 0$, for all $f \in D(\mathcal{A}^H)_{L_1} \cap C_o(N)$. These facts imply that A^H is a generator of a semigroup of probability measures that are concentrated on H. Now, if $\nu(G^{\infty}) < 1$ then there exist c', $c'' > 0$ and a probability measure γ such that

$$c(\nu - \delta_e) = c'(\gamma - \delta_e) - c''\delta_e .$$

Then, however, we would have

$$A = -c''\delta_e + c'(\gamma - \delta_e) + A^H$$

which would immediately imply that A is a generator of a semigroup of subprobability measures with the masses < 1, which is not true. Therefore ν is a probability measure on G^∞. Thus, the formula (4) holds for $(\mu_t)_{t>0}$ considered as a strongly continuous semigroup on $C_u(N)$. We may always assume that $\nu(H) = 0$, otherwise we may decompose $\nu = \nu|H + \nu|H^c$ and change A^H by adding $\nu|H$.

Now, a semigroup $(\mu_t)_{t>0}$ of probability measures on G^∞ is called Gaussian if

$$\lim_{t\to 0+} \frac{1}{t} \mu_t(U^c) = 0$$

for every open neighbourhood U of the identity e of G^∞. It is not difficult to derive from Theorem 1 the following:

Corollary. Assume that $(\mu_t)_{t>0}$ is a Gaussian semigroup on G^∞. If H is a Borel measurable normal subgroup of G^∞ such that $\mu_t(H)>0$, for all $t>0$, then $\mu_t(H) = 1$, for every $t>0$. If μ_t are symmetric then for a normal Borel subgroup H $\mu_t(H) > 0$ for a single $t > 0$ implies $\mu_t(H) = 1$ for all $t>0$.

Remark. The above decomposition is closely related to results of [7] and can be used to obtain several versions of 0-1 law also for nongaussian measures on G^∞, as it was done in [7], using different techniques, for measures on a locally compact group.

3. Lévy - Khintchine decomposition on abelian groups. Throughout the rest of this note G will be polish abelian group. Here we also use the semigroup technique. Our basic function space is now C_u, that is the space of all uniformly continuous and bounded real-valued functions on G, equipped with the supremum norm.
The following lemma is well-known (see [5]).

Lemma 1. Let (A_n) be a sequence of bounded generators of semigroups $(\mu_t^{(n)})_{t>0}$, n=1,2,..., which commute with one another.

If the sequence (A_n) is convergent (in the strong topology) to a generator A of a semigroup $(\mu_t)_{t>0}$ on $D(A)$ then for every $t>0$ the sequence $(\mu_t^n)_n$ is weakly convergent to μ_t.

The second lemma is taken from Parthasarathy's book [9]. It deals with shift compactness.

Lemma 2. Let G be a polish abelian group and let $\{\lambda_n\}$, $\{\mu_n\}$, $\{\nu_n\}$ be three sequences of measures on G such that for each $n \in N$ $\lambda_n = \mu_n * \nu_n$. If the sequences $\{\lambda_n\}$ and $\{\mu_n\}$ are conditionally compact so is the sequence $\{\nu_n\}$. If the sequence $\{\lambda_n\}$ is conditionally compact then the sequences $\{\mu_n\}$ and $\{\nu_n\}$ are shift compact.

We need the following definition.

Definition. Let ν be a finite Borel measure on G, $\nu\{e\} = 0$, where e stands for the identity of G. The Poisson measure associated with ν is defined as follows:

$$\exp(\nu) = e^{-\nu(G)} \sum_{k=0}^{\infty} \frac{\nu^{*k}}{k!} \qquad \text{where} \qquad \nu^{*0} = \delta_e.$$

Let now ν be a Borel (not necessarily finite) measure on G. If there exists a sequence of finite measures $(\nu_n)_n$ such that $\nu = \sup_n \nu_n$ and $(\exp(\nu_n))_n$ is shift compact then every accumulation point of $(\exp(\nu_n) * \delta_{x_n})_n$ is called the generalized Poisson measure associated with ν.
If λ_1 and λ_2 are such accumulation points then there exists $x \in G$ such that $\lambda_1 = \lambda_2 * \delta_x$ [12].
It is obvious that if ν and θ are finite measures then $\exp(\nu) * \exp(\theta) = \exp(\nu+\theta)$. What is more $(\exp(t\nu))_{t>0}$ is a continuous semigroup of probability measures and a bounded operator $c[T_{\frac{1}{c}\nu} - I]$, where $c = \nu(G)$, is the generator of $(\exp(t\nu))_{t>0}$. The domain of this generator is equal to C_u.

The third lemma can also be found in [9].

Lemma 3. Let $(\nu_n)_n$ be a sequence of finite measures. If the sequence $(\exp(\nu_n))_n$ is shift compact then for every neighbourhood of the identity U the family of measures $(\nu_n|U^c)_n$ is conditionally compact.

In the paper [12] Tortrat proved that every infinitely divisible probability measure μ without idempotent factors on polish abelian group G can be represented as a convolution: $\mu = \rho * \exp(\nu)$, where $\exp(\nu)$ is a generalized Poisson measure and ρ is the measure without factors of Poisson type.

Modyfying Tortrat's method we can prove that every symmetric semigroup of measures without idempotent factors on G can be represented in the same way. Namely we prove the following theorem.

Theorem 2. Let $(\mu_t)_{t>0}$ be an e-continuous semigroup of measures on abelian polish group G and let U denote a neighbourhood of e. There exist a finite measure ν_U with the support contained in U^c and a continuous semigroup of measures $(\rho_t^U)_{t>0}$ such that

$$\forall\ t>0 \qquad \mu_t = \rho_t^U * \exp(t\,\nu_U).$$

The measures ρ_t^U, $t>0$, have the following property: if $\exp(\theta)$ is a factor of ρ_t^U then $\mathrm{supp}(\theta) \subseteq \overline{U}$.

If $(\mu_t)_{t>0}$ is a semigroup of symmetric measures then

$$\forall\ t>0 \qquad \mu_t = \rho_t * \exp(t\,\nu),$$

where $(\exp(t\nu))_{t>0}$ is the semigroup of generalized Poisson measures and $(\rho_t)_{t>0}$ is a continuous semigroup of measures without factors of Poisson type.

Proof. Let $(\mu_t)_{t>0}$ be an e-continuous semigroup of measures on G. The measure $\frac{1}{t}\mu_t$ is finite for every $t>0$ hence for all $t>0$ and for every neighbourhood U of the identity the following equality holds:

$$(5) \qquad \exp\left(s\left(\tfrac{1}{t}\mu_t|U^c\right)\right) * \exp\left(s\left(\tfrac{1}{t}\mu_t|U\right)\right) = \exp\left(s\left(\tfrac{1}{t}\mu_t\right)\right).$$

Let us denote the generators of the above semigroups by $A_{1,t}$, $A_{2,t}$ and A_t, respectively. These generators are bounded hence their domains are equal C_u. For every $f \in C_u$ we have $A_t f(x) = \frac{1}{t}(T_t f - f)(x)$ and the limit $\lim_{t\to 0+} A_t f(x)$ exists for $f \in D(A)$, where A denote the generator of $(\mu_t)_{t>0}$.

Lemma 1 implies that the family of semigroups $\left\{\exp\left(s\left(\tfrac{1}{t}\mu_t\right)\right)_{s>0}\right\}_{t>0}$ converges if $t\to 0+$ (to the semigroup $(\mu_s)_{s>0}$).

This fact and Lemma 2 imply that the families of measures

$\left\{\exp\left(\frac{1}{t}\mu_t|U\right)\right\}_{t>0}$ and $\left\{\exp\left(\frac{1}{t}\mu_t|U^c\right)\right\}_{t>0}$ are shift compact.
Using Lemma 3 we can see that the family of measures $\left\{\frac{1}{t}\mu_t|U^c\right\}_{t>0}$ is conditionally compact.
Let us choose now a measure ν_U which is an accumulation point of this family. It is obvious that $\text{supp}(\nu_U) \subseteq U^c$.
Let $t_n \to 0$ be such a sequence that the sequence of measures $\left(\frac{1}{t_n}\mu_{t_n}|U^c\right)_n$ converges weakly to ν_U. The measure ν_U is finite hence the generator of the semigroup $(\exp(s\nu_U))_{s>0}$ is a bounded operator. We denote it by A_1.
The weak convergence of the sequence $\left(\frac{1}{t_n}\mu_{t_n}|U^c\right)_n$ implies that the generators A_{1,t_n} of the semigroups $\left(\exp s\left(\frac{1}{t_n}\mu_{t_n}|U^c\right)\right)_{s>0}$ $n = 1,2,\ldots$ converge to A_1 for every $f \in C_u$.
If we denote A_{2,t_n} the generator of the semigroup $\left(\exp s\left(\frac{1}{t_n}\mu_{t_n}|U\right)\right)_{s>0}$ we have the following equality:

$$\forall\, n \in N \qquad A_{t_n} = A_{1,t_n} + A_{2,t_n} .$$

The limit $\lim_n A_{2,t_n}$ exists for $f \in D(A)$ because for these functions $\lim_n A_{t_n} f = Af$ and $\lim_n A_{1,t_n} f = A_1 f$.
The convergence of commuting generators on a subspace which is dense in C_u implies the weak convergence of the semigroups [5].
The equality (5) implies that for every $s > 0$ the sequence $\left(\exp s\left(\frac{1}{t_n}\mu_{t_n}|U\right)\right)_n$ is conditionally compact hence the limiting semigroup is a continuous semigroup of measures.
Let us define, for every $s > 0$

$$\exp(s\nu_U) = \lim_n \exp\left(s\left(\frac{1}{t_n}\mu_{t_n}|U^c\right)\right) \text{ and } \quad \rho_s^U = \lim_n \exp\left(s\left(\frac{1}{t_n}\mu_{t_n}|U\right)\right).$$

The continuity of convolution implies that for every $s > 0$

$$\mu_s = \rho_s^U * \exp(s\nu_U) .$$

Let V be a neighbourhood of the identity such that $\overline{V} \subset U$.
For every $t > 0$ we have

$$(6) \quad \exp\left(s\left(\tfrac{1}{t}\mu_t|U\right)\right) = \exp\left(s\left(\tfrac{1}{t}\mu_t|V\right)\right) * \exp\left(s\left(\tfrac{1}{t}\mu_t|V^c \cap U\right)\right).$$

If there exist the measures ρ_s' and $\exp(\theta)$ such that

$$\lim_n \exp\left(s\left(\frac{1}{t_n}\mu_{t_n}\middle| U\right)\right) = \rho_s^U = \rho'_s * \exp(\theta|V + \theta|V^c)$$

then (6) and Lemmas 2 and 3 imply that

$$\mathrm{supp}(\theta) \cap V^c \subseteq \overline{V^c \cap U} .$$

Now, let $(\mu_t)_{t>0}$ be a semigroup of symmetric measures. Let us choose and fix a sequence of symmetric neighbourhoods of the identity $U_k \searrow \{e\}$. Repeating the above arguments for the sets $U_1, U_1 \setminus U_2, U_2 \setminus U_3, \ldots$ we obtain:

(a) a sequence of finite measures $(\nu_k)_k$ such that $\mathrm{supp}(\nu_1) \subseteq U_1^c$ and $\mathrm{supp}(\nu_k) \subseteq \overline{U_{k-1} \setminus U_k}$, $k=2,3,\ldots$.

(b) a sequence of continuous semigroups $(\rho_t^{(k)})_{t>0}$ $k=1,2,\ldots$ with the following property:

(7) $\forall k \in N \ \ \forall t > 0 \qquad \mu_t = \rho_t^{(k)} * \exp(t(\nu_1 + \ldots + \nu_k)).$

Lemma 2 implies that the sequences of measures $(\rho_{t/2}^{(k)})_k$ and $\exp(\frac{t}{2}(\nu_1 + \ldots + \nu_k))_k$ are shift compact.

All the measures ν_k and semigroups $(\rho_t^{(k)})_{t>0}$ are symmetric so for every $t > 0$ the sequences $(\rho_t^{(k)})_k$ and $\exp(t(\nu_1 + \ldots + \nu_k))_k$ are conditionally compact.

Let us denote by D the set of all positive dyadic-rational numbers. By the diagonal method we can choose a subsequence (k') such that for every $t \in D$ the sequences of measures $(\rho_t^{(k')})_{k'}$ and $\exp(t(\nu_1 + \ldots + \nu_{k'}))_{k'}$ are weakly convergent.

Let us define for every $t \in D$

$$\rho_t = \lim_{k'} \rho_t^{(k')} \quad \text{and} \quad \exp(t\nu) = \lim_{k'} \exp(t(\nu_1 + \ldots + \nu_{k'})).$$

A semigroup of symmetric measures is continuous (Proposition 1) so there exist the unique continuous extensions $(\rho_t)_{t>0}$ and $(\exp(t\nu))_{t>0}$ (see [11]).

The formula (7) is valid for all $k \in N$ and all $t > 0$ so we have :

$$\forall\ t > 0 \qquad \mu_t = \rho_t * \exp(t\nu) .$$

It is easy to see that the semigroup $(\rho_t)_{t>0}$ has no factors of Poisson type (compare the first part of the proof of this theorem).

Remark. The semigroup $(\rho_t)_{t>0}$ is called Gaussian part of $(\mu_t)_{t>0}$. In the case of separable Banach space of cotype 2 one can show that the space of twice Fréchet differentiable functions is always contained in the domain of the generator of any continuous semigroup of measures. The representation of the generator in these spaces was obtained in [13]. In the case of semigroups of p-stable measures, $0 < p \leq 2$, the representation is valid without any assumption on the geometry of Banach space.

References

[1] Billingsley P. Convergence of Probability Measures. Wiley, N.Y.

[2] Byczkowski T. Zero-one laws for Gaussian measures on metric abelian groups. Studia Math. 69 (1980) 159-189 .

[3] Byczkowski T. and Hulanicki A. Gaussian measure of normal subgroups. Ann. Probability 11 (1983) 685-691

[4] Dunford N. and Schwartz J. Linear Operators. Part I, Interscience Publishers, N.Y. London (1958)

[5] Feller W. An Introduction to Probability Theory and its Applications 2. 2nd ed. Wiley, N.Y. (1971)

[6] Heyer H. Probability Measures on Locally Compact Groups. Springer, Berlin Heidelberg New York (1977)

[7] Janssen A. Zero-one laws for infinitely divisible probability measures on groups. Z. Wahr. verw. Gebiete 60 (1982) 119-138

[8] Kloss B.M. Probability measures on compact topological groups. Prob. Theory and Appl. 4 (1959) (in Russian) 255-290

[9] Parthasarathy K.R. Probability Measures on Metric Spaces. Acad. Press, N.Y. (1967)

[10] Siebert E. Einbettung unendlich teilbarer Wahrscheinlichkeitsmasse auf topologischen Gruppen. Z. Wahr. verw. Gebiete 28 (1974)

[11] Siebert E. Convergence and convolutions of probability measures on a topological group. Ann. Probability 4 (1976) 433-443

[12] Tortrat A. Structure des lois indéfiniment divisible dans un espace vectoriel topologique (separe) X Lecture Notes in Math. 31 Springer (1967) 299-328

[13] Żak T. A representation of infinitesimal operators of semigroups of measures on Banach spaces of cotype 2. Bull. Acad. Polon. Sci. 31 (1983) 71-74

Institute of Mathematics
Technical University
Wybrzeże Wyspiańskiego 27
Wrocław 50-370 Poland

BROCOT SEQUENCES AND RANDOM WALKS IN SL(2,$\mathbb{R}$)

Philippe Chassaing, Gérard Letac, Marianne Mora.

1. Introduction. To every (2,2) matrix $A = \begin{pmatrix} a & b \\ c & d \end{pmatrix}$ with real coefficients such that $ad-bc \neq 0$, we associate the projectivity on $\mathbb{R} \cup \infty$ defined by

$$x \longmapsto A(x) = \frac{ax+b}{cx+d}$$

with the usual understanding $A(-d/c) = \infty$ and $A(\infty) = a/c$ if $c \neq 0$, $A(\infty) = \infty$ if $c=0$. Obviously $(AB)(x) = A[B(x)]$ for all x in $\mathbb{R} \cup \infty$.

Suppose that $(A_n)_{n=1}^{\infty}$ is a sequence of independent random matrices of SL(2,$\mathbb{R}$) (that is (2,2), real, and with determinant 1), with the same distribution ν, it is known that (under fairly general conditions, and we won't go into the details) the limit $Z = \lim_{n\to\infty} (A_1 \dots A_n)(x)$ exists almost surely for any x in $\mathbb{R} \cup \infty$ and does not depend on x. Note that, unless the law of Z is the Cauchy one, the exact distribution of Z is seldom known, and this is the aim of this paper to give examples where this distribution can be made reasonably explicit.

Actually, we shall concentrate on a simpler case, the random continued fractions. Let us start from a sequence $(X_n, Y_n)_{n=1}^{\infty}$ of independent pairs of strictly positive random variables with the same distribution γ in $(0,+\infty)^2$, and consider :

$$A_n = \begin{pmatrix} X_n & 1 \\ 1 & 0 \end{pmatrix} \begin{pmatrix} Y_n & 1 \\ 1 & 0 \end{pmatrix} = \begin{pmatrix} X_nY_n + 1 & X_n \\ Y_n & 1 \end{pmatrix}$$

In this case $A_1(x) = X_1 + \cfrac{1}{Y_1 + \cfrac{1}{x}}$, and more generally :

$(A_1A_2,\dots,A_n)(x) = X_1 + \cfrac{1}{Y_1 + \cfrac{1}{X_2 + \cfrac{1}{\dots + \cfrac{1}{x}}}}$ and Z can be written as a continued fraction.

Example 1.1 (G. Letac and V. Seshadri [6]). Let λ, a, b positive numbers ; the gamma distribution $\gamma_{\lambda,2/a}$ on $(0,+\infty)$ is :

$$\gamma_{\lambda,2/a}(dx) = \frac{2^{-\lambda}a^{\lambda}}{\Gamma(\lambda)} x^{\lambda-1} \exp(-ax/2)\, \mathbb{1}_{(0,\infty)}(x)dx .$$

$(X_n)_{n=1}^{\infty}$ and $(Y_n)_{n=1}^{\infty}$ being two independent sequences of independent random variables, such that $\mathcal{L}(X_n) = \gamma_{\lambda,2/a}$ and $\mathcal{L}(Y_n) = \gamma_{\lambda,2/b}$, the distribution of Z is the generalized inverse-gaussian law :

$$\mu_{\lambda,a,b}(dz) = \frac{a^{\lambda/2}\, b^{-\lambda/2}}{2K_\lambda(\sqrt{ab})}\, z^{\lambda-1} \exp -\frac{1}{2}(az + \frac{b}{z})\, \mathbb{1}_{(0,\infty)}(z)dz,$$

where K_λ is a Bessel function.

Furthermore, we specialize our study to the case where (X_n, Y_n) are taking values in $\mathbb{N}^2$ ($\mathbb{N}$ is the set of integers, $\mathbb{N}^* = \mathbb{N}\setminus\{0\}$). A motivation of such a choice is the fact that, using the familiar representation of a positive number by a continued fraction with coefficients in $\mathbb{N}^*$ (details in §2), one can get a representation of the distribution of Z(Th.3.1). Replacing $\mathbb{N}^*$ by $\mathbb{N}$ gives some difficulties in interpreting Z as a continued fraction, since X_n or Y_n could be zero. The rewarding is that we can cope with simple and interesting cases like

$$P\left(A_n = \begin{bmatrix} 1 & 1 \\ 0 & 1 \end{bmatrix}\right) = \alpha\,, \quad P\left(A_n = \begin{bmatrix} 1 & 0 \\ 1 & 1 \end{bmatrix}\right) = \beta = 1-\alpha \quad \text{, with } \alpha \text{ in } (0,1).$$

2. <u>The Brocot sequences and the Denjoy-Minkowski measure</u>. Before stating the lengthy Theorem 3.1, we have to prepare it by recalling a few facts on continued fractions with integer coefficients, and by building the Denjoy-Minkowski measure $\mu^{(\alpha)}$ on $(0,+\infty)$, which will be related to our best example.

Define $E_1 = E_1^* = \mathbb{N}$ and, if n is an integer >1, E_n as the set of sequences $a = (a_1,a_2,\dots,a_n)$ included in $\mathbb{N}^n$ such that $a_1 \geq 0$ $a_j \geq 1$ if $j=1,\dots,n-1$, $a_n \geq 2$, and E_n^* as the set of sequences a of $\mathbb{N}^n$ such that $a_1 \geq 0$, $a_j \geq 1$ if $j=1,\dots,n-1$ and $a_n = 1$. On E_n and E_n^*, one defines the rational number $[a] = [a_1,a_2,\dots,a_n]$ by induction on n, with :

$$[a_1] = a_1\,, \qquad [a_1,\dots,a_n] = a_1 + \frac{1}{[a_2,\dots,a_n]}\,.$$

A standard result (see e.g. C.D. Olds [9] page 14) is that if a and a' are in $\bigcup_{n=1}^{\infty} E_n$ or $\bigcup_{n=1}^{\infty} E_n$ such that $[a] = [a']$ then either a=a', or there exists $n \geq 1$ such that $a \in E_n$, $a' \in E_{n+1}^*$, with $a_j = a'_j$ for $j=1,\dots,n-1$ and $a'_n = a_n - 1$. Furthermore, the maps $a \longrightarrow [a]$ from $\bigcup_{n=1}^{\infty} E_n$ and $\bigcup_{n=1}^{\infty} E_n^*$ to the set $\mathbb{Q}^+$ of non negative rational numbers are bijective. If r is in $\mathbb{Q}^+$, write $r = [a] = [a']$, with a in E_n and a' in E_{n+1}^*. Then $a_1 + \dots + a_n = a'_1 + \dots + a'_{n+1}$: this integer will be called the <u>mass of the rational r</u> (like in [15]).

Define now the set E of infinite sequences of integers $a = (a_1, a_2, \ldots, a_n, \ldots)$ such that $a_1 \geq 0$ and $a_n > 0$ if $n > 1$. One shows that for a in E, $[a] = \lim_{n\,\infty} [a_1, a_2, \ldots, a_n]$ exists (see [9]) and that the map $a \longmapsto [a]$ from E to the set $\mathbb{I}^+$ of positive irrational numbers is bijective. If x is $\mathbb{I}^+$, the element a of E such that $x = [a]$ is called <u>the</u> development in continued fraction of x.

Let us explain now what <u>Brocot sequences</u> $(B_n)_{n=o}^{\infty}$ are. We decide $1/0 = +\infty$, and define the first $B_n \subset [0,\infty]$ by :

$$B_o = \left\{ \frac{0}{1}, \frac{1}{0} \right\} \quad , \quad B_1 = \left\{ \frac{0}{1}, \frac{1}{1}, \frac{1}{0} \right\} \quad , \quad B_2 = \left\{ \frac{0}{1}, \frac{1}{2}, \frac{1}{1}, \frac{2}{1}, \frac{1}{0} \right\} .$$

To define B_n for all n in $\mathbb{N}$, let us adopt the following definition . If $r = p/q$ and $r' = p'/q'$ are in $\mathbb{Q}^+$ such that the Greatest Common Divisors (p,q) and (p',q') are 1, define their mediant $m(r,r') = (p+p')/(q+q')$. One easily sees that $pq' - p'q = \pm 1$ implies $(p+p, q+q') = 1$.

Suppose now that the Brocot sequence B_n is an increasing sequence of 2^n elements of $\mathbb{Q}^+$: $r_o^{(n)} < r_1^{(n)} < \ldots < r_{2^n-1}^{(n)}$ completed by $+\infty = r_{2^n}^{(n)}$ such that, for all $0 \leq j < 2^n$, if $r_j^{(n)} = \frac{p}{q}$ and $r_{j+1}^{(n)} = \frac{p'}{q'}$, with $(p,q) = (p',q') = 1$ one has $pq' - p'q = 1$ (it is indeed the case for n=0). The sequence B_{n+1} will be now defined by $r_k^{(n+1)} = r_k^{(n)}$ if $k = 0,1,\ldots,2^n$, and $r_{2k+1}^{(n+1)} = m(r_k^{(n)}, r_{k+1}^{(n)})$ if $k = 0,1,\ldots,2^n-1$. We have now a simple caracterisation of B_n :

THEOREM 2.1. (see [15]) $B_n \backslash \infty$ is the set of non negative rationals with mass $\leq n$.

<u>Sketch of the proof</u> : One proves by induction on n that elements of $B_n \backslash \infty$ have masses $\leq n$ (see [15] for a neat proof). Conversely we show that the number of r in $\mathbb{Q}^+$ with mass $\leq n$ is 2^n. To see this, we consider formal variables $(x_j)_{j=1}^{\infty}$, and define for a in E_n : $x^a = x_1^{a_1} x_2^{a_2} \ldots x_n^{a_n}$. Hence

$$\sum_{a \in \cup_n E_n} x^a = \frac{1}{1-x_1} + \frac{x_2^2}{(1-x_1)(1-x_2)} + \frac{x_3^2}{(1-x_1)(1-x_2)(1-x_3)} + \ldots$$

Taking $t = x_1 = x_2 = \ldots$, $x^a = t^{w(r)}$ where w(r) is the mass of r=[a], one gets $\sum_{r \in Q^+} t^{w(r)} = \sum_{a \in \cup E_n} x^a = 1 + \frac{t}{1-2t}$, which gives $\#\{r \; ; \; w(r)=n\} = 2^{n-1}$ if $n \geq 1$, and 1 if n=0. Thus $\#\{r \; ; \; w(r) \leq n\} = 1+1+2+4+\ldots+2^{n-1} = 2^n$. Since $\# B_n \backslash \infty = 2^n$, the proof is done. □

Let us introduce the Denjoy-Minkowski measure $\mu^{(\alpha)}$. We fix α in $(0,1)$, and $\beta = 1-\alpha$, and we consider the probability measure $\mu_n^{(\alpha)}$ on $B_n\setminus\infty$ defined by induction as follows :

$$\mu_o^{(\alpha)} = \delta_o$$

$$\mu_{n+1}^{(\alpha)}(r_{2k}^{(n+1)}) = \beta\, \mu_n^{(\alpha)}(r_k^{(n)}) \qquad 0\le k\le 2^n$$

$$\mu_{n+1}^{(\alpha)}(r_{2k+1}^{(n+1)}) = \alpha\, \mu_n^{(\alpha)}(r_k^{(n)}) \qquad 0\le k<2^n .$$

Such a definition can perhaps be enlightened by the following table :

B_o	$\frac{0}{1}$								$\frac{1}{0}$
$\mu_o^{(\alpha)}$	1								0
B_1	$\frac{0}{1}$				$\frac{1}{1}$				$\frac{1}{0}$
$\mu_1^{(\alpha)}$	β				α				0
B_2	$\frac{0}{1}$		$\frac{1}{2}$		$\frac{1}{1}$		$\frac{2}{1}$		$\frac{1}{0}$
$\mu_2^{(\alpha)}$	β^2		$\alpha\beta$		$\beta\alpha$		α^2		0
B_3	$\frac{0}{1}$	$\frac{1}{3}$	$\frac{1}{2}$	$\frac{2}{3}$	$\frac{1}{1}$	$\frac{3}{2}$	$\frac{2}{1}$	$\frac{3}{1}$	$\frac{1}{0}$
$\mu_3^{(\alpha)}$	β^3	$\alpha\beta^2$	$\beta\alpha\beta$	$\alpha^2\beta$	$\beta^2\alpha$	$\alpha\beta\alpha$	$\beta\alpha^2$	α^3	0

The measure $\mu_n^{(\alpha)}$ can be seen in terms of a non homogeneous Markov chain $(Z_n)_{n=o}^{\infty}$ on $\mathbb{Q}^+$: Z_n is valued in B_n and has the following transition probability :

$$P[Z_{n+1} = r_{2k}^{(n+1)} \mid Z_n = r_k^{(n)}] = \beta$$

$$P[Z_{n+1} = r_{2k+1}^{(n+1)} \mid Z_n = r_k^{(n)}] = \alpha \qquad \text{if } 0\le k<2^n .$$

$\mu_n^{(\alpha)}$ is the distribution of Z_n (recall that $Z_o = 0$) ; $(Z_n)_{n=o}^{\infty}$ is increasing and bounded with probability one : actually $Z_n = r_k^{(n)}$ implies $Z_n < r_{k+1}^{(n)}$ for all m>n. Hence, the sequence $(\mu_n^{(\alpha)})_{n=o}^{\infty}$ converges tightly to the probability $\mu^{(\alpha)}$ on $(0,+\infty)$, which is the distribution of $Z = \lim_{n\to\infty} Z_n$. The measure $\mu^{(\alpha)}$ is called the Denjoy-Minkowski measure (with parameter α in $(0,1)$). Denote, with Denjoy, $\chi_\alpha(x) = \mu^{(\alpha)}((x,+\infty))$ for $x>0$. Here is a list of the properties of χ_α

THEOREM 2.2. ① $\chi_\alpha(r) = \mu_n^{(\alpha)}([r,+\infty))$ if r in $\mathbb{Q}^+$ has a mass $\leq n$.

② χ_α is continuous on $[0,+\infty)$.

③ $\chi_\alpha(x) + \chi_\beta(1/x) = 1$ for all $x\geq 0$.

④ $\chi_\alpha(a+x) = \alpha^a \chi_\alpha(x)$ for all $x\geq 0$ and all a in $\mathbb{N}$.

⑤ If x is in $\mathbb{I}^+$, with $x = [a]$, then

$$\chi_\alpha(x) = \alpha^{a_1} - \alpha^{a_1}\beta^{a_2} + \alpha^{a_1+a_3}\beta^{a_2} - \alpha^{a_1+a_3}\beta^{a_2+a_4} + \dots$$

or, denoting $\sigma_k = \sum_{j=1}^{k} a_{2j-1}$ and $\sigma'_k = \sum_{j=1}^{k-1} a_{2j}$.

$$\chi_\alpha(x) = \sum_{k=1}^{\infty} \alpha^{\sigma_k} \beta^{\sigma'_k}(1 - \beta^{a_{2k}}) .$$

If x is in $\mathbb{Q}^+$ with $x = [a]$ with a in some E_n,

$$\chi_\alpha(x) = \sum_{1\leq k\leq n/2} \alpha^{\sigma_k} \beta^{\sigma'_k}(1 - \beta^{a_{2k}}) .$$

Proof. ① Since $Z_n \uparrow Z$, $\chi_\alpha(x) = P(Z>x) = \lim_{n\,\infty} P(Z_n > x)$. Now from the definition of $\mu_n^{(\alpha)}$, for fixed r in $\mathbb{Q}^+$ the sequence in $n:\mu_n^{(\alpha)}([r,+\infty))$ is constant on $\{n\ ;\ n\geq w(r)\}$ where $w(r)$ is the mass of r.

But $P[Z_n \geq r] - P[Z_n > r] = P(Z_n = r) \leq [\max(\alpha,\beta)]^n \xrightarrow[n\,\infty]{} 0$.

Hence $P[Z > r] = P[Z_n \geq r]$ for $n \geq w(r)$.

② If I is an interval of $[0,+\infty]$, n is the largest integer such that there exists k in $\{0,\dots,2^n -1\}$ such that $I \subset [\,r_k^{(n)}, r_{k+1}^{(n)}) = J$. If $I_\varepsilon = (x-\varepsilon,x]$ with fixed $x>0$, we denote $n=n(\varepsilon)$ and $J = J_\varepsilon$; $n(\varepsilon)$ is increasing when $\varepsilon\downarrow 0$ and cannot be bounded (from Th.2.1 and the density of $\mathbb{Q}^+$ in $(0,+\infty)$). Hence :

$$\chi_\alpha(x-\varepsilon) - \chi_\alpha(x) = \mu^{(\alpha)}(I_\varepsilon) \leq \mu^{(\alpha)}(J_\varepsilon) = \mu_{n(\varepsilon)}^{(\alpha)}(J_\varepsilon) \quad ,$$

from ① . We have $\chi_\alpha(x-\varepsilon) - \chi_\alpha(x) \leq [\max(\alpha,\beta)]^{n(\varepsilon)} \xrightarrow[\varepsilon\to o]{} 0$, and χ_α is left continuous. Right continuity was obvious from definition.

③ Define $\mu_n'^{(\beta)}$ as the distribution of $1/Z_n$ on $(0,+\infty]$, and $\mu'^{(\beta)}$ as the distribution of $1/Z$, i.e. $\mu'^{(\beta)} = \lim_{n\,\infty} \mu_n'^{(\beta)}$. Differences between $\mu_n'^{(\alpha)}$ and $\mu_n^{(\alpha)}$ are small : the following table gives $\mu_3'^{(\alpha)}$:

B_3	$\frac{0}{1}$	$\frac{1}{3}$	$\frac{1}{2}$	$\frac{2}{3}$	$\frac{1}{1}$	$\frac{3}{2}$	$\frac{2}{1}$	$\frac{3}{1}$	$\frac{1}{0}$
$\mu_3'^{(\alpha)}$	0	β^3	$\alpha\beta^2$	$\beta\alpha\beta$	$\alpha^2\beta$	$\beta^2\alpha$	$\alpha\beta\alpha$	$\beta\alpha^2$	α^3

For all r in B_n, we have : $\mu_n^{(\alpha)}([r,+\infty)) = \mu_n'^{(\alpha)}((r,+\infty))$.
Since the first member is $\chi_\alpha(r)$ from ①, taking the limit when $n \to \infty$ and using the continuity of χ_α, one deduces that $\mu'^{(\alpha)} = \mu^{(\alpha)}$. Hence $\mu^{(\beta)}$ is the distribution of $1/Z$, which is ③.

④ Let r in $\mathbb{Q}^+$, and a in $\mathbb{N}$. Obviously $P[Z_a = a] = \alpha^a$ and $\{Z_{a+n} \geq a+r\} \subset \{Z_n = a\}$. Also $(B_{a+n} - a) \cap [0,+\infty) = B_n$.

One deduces $P[Z_{a+n} \geq a+r \mid Z_a = a] = P[Z_n \geq r]$.
Taking now $n \geq w(r)$, the mass of r, we get from ①

$$\chi_\alpha(a+r) = \alpha^a \chi_a(r) \text{ , which gives ④ .}$$

⑤ If x in $\mathbb{I}^+$ is equal to $[a_1, a_2, \ldots] = [a]$, with a in E, denote $x_n = [a_{2n+1}, a_{2n+2}, \ldots]$. We have

$$x_n = a_{2n+1} + \cfrac{1}{a_{2n+2} + \cfrac{1}{x_{n+1}}} \text{ ,}$$

and using ③ and ④ we get

$$\chi_\alpha(x_n) = \alpha^{a_{2n+1}} [1 - \beta^{a_{2n+2}}(1 - \chi_\alpha(x_{n+1}))] .$$

Since $(\chi_\alpha(x_n))_{n=1}^\infty$ is bounded, one is easily led to ⑤. The proof is pretty much the same when x is in $\mathbb{Q}^+$. □

Some bibliographical comments are in order. According to Lucas [7] , Brocot was a watchmaker. His sequences should not be confused with the Farey sequences (see [7] pages 469-475). The second author met them for the first time when he was a student in a Problème d'Agrégation [10] .
For Denjoy ([2] page 135) Farey sequences are our Brocot sequences. We have been using here the presentation given by M. Shrader-Frechette [15], who apparently rediscovered the Brocot sequences under the name of "modified Farey sequences".

Taking $\alpha = 1/2$, the function ?(x) defined on [0,1] by $?(x) = 2(1-\chi_{1/2}(x))$ was introduced by Minkowski [8] . Th.2.2. ⑤ provides, for $a_j \in N^*$ j=1,2,...

$$?([0,a_1,\ a_2,\dots\ ) = \sum_{k=1}^{\infty} (-1)^k\ 2^{-(a_1+a_2+\dots+a_k)}$$

? gives a bijection between quadratic numbers (i.e. numbers $[a]$ of $\mathbb{I}^+$ such that a is ultimately periodic) on $[0,1]$ and $\mathbb{Q}^+\cap\ [0,1]$; $\chi_\alpha(x)$ was introduced by Denjoy as a solution of the functional equation

$$\chi_\alpha(r_{2k+1}^{(n+1)}) = \alpha\chi_\alpha(r_k^{(n)}) + \beta\chi_\alpha(r_{k+1}^{(n)})$$

for all $n\in N$ and all $k=0,1,\dots,2^n-1$.

Denjoy does not really consider $\mu^{(\alpha)} = -d\chi_\alpha$, and our definition of χ_α seems to be new. Various papers by G. de Rham ([11], [12], [13]) mention the functions ? and χ_α ; it seems clear to these authors that χ_α is a singular function.

3. The main theorem.

THEOREM 3.1. Consider a distribution γ in $\mathbb{N}^2$ such that $\gamma(\mathbb{N}\times\mathbb{N}^*)$ and $\gamma(\mathbb{N}^*\times\mathbb{N})$ are not 0. Consider also a sequence $(X_n,Y_n)_{n=1}^{\infty}$ of independent random variables, γ distributed, and the random matrices

$$A_n = \begin{bmatrix} X_nY_n + 1 & X_n \\ Y_n & 1 \end{bmatrix}$$

and $Z = \lim_{n\to\infty} A_1,\dots,A_n(x)$. Then there exist two applications A and B from $\mathbb{N}\times\mathbb{N}^*$ to $(0,1)$ such that if a is in E, with $a = (a_1,a_2,\dots,a_n,\dots)$, we have :

$$\boxed{P\,[Z \geq [a]] = \sum_{n=1}^{\infty} B(a_{2n-1},\ a_{2n}) \prod_{k=1}^{n-1} A(a_{2k-1},\ a_{2k})}$$

Furthermore, denoting $\varphi(x,y) = \sum_{a_1=o}^{\infty}\ \sum_{a_2=o}^{\infty} \gamma(a_1,a_2)x^{a_1}\ y^{a_2}$,

$k = P[X_1 > 0]$ and $K(x) = \frac{1}{1-x}\ (1 - \mathbb{E}(x^{X_1}))$, A and B can be computed from the formulae :

$$\tilde{A}(x,y) = \sum_{a_1=o}^{\infty}\ \sum_{a_2=1}^{\infty} A(a_1,a_2)x^{a_1}\ y^{a_2} = \frac{[1-\gamma(o,o)][\varphi(x,y - \varphi(x,o)]}{[1-\varphi(x,o)][1-\varphi(o,y)]}$$

$$\tilde{B}(x,y) = \sum_{a_1=o}^{\infty}\ \sum_{a_2=1}^{\infty} B(a_1,a_2)x^{a_1}y^{a_2} = \frac{y}{1-y} \times \frac{K(x)[1-\varphi(o,y)]+k[\varphi(x,y)-\varphi(x,o)]}{1-[\varphi(x,o)][1-\varphi(o,y)]}$$

To tame this statement, let us consider a few examples :

Example 3.1. $\gamma(1,0) = \alpha \in (0,1)$, $\gamma(0,1) = \beta = 1-\alpha$; thus

$$P\left[A_n = \begin{bmatrix} 1 & 1 \\ 0 & 1 \end{bmatrix}\right] = \alpha \text{ and } P\left[A_n = \begin{bmatrix} 1 & 0 \\ 1 & 1 \end{bmatrix}\right] = \beta \quad , \varphi(x,y) = \alpha x + \beta y$$

$$\mathbb{E}(x^{X_1}) = \alpha x + \beta \qquad k = K(x) = \alpha ,$$

$$\tilde{A}(x,y) = \frac{\beta y}{(1-\alpha x)(1-\beta y)} \quad , \quad \tilde{B}(x,y) = \frac{y}{1-y} \times \frac{\alpha}{(1-\alpha x)(1-\beta y)}$$

$$A(a_1,a_2) = \alpha^{a_1} \beta^{a_2} \quad , \quad B(a_1,a_2) = (1-\beta^{a_2})\alpha^{a_1} \quad ,$$

$$P[Z \geq [a]] = \sum_{n=1}^{\infty} (1-\beta^{a_{2n}})\alpha^{a_{2n-1}} \prod_{k=1}^{n-1} \alpha^{a_{2k-1}} \beta^{a_{2k}} = \chi_\alpha([a])$$

from theorem 2.2 ⑤ . Hence Z has the Denjoy-Minkowski distribution $\mu^{(\alpha)}$.

Example 3.2. Suppose that X_n and Y_n are independent and Bernoulli distributed, i.e. $\varphi(x,y) = (px+q)(\alpha y + \beta)$, with $p=1-q$ and $\alpha = 1-\beta$ in $(0,1)$. Taking $r = \beta p/(1-\beta q)$ and $R = \alpha q/(1-\beta q)$, we get :

$$\tilde{A}(x,y) = \frac{(\frac{p}{q}x+1)Ry}{(1-rx)(1-Ry)} \quad , \quad \tilde{B}(x,y) = \frac{y}{1-y} \; \frac{(1-R)(\frac{p}{q}x.Ry+1)}{(1-rx)(1-Ry)}$$

$$A(0,a_2) = R^{a_2} \quad , \quad A(a_1,a_2) = \frac{r^{a_1}}{\beta}\,\frac{R^{a_2}}{q} \quad \text{if } a_1 > 0$$

$$B(0,a_2) = 1-R^{a_2} \quad , \quad B(a_1,a_2) = \frac{r^{a_1}}{\beta}\left(1 - \frac{R^{a_2}}{q}\right) \quad \text{if } a_1 > 0 .$$

The distribution of Z is given by the formulae :

if $a_1 = 0$ (i.e. $[a] < 1$) :

$$P[Z \geq [a]] = 1 - R^{a_2} + R^{a_2} \sum_{n=3}^{\infty} \frac{r^{a_3+\dots+a_{2n-1}}}{\beta^{n-1}} \times \frac{R^{a_4+\dots+a_{2n-2}}}{q^{n-2}} \left(1 - \frac{R^{a_{2n}}}{q}\right)$$

if $a_1 > 0$ (i.e. $[a] \geq 1$)

$$P[Z \geq [a]] = \sum_{n=1}^{\infty} \frac{r^{a_1+\dots+a_{2n-1}}}{\beta^n} \times \frac{R^{a_2+\dots+a_{2n-2}}}{q^{n-1}} \left(1 - \frac{R^{a_{2n}}}{q}\right)$$

For instance, if $p = \alpha = 1/2$:

if $a_1 = 0$ $\qquad P[Z \geq [a]] = 1 - \sum_{n=2}^{\infty} (-1)^n\, 2^{n-2}\, 3^{-(a_2+\dots+a_n)}$

if $a_1 > 0$ $$P[Z \geq [a]] = \sum_{n=1}^{\infty} (-1)^{n+1}\, 2^n\, 3^{-(a_1+\ldots+a_n)}$$

In this example, consider now what happens if we impose to p/α to be a constant number $\rho > 0$ and if we let $\alpha \to 0$. Then $r \xrightarrow[\alpha\to 0]{} \rho/(1+\rho) = \alpha_1$, $R \to 1 - \alpha_1 = \beta_1$, and the limiting distribution of Z is nothing but the Denjoy-Minkowski measure $\mu^{(\alpha_1)}$. So we have $\gamma \to \delta_{0,0}$ weakly, i.e. the distribution of A_1 tends to be concentrated on the origin of $SL(2,\mathbb{R})$, and the correspondence between γ and the distribution of Z is by no means continuous.

Example 3.3. Suppose that X_n and Y_n are independent such that $\varphi(x,y) = \frac{q}{1-px}(\alpha y+\beta)$, with $p = 1-q$ and $\alpha = 1-\beta$ in $(0,1)$. Taking $R = \alpha q/(1-\beta q)$ and $r = 1-R$ (they are in $(0,1)$) one gets again the Denjoy-Minkowski measure $\mu^{(r)}$ as the distribution of Z.

Let us do now some comments on Theorem 3.1.

① Letting $\gamma(\mathbb{N}\times\{0\}) = 1$ would imply :

$$A_1,\ldots,A_n = \begin{bmatrix} 1 & X_1+\ldots+X_n \\ 0 & 1 \end{bmatrix}$$

a case where clearly Z would be ∞ a.s.

$\gamma(\{0\} \times \mathbb{N}\} = 1$ would give $Z = 0$ a.s.

② If $\gamma(\mathbb{N} \times \{0\}) = \gamma(\{0\} \times \mathbb{N}) = 0$, A and B have a simple form given by

$$A = \gamma , \quad \tilde{B} = \frac{y}{1-y} \times [K(x) + k\,\varphi(x,y)]$$

③ If X_n and Y_n are independent, A and B have a somewhat simpler form ; in particular $A(a_1,a_2) = A_1(a_1)\,A_2(a_2)$ where $A_1 : \mathbb{N} \to (0,1)$ and $A_2 = \mathbb{N}^* \to (0,1)$.. It worths to mention that if furthermore $\varphi(x,y)$ is a rational function, then A_1, A_2 and B are exponential polynomials with respect to a_1 and a_2.

4. Proof of Theorem 3.1. For fixed $a = (a_1,a_2,\ldots)$ in E, denote

$$v = P[Z \geq [a_3,a_4,\ldots]] \quad \text{and} \quad G(a_1,a_2) = P[Z \geq [a]]$$

The basic relation is :

$$G(a_1,a_2) = \sum_{k_2=0}^{\infty} \sum_{k_1=a_1+1}^{\infty} \gamma(k_1,k_2) + \sum_{k_1=0}^{a_1-1} \gamma(k_1,0)\, G(a_1-k_1,a_2)$$
$$+ \sum_{k_2=0}^{a_2-1} \gamma(a_1,k_1)\, G(0,a_2-k_2) + \gamma(a_1,a_2)v \quad (1)$$

Define $\tilde{G}(x,y) = \sum_{a_1=o}^{\infty} \sum_{a_2=1}^{\infty} G(a_1,a_2)x^{a_1} y^{a_2}$.

We compute first $\tilde{G}(0,y)$ from (1) by doing $a_1 = 0$:

$$G(0,a_2) = k + \sum_{k_2=o}^{a_2-1} \gamma(0,k_2)\, G(0,a_2-k_2) + \gamma(0,a_1)v \ . \tag{2}$$

Multiplying by y^{a_2} in (2) and adding on $\mathbb{N}^*$, we get

$$[1-\varphi(0,y)]\, \tilde{G}(x,y) = \frac{ky}{1-y} + [\varphi(0,y) - \gamma(0,0)]\, v \tag{3}$$

Coming back to (1), multiplying by $x^{a_1}\, y^{a_2}$ and adding on $\mathbb{N} \times \mathbb{N}^*$, we get :

$$\tilde{G}(x,y) = \frac{y}{1-y} K(x) + \varphi(x,0)\, [\tilde{G}(x,y) - \tilde{G}(0,y)] + \varphi(x,y)\, \tilde{G}(0,y) + [\varphi(x,y) - \varphi(x,0)]\, v \tag{4}$$

Putting (3) and (4) together, one obtains

$$\tilde{G}(x,y) = \tilde{A}(x,y)v + \tilde{B}(x,y) \tag{5}$$

where $\tilde{A}$ and $\tilde{B}$ are described in the statement of the theorem. Identity (5) says that

$$P[Z \geq [a_1,a_2,a_3,\ldots]] = B(a_1,a_2) + A(a_1,a_2)\, P\,[Z \geq [a_3,a_4,\ldots]]$$

Iterating, one gets the theorem. □

5. Singularity.

Theorem 5.1. With the same hypothesis as in Theorem 3.1. and with the further assumption that γ is not concentrated on one point, the distribution of Z is purely singular.

We are not going to prove this theorem, whose the proof would be a mere adoption of the proofs given by S.J. Chatterji [1] and F. Schweiger [14] in the particular case where X_n, Y_n are in $\mathbb{N}^*$ and independent. Such a result seems sometimes ignored : A. Kirillow and A. Gvichiani [4] through the solution of their exercise 130b) claim that Z could have the Gauss distribution

$$\mathbb{1}_{(0,1)}(x)\, \frac{dx}{(\mathrm{Log}2)(1+x)}$$, certainly a not singular one.

6. Characteristic exponents. Take the euclidean norm in $\mathbb{R}^2$; if A in the linear group GL(2,R) $\|A\|$ is the norm of the corresponding linear endomorphism. From [3] we know that, under general conditions, if $(A_n)_{n=1}^{\infty}$ is a sequence of i.i.d. random matrices of GL(2,R), then

$$\lambda = \lim_{n\to\infty} \frac{1}{n} \operatorname{Log} \|A_1, A_2 \dots A_n\|$$

exists almost surely ; λ is called the characteristic exponent. It can be expressed as an integral :

$$\lambda = \int_{GL(2,\mathbb{R})} \int_{\mathbb{R}\cup\infty} \frac{1}{2} \operatorname{Log} \frac{(az+b)^2 + (cz+d)^2}{z^2+1} \, d\nu(a,b,c,d)\,d\mu(z)$$

where ν is the distribution of A_1 and μ is the distribution of $Z = \lim_{n\,\infty} A_1 A_2, \dots A_n(x)$. For a beautiful account of this theory, the lectures notes of F. Ledrappier [5] should be consulted.

Unfortunately, λ can be seldom computed, even in the simplest circumstance of Example 3.1. We shall content ourselves to compute λ in the case of the example 1.1 of the introduction for special values of the parameters.

THEOREM 6.1. Let $(X_n)_{n=1}^{\infty}$ a sequence of i.i.d. random variables with distribution $\gamma_{1,2/a}(dx) = \frac{a}{2}\exp(-\frac{ax}{2})\, 1_{(o,\infty)}(x)dx$, $a > 0$ and $A_n = \begin{bmatrix} 0 & 1 \\ 1 & X_n \end{bmatrix}$. The characteristic exponent λ is $\lambda = \frac{2K_o(a)}{a\,K_1(a)}$, where

$$K_\lambda(a) = \frac{1}{2}\int_0^\infty u^\lambda \exp\left[-\frac{a}{2}\left(u+\frac{1}{u}\right)\right]\frac{du}{u} \quad .$$

Proof. From the Example 1.1, we know that

$$\mu(dz) = \frac{1}{2K_1(a)} \exp - \frac{a}{2}\left(z+\frac{1}{z}\right) 1_{(0,\infty)}(z)\,\frac{dz}{z^2} ,$$

and we get

$$\lambda(a) = \frac{a}{4K_1(a)} \int_0^\infty\int_0^\infty \operatorname{Log}\frac{(1+(z+x)^2}{1+z^2} \exp\left[-\frac{a}{2}\left(z+x+\frac{1}{z}\right)\right]\frac{dz}{z^2}\,dx$$

Integration by parts with respect to x yields :

$$\lambda(a) = \frac{1}{K_1(a)}\int_0^\infty \exp\left[-\frac{a}{2}\left(z+\frac{1}{z}\right)\right]\frac{dz}{z^2}\int_0^\infty \frac{z+x}{1+(z+x)^2}\exp\left(-\frac{ax}{2}\right)dx$$

Then, we change the variables $u = x+z$, $v = 1/z$:

$$\lambda(a) = \frac{1}{K_1(a)} \int_0^\infty \frac{u}{1+u^2} \exp(-\frac{au}{2})du \int_{1/u}^\infty \exp(-\frac{av}{2})dv$$

$$= \frac{2}{aK_1(a)} \int_0^\infty \frac{u}{1+u^2} \exp[-\frac{a}{2}(u+\frac{1}{u})]du$$

Splitting the integral in $\int_0^1$ and $\int_1^\infty$ and changing u in 1/u in the first, one gets :

$$\lambda(a) = \frac{2}{aK_1(a)} \int_0^\infty \exp[-\frac{a}{2}(u+\frac{1}{u})]\frac{du}{u} = \frac{2K_0(a)}{aK_1(a)} . \qquad \square$$

REFERENCES

[1] S.D. Chatterji. Masse, die von regelmässigen Kettenbrüchen induziert sind, Math. Ann. vol.164 (1966), 113-117.

[2] A. Denjoy. Sur une fonction réelle de Minkowski, J. Math. pures et appl., vol.17 (1938), 105-151.

[3] H. Furstenberg. Non commuting random products, Trans. Amer. Math. Soc. vol.108 (1963), 377-428.

[4] A. Kirillov, A. Gvidiani. Théorèmes et Problèmes d'Analyse fonctionnelle, Editions Mir (1982 for the French translation), Moscow.

[5] F. Ledrappier. Quelques propriétés des exposants caractéristiques, Cours de l'Ecole d'Eté de Saint Flour (to appear in Springer Verlag Lectures Notes).

[6] G. Letac, V. Seshadri. A characterization of the generalized inverse Gaussian distribution by continued fractions, Z. Wahrscheinlichkeits-theorie und verw. Gebiete, Vol.62 (1983), 485-489.

[7] E. Lucas. Théorie des Nombres, Tome premier (1891), Librairie Joseph Gibert, Paris.

[8] H. Minkowski. Zur Geometrie der Zahlen, in Verhand lungen des III. Internationalen Mathematiker Krongresses in Heidelberg (1904), 171-172.

[9] C.D. Olds. Continued fractions, Vol.9 (1963), New Mathematical Library, Mathematical Association of America, Yale.

[10] Problèmes du concours 1962 d'Agrégation Masculine de Mathématiques, épreuve de Mathématiques élémentaires et spéciales, published in Bulletin de l'Association des Professeurs de Mathématiques de l'Enseignement Public, Vol.42, n°225 (1962), 103-106.

[11] G. de Rham. Sur une courbe plane, J. Math. pures et appl., Vol.35 (1956), 25-42.

[12] G. de Rham. Sur quelques courbes définies par des équations fonctionnelles, Univ. c. Politec. Torino, Rend. Sem. Math. Vol.16 (1956-57), 101-103.

[13] G. de Rham. Sur les courbes limites de polygônes définies par trisection, L'enseignement Mathématique, Vol.2 n°5 (1959), 29-43.

[14] F. Schweiger. Eine Bemerkung zu einer Arbeit von D.S. Chatterji Matematicky Časopis, Vol.19 (1969), 89-91.

[15] M. Shrader-Frechette. Modified Farey sequences and continued fractions, Mathematics Magazine, Vol.54, n°2 (1981), 60-63.

Laboratoire de Statistique et Probabilités
E.R.A.-C.N.R.S. 591
Université Paul Sabatier
118, route de Narbonne
31062 Toulouse Cedex, France.

UNE CONDITION SUFFISANTE DE RECURRENCE POUR DES CHAINES DE MARKOV SUR LA DROITE

Y. DERRIENNIC

Université de Bretagne Occidentale
Département de Mathématiques et Informatique
BREST

Abstract : A direct, elementary, proof is given to the following result :

"Let $P(x,dy)$ a transition probability on the real line. Assume that :

i) for f continuous and bounded, Pf is continuous.

ii) P is irreductible, with respect to open sets.

iii) for some constant K, $P(x,]-\infty, -K[) = 0$ for x close enough to $+\infty$ and $P(x,]+K, +\infty[) = 0$ for x close enough to $-\infty$.

iv) for x outside a compact set, $\int y\, P(x,dy) = x$

Then the Markov chain associated to P is topologically recurrent on the line"

A similar result is given on $\mathbb{Z}$.

INTRODUCTION

Une marche aléatoire ayant un moment d'ordre 1 est récurrente sur la droite réelle si et seulement si elle est centrée. Cette propriété classique suggère la question suivante : existe t'il des conditions suffisantes et/ou nécessaires de récurrence pour des chaînes de Markov "adaptées" à la structure de la droite ? Parmi les méthodes connues pour démontrer la récurrence des marches aléatoires centrées, aucune ne semble s'étendre au cas de chaînes de Markov plus générales, car toutes utilisent la commutativité des translations et de l'opérateur de transition de la marche.

Le présent article donne une condition suffisante de récurrence pour des chaînes de Markov qui sont aussi des martingales, autrement dit dont les accroissements sont conditionnellement centrés. La démonstration repose sur une étude élémentaire de fonctions réelles vérifiant une propriété de "surmoyenne". Le résultat obtenu est à rapprocher de résultats voisins concernant le comportement asymptotique des martingales (voir le livre de Neveu, chapître VII, [4]) et aussi de résultats récents de Cocozza-Thivent, Kipnis et Roussignol ([1]). Pour des diffusions cette question a fait l'objet de travaux bien connus (voir, par exemple, le livre de Friedman ([3]). Pour les propriétés classiques des chaînes de Markov et des marches aléatoires on renvoie au livre de Revuz ([5]).

ENONCE DU RESULTAT

Soit $P(x,dy)$ une probabilité de transition sur $\mathbb{R}$. On note P_x la loi de la chaîne (canonique) X_n issue de x et de transition P. L'action de P sur les fonctions mesurables bornées ou positives est notée $Pf(x) = \int f(y)\, P(x,dy)$.

Par définition, la chaîne est récurrente topologiquement si, pour tout $x \in \mathbb{R}$ et tout voisinage V_x de x, on a

$$P_x\ (\limsup\ (X_n \in V_x)) = 1$$

(d'autres notions de récurrence, par exemple la récurrence-Harris, pourraient aussi être envisagées mais ne le seront pas).

Pour que le problème soit bien posé il faut que P possède certaines propriétés de régularité liées à la structure de $\mathbb{R}$. Dans la suite on utilisera les hypothèses suivantes :

- P possède la propriété de Feller, i.e. pour toute fonction f continue et bornée, Pf est continue.

- P est topologiquement irréductible, i.e. pour tout $x \in \mathbb{R}$ et tout ouvert V il existe n tel que $P_x\ (X_n \in V) > 0$.

Le résultat qu'on se propose de démontrer s'énonce alors ainsi :

<u>Théorème</u> : *Si $P(x, dy)$ est une probabilité de transition sur $\mathbb{R}$ satisfaisant aux hypothèses :*

1) P a la propriété de Feller et est topologiquement irréductible,

2) Il existe une constante K positive telle que $P(x,]-\infty\ , -K]) = 0$ pour x assez voisin de $+\infty$ et $P(x, [+K, +\infty[) = 0$ pour x assez voisin de $-\infty$,

3) pour tout x hors d'un compact, $P(x, dy)$ est centrée en x, i.e.

$$\int y\ P(x,dy) = x,$$

alors la chaîne de Markov de transition P est topologiquement récurrente.

Les hypothèses du théorème sont vérifiées par une marche aléatoire adaptée de loi centrée et à support compact. Parmi les nombreux autres exemples qu'il est facile de construire, il est intéressant de considérer particulièrement celui défini par :

$$P(x,dy) = \frac{1}{\delta(x)}\ 1_{[x - \frac{1}{2}\delta(x),\ x + \frac{1}{2}\delta(x)]}(y)\ dy$$

où $\delta(x)$ est une fonction continue et strictement positive vérifiant $\delta(x) \leq |x| + K$. La continuité donne la propriété de Feller, la stricte positivité donne l'irréductibilité ; les hypothèses 2 et 3 sont évidentes. La chaîne associée s'écrit $X_{n+1} = X_n + \frac{1}{2}\delta(X_n)\ Y_{n+1}$, avec $(Y_n)_n$ une suite de variables indépendantes uniformément distribuées sur $[-1, +1]$; c'est évidemment une martingale.

La démonstration du théorème repose sur la caractérisation des fonctions positives et sur-invariantes pour P, donnée par la proposition suivante :

Proposition : Soit $P(x,dy)$ une probabilité de transition sur $\mathbb{R}$, topologiquement irréductible, et vérifiant les hypothèses 2 et 3 du théorème. Si une fonction f positive semi-continue inférieurement (s.c.i.) est sur-invariante (i.e. pour tout x, $f(x) \geq \int f(y)\, P(x,dy)$) alors elle est constante.

On verra dans la démonstration que, pour cet énoncé, la condition de Feller et même la mesurabilité en x de $P(x,dy)$ ne sont pas requises.

DEMONSTRATIONS

Tout d'abord indiquons brièvement comment le théorème résulte de la proposition. Ceci est tout à fait classique (voir, par exemple, Foguel [2]). Si V est ouvert, 1_V est s.c.i. et d'après la condition de Feller, la fonction $h(x) = P_x\,(\bigcup_{n\geq 0} (X_n \in V))$ est aussi s.c.i. Sur V elle vaut 1. Comme elle est sur-invariante la proposition dit qu'elle est constamment égale à 1. Alors $P_x(\limsup (X_n \in V)) = \lim_n (P^n h)(x)$ est aussi égale à 1 et la chaîne est récurrente.

La démonstration de la proposition résulte des lemmes suivants :

Lemme 1 : (principe du minimum). Supposons P irréductible topologiquement. Toute fonction s.c.i., sur-invariante, qui atteint son minimum en un point est constante.

Démonstration : soit f s.c.i. positive, non nulle. Pour $\varepsilon > 0$ assez petit, $\{f > \varepsilon\}$ est un ouvert V non vide. Par irréductibilité, pour tout x il existe n tel que $P^n(x,V) > 0$, donc $P^n f(x) > 0$. Si f est sur-invariante, $f(x) \geq P^n f(x) > 0$, donc f ne peut pas s'annuler.

D'après ce lemme il suffit, pour démontrer la proposition, de prouver que toute fonction positive s.c.i. sur-invariante atteint son minimum.

Lemme 2 : (variation sur l'inégalité de Jensen). Soit μ une probabilité centrée sur $\mathbb{R}$, telle que $\mu\{0\} = 0$. Etant donnée f mesurable et positive, il existe s et t dans le support de μ, avec $s < 0 < t$ tels que

$$\int f d\mu \geq \frac{s}{s-t} f(t) + \frac{t}{t-s} f(s)$$

(si f n'est pas linéaire μp.p., on a l'inégalité stricte).

Démonstration : Seul le cas où $\int f d\mu < \infty$ est à considérer. Notons $M_{s,t}(x)$ la fonction linéaire qui interpole f entre s et t. L'inégalité cherchée s'écrit alors $\int f d\mu \geq M_{s,t}(0)$. Notons $\ell_t(x)$ la fonction linéaire définie par

$\ell_t(x) = \alpha(t)\, x + \int f d\mu$ où $\alpha(t) = \frac{1}{t}\,(f(t) - \int f d\mu\,)$ avec $t > 0$. Soit S le support de μ, $S^+ = S \cap]0, +\infty[$, $S^- = S \cap]-\infty, 0[$; ces deux ensembles sont non vides. Observons que l'inégalité $\ell_t(s) \geq f(s)$ entraîne $\int f d\mu \geq M_{s,t}(0)$ $(s < 0 < t)$. Dans le cas $\inf_{t \in S^+} \alpha(t) = -\infty$ l'inégalité cherchée est donc une conséquence directe du fait que f est finie μp.p. sur S^-. Dans le cas $\inf_{t \in S^+} \alpha(t) = \alpha > -\infty$ on a $f(x) \geq \alpha x + \int f d\mu$ pour tout $x \in S^+$. S'il existe $s \in S^-$ tel que $f(s) < \alpha s + \int f d\mu$, il existe $t \in S^+$ tel que $f(s) < \ell_t(s)$ et le résultat s'en suit. Sinon $f(x) \geq \alpha x + \int f d\mu$ en tout $x \in S^-$ et comme $\mu\{0\} = 0$,

$f(x) \geq \alpha x + \int f d\mu$ μp.p. Or, μ étant centrée, $\int f(x)\, d\mu\,(x) = \int (\alpha x + \int f d\mu)\, d\mu(x)$ et donc $f(x) = \alpha x + \int f d\mu$ μp.p. Ceci achève la démonstration.

Pour alléger la rédaction posons la :

Définition : *Une fonction f définie sur un intervalle I est dite partiellement sur-moyenne en x s'il existe un intervalle d'extrémités a et b dans I, dont l'intérieur contient x et tel que*

$$f(x) \geq \frac{x-a}{b-a}\, f(b) + \frac{b-x}{b-a}\, f(a).$$

Le lemme 2 va nous permettre de montrer qu'une fonction sur-invariante est partiellement sur-moyenne au voisinage de l'infini.

Lemme 3 : *Supposons P irréductible et vérifiant les hypothèses 2 et 3 du théorème Il existe $A > 0$ tel que toute fonction f définie sur $[-K, +\infty[$ (resp. $]-\infty, K]$), positive et sur-invariante en tout point de $[A, +\infty[$ (resp. $]-\infty, -A]$) soit partiellement sur-moyenne en tout $x \geq A$ (resp. $x \leq -A$).*

Démonstration : D'après les hypothèses 2 et 3, il existe $A > 0$ tel que $P(x,dy)$ est centrée en x pour $|x| \geq A$, $P(x,]-\infty, -K]) = 0$ pour $x \geq A$, $P(x, [K, +\infty[\,) = 0$ pour $x \leq -A$. Soit f définie sur $[-K, +\infty[$, positive et sur-invariante en tout $x \geq A$ i.e. $f(x) \geq \int f(y)\, P(x,dy)$. Comme P est topologiquement irréductible, $P(x, \{x\}\,) \neq 1$ et on peut considérer la probabilité

$$\mu_x(dy) = \frac{1}{1 - P(x,\{x\})}\,(P(x,dy) - P(x,\{x\})\, \varepsilon_x\,(dy))$$ où ε_x est la mesure de Dirac en x.

Elle est centrée en x, ne charge pas x et

$$f(x) \geq \int f(y)\, \mu_x(dy), \quad (x \geq A).$$

Le lemme 2 dit alors que f est partiellement sur-moyenne en x relativement à un intervalle contenu dans $[-K, +\infty[$ car $P(x,]-\infty, -K]) = 0$. Le lemme est ainsi démontré.

Lemme 4 : Soit f une fonction définie sur une demi-droite $[B, +\infty[$ (resp. $]-\infty, B]$), positive, s.c.i., nulle en un point $s > B$ (resp. $s < B$). Si f est partiellement sur-moyenne en tout $x \geq s$ (resp. $x \leq s$) alors l'ensemble des x où $f(x) = 0$ est non borné.

Démonstration : Si cet ensemble était borné il aurait une borne supérieure z. Comme f est s.c.i. et positive on aurait $f(z) = 0$, et f ne pourrait pas être partiellement sur-moyenne en z.

Lemme 5 : Soit f une fonction définie sur $[-K, +\infty[$ (resp $]-\infty, K]$) positive, s.c.i. et partiellement sur-moyenne en tout $x \geq A$ (resp. $x \leq -A$). Si $\liminf_{x\to+\infty} f(x) = 0$ (resp. $\liminf_{x\to-\infty} f(x) = 0$) alors f s'annule en un point $t \in [-K, A]$ (resp. $[-A, K]$).

Démonstration : Supposons que f ne s'annule pas sur $[-K, A]$. Comme f est s.c.i. posons $\inf_{-K\leq x\leq A} f(x) = \alpha > 0$. Notons ℓ_λ la fonction linéaire $\ell_\lambda(x) = \lambda x + \alpha + \lambda K$. A l'aide de l'hypothèse $\liminf_{x\to+\infty} f(x) = 0$, c'est alors un exercice sur la semi-continuité inférieure de montrer qu'il existe, parmi les fonctions ℓ_λ qui vérifient $\ell_\lambda \leq f$ sur $[A, +\infty[$, une fonction ℓ_{λ_0} plus grande que toutes les autres "tangente" à f avec $\lambda_0 < 0$ (on pose $\lambda_0 = \sup(\lambda;\ \ell_\lambda \leq f$ sur $[A, +\infty[$); alors $\lambda_0 < 0$, $f - \ell_{\lambda_0}$ est s.c.i. et $\lim_{x\to+\infty} (f(x) - \ell_{\lambda_0}(x)) = +\infty$; $f - \ell_{\lambda_0}$ atteint donc son minimum (relatif à $[A, +\infty[$) en un point $s > A$ et ce minimum est nécessairement nul). La fonction $f - \ell_{\lambda_0}$ vérifie alors les hypothèses du lemme 4 sur $[A, +\infty[$, ce qui est contradictoire avec $\lim_{x\to+\infty} (f(x) - \ell_{\lambda_0}(x)) = +\infty$ Le lemme est donc démontré.

Conclusion : Considérons une fonction f positive, s.c.i., sur-invariante sur IR, la probabilité de transition P étant irréductible topologiquement et vérifiant les hypothèses 2 et 3 du théorème. D'après le lemme 3, f satisfait aux hypothèses du lemme 5. Donc, si $\liminf_{x\to+\infty} f(x) = 0$ (resp. $\liminf_{x\to-\infty} f(x) = 0$) f s'annule en un point de $[-K, A]$ (resp. $[-A, K]$). Dans tous les cas, si $\inf_{x\in\mathbb{R}} f(x) = 0$, f doit s'annuler en un point. D'après le lemme 1, f est alors constamment nulle. Ceci achève la démonstration de la proposition et donc du théorème.

LE RESULTAT SUR $\mathbb{Z}$

En suivant exactement la même méthode on peut démontrer un résultat analogue sur l'ensemble des entiers, dont l'énoncé précis est le suivant.

Théorème : *Si $P(i, j)$ est une probabilité de transition sur $\mathbb{Z}$ satisfaisant aux hypothèses :*

1) P est irréductible, i.e. pour tout $i, j \in \mathbb{Z}$ il existe n tel que $P^n(i,j) > 0$,

2) il existe $K > 0$ tel que $\sum_{j<-K} P(i,j) = 0$ pour i assez voisin de $+\infty$, et $\sum_{j>K} P(i,j) = 0$ pour i assez voisin de $-\infty$,

3) pour tout i hors d'un ensemble fini, $P(i,j)$ est centrée en i, i.e. $\sum_j j\, P(i,j) = i$,

alors la chaîne de Markov de transition P est récurrente sur $\mathbb{Z}$.

REMARQUES SUPPLEMENTAIRES

D'après Skorokhod toute probabilité centrée sur $\mathbb{R}$ est la loi du mouvement brownien arrêté à un instant (aléatoire) convenable. Ce résultat classique permet d'interpréter le théorème démontré ci-dessus comme une condition sur une suite croissante de temps d'arrêt T_n suffisante pour que la chaîne $X_n = W_{T_n}$ obtenue en arrêtant le mouvement brownien aux instants T_n, conserve la propriété de récurrence que le mouvement brownien lui-même possède en dimension 1. Cette interprétation conduit à un résultat analogue pour le plan, où le mouvement brownien est encore récurrent. Cependant l'énoncé exact et sa démonstration posent des problèmes délicats ; la condition de centrage n'est plus suffisante. Ceci fera l'objet d'une autre rédaction. Indiquons seulement que le résultat sur les fonctions sur-moyennes s.c.i., (lemme 5) peut être démontré par une méthode différente dans laquelle on considère l'enveloppe inférieure des fonctions linéaires intervenant dans les inégalités de sur-moyenne. Cette méthode est moins élémentaire en dimension 1 mais a l'avantage de s'adapter à la dimension 2.

R E F E R E N C E S

[1] C. Cocozza-Thivent, C. Kipnis, M. Roussignol (1982)
Stabilité de la récurrence nulle pour certaines chaînes de Markov perturbées.
(à paraître)

[2] S.R. Foguel (1973). The ergodic theory of positive operators on continuous functions.
Annali della Scuola Normale Superiore di Pisa. Classe di Scienze. Vol XXVII
Fasc. 1, 19-51

[3] A. Friedman (1975) Stochastic differential equations and applications. Academic Press.

[4] J. Neveu (1972). Martingales à temps discret. Masson et Cie

[5] D. Revuz (1975). Markov chains. North Holland pub.

-o-o-o-O-o-o-o-

Stable Laws on the Heisenberg Groups.

Thomas Drisch / Léonard Gallardo

Abstract: We determine the generating distributions of the full continuous convolution semigroups of probabilities on the Heisenberg groups which are stable in the sense of Hazod. We obtain a classification of the limit distributions on the Heisenberg groups for the case of identically distributed random variables without centering.

0 Introduction

A probability measure μ on a locally compact group G is called embeddable iff there exist a continuous convolution semigroup of probability measures μ_t on G with $\mu_1 = \mu$. An embeddable probability μ is called stable with associated semigroup $(\mu_t : t \in \mathbb{R}_+^*)$ iff there exist a (multiplicative) continuous one-parameter group $(\tau(t) : t \in \mathbb{R}_+^*)$ of automorphisms of G with $\mu_t = \tau(t)(\mu)$. This notion of stability which generalises M. Sharpe's notion of operator-stable measures on vector groups as well as the notion of dilation-stable measures on locally compact groups, was introduced by W. Hazod in [3].
We determine all one-parameter automorphism groups of the n-th Heisenberg groups $\mathbb{H}_n$ which allow full stable continuous convolution semigroups of probabilities. They are given by conjugations of the mappings $t \mapsto \left(\begin{array}{c|c} t^{m+M} & o \\ \hline o & t^{2m} \end{array}\right)$ with an inner automorphism of $\mathbb{H}_n$,

where $m \in [1/2, \infty)$, and where M is a non-singular operator of the symplectic Lie algebra of the 2n - dimensional "plane" of $\mathbb{H}_n$ with certain spectral conditions (theorem 1). Therefore the generating distributions of the full stable continuous convolution semigroups are certain generating functionals of full operator - stable probabilities on the vector space underlying the Lie algebra $\mathfrak{h}_n$ of $\mathbb{H}_n$ or on an affine hyperplane of $\mathfrak{h}_n$ (which are completely known). The 4 th paragraph shows that in our case Hazod's notion is the same as the notion of P. Baldi introduced for the groups of n - dimensional motions in [1], characterising stable probabilities as weak limits of the distributions of automorphic transformations of partial products of independent identically distributed group - valued random variables (theorem 2). As a by - product of the proof, we obtain the uniqueness of the associated semigroup of a stable probability on $\mathbb{H}_n$.

1 Automorphisms of the Heisenberg groups

Let W be a $(2n+1)$ - dimensional real vector space; let V be a $2n$ - dimensional subspace of W; and let P be a projection from W on V. We write $Q := 1-P$, $Z := Q(W)$, and - for simplicity -

$$\bar{w} := Pw, \quad w' := Qw, \quad w =: (\bar{w}, w')$$

for $w \in W$. Let β be a nondegenerate symplectic form on V; let e be a nontrivial element of $Q(W)$. Then, W with the bracket

$$[u,w] := \beta(\bar{u},\bar{w})\, e \qquad u,w \in W$$

is a realisation of the n - th Heisenberg Lie algebra $\mathfrak{h}_n$.
The group structure of the associated simply connected real Lie group $\mathbb{H}_n$, the n - th Heisenberg group, realised on W, is given by the Campbell - Hausdorff formula:

$$uw = u + w + \frac{1}{2}[u,w] \qquad u,w \in W.$$

With respect to these special realisations, the exponential $\exp_{\mathbb{H}_n} : \mathfrak{h}_n \mapsto \mathbb{H}_n$ is the identity id_W of W, the center of $\mathbb{H}_n$ is identified with Z, and we have $o = e_{\mathbb{H}_n}$ (the neutral element of $\mathbb{H}_n$) and $u^{-1} = -u$ for $u \in \mathbb{H}_n$.
Let $Sp(V,\beta)$ resp. $\widetilde{Sp}(V,\beta)$ denote the group resp. the set of all $A \in Gl(V)$ which are β- symplectic resp. β- skewsymplectic.
$S(V,\beta) := Sp(V,\beta) \cup \widetilde{Sp}(V,\beta)$ operates as automorphism group on $\mathbb{H}_n$ by

$$(A,w) \mapsto \alpha_A w := \begin{cases} (A\bar{w}, w') & A \in Sp(V,\beta) \\ (A\bar{w}, -w') & A \in \widetilde{Sp}(V,\beta). \end{cases} \quad \text{for}$$

Further, $\mathbb{R}_+^*$ operates as group of dilations on $\mathbb{H}_n$ by

$$(s,w) \mapsto \delta_s w := (s\bar{w}, s^2 w') \qquad s \in \mathbb{R}_+^*.$$

$S(V,\beta)$ and $\mathbb{R}_+^*$ will be identified with their corresponding subgroups of the automorphism group $Aut(\mathbb{H}_n)$ of $\mathbb{H}_n$.

Let $Inn(\mathbb{H}_n)$ denote the group of inner automorphisms

$$inn\, u : w \mapsto uwu^{-1} \qquad u,w \in W$$

of $\mathbb{H}_n$. Using the identification of $\mathbb{H}_n$ and $\mathfrak{h}_n$, the formula

$$inn\, u(\exp w) = \exp(Ad\, u)w \qquad u,w \in W$$

shows inn u = Ad u for all $u \in W$. The formula

1.1 $$\operatorname{inn} u\,(w) = w + \beta(\bar{u},\bar{w})e \qquad u,w \in W$$

shows with the nondegeneracy of β that inn u_1 = inn u_2 is equivalent to $\bar{u}_1 = \bar{u}_2$ in $\operatorname{Aut}(\mathbb{H}_n)$ and $\operatorname{Aut}(\mathfrak{h}_n)$.

Therefore, we identify V with his additive structure via

$$v \longmapsto \operatorname{inn}(v,o) = \operatorname{Ad}(v,o) \qquad v \in V$$

with $\operatorname{Inn}(\mathbb{H}_n)$ resp. $\operatorname{Ad}(\mathbb{H}_n)$, and we denote inn(v,o) simply by inn v.

1.2 Prop.: $\operatorname{Aut}(\mathbb{H}_n)$ is the semidirect product of V by $S(V,\beta) \oplus \mathbb{R}_+^*$, the defining homomorphism $\gamma : S(V,\beta) \oplus \mathbb{R}_+^* \longrightarrow \operatorname{Aut}(V)$ being given by

$$\gamma(A,s)\,v := Asv \qquad v \in V.$$

1.3 Cor.: The identity component of $\operatorname{Aut}(\mathbb{H}_n)$ is $V \circledS (Sp(V,\beta) \oplus \mathbb{R}_+^*)$.

1.4 Cor.: $\operatorname{Aut}(\mathfrak{h}_n) = V \circledS (S(V,\beta) \oplus \mathbb{R}_+^*)$.

Convention: We denote inn v o α_A o $\delta_s \in \operatorname{Aut}(\mathbb{H}_n)$ by (v;A,s).

Proof: Obviously, $S(V,\beta)$ and $\mathbb{R}_+^*$ are commuting. As for every group, $V = \operatorname{Inn}(\mathbb{H}_n)$ is normal in $\operatorname{Aut}(\mathbb{H}_n)$. Because $P \circ \operatorname{Inn}(\mathbb{H}_n)$ operates identically on V (see 1.1),

$$V \cap (S(V,\beta) \oplus \mathbb{R}_+^*) = \{\operatorname{id}_{\mathbb{H}_n}\}.$$

Because the operation in V as subgroup of $\operatorname{Aut}(\mathbb{H}_n)$ is the vector addition, $\gamma(A,s)$ is an automorphism of $V = \operatorname{Inn}(\mathbb{H}_n)$, and γ is homomorphic. A short calculation shows, using the notational convention:

1.5 $$(u;A,s)\,(w;B,t) = (u(\gamma(A,s)w);\, AB,\, st).$$

Let $\xi \in \operatorname{Aut}(\mathbb{H}_n)$. Then, $\xi(Z) \subset Z$, and there exist $s \in \mathbb{R}_+$, $\varepsilon \in \{1,-1\}$ with $\xi e = s^2 \varepsilon e$; clearly we have $s > o$. Define an operator $A: V \longrightarrow V$ by

$$A\bar{w} := s^{-1} P\,\xi(\bar{w},o) \qquad w \in V.$$

Identifying $\mathfrak{h}_n$ with the tangent space of $\mathbb{H}_n$ at the identity, and denoting the differential of ξ at $e_{\mathbb{H}_n}$ by $D\xi$, the equation

$$\xi(\exp u) = \exp((D\xi)u) \qquad u \in \mathfrak{h}_n = W$$

shows that ξ operates as a linear transformation on W. Especially, A is linear. Obviously, A is surjective on V. Comparing the equations

$$\xi(\bar{u},o)\,\xi(\bar{w},o) = \xi(\bar{u},o) + \xi(\bar{w},o) + \tfrac{1}{2}\beta(sA\bar{u},sA\bar{w})e,$$

$$\xi((\bar{u},o)(\bar{w},o)) = \xi(\bar{u} + \bar{w},o) + \tfrac{1}{2}\beta(\bar{u},\bar{w})\; s^2\,\varepsilon\, e,$$

we obtain the symplectic resp. skewsymplectic nature of A.
Define v as the unique solution of

$$\beta(v,sA\bar{w}) = Q\xi(\bar{w},o) \qquad\qquad \bar{w}\in V.$$

Then a short calculation using the linearity of ξ shows $\xi = (v;A,s)$, and the algebraic part of the assertion 1.2 is proven.

$\mathbb{H}_n$ being simply connected, $id_W = D_{\mathbb{H}_n} : \xi \longmapsto D\xi$ is an analytic isomorphism of Aut($\mathbb{H}_n$) on Aut($\mathfrak{h}_n$). Finally, considering Aut($\mathfrak{h}_n$) as subgroup of Gl(W), we see that Aut($\mathfrak{h}_n$) = $V \circledS (S(V,\beta) \oplus \mathbb{R}_+^*)$ (and also Aut($\mathbb{H}_n$)) inherits the product topology of the natural topologies of V, $S(V,\beta)$, $\mathbb{R}_+^*$. □

2 One-parameter automorphism groups on the Heisenberg groups

A (one-parameter) group of automorphisms on a locally compact group G is a continuous homomorphism τ: $\mathbb{R}_+^* \longrightarrow \mathrm{Aut}(G)$. τ is called contracting iff for any compact subset K of G and any neighbourhood V of the identity of G there exists $x>0$ such that $\tau(y)(K) \subset V$ for all $y<x$.

Convention: In the following, a (contracting) automorphism group means a multiplicative one-parameter continuous (contracting) automorphism group.

2.1 Prop.: Let τ be an automorphism group on $\mathbb{H}_n$. Then:

2.2 $$\tau(x) = ((v - x^{m+M}v) - v_x;\, x^M,\, x^m),$$

where $v\in V$, $m\in\mathbb{R}$, $M\in sp(V,\beta)$ (the symplectic Lie algebra of β), and $x \longmapsto v_x \in V$ is a solution of the (non-unitary) firstcocycle equation

2.3 $$v_x + x^{m+M}\, v_y = v_{xy}$$

where all v_x are elements of the weight space V_{-m} of M to the characteristic value $-m$ of M.

Proof: 1. Denote $\tau(x)$ by $(v_x; A_x, s_x)$. By continuity of τ, $\tau(\mathbb{R}_+^*) \subset V \circledS (Sp(V,\beta) \oplus \mathbb{R}_+^*)$ (Cor. 1.4). 1.5 is equivalent to

2.4 $$A_x A_y = A_{xy},\ s_x s_y = s_{xy},\ v_x + A_x s_x v_y = v_{xy},$$

giving continuous homomorphisms $A: \mathbb{R}_+^* \to Sp(V,\beta)$, $s: \mathbb{R}_+^* \to \mathbb{R}_+^*$ with the solutions

$$A_x = e^{(\log x)M} =: x^M,\ s_x = x^m$$

where $M \in sp(V,\beta)$, $m \in \mathbb{R}$, and the third equation of 2.4 takes then the form 2.3.

2. To solve equation 2.3, denote by $V_\lambda^{\mathbb{C}}$ the weight space of M with respect to the characteristic value λ of M. Let $W_{-m}^{\mathbb{C}}$ be the direct sum of all $V_\lambda^{\mathbb{C}}$ with $\operatorname{Re}\lambda = -m$; let $\widetilde{W}^{\mathbb{C}}$ be the direct sum of all $V_\lambda^{\mathbb{C}}$ with $\operatorname{Re}\lambda \neq -m$. Then, with $W_{-m} := W_{-m}^{\mathbb{C}} \cap V$, $\widetilde{W} := \widetilde{W}^{\mathbb{C}} \cap V$: $V = W_{-m} \oplus \widetilde{W}$; and the restriction of $id_V - x^{m+M}$ on $\widetilde{W}$ is an isomorphism $A_x : \widetilde{W} \to \widetilde{W}$.

3. Let $x \mapsto w_x$ be a solution of 2.3 with $w_{x_0} = 0$ for a positive real $x_0 \neq 1$. Inserting $x := x_0$ resp. $y := x_0$ in 2.4 we obtain

$$x_0^{m+M} w_x = w_{xx_0},\quad w_x = w_{xx_0} \qquad x \in \mathbb{R}_+^*.$$

Therefore $w_x \in W_{-m}$ for all $x \in \mathbb{R}_+^*$.

4. Let $x \mapsto u_x$ be a solution of 2.3 with $u_{x_1} \in \widetilde{W}$ for a $x_1 \in \mathbb{R}_+^*$ with $x_1 \neq 1$. Define:

$$v := A_{x_1}^{-1} u_{x_1},\quad v_x := (v - x^{m+M} v) - u_x.$$

Then $x \mapsto v_x$ is a solution of 2.3 with $v_{x_1} = 0$. Obviously, $v \in \widetilde{W}$, and

$$u_x = (v - x^{m+M} v) + v_x \quad \text{with} \quad v_x \in W_{-m} \text{ for all } x \in \mathbb{R}_+^*.$$

5. Denote the weight space of $-m+ri$ by $V_{-m+ri}^{\mathbb{C}}$ $(i := \sqrt{-1})$ and the real weight space of $-m$ by V_{-m}. Assume $V_{-m} \neq W_{-m}$. Consider the linear transformation B_x, induced by $id_V - x^{m+M}$ on W_{-m}. Let $r_1,\dots,r_{2k}$ be the imaginary parts $\neq 0$ of the characteristic values λ of M with $\operatorname{Re}\lambda = -m$. If there exist $x_0 \in \mathbb{R}_+^*$ such that $(\log x_0) r_j \notin 2\pi\mathbb{Z}$ for $j = 1, \dots, 2k$ and such that $x_0^{m+M} v_{x_0} = 0$, then the argument of the third step shows:

$$e^{(\log x_0)\, r_j} (v_x)_j = (v_x)_j$$

where $(v_x)_j$ denotes the component of v_x in $V^{\mathbb{C}}_{-m+r_j i}$ $(j = 1, \ldots, 2k)$; therefore $(v_x)_j = 0$ and $v_x \in V_{-m}$.

6. In the other case, we can use the argument of the forth step showing

$$v_x = (\tilde{v} - x^{m+M}\tilde{v}) + \tilde{\tilde{v}}_x$$

with $\tilde{v} \in (\bigoplus V^{\mathbb{C}}_{-m+r_i i}) \cap V$ and $\tilde{\tilde{v}}_x \in V_{-m}$. □

Let $\sigma_{M,m}$ denote the continuous automorphism group $x \mapsto (0;\, x^M,\, x^m)$; let τ be given by 2.2 with $v_x = 0$ for all $x \in \mathbb{R}^*_+$. Then a short calculation shows:

2.5 $$\tau(x) = \operatorname{inn}(v) \circ \sigma_{M,m}(x) \circ \operatorname{inn}(-v),$$

i.e. a conjugation of τ by an inner automorphism "separates the variables".

2.6 Cor.: Let τ_0 be a contracting automorphism group of $\mathbb{H}_n$. Then:

$$\tau_0(x) = (v - x^{m+M} v;\; x^M;\; x^m)$$

where $v \in V$, $m \in \mathbb{R}^*_+$, $M \in sp(V,\beta)$, and where the characteristic values λ of M fulfill $\operatorname{Re}\lambda > -m$.

Proof: The conditions on m resp. M are necessary for the contracting property on Z resp. V. Especially, $V_{-m} = \{0\}$, implying $v_x = 0$ for all $x \in \mathbb{R}^*_+$. Conversely, the formula

$$\tau_0(x)\, u = (x^{m+M}\bar{u},\; x^{2m} u' + x^m \beta(v - x^{m+M} v,\, x^M \bar{u}))$$

shows that the conditions are sufficient. □

For $j \in \mathbb{N}$, let $j(x)$ denote the vector

$$\left(\frac{\log^j x}{j!},\; \frac{\log^{j-1} x}{(j-1)!},\; \ldots,\; \log x\right) \in \mathbb{R}^j.$$

2.7 Prop.: There exists an ordered basis of V_{-m} - the weight space of M to $-m$ - such that the coordinate vector of v_x in the formula 2.2 is given by $(r_1 j_1(x), \ldots, r_k j_k(x)) \in \mathbb{R}^{j_1 + \ldots j_k}$.

Proof: Let A_j resp. $B_j(x)$ denote the (j,j) - matrices

$$\begin{pmatrix} 0 & 1 & & 0 \\ & \ddots & \ddots & \\ & & \ddots & 1 \\ 0 & & & 0 \end{pmatrix} \quad \text{resp.} \quad \begin{pmatrix} 1 & \log x & \cdots & \frac{\log^j x}{j!} \\ & \ddots & \ddots & \vdots \\ & & \ddots & \log x \\ 0 & & & 1 \end{pmatrix}.$$

The restriction of $m + M$ on V_{-m} defines a nilpotent linear transformation $A : V_{-m} \to V_{-m}$. There exists an ordered basis of V_{-m} such that the matrix of A with respect to this basis is given by

$$\begin{pmatrix} A_{j_1} & & 0 \\ & \ddots & \\ 0 & & A_{j_k} \end{pmatrix}.$$ Then the matrix of x^A to the same ordered basis is $$\begin{pmatrix} B_{j_1}(x) & & 0 \\ & \ddots & \\ 0 & & B_{j_k}(x) \end{pmatrix}.$$

Obviously, $(r_1 j_1(x), \ldots, r_k j_k(x))$ is a solution of

$v_x + x^A v_y = v_{xy}$, and a standard argument shows, that this solution is the only continuous one. □

We say that τ <u>contracts</u> $F \subset G$ iff for every neighbourhood U of e there exist $t_U \in \mathbb{R}_+^*$ such that $\tau_t(F) \subset U$ for $t \geq t_U$. We say that τ is <u>contracting</u> on F iff τ contracts every compact subset of F. We define:

$$C(\tau) := \{g \in G : \lim_{x \to 0} \tau(x) g = e\}.$$

Obviously, $C(\tau)$ is a connected subgroup of G which is, for all $x \in \mathbb{R}_+^*$, $\tau(x)$ - invariant.

Now let G be a Lie group with Lie algebra $\mathfrak{g}$. We denote the differential of $\tau(x)$ at the identity e_G of G bei $D\tau(x)$ and define:

$$C(D\tau) := \{X \in \mathfrak{g} : \lim_{x \to 0} D\tau(x) X = 0\}.$$

Then $(D\tau(x) : x \in \mathbb{R}_+^*)$ is a continuous automorphism group of $\mathfrak{g}$, and $C(D\tau)$ is a Lie subalgebra of $\mathfrak{g}$ which is, for all $x \in \mathbb{R}_+^*$, $D\tau(x)$ - invariant. From [3], 2.7 we obtain:

2.8 <u>Lemma:</u> $C(\tau)$ is nilpotent simply connected, and $\exp_G$ induces a diffeomorphism $C(D\tau) \to C(\tau)$.

2.9 Prop.: τ is contracting on $C(\tau)$.

Proof: As a continuous one-parameter group of $\mathfrak{g}$, $\tau(x) = x^M$ with a derivation M of $\mathfrak{g}$. By the invariance of $C(D\tau)$, $\tau(x)$ induces a continuous automorphism group $\sigma(x) = x^{\tilde{M}}$ of $C(D\tau)$.
Because $\lim_{x\to 0} x^{\tilde{M}}(X) = o$ for all $X \in C(D\tau)$, the spectrum of M is a subset of $\{z \in \mathbb{C} : \operatorname{Re} z > o\}$. Therefore $(D\tau(x) : x \in \mathbb{R}_+^*)$ is contracting on $C(D\tau)$. By 2.8. we have $\tau(x) = \exp_G \circ D(x) \circ \log_G$; this equation implies, again with 2.8., that τ is contracting on $C(\tau)$. □

2.10 Example: The Heisenberg groups. 2.1 and 2.3 show:

$$\tau(x)u = (x^{m+M}\bar{u}, x^{2m}u' + \beta(v - x^{m+M}v + v_x, x^{m+M}\bar{u})),$$

$$\beta(v_x, x^{m+M}\bar{u}) = x^{2m}\,\beta(x^{-(m+M)}v_x, \bar{u}) = -x^{2m}\,\beta(v_{x^{-1}}, \bar{u}) .$$

For the aim of determining the subgroups $C(\tau) \subset \mathbb{H}_n$, denote by $\tilde{W}$ the intersection of V with the direct sum of the weight spaces of M to the characteristic values λ with $\operatorname{Re}\lambda > -m$.
Then, by 2.7:

$$\begin{aligned}
v_x &\neq o,\ m > o: & C(\tau) &= \tilde{W} \oplus \\
v_x &\neq o,\ m \leq o: & C(\tau) &= \{o\} = \{e_{\mathbb{H}_n}\}, \\
v_x &= o,\ m > o: & C(\tau) &= \tilde{W} \oplus Z\ , \\
v_x &= o,\ m \leq o: & C(\tau) &= \tilde{W}\ .
\end{aligned}$$

3 Generators of stable semigroups on the Heisenberg groups

Let $\tau : \mathbb{R}_+^* \to G$ be an automorphism group of the locally compact group G. Let $(\mu_t : t \in \mathbb{R}_t^*)$ be a (weakly) continuous convolution semigroup of probability measures μ_t on G (abbreviated c.c.s.) enjoying

3.1 $$\lim_{t\downarrow 0} \mu_t = \varepsilon_e$$

(ε_g denotes the point measure in $g \in G$). (μ_t) is called τ- stable resp. strictly τ - stable iff there exist a one-parameter subgroup $(x_t : t \in \mathbb{R}_+^*)$ of G (see [5], 4.4.5) such that

3.2 $\tau_t \mu_1 = \lim_k (\mu_{\frac{t}{k}} * \varepsilon_{x_{\frac{t}{k}}})^k$ $t \in \mathbb{R}_+^*$

resp. iff

3.2* $\tau_t \mu_1 = \mu_t$ $t \in \mathbb{R}_+^*$.

The motivation for this notion, obviously generalising M. Sharpe's definition of operator-stable probabilities on vector groups ([9],§3), is given by W. Hazod in [3], § 1.

Let (μ_t) be a c.c.s. on G with property 3.1 and let A be the generating distribution of (μ_t). Then the conditions 3.2 resp. 3.2* of τ - stability of (μ_t) means that there exist $X_t \in \mathfrak{g}$ (the Lie algebra of G, see [5], 4.4.5) such that

3.2a $\tau_t A = tA + X_t$ $t \in \mathbb{R}_+^*$,

3.2a* $\tau_t A = tA$ $t \in \mathbb{R}_+^*$.

In this case A is also called <u>τ- stable</u> resp. <u>strictly τ- stable</u>. The generating distributions of the continuous semigroups are the real functionals $B \in D'(G)$ which are <u>normed</u> and <u>almost positive</u> ([5],4.4.18): there exist a neighbourhood U of e with $\sup(\langle B,f\rangle : f \in D(G),\ 1_U \le f \le 1_G) = o$,

and for $f \in D(G), f \ge o, f(e) = o$: $\langle B,f\rangle \ge o$.

These properties remain valid by transporting the distributions in the case of an exponential Lie group G via $\log_G$ on the Lie algebra $\mathfrak{g}$ (see also [3], 3.6). In this case, denoting the semigroup generated by $\log_G(A)$ on the underlying vector group by (ν_t), we have the relations ([3], 3.7):

3.3 $\mu_t = \lim_{n\to\infty} (\exp_G \nu_{\frac{t}{n}})^n$ $t \in \mathbb{R}_+^*$.

Now, let τ be an automorphism group with $v_x = o$ on $\mathbb{H}_n$ and let (μ_t) be a τ - stable c.c.s. on $\mathbb{H}_n$ with generating distribution A. Then, τ given by Cor. 2.6, and setting

$$\mu_t^v := \mathrm{inn}(-v)\mu_t,\quad y_t := \mathrm{inn}(-v)x_t,$$

$$A_v := \mathrm{inn}(-v)A,\quad Y_t := \mathrm{Ad}(-v,o)X_t = \mathrm{inn}(-v)X_t,$$

we obtain the <u>characterising equations</u>:

3.4 $\sigma_{M,m}(t)\mu_1^v = \mu_t^v * \varepsilon_{y_t}$, 3.4a $\sigma_{M,m}(t)A_v = tA_v + Y_t$ $t \in \mathbb{R}_+^*$.

By our identification $\exp_{\mathbb{H}_n} = id_W$, the automorphism group underlies by passing from $\mathbb{H}_n$ to $\mathfrak{h}_n$ no transformation; therefore, equation 3.4a can be considered as an equation on the vector group underlying $\mathfrak{h}_n = W$. We denote the measures resp. distributions which are given by the marginals

$$P(\mu_t^V) = P(\mu_t), \; P(A_V) = P(A) \quad \text{and} \quad Q(A_V)$$

on the vector groups underlying V resp. Z, by $\bar{\mu}_t$, $\bar{A}$ and A_V'. P identifies the vector group underlying V with $\mathbb{H}_n/_Z$. Therefore, using

$$\sigma_{M,m}(t) \circ P = P \circ \sigma_{M,m}(t) \qquad t \in \mathbb{R}_+^*$$

we obtain by 3.4:

$$\begin{aligned} \bar{\mu}_t &= (\sigma_{M,m}(t) \circ P)(\mu_1^V) * P(\varepsilon_{y_t^{-1}}) \\ &= \sigma_{M,m}(t)(\bar{\mu}_1) * \varepsilon_{P(y_t^{-1})} = t^{m+M}\bar{\mu}_1 * \varepsilon_{(-Px_t)} \qquad t \in \mathbb{R}_+^* . \end{aligned} \tag{3.5}$$

We see that $(\bar{\mu}_t)$ is a (t^{m+M}) - stable c.c.s. on the vector group underlying V which is generated by $\bar{A}$. Also, A_V' is a generating distribution (i.e. real, normed and almost positive) on the center $Z = \mathbb{R}$, and 3.4a implies:

3.6 $$t^{m+M}\,\bar{A} = t\bar{A} + \bar{Y}_t ; t^{2m}A_V' = tA_V' + Y_t' \qquad t \in \mathbb{R}_+^* ,$$

$\bar{Y}$ resp. Y_t' being first order differential forms on V resp. Z.

A probability measure μ on $\mathbb{H}_n$ will be called <u>full</u> iff $\bar{\mu} := P(\mu)$ is not concentrated on a proper affine hyperplane of V. If (μ_t) is a τ- stable c.c.s. on $\mathbb{H}_n$ such that μ_1 is full, then $\bar{\mu}_1$ is an operator-stable full measure in the sense of Sharpe (see [9], th. 1) on the vector group underlying V (by 3.5), implying by [9], th. 2:

3.7 $$\bar{\mu}_t \text{ is full} \qquad t \in \mathbb{R}_+^* .$$

Especially, all μ_t are full, allowing the notion of a full c.c.s. on $\mathbb{H}_n$.

Conversely, let $\bar{A}$ resp. A' be generating distributions on the groups underlying V resp. Z, $\bar{A}$ generating a full c.c.s., $\bar{A}$ and A' enjoying 3.6. Then $\bar{A} \otimes \varepsilon_o + \varepsilon_o \otimes A'$ defines a real almost positive distribution A on $\mathbb{H}_n$ with marginals $\bar{A}$ and A' (because $\bar{A}$ and A' annihilates constants), and A generates a $\sigma_{M,m}$-stable c.c.s. on $\mathbb{H}_n$ (because 3.2a is verified) being full (because $P(\nu_t)$ is generated by $P(A) = \bar{A}$ and, by 3.3

3.5a $$\bar{\mu}_t = \lim P(\nu_{\frac{t}{n}}))^n = P(\nu_t)).$$

3.8 Prop.: A probability μ on $\mathbb{H}_n$ is full if and only if μ is not concentrated on a conjugacy class of a proper closed connected subgroup of $\mathbb{H}_n$.

Proof: Assume that μ is not full. Then there exist a $v \in V$ and a proper linear subspace V' of V such that $P\mu$ is concentrated on $v + V'$. But then

$$\operatorname{supp} \mu \subset P^{-1}(v+V') = v + (V'+Z),$$

and, because $V' + Z$ is a Lie subalgebra of $\mathfrak{h}_n$, $V' + Z$ is a proper connected subgroup of $\mathbb{H}_n$ (obviously closed). Finally,

$$v + (V' + Z) = v(V' + Z).$$

Conversely, let μ be a full probability on $\mathbb{H}_n$. Assume that there exist a conjugacy class $(\bar{v}, v')\,H$ of a closed connected subgroup H on which μ is concentrated. Identify H with his Lie algebra $\mathfrak{h}$. Denote the linear transformation $w \mapsto \beta(\bar{v},\bar{w})e$ on W by η. Then:

$$(\bar{v}, v')\,H = (\bar{v}, v') + (\mathrm{id}_W + \eta)\,\mathfrak{h} \quad ;$$

therefore, $(\bar{v}, v')\,H$ is an affine subspace W' of W. Using

$$\operatorname{supp}(P\mu) = \overline{P(\operatorname{supp} \mu)} \subset P(W')$$

we see by the fullness of $P\mu$ that $P(W') = V$. Therefore, $\dim \mathfrak{h} = \dim W' \geq 2n$, and because $\mathfrak{h}_n$ contains no proper Lie subalgebra $\mathfrak{k}$ of dimension $2n$ without $Z \subset \mathfrak{k}$, $\mathfrak{h} = \mathfrak{h}_n$. □

3.9 Lemma: Let τ be an automorphism group on $\mathbb{H}_n$ such that there exists a full τ - stable c.c.s. Then τ is contracting.

Proof: Denote $\tau(t)$ by $(v_t;\, t^M,\, t^m)$ (see 2.2). Let (μ_t) be a full τ - stable c.c.s. on $\mathbb{H}_n$. Then the calculation 3.5 shows that $\bar{\mu}_t$ is a full (t^{m+M}) - stable c.c.s. on V. By [9], th. 2 and th. 3, the eigenvalues of $m+M$ have real parts $\geq \frac{1}{2}$. By $M \in sp(V,\beta)$, trace $M=o$. Especially, $m > o$ and for all eigenvalues λ of M, $\operatorname{Re} \lambda > -m$. By 2.1, for all $x \in \mathbb{R}_+^*$ $v_x = o$ in formula 2.2, and by 2.6, τ is contracting. □

Let τ be an automorphism group with $v_x = o$ on $\mathbb{H}_n$, let $\sigma_{M,m}$ given by 2.5. We will call the pair (M,m) the exponent of τ. Denote the set of complex eigen-values of M by spec M . From the work of Sharpe on operator-stable laws on $\mathbb{R}^n$ ([9], th. 2, th.3) we obtain as a summary of the afore-said:

Theorem 1: Let τ be on automorphism group on $\mathbb{H}_n$. There exists a full τ - stable c.c.s. on $\mathbb{H}_n$ if and only if $v_x = o$ for all $x \in \mathbb{R}_+^*$ and if the the exponent (M,m) of τ has the following properties:

3.10 $\quad m \in [\frac{1}{2}, \infty)$, $\quad M \in Gl(V) \cap sp(V,\beta)$

3.11 $\quad \text{spec } M \subset \{ z \in \mathbb{C} : \ \text{Re} z \geq -m + \frac{1}{2} \}$,

3.11* $\quad (z \in \text{spec } M, \text{ Re} z = -m + \frac{1}{2}) \Rightarrow z$ is a simple zero of the minimal polynomial of M.

If the conditions 3.10, 3.11, 3.11* are fulfilled, then the τ - stable generating functionals are given by (inn v)B ,v being a vector of V, B being the generating distribution of a $(\sigma_{M,m})$-stable c.c.s. on the vector group underlying $W = \mathbb{H}_n$ such that the marginal $\bar{B}$ on V generates a full (t^{M+m}) - stable c.c.s. on the vector group of V.

Proof: As we have seen, the existence of a full τ - stable c.c.s. on $\mathbb{H}_n$ implies the existence of a (t^{m+M}) - stable c.c.s. on V which is equivalent to $M \in Gl(V)$ and 3.11/3.11*. By comparing trace M = o (because $M \in sp(V,\beta)$, β non - degenerate) with trace $M \geq 2n(-m+\frac{1}{2})$ we obtain $m \geq \frac{1}{2}$. The proof of 3.9 shows $v_x = o$ for all $x \subset \mathbb{R}_+^*$.

Conversely, given (m,M) with the properties 3.10, 3.11, 3.11* there exists a generator A of a full (t^{m+M}) - stable c.c.s. on V, and $B := A \otimes \varepsilon_o$ is a generator of a full $\sigma_{M,m}$ - stable c.c.s. on $\mathbb{H}_n$. $\square$

Let (μ_t) be a full τ - stable c.c.s. on $\mathbb{H}_n$. Then the exponent (M,m) of τ will be also called the exponent of (μ_t) or of the generating functional A. Let $\mathcal{E}(\mu_t) = \mathcal{E}(A)$ denote the set of exponents of (μ_t) (obviously, $\mathcal{E}(A) = \mathcal{E}(\text{inn}(v)\ A))$. We fix an ordered β - symplectic base $\mathcal{B} := (\nu_1, \ldots, \nu_n, \tilde{\nu}_1, \ldots, \tilde{\nu}_n)$ of V (i.e.

3.12 $\beta(\nu_j,\tilde{\nu}_k) = -\beta(\tilde{\nu}_j,\nu_k) = \delta_{jk}$, $\beta(\nu_j,\nu_k)=\beta(\tilde{\nu}_j,\tilde{\nu}_k) = o$
$j,k = 1, \ldots, n)$,

and we define an inner product γ on V by

$$\gamma(\nu_j,\nu_k) = \gamma(\tilde{\nu}_j,\tilde{\nu}_k) = \delta_{jk} \qquad j,k = 1, \ldots, n .$$

We denote by $so(V,\gamma)$ the Lie algebra of γ- skewsymmetric linear operators on V.

3.13 Cor.: There exist a Lie subalgebra $\mathfrak{h}$ of $so(V,\gamma)$ and a γ- positive definite symmetric operator L on V such that

$$\mathcal{E}(A) = \{(m,M + LNL^{-1}) : N \in \mathfrak{h} \cap sp(V,\beta)\}.$$

Proof: It is known that

$$\mathcal{E}(\bar{A}) = m + M + L\mathfrak{h}L^{-1}$$

([6], th. 1 and proof of cor. 3). Therefore, if $(m',M') \in \mathcal{E}(A)$,

$$\begin{aligned} 2n(m-m') &= \text{trace } (m-m'+M-M') \\ &= \text{trace } m-m'+LHL^{-1} = \text{trace } m-m'+H = 0 \end{aligned}$$

for an operator $H \in \mathfrak{h}$, implying $m = m'$ and $M - M' \in L\mathfrak{h}L^{-1} \cap sp(V,\beta)$. □

3.14 Remark: The last argument of the preceding proof shows the following for a symplectic operator M on V:

$$(m,M) \in \mathcal{E}(A) \iff m + M \in \mathcal{E}(\bar{A}) .$$

Now, let A be a generating distribution of local type on $\mathbb{H}_n$ (or equivalently: let (μ_t) be a Gaussian semigroup (see [5],6.2.1)). Then $(\bar{\mu}_t)$ is a Gaussian semigroup on V and has therefore $\frac{1}{2}\, id_V$ as exponent, and the remark shows that (μ_t) has $(\frac{1}{2},o)$ as exponent: Every Gaussian semigroup is stable with respect to a dilation group.

3.15 Example: In the case $n = 1$, $sp(V,\beta)$ has no proper non-trivial Lie subalgebra, and $L\, sp(V,\beta)L^{-1} = \mathbb{R}C$ with a symplectic operator C (depending on L). Therefore, $\mathcal{E}(A) = (m,\mathfrak{k})$ where $\mathfrak{k}$ is an affine subspace of $sp(V,\beta)$ of dimension 0 or 1. In the second case, we see by

3.9 that the eigenvalues of M with $(m,M) \in \mathcal{E}(A)$ are all purely imaginary. Further we know in the second case that there exist $(r = m,o) \in \mathcal{E}(A)$ ([6], th.2): If a stable c.c.s. (μ_t) on $\mathbb{H}_1$ has different exponents, then (μ_t) is stable with respect to a dilation group.

Theorem 1 allows the description of the generating distributions of the full stable c.c.s. on $\mathbb{H}_n$ by reduction on the Lévy - Khintchine representation of the full stable probabilities on $\mathbb{R}^{2n+1}$ or on $\mathbb{R}^{2n}$ (which is completely known; see, for example [8], th. 1/3):

3.16 Cor.: The generating distributions of full stable c.c.s. (μ_t) on $\mathbb{H}_n$ are given by (inn v) B, $v \in V$, B being the generating distribution of a $(\sigma_{M,m})$ - stable c.c.s. (ν_t) on the vector group of W such that either (ν_t) is full or $B = \bar{B} \otimes \varepsilon_o + Y$, where $\bar{B}$ generates a full c.c.s. on the vector group of V and Y is a central derivation.

Proof: Assume that (ν_t) is not full. Let S be a proper affine subspace of W with $\operatorname{supp} \nu_1 \subset S$. Then the c.c.s. $(P\nu_t)$ on V is generated by P(A); therefore $P\nu_t = \bar{\mu}_t$ and, by

$$P(S) \supseteq \overline{P(\operatorname{supp} \mu_1)} = \operatorname{supp} \bar{\mu}_1$$

and the fullness of (μ_t), $P(S) = V$ (implying $\dim S \geq 2n$; by assumption, $\dim S \leq 2n$). Therefore, $S \cap Z = \{v'\}$ for a unique $v' \in Z$. $(\tilde{\mu}_t := \mu_t * \varepsilon_{-tv'})$ is a full $(\sigma_{M,m})$ - stable c.c.s. on $\mathbb{H}_n$ with generating distribution B - Y for a central derivation Y, and the c.c.s.$(\tilde{\nu}_t)$, generated by B - Y on W, is given by $\tilde{\nu}_t = \nu_t * \varepsilon_{-tv'}$, and $\tilde{\nu}_t$ is concentrated on $V \times \{o\}$; finally $P\tilde{\mu}_t = P\tilde{\nu}_t$ shows that $P\tilde{\nu}_t$ is full on V. The converse is obvious. □

We conclude this paragraph with a short discussion of the case of non-full strictly stable probabilities on $\mathbb{H}_n$:

3.17 Prop.: Let (μ_t) be a strictly τ - stable c.c.s. on a exponential Lie group G. Let S be the support $\bigcup (\mu_t : t \in \mathbb{R}_+^*)$ of (μ_t). Then τ is contracting on the closed subgroup generated by S.

Remark: Prop.3.17 can not be generalised to (not necessary strictly) stable c.c.s. (in contrast to the situation in the full case (3.8)).

Proof: Let A be the generating distribution of (μ_t), let Q be the (symmetric) Gaussian part of A, let η be the Lévy measure of A ([5], 4.4.18), let (ν_t) resp. (ρ_t) the c.c.s. generated by Q resp. by the Lévy part of A. Then, by [4], 2.2 and 2.3, τ contracts every element of $\bigcup(\text{supp}\ \mu_t : t \in \mathbb{R}_+^*) =: U$ and of supp η.

Let H denote the closed subgroup generated by U and supp η. Then $\bigcup \text{supp}\ \rho_t \subset H$ and, by

$$\mu_t = \lim_n (\nu_{\frac{t}{n}} * \rho_{\frac{t}{n}})^n \qquad t \in \mathbb{R}_+^*$$

$\bigcup \text{supp}\ \mu_t \subset H$. The subgroup generated by $U \cup \text{supp}\ \eta$, is contained in $C(\tau)$. As a (simply) connected subgroup of the exponential group G, $C(\tau)$ is closed. Therefore $H \subset C(\tau)$, and the proof is complete by 2.9. $\square$

3.18 Example: The Heisenberg groups.

Let (μ_t) be a strictly τ- stable c.c.s on $\mathbb{H}_n$. Let T be the support of $\bar{\mu}_1$ $(\bar{\mu}_t)$ is a full strictly stable c.c.s. on the linear subspace [T] with $\bar{\mu}_t = (P \circ \tau(t))\bar{\mu}_1$. It follows that (μ_t) is a full c.c.s. on the closed subgroup of $\mathbb{H}_n$ supported by $[T] \oplus Z$. We show:

3.19 $\qquad [T] \oplus Z$ is $\tau(t)$-invariant for all $t \in \mathbb{R}_+^*$.

By 2.10 and 3.17, we see that the one-parameter automorphism group, induced by τ on the subgroup $[T] \oplus Z$ is of the form 2.6. Therefore the problem is reduced to the case of a full stable c.c.s. on a"Heisenberg-like" subgroup or on a vector subgroup, giving the possibility of generalising theorem 1 and corollar 3.16 to the non-full strict stable case.

Proof of 3.19: Denote $\tau(t)$ by $(v_t; B_t, s_t)$. It is sufficient to show that [T] is B_t - invariant. $\bar{\mu}_1$ has a continuous density φ. The set $\{\varphi \neq 0\}$ generates [T] by the fullness of $\bar{\mu}_1$ as a measure on [T]. Choose a basis $\{e_1,\dots,e_k\}$ of [T] with $\varphi(e_j) \neq 0$ for $j=1,\dots,k$. Then the density of $\bar{\mu}_1$ is given by

$$\varphi_t := s_t^{-2n}(\det B_t)^{-1}\varphi(s_t^{-1} B_t^{-1}) \qquad t \in \mathbb{R}_+^* .$$

If [T] is not B_t - invariant, there exist $j \in \{1,\dots,k\}$ with $B_t e_j \notin [T]$, implying by $\varphi_t(s_t B_t e_j) \neq 0$ and the continuity of φ_t that supp $\bar{\mu}_t \not\subset [T]$ which is impossible. $\square$

§ 4 Strict stability equals non-emptiness of the domain of attraction

In this paragraph, we will clarify the meaning of fullness in the context of stable probability measures, and we will give the probabilistic interpretation of strict stability of full probability measures.

4.1 The Compactness Lemma: Given probabilities λ_k on $\mathbb{H}_n$ and automorphisms τ_k of $\mathbb{H}_n$ such that λ_k resp. $\tau_k\lambda_k$ converge weakly to full probabilities λ resp. μ, then the set $\{\tau_n\}$ is precompact in Aut($\mathbb{H}_n$), and if τ is an adherence point of $\{\tau_n\}$, then $\tau\lambda = \mu$.

Proof: Denote τ_k by $(v_k; B_k, s_k)$. With

$$P\lambda_k =: \bar{\lambda}_k, \; P\lambda = \bar{\lambda}\,, \; P\mu =: \bar{\mu}, \quad \text{we obtain:}$$

$$\bar{\lambda}_k \rightarrow \bar{\lambda}\,, \quad s_k B_k \bar{\lambda}_k \longrightarrow \bar{\mu}\,.$$

$\bar{\lambda}$, $\bar{\mu}$ being full on V, and $s_k B_k \in Gl(V)$, the commutative version of the Compactness Lemma ([9],Prop.4) shows that $\{s_k B_k : k \in \mathbb{N}\}$ is precompact in Gl(V); by $\det s_k B_k = s_k^{2n}$ it follows that $\{s_k : k \in \mathbb{N}\}$ is bounded. Let s be an adherence point of $\{s_k : k \in \mathbb{N}\}$; let$(s_{k_1})$ be a subsequence converging to s; let$(\tilde{s}_m \tilde{B}_m)$ be a convergent subsequence of $(s_{k_1} B_{k_1})$ with limit C. Then:

$$\tilde{s}^{2n} = \lim \tilde{s}_m^{2n} = \lim \det \tilde{s}_m \tilde{B}_m = \det C \neq o.$$

Let B be an adherence point of $\{B_k : k \in \mathbb{N}\}$; the same argumentation shows that there exists a converging sequence $(s_1 B_1)$ with $\lim B_1 = B$, $\lim s_1 =: s > o$ and $\lim s_1 B_1 =: D \in GL(V)$, and we conclude $B \in Sp(V, \beta)$. Therefore, $\{(o; B_k, s_k)\}$ is precompact in Aut($\mathbb{H}_n$).

Assume now that $\{v_k\}$ is not bounded. Choose a subsequence (v_{k_1}) with $\lim_1 \|v_{k_1}\| = \infty$. Choose a subsequence $(s_{k_{1_m}} B_{k_{1_m}})$ of $(s_{k_1} B_{k_1})$ such that $(s_{k_{1_m}})$ and $(B_{k_{1_m}})$ converge to s resp. B. By joint continuity of $(\tau, \mu) \mapsto \tau\mu$ (from $Aut(\mathbb{H}_n) \times M^1(\mathbb{H}_n)$ in $M^1(\mathbb{H}_n)$ where $M^1(\mathbb{H}_n)$ is the semigroup of probabilities on $\mathbb{H}_n$ furnished with the weak topology) we obtain with $(o;B,s)\lambda =: \lambda'$:

$$\lim_m (\mathrm{inn}\, v_{k_{l_m}})\,\lambda' = \lim_m (\mathrm{inn}\, v_{k_{l_m}})\ (o;\, B_{k_{l_m}}, s_{k_{l_m}})\ \lambda_{k_{l_m}} = \mu .$$

Choose an ordered β - symplectic base $\mathfrak{B} = (u_1,\dots,u_n, \tilde{u}_1,\dots,\tilde{u}_n)$ (see 3.12); let $(v_k^1,\dots,v_k^n,\tilde{v}_k^1,\dots,\tilde{v}_k^n)$ denote the coordinates of v_k with respect to $\mathfrak{B}$. We can assume: $\lim_k v_k^1 = \infty$. Denote $[\{u_1,\dots,u_n, u_{n+2},\dots,u_n\}]$ by V'. Let $\bar{K}$ resp. K' be compact subsets of $V \setminus V'$ resp. Z; let f be a continuous function $\mathbb{H}_n \longrightarrow \mathbb{R}_+$ with supp $f \subset \bar{K} \times K'$. Then a short calculation shows: To $n \in \mathbb{N}$ there exists $m(n)$ such that for $m \geq m(n)$:

$$\mathrm{inn}(-v_{k_{l_m}})(\bar{K} \times K') \subset \bar{K} \times (Z \setminus [-n,n]).$$

Therefore $\langle \mu, f \rangle = \lim_m \langle \mathrm{inn}\, v_{k_{l_m}} \lambda', f \rangle = o$, implying supp $\mu \subset V' \times Z$ or supp $\bar{\mu} \subset V'$ contradicting the fullness of μ. The last assertion is trivial. □

We call a sequence $(\mu_k : k \in \mathbb{N})$ of probabilities on $\mathbb{H}_n$ <u>infinitesimal</u> iff (μ_k) converges weakly to ε_e. If $\mu_k = \tau_k\, \mu$ with $\tau_k \subset \mathrm{Aut}(\mathbb{H}_n)$, $\mu \in M^1(\mathbb{H}_n)$ and if μ is full, then the infinitesimality of (μ_k) is equivalent to the contracting property of

$$\tau : \mathbb{N} \longrightarrow \mathrm{Aut}(\mathbb{H}_n), \quad k \longmapsto \tau_k$$

e.g. to $\lim \tau_k(g) = e$ for all $g \in \mathbb{H}_n$ (be cause, by the discrete analogue of 2.8, the last property implies $\lim \langle \tau_k \mu, f \rangle = o$ for all continuous real valued f on $\mathbb{H}_n$ with compact support). The <u>domain of attraction</u> of a full probability μ on $\mathbb{H}_n$ is defined as

4.3 $\quad A(\mu) := \{\nu \in M^1(\mathbb{H}_n):$ there exist $\tau_k \in \mathrm{Aut}(\mathbb{H}_n)$ such that $\lim \tau_k \nu^k = \mu$ and (τ_k) is contracting$\}$.

We call a probability measure on $\mathbb{H}_n$ <u>Baldi - stable</u> iff for every $k \in \mathbb{N}$ there exist $\tau_k \in \mathrm{Aut}(\mathbb{H}_n)$ such that

4.2 $\quad \mu^k = \tau_k\, \mu$.

This notion was introduced by P. Baldi for the groups $\mathbb{R}^n \circledS SO(n)$ of n-dimensional motions ([1], Def. 2).

The motivation for Baldi's definition is given by the following result:

4.4 Prop.: Let μ be a full probability on $\mathbb{H}_n$. Then μ is Baldi - stable if and only if the domain of attraction of μ is nonempty.

Proof: One implication can be proved as in [1], proof of th.8, using the Compactness Lemma. The other implication follows from lemma 3.9 and the next theorem.

Theorem 2: Let (μ_t) be a strictly stable c.c.s. on $\mathbb{H}_n$. Then every μ_t is Baldi - stable. Let μ be a full Baldi - stable probability on $\mathbb{H}_n$. Then there exists a strictly stable c.c.s. (μ_t) on $\mathbb{H}_n$ with $\mu_1 = \mu$.

Proof: The first assertion is trivial.

Let μ be a Baldi - stable probability on $\mathbb{H}_n$. Define for $k,m \in \mathbb{N}$:

4.5 $$\tilde{\mu}_{k \cdot 2^{-m}} := \tau_2^{-m} \tau_k \mu \quad (= (\tau_2^{-m} \mu)^k) ,$$

using the automorphisms of definition 4.2. Denote by D the additive semigroup of strict positive dyadic numbers.

4.5a $\tilde{\mu}_{k \cdot 2^{-m}}$ is well-defined $\qquad k,m \in \mathbb{N}$.

4.6 $\tilde{\mu}_s * \tilde{\mu}_t = \tilde{\mu}_{s+t} \qquad s,t \in D$.

Proof of 4.5a: It is sufficient to prove $\tilde{\mu}_{\frac{2k}{2^{m+1}}} = \tilde{\mu}_{\frac{k}{2^m}} \qquad k,m \in \mathbb{N}$.

We use that τ_2^{-1}, τ_k are convolution homomorphisms:

$$\tilde{\mu}_{\frac{2k}{2^{m+1}}} = \tau_2^{-(m+1)} \mu^{2k} = (\tau_2^{-m} \tau_2^{-1} \mu^2)^k$$
$$= (\tau_2^{-m} \mu)^k = \tau_2^{-m} \mu^k = \tilde{\mu}_{\frac{k}{2^m}} . \quad \square$$

Proof of 4.6: By 4.5a, it is sufficient to prove

$$\tilde{\mu}_{\frac{k}{2^m}} * \tilde{\mu}_{\frac{l}{2^m}} = \tilde{\mu}_{\frac{k+l}{2^m}} \qquad k,l,m \in \mathbb{N}.$$

This is done in the same way as in the proof of 4.5a. $\square$

By [2], lemma 6, there exist a c.c.s. (μ_t) on $\mathbb{H}_n$ such that $\mu_t = \widetilde{\mu}_t$ for $t \in D$. Define

$$F_t := \{\tau \in \mathrm{Aut}(\mathbb{H}_n) : \mu_t = \tau\mu\} \qquad t \in \mathbb{R}_+^* .$$

By 4.5, $F_t \neq \emptyset$ for $t \in D$. The calculation 3.5 shows that with $\bar{\mu}_t := P\mu_t, t \in \mathbb{R}_+^*$, $\bar{\tau}(k) := P \circ \tau(k)\big|_V : V \longrightarrow V$ $(k \in \mathbb{N})$ we obtain a measure $\bar{\mu} := \bar{\mu}_1$ on V with associated c.c.s $(\bar{\mu}_t)$ such that $\bar{\tau}(k)\,\bar{\mu} = \bar{\mu}^k$.
Assume now that μ is full; then $\bar{\mu}$ is a full stable probability on V ([9], th. 1) implying supp $\bar{\mu}_t = V$ for all $t \in \mathbb{R}_+^*$ (3.7).
Therefore, by definition:

4.7 Every μ_t is full $\qquad t \in \mathbb{R}_+^*$.

Now, the Compactness Lemma shows $F_t \neq \emptyset$ for all $t \in \mathbb{R}_+^*$. Define

$$G_t := \{\tau \in Gl(V) : \bar{\mu}_t = \tau\bar{\mu}\},$$

$$H_t := G_t \cap \mathbb{R}_+^* \cdot Sp(V,\beta) \qquad t \in \mathbb{R}_+^* .$$

Choosing $\tau(t) \in F_t$, and defining $\bar{\tau}(t) := P \circ \tau(t)\big|_V : V \longrightarrow V$ we see that $\bar{\tau}(t) = s_t B_t$, if $\tau(t) =: (v_t; B_t, s_t)$ and again by 3.5, $\bar{\tau}(t)\bar{\mu} = \bar{\mu}_t$. Therefore $H_t \neq \emptyset$. It is known that $\bigcup(G_t : t \in \mathbb{R}_+^*)$ is a closed subgroup of $Gl(V)$ ([9], § 6, lemma 4). By $\det C = 1$ for $C \in Sp(V,\beta)$, $\mathbb{R}_+^* \cdot Sp(V,\beta)$ is the direct product of the closed subgroups $\mathbb{R}_+^* \cdot id_V$ and $Sp(V,\beta)$ of $Gl(V)$. Therefore

$$H := \bigcup(H_t : t \in \mathbb{R}_+^*) = (\bigcup(G_t : t \in \mathbb{R}_+^*)) \cap (\mathbb{R}_+^* \cdot Sp(V,\beta))$$

is a closed subgroup of $Gl(V)$. By $G_s \cap G_t = \emptyset$ for $s \neq t$ ([9], § 6, lemma 3), the mapping $\xi : H \longrightarrow \mathbb{R}$, given by $H_t \ni C \longmapsto \log t$ is well-defined and is a continuous homomorphism from H in $(\mathbb{R},+)$ ([9], § 6, lemma 5) which is surjective by $H_t \neq \emptyset$. Now, we follow the proof of [9], theorem 2, obtaining an invertible operator C of the Lie algebra of $\mathbb{R}_+^* \cdot Sp(V,\beta)$ such that $e^{tC} \in H_t$ for all $t \in \mathbb{R}_+^*$. Obviously, $C = m+M$ with $m \in \mathbb{R}$, $M \in sp(V,\beta)$. By definition, $t^{m+M}\bar{\mu} = s_t B_t \bar{\mu}$.

4.8 $\qquad s_t = t^m \qquad t \in \mathbb{R}_+^*$.

Proof of 4.8: By 4.7, $\bar{\mu}$ has a continuous bounded density φ ([8], th.1). Assume $s_{t_o} > t_o^m$ for a $t_o \in \mathbb{R}_+^*$. Choose a $v_o \in V$ with $\varphi(v_o) > o$. Choose a regularisation of ε_v, $v \in V$ fixed, by continuous functions γ_r ($r \in \mathbb{R}_+^*$) with compact support. By

$$\det t^{m+M} = t^{2nm}, \ \det s_t B_t = s_t^{2n} ,$$

the transformation theorem shows, λ being the Lebesgue measure of V:

$$t^{2nm} \int \gamma_r (\varphi \circ t^{-m-M})\, d\lambda = \int (\gamma_r \circ t^{m+M}) \varphi \, d\lambda$$

$$= \int \gamma_r \, d(t^{m+M} \bar{\mu}) = \int \gamma_r \, d(s_t B_t \bar{\mu})$$

$$= \int (\gamma_r \circ (s_t B_t)) \varphi \, d\lambda = s_t^{2n} \int \gamma_r (\varphi \circ (s_t B_t)^{-1})\, d\lambda .$$

For $r \to o$, we arrive at

$$t^{2nm} \varphi(t^{-m-M} v) = s_t^{2n} \varphi(s_t^{-1} B_t^{-1} v).$$

Especially, for $t = t_o$ and with $C := t_o^{-m-M} s_{t_o} B_{t_o}$, $c := \dfrac{s_{t_o}^{2n}}{t_o^{2nm}}$, $w := s_t B_t v$:

$$\varphi(C_w) = c \cdot \varphi(w), \ \varphi(C^k w) = c^k \varphi(w) \qquad k \in \mathbb{N}.$$

By $\varphi(v_o) > o$, $c > 1$, we obtain a contradiction to the boundedness of φ. (The argumentation with $s_{t_o} < t_o^m$ is analogue.) □

Define $\sigma(t)$ by $(v_t; t^M, t^m) \in \mathrm{Aut}(\mathbb{H}_n)$, the v_t being given by $\tau(t) = (v_t; B_t, s_t) \in F_t$.

4.10 $\tau(t)\mu = \sigma(t)\mu$.

<u>Proof:</u> It is sufficient to show

$$(o; B_t, s_t)\mu = (o; t^M, t^m)\mu .$$

Choose a desintegration of μ:

$$\mu = \int_V \nu_v \, d\bar{\mu}(v)$$

with a regular conditional distribution

$$\nu : V \times \mathcal{B}_V \longrightarrow \mathbb{R}_+, (v; U) \mapsto \nu_v(U) \qquad U \in \mathcal{B}_V,$$

where $\mathcal{B}_V$ denotes the Borel subsets of V.
Let U' be a Borel subset of Z. Then:

$$(o; B_t, s_t)\mu(U \times U') = \int_U \nu_v(s_t^{-2} U')d((s_t B_t)\bar{\mu})$$

$$= \int_U \nu_v(t^{-2m} U') \, d(t^{m+M} \bar{\mu}) = (o; t^M, t^m)\mu(U \times U'). \quad \square$$

4.11 $\qquad v_k + k^{m+M} v_l = v_{kl} \qquad k,l \in \mathbb{N}.$

Proof of 4.11: By

$$\sigma_k \sigma_l \mu = \mu^{kl} = \sigma_{kl} \mu \quad \text{and}$$

4.11a $\qquad \sigma_k \sigma_l = (v_k + k^{m+M} v_l; (kl)^M, (kl)^m)$

we arrive with $\tilde{v} := v_k + k^{m+M} v_l$ and $\nu := (\alpha_M \circ \delta_m)\mu$ at

4.11b $\qquad (\text{inn } v_{kl})\nu = (\text{inn } \tilde{v})\nu$

or $\nu = (\text{inn } u)\nu$ for $u := (v_{kl} - \tilde{v})$; by iteration:

4.11c $\qquad \nu = (\text{inn } u)^j \nu = (\text{inn}(ju)) \qquad j \in \mathbb{N}.$

The Compactness Lemma shows, that inn(ju) has an adherence point in Inn($\mathbb{H}_n$), but by Inn($\mathbb{H}_n$) = V, this means that $\{ju: j \in \mathbb{N}\}$ has an adherence point in V, implying $u = o$. $\square$

4.11d <u>Remark:</u> The last argument (beginning with 4.11b) shows that, for a full probability ρ on H_n, $(\text{inn } v_1)\rho = (\text{inn } v_2)\rho$ implies $v_1 = v_2$.

By 4.11

$$v_l + l^{m+M} v_k = v_{lk} = v_{kl} = v_k + k^{m+M} v_l ,$$

which implies by 4.11a:

4.12 $$\sigma_k \sigma_l = \sigma_{kl} = \sigma_l \sigma_k \qquad k,l \in \mathbb{N}.$$

Define now

$$\tilde{\sigma}_{\frac{k}{l}} := \sigma_l^{-1} \sigma_k , \ \tilde{\tilde{\mu}}_{\frac{k}{l}} := \tilde{\sigma}_{\frac{k}{l}} \mu \qquad k,l \in \mathbb{N}.$$

By 4.12, the definition is consistent, and we have for $k,k',l,l' \in \mathbb{N}$:

$$\tilde{\tilde{\mu}}_{\frac{k}{l}} * \tilde{\tilde{\mu}}_{\frac{k'}{l'}} = \sigma_{ll'}^{-1}(\sigma_k \sigma_{l'} \mu * \sigma_{k'} \sigma_l \mu)$$
$$= \sigma_{ll'}^{-1} \mu^{kl' + k'l} = \tilde{\tilde{\mu}}_{(\frac{k}{l} + \frac{k'}{l'})} ,$$

4.13 $$\tilde{\sigma}_{\frac{k}{l}} \tilde{\tilde{\mu}}_{\frac{k'}{l'}} = \sigma_l^{-1} \sigma_{l'}^{-1} \sigma_k \sigma_{k'} \mu = \tilde{\tilde{\mu}}_{\frac{kk'}{ll'}} .$$

Let (ρ_t) be the c.c.s. on $\mathbb{H}_n$ with $\rho_{\frac{k}{l}} = \tilde{\tilde{\mu}}_{\frac{k}{l}}$ ([5],3.5.4, 3.5.8, 3.5.11). 4.1 defines $\tilde{\sigma}_t \in \mathrm{Aut}(\mathbb{H}_n)$ with $\tilde{\sigma}_t \mu = \rho_t$, and by 4.13, we obtain $\tilde{\sigma}_t \rho_s = \rho_{ts} (t,s \in \mathbb{R}_+^*)$. Taking sequences $\tilde{\sigma}_{s_n} \to \tilde{\sigma}_{s_n}$, $\tilde{\sigma}_{t_n} \to \tilde{\sigma}_t$ with rational indexes, we see by 4.12 that $(\tilde{\sigma}_t)$ is a (multiplicative) group of automorphisms. Now the remark 4.11 of shows that every v_t is uniquely determined.

Therefore, given a sequence (t_n) of positive reals converging against t, we see that $\{\tilde{\sigma}_{t_n}\}$ has at most one adherence point $\tilde{\sigma}$ (which exist by the Compactness Lemma), and the second assertion of the lemma gives $\tilde{\sigma} = \tilde{\sigma}_t$. Now, the "rational versions" of prop. 2.1, cor. 2.6 and lemma 3.9 shows

4.14 $$\tilde{\sigma}_t = (v - t^{m+M} v;\ t^M, t^m) \qquad t \in \mathbb{Q}_+^* .$$

Then, by the uniqueness of the adherence points, 4.14 is valid for all $t \in \mathbb{R}_+^*$. Therefore $(\tilde{\sigma}_t : t \in \mathbb{R}_+^*)$ is continuous. □

4.14 Remark: The example 3.18 shows that it is possible to give also a version of 4.4 and theorem 2 for the non-full case.

§ 5 The uniqueness of the associated semigroup

Prop.: Let (μ_t) and (ν_t) be full τ- stable resp. τ' - stable c.c.s. on $\mathbb{H}_n$ with $\mu_1 = \nu_1$. Then $\mu_t = \nu_t$ for all $t \in \mathbb{R}_+^*$.

Proof: Consider the c.c.s. $(\bar{\mu}_t)$ and $(\bar{\nu}_t)$ on V. By prop. 2.1, we know that there exist $m, r \in \mathbb{R}$, $M, R \in sp(V, \beta)$ with

$$t^{m+M}\bar{\mu}_1 = \tau(t)\bar{\mu}_1 = \bar{\mu}_t, \qquad t^{r+R}\bar{\mu}_1 = \tau'(t)\bar{\mu}_1 = \bar{\nu}_t .$$

The uniqueness of the roots of $\bar{\mu}_1$ shows $t^{m+M}\bar{\mu}_1 = t^{r+R}\bar{\mu}_1$ and by [6], th. 1 : $m = r$.

Denote $\tau(t)$ by $(v_t; B_t, s_t)$, $\tau'(t)$ by $(v_t'; B_t', s_t')$.

Define $\sigma(t)$ by $(v_t; t^M, t^m)$, $\sigma'(t)$ by $(v_t'; t^R, t^r)$.

The argument of 4.10 shows:

$$\text{5.1} \qquad \operatorname{inn}(-v_t)\mu_t = \operatorname{inn}(-v_t)\,\sigma(t)\mu_1 = \operatorname{inn}(-v_t')\,\sigma'(t)\mu_1 = \operatorname{inn}(-v_t')\,\nu_t .$$

Especially, with $u_{kl} := v'_{\frac{k}{l}} - v_{\frac{k}{l}}$ $\qquad k, l \in \mathbb{N}$:

$$\operatorname{inn} u_{kl}\,\mu_1^k = \operatorname{inn} u_{kl}\,\mu_{\frac{k}{l}}^l = (\operatorname{inn} u_{kl}\,\mu_{\frac{k}{l}})^l = \nu_{\frac{k}{l}}^l = \nu_1^k = \mu_1^k .$$

Now the remark 4.11. d shows $u_{kl} = 0$.

Therefore, by 5.1:

$$\mu_{\frac{k}{l}} = \nu_{\frac{k}{l}} \qquad k, l \in \mathbb{N},$$

and by continuity: $\mu_t = \nu_t$ for all $t \in \mathbb{R}_+^*$. □

References:

[1] P. Baldi: Lois stables sur les déplacements de R^d. In: Probability Measures on Groups. Lecture Notes Math. 706. Berlin-Heidelberg-New York: Springer 1979.

[2] Q.L. Burrell, M. McCrudden: Infinitely divisible distributions on connected nilpotent Lie groups. J. London Math. Soc. 7, 584 - 588 (1974).

[3] W. Hazod: Stable Probabilities on Locally Compact Groups. In: Probability Measures on Groups. Lecture Notes Math. 928. Berlin-Heidelberg-New York: Springer 1982.

[4] W. Hazod: Remarks on (Semi-) stable Probabilities. In: Probability Measures on Groups. Lecture Notes Math. Berlin-Heidelberg-New York: Springer 1984.

[5] H. Heyer: Probability Measures on Locally Compact Groups. Berlin-Heidelberg-New York: Springer 1977.

[6] J.P. Holmes, W.N. Hudson, J.D. Mason: Operator Stable Laws: Multiple Exponents and Elliptical Symmetry. Annals Prob. 10, 602-612 (1982).

[7] W.N. Hudson: Operator-Stable Distributions and Stable Marginals. J. Multivariate Analysis 10, 26-37 (1980).

[8] W.N. Hudson, J.D. Mason: Operator - Stable Laws. J. Multivariate Analysis 11, 434-447 (1981).

[9] M. Sharpe: Operator-Stable Probability Measures on Vector Groups. Trans. Amer. Math. Soc. 136, 51-65 (1969).

Thomas Drisch
Abteilung Mathematik
Univ. Dortmund
Postfach 500 500
D-4600 Dortmund

Léonard Gallardo
UER Math. et Inform.
Univ. de Nancy II
23 Boulevard Albert 1er
F-5400 Nancy

AN ANALOGUE OF THE LÉVY-KHINTCHIN FORMULA ON SL(2;ℂ)

B.-J. Falkowski
Hochschule der Bundeswehr München
Fachbereich Informatik
Werner-Heisenberg-39

D-8014 Neubiberg

INTRODUCTION

The interest in an analogue of the classical Lévy-Khintchin formula for non-abelian groups arises on the one hand from the construction of "continuous tensor products", cf. [9], [18], [4] and on the other hand from Quantum Mechanics, cf. [15], [2]. The purely mathematical interest in the construction of a continuous tensor product is intimately connected with applications in Quantum Mechanics via the construction of irreducible representations of so-called "current groups", cf. [8], [16], [17], [4], [2]. In [15] cohomology groups (first and second order) are used to derive the classical Lévy-Khintchin formula. This method generalizes to non-abelian groups: In [14] all infinitely divisible positive definite functions on a compact group are described. Further partial results may be obtained from [15]. For connected semi-simple Lie groups quite a bit is known about the solution since a large class of groups is known to admit only trivial first cohomology groups, cf. [3]. However, for two important classical groups, namely SO(n;1) and SU(n;1) the complete solution is not yet known. A partial solution for SU(1;1) was given in [5]. Here we obtain a complete solution for SL(2;ℂ), which is the double cover of $SO_e(3;1)$, the connected component of the identity of the homogeneous Lorentz group. It seems likely that similar results can be proved for $SO_c(n;1)$ (n>3) and for SU(n;1). Partial results are known in this case as well, cf. [8].

For notation and definitions we refer the reader to [5]. In this paper a representation will always be a continuous homomorphism from a topological group G into the group of unitary operators on some Hilbert space endowed with the weak (or equivalently strong) topology.

§ 1 I.D.P. FUNCTIONS AND 1-COHOMOLOGY

We start with the following

(1.1) Theorem: Let G be a connected, semi-simple Lie group. Then for every pair (δ,b) consisting of a continuous 1-cocycle δ and a continuous function $b:G \to \mathbb{R}$ satisfying

$$\operatorname{Im} < \delta(g_2),\ \delta(g_1^{-1}) > = b(g_1) + b(g_2) - b(g_1g_2)$$

we obtain an I.D.P. function f given by

$$f(g) := \exp-[\tfrac{1}{2} < \delta(g),\ \delta(g) > + i\ b(g)] \qquad (*)$$

Conversely: Given an I.D.P function f on G there exists a pair (δ,b), as described above, such that equation (*) is satisfied. Moreover b is uniquely determined and δ up to unitary equivalence, if $\{\delta(g) : g \in G\}$ is total in the relevant Hilbert space.

Proof: We note that theorem (1.7) in [5] holds for a connected, semi-simple Lie group since the same proof as for SU(1;1) goes through verbatim. The rest follows easily from theorems (1.2), (1.4) in [4] p. 78-80. Q.E.D.

Thus it is clear that the solution of our problem depends on a complete description of the 1-cohomology for $SL(2;\mathbb{C})$. Moreover (since $\operatorname{Im} < \delta(g_2), \delta(g_1^{-1}) >$ describes a 2-cocycle) some information about the 2-cohomology is needed as well. For technical convenience, however, let us first investigate how changing the 1-cocycle δ in (1.1) by a trivial cocycle (i.e. a coboundary) alters the corresponding I.D.P. function. The answer is provided by

(1.2) Lemma: Suppose f,δ,b are as in (1.1) and satisfy equation (*). Let $\delta'(g) := \delta(g) + U_g v - v$, where δ is associated with the representation $g \mapsto U_g$ and v is a fixed vector in the Hilbert space where U_g acts. Then we obtain (via (1.1)) an I.D.P function f' given by

$$f'(g) = \exp-[\tfrac{1}{2} < \delta'(g),\ \delta'(g) > + i\ b'(g)]$$

where $b'(g) = b(g) + \operatorname{Im}[< v, U_g v > + < v,\delta(g) > - < v,\delta(g^{-1}) >]$

and thus

$$f'(g) = f(g)\ \exp[< U_g v-v,v > + < \delta(g),v > + < v,\delta(g^{-1}) >]$$

Proof: This is a straightforward computation using the cocycle identity.
Q.E.D.

Let us proceed to describe the nontrivial 1-cocycles for $SL(2;\mathbb{C})$ associated with irreducible unitary representations:
Consider all functions $f : \mathbb{C} \to \mathbb{C}$ which are square integrable with respect to Lebesgue measure. These form a Hilbert space in the usual way and we define an irreducible unitary representation of $SL(2;\mathbb{C})$ in this Hilbert space by setting

$$(U_g f)(z) := (\alpha-\gamma z)^{-2}\ f\left(\frac{\delta z-\beta}{\alpha-\gamma z}\right)$$

where $g \in SL(2;\mathbb{C}) := \left\{ \begin{bmatrix} \alpha & \beta \\ \gamma & \delta \end{bmatrix} : \alpha\delta-\beta\gamma = 1;\ \alpha,\beta,\gamma,\delta \in \mathbb{C} \right\}$, $z \in \mathbb{C}$.

Then we have

(1.3) Theorem: There is exactly one nontrivial 1-cocycle associated with an irreducible representation of $SL(2;\mathbb{C})$ (up to scalar multiples). It is associated with the representation described above and explicitly given by

$$\delta(g)(z) := \frac{(\alpha\bar{z}+\gamma)(\overline{\alpha}-\overline{\gamma z}) + (\beta\bar{z}+\delta)(\overline{\beta}-\overline{\delta z})}{(1+|z|^2)(|\alpha-\gamma z|^2 + |\delta z-\beta|^2)}$$

where g is as above.

Proof: See [6].

We are now ready to describe the main result.

§ 2 THE MOST GENERAL 1-COCYCLE AND THE LÉVY-KHINTCHIN FORMULA

$SL(2;\mathbb{C})$, being a semi-simple, connected Lie group, is of type I. Thus we have a perfectly good direct integral decomposition of unitary representations into irreducibles, cf. [10]. The associated 1-cocycles decompose equally well by the theory given in [15]. Thus given an arbitrary unitary representation V and a 1-cocycle Δ associated with it, we may find a measure space (Ω,μ) such that

$$V(g) = \int_{\Omega}^{\oplus} U_g^{\omega} \, d\mu(\omega)$$

in the sense of direct integrals.

$$\Delta(g) = \int_{\Omega}^{\oplus} \delta^{\omega}(g) \, d\mu(\omega)$$

Here $g \mapsto U_g^{\omega}$ is an irreducible representation in a Hilbert space H^{ω} and δ^{ω} is a 1-cocycle associated with U^{ω} for all $\omega \in \Omega$.

Let $\Omega_1 := \{\omega \in \Omega : U^{\omega}$ is the representation which admits the non-trivial cocycle δ appearing in (1.3)$\}$ then it follows from the theory in [10] that Ω_1 must be measurable. Thus we may write Δ as a direct sum

$$\Delta(g) = \int_{\Omega_1}^{\oplus} \varsigma(\omega)\delta(g) \, d\mu(\omega) \oplus \int_{\Omega\setminus\Omega_1}^{\oplus} [U_g^{\omega}v^{\omega} - v^{\omega}] d\mu(\omega)$$

where $\varsigma : \Omega_1 \to \mathbb{C}$ is an element in $L^2(\Omega_1,\mu)$ and $v^{\omega} \in H^{\omega}$ for all $\omega \in \Omega\setminus\Omega_1$. In order to calculate the logarithm appearing in (1.1) it will be convenient to consider a 1-cocycle Δ satisfying $\Delta(k) = 0$ for all $k \in SU(2)$. (This is the compact group appearing in the Iwasawa decomposition of $SL(2;\mathbb{C})$). Then the most general 1-cocycle will differ from this only by a coboundary since any 1-cocycle on a compact group is a coboundary [15]. Moreover, in view of (1.2), this will represent no loss of generality. In terms of the direct integral decomposition of Δ given above $\Delta(k) = 0$ for all $k \in SU(2)$ means simply that

$U_k^{\omega} v^{\omega} = v^{\omega}$ for all $(\omega,k) \in \Omega\setminus\Omega_1 \times SU(2)$ without loss of generality.

For such a Λ we then calculate

$$\frac{1}{2} < \Delta(g), \Delta(g) > = \|\varsigma\|^2 [2t \coth 2t - 1] + \frac{1}{2} \int_{\Omega\setminus\Omega_1} \| U_g^{\omega} v^{\omega} - v^{\omega} \|^2 \, d\mu(\omega)$$

where

(i) $t = \log \frac{1}{2}\{[\text{trace } g^*g+2]^{1/2} + [\text{trace } g^*g-2]^{1/2}\}$

(the nonnegative square root being taken)

(ii) $\|\varsigma\|^2 = \int_{\Omega_1} |\varsigma(\omega)|^2 \, d\mu(\omega)$

From the triviality of the relevant 2-cohomology, cf. e.g. [13], it follows that a continuous function

$B : G \to \mathbb{R}$ which satisfies

$$\mathrm{Im} < \Delta(g_2), \Delta(g_1^{-1}) > = B(g_1) + B(g_2) - B(g_1 g_2)$$

exists. Thus we have to solve the following functional equation:

$$\mathrm{Im}\,[c^2 < \delta(g_2), \delta(g_1^{-1}) > + \int_{\Omega\setminus\Omega_1} < (U^{\omega}_{g_1 g_2} - U^{\omega}_{g_1} - U^{\omega}_{g_2}) v^{\omega}, v^{\omega} > d\mu(\omega)] =$$

$$B(g_1) + B(g_2) - B(g_1 g_2)$$

However, it is shown in [7], that $\mathrm{Im} < \delta(g_2), \delta(g_1^{-1}) > \equiv 0$. Moreover an inspection of the relevant spherical functions $< U^{\omega}_g v^{\omega}, v^{\omega} >$ in [11] shows that these are real valued and thus we see that B must be a homomorphism and hence identically zero.
An application of (1.1) now yields

<u>(2.1) Lemma:</u> Let Δ be a 1-cocycle which is a direct integral as described above and satisfies $\Delta(k) = 0$ for all $k \in SU(2)$. Then the associated I.D.P. function f is (via (1.1)) given by

$$f(g) = \exp \Psi(g)$$

where $\Psi(g) = - \Pi c^2 [2t \coth 2t-1] + \int_{\Omega\setminus\Omega_1} < U^{\omega}_g v^{\omega} - v^{\omega}, v^{\omega} > d\mu(\omega)$

All the relevant terms are described above.

<u>(2.2) Remarks:</u>

a) Combining (2.1) with (1.2) we obtain the complete analogue of the classical Lévy-Khintchin Formula for $SL(2;\mathbb{C})$. Apart from the consequences mentioned in the introduction this should also be an interesting result in the context of [19].

b) There is a whole family of I.D.P. functions determined by the analogue of the "Gaussian part" of Ψ appearing in (2.1) namely

$$f(g) = \exp - \Pi c^2 [2t \coth 2t-1].$$

This leads to the construction of irreducible representations of current groups, cf. [7].

REFERENCES

[1] *Albeverio, S.; Høegh-Krohn, R.; Testard, D.:* Irreducibility and Reducibility for the Energy Representation of the Group of Mappings of a Riemannian Manifold into a Compact Semi-simple Lie Group, J. Functional Analysis 41 (378-396) (1981)

[2] *Araki, H.:* Factorizable Representation of Current Algebra - Non commutative extension of the Lévy-Khintchin formula and cohomology of a solvable group with values in a Hilbert Space -, Publ. RIMS, Kyoto Univ. Vol. 5 (1969/70) p. 361-422

[3] *Delorme, P.:* 1-COHOMOLOGIE DES REPRÉSENTATIONS UNITAIRES DES GROUPES DE LIE SEMI-SIMPLES ET RESOLUBLES, PRODUITS TENSORIELS CONTINUS DE REPRESENTATIONS, Centre de Mathématiques de l'Ecole Polytechnique, Plateau de Palaiseau - 91120 Palaiseau (France) (1976)

[4] *Erven, J.; Falkowski, B.-J.:* Low Order Cohomology and Applications, Springer Lecture Notes in Mathematics, Vol. 877 (1981)

[5] *Erven, J.; Falkowski, B.-J.:* Continuous Cohomology, Infinitely Divisible Positive Definite Functions and Continuous Tensor Products for SU(1;1), in "Probability Measures on Groups" (Ed. H. Heyer), Springer Lecture Notes in Mathematics, Vol. 928 (1982)

[6] *Falkowski, B.-J.:* First Order Cocycles for SL(2;ℂ), J. Ind. Math. Soc. 41 (1977) 245-254

[7] *Falkowski, B.-J.:* Current Group and 1-Cohomology for SL(2;ℂ), submitted for publication to J. Ind. Math. Soc.

[8] *Gelfand, I.M.; Graev, I.M.; Vershik, A.M.:* Irreducible Representations of the Group G^X and Cohomologies, Funct. Analy. and its Appl. 8, no. 2 (1974)

[9] *Guichardet, A.:* Symmetric Hilbert Spaces and Related Topics, Springer Lecture Notes in Mathematics, Vol. 261 (1972)

[10] *Mackey, G.W.:* The Theory of Unitary Group Representations, Univ. of Chicago Press (1976)

[11] *Naimark, M.A.:* Normed Algebras, Wolters-Noordhoff Publishing, Groningen, The Netherlands (1972)

[12] *Newman, C.M.:* Ultralocal Quantum Field Theory in Terms of Currents, Comm. in Math. Phys., 26, (1972)

[13] *Parthasarathy, K.R.:* Multipliers on Locally Compact Groups, Springer Lecture Notes in Mathematics, Vol. 93 (1969)

[14] *Parthasarathy, K.R.:* Infinitely Divisible Representations and Positive Definite Functions on a Compact Group, Comm. Math. Phys. Vol. 16 (1970)

[15] *Parthasarathy, K.R.; Schmidt, K.:* Positive Definite Kernels, Continuous Tensor Products, and Central Limit Theorems of Probability Theory, Springer Lecture Notes in Mathematics, Vol. 272 (1972)

[16] *Parthasarathy, K.R.; Schmidt, K.:* A New Method for constructing Factorizable Representations for Current Groups and Current Algebras, Comm. in Math. Phys., 50, (1976)

[17] *Segal, G.:* Unitary Representations of some Infinite Dimensional Groups, Comm. in Math. Phys. 80 (1981)

[18] *Streater, R.F.;* Current Commutation Relations, Continuous Tensor Products, and Infinitely Divisible Group Representations, Rend. Sci. Int. Fisica E. Fermi, XI, (1969)

[19] *Walter, M.E.:* Differentiation on the Dual of a Group: An Introduction, Rocky Mountain Journal of Math. Vol. 12, No. 3, (1982)

BERNOULLI SYSTEMS IN SEVERAL VARIABLES

Philip Feinsilver
Department of Mathematics
Southern Illinois University
Carbondale, Illinois 62901/USA

I. Introduction

Let $w(t)$ be a process with values in $\mathbb{R}^N$ having stationary independent increments. Assume that the generator $L(z)$, $z = (\frac{\partial}{\partial x_1},\ldots,\frac{\partial}{\partial x_N})$, is, as a function of z, analytic in a neighborhood of 0 in $\mathbb{C}^N$. We center and normalize L in general so that $L'(0) = 0$, $L''(0) = I$ where $L' = \text{grad } L$, $L'' = (\frac{\partial^2 L}{\partial z_j \partial z_k})$.

The exponential martingale

$$e^{a\cdot x - tL(a)} \quad , \quad a\in\mathbb{C}, \quad x = w(t)$$

has an expansion around zero

$$\sum_n \frac{a^n}{n!} h_n(x,t) \quad , \quad n = (n_1,\ldots,n_N)$$

In the case $L(z) = \frac{1}{2}\sum z_j^2$, the h_n's are an orthogonal family, the Hermite polynomials. We can ask the question: If we allow for a smooth change of variables $a \to V(a)$, can we get a system of orthogonal polynomials (in cases other than the Hermite case)?

That is, we want

$$\exp(a\cdot x - tL(a)) = \sum \frac{(V(a))^n}{n!} J_n(x,t)$$

where $J_n(x,t)$ are to be orthogonal with respect to the underlying measure $p_t(x)$, satisfying

$$\int_{\mathbb{R}^N} e^{a\cdot x} p_t(x) = e^{tL(a)}.$$

With $\langle\ \rangle$ denoting expectation with respect to p_t the condition for orthogonality is

$$\langle e^{a\cdot x - tL(a)} e^{b\cdot x - tL(b)}\rangle = \Phi(V_1(a)V_1(b),\ldots,V_N(a)V_N(b))$$

for some smooth Φ. This gives us

$$L(a+b) - L(a) - L(b) = F(V_1(a)V_1(b),\ldots,V_N(a)V_N(b))$$

with $F = t^{-1}\log\Phi$.

This relation is the basis for the theory. A function L satisfying the above equation is a <u>Bernoulli generator</u> and the orthogonal system $\{J_n\}$ is a <u>Bernoulli system</u>.

II. The Characteristic Equations

The first step is to expand around $a = 0$:

$$L(a+b)-L(a)-L(b) = a_\lambda(L_\lambda(b)-L_\lambda(0))+\tfrac{1}{2}a_\lambda a_\mu(L_{\lambda\mu}(b)-L_{\lambda\mu}(0))+\ldots$$

$$F(V(a)V(b)) = a_\lambda V_{\epsilon\lambda}(0)V_\epsilon(b)f_\epsilon+\tfrac{1}{2}a_\lambda a_\mu(V_{\epsilon\lambda\mu}(0)V_\epsilon(b)f_\epsilon + V_{\epsilon\lambda}(0)V_\epsilon(b)V_{\zeta\mu}(0)V_\zeta(b)f_{\epsilon\zeta})+\ldots,$$

with the f's denoting derivatives of F at zero.

Note.

1. Subscripts denote partial derivatives: $L_j = \partial L/\partial z_j$ and so on.

2. The summation convention used is: repeated Greek indices are summed from 1 to N. (In explicitly indicated sums, Latin indices will be used.)

The next step is to compare terms:

1st order: Since grad $L(0) = 0$, we have

$$L_j(z) = V_{\epsilon j}(0)f_\epsilon V_\epsilon(z) = O_{\epsilon j}V_\epsilon, \text{ with } O_{jk} = V_{jk}(0)f_j$$

2nd order:

$$O_{\epsilon j}V_{\epsilon k} = L_{jk} = \delta_{jk} + V_{\epsilon jk}(0)V_\epsilon f_\epsilon + O_{\epsilon j}O_{\zeta k}\, f_{\epsilon\zeta}f_\epsilon^{-1}f_\zeta^{-1}V_\epsilon V_\zeta$$

Multiply by o^t, o transpose, with $o = O^{-1}$,

$$V_{jk} = o_{kj} + a^\lambda_{jk}V_\lambda + O_{\zeta k}f_{j\zeta}f_j^{-1}f_\zeta^{-1}V_jV_\zeta$$

Here and in subsequent steps the a^λ_{jk} terms may vary; they are "generic" constants. Next, change coordinates, setting

$$V_j(z) = v_j(o^tz)$$

$$V_{jk}(z) = o_{k\lambda}v_{j\lambda}(o^tz)$$

And

$$o_{k\lambda}v_{j\lambda} = o_{kj} + a^\lambda_{jk}v_\lambda + f_{j\zeta}f_j^{-1}v_jo_{k\zeta}v_\zeta$$

The last term arises as follows: The equations for V_{jk} give us $V_{jk}(0) = o_{kj}$. The definition of O says then that

$O_{jk} = V_{jk}(0)f_j = o_{kj}f_j$. That is, $f_j^{-1}O_{jk} = o_{kj}$.

Multiplying by O yields

$$v_{jk} = \delta_{jk} + a^\lambda_{jk}v_\lambda + f_{jk}f_j^{-1}v_jv_k$$

Now differentiate:

$$v_{jk\ell} = a^\lambda_{jk}v_{\lambda\ell} + f_{jk}f_j^{-1}(v_{j\ell}v_k + v_jv_{k\ell})$$

Resubstituting yields

$$v_{jk\ell} = a^\ell_{jk}+(1^{st} \text{ and } 2^{nd} \text{ order terms}) + f_{jk}f_j^{-1}(f_{k\ell}f_k^{-1}+f_{j\ell}f_j^{-1})v_jv_kv_\ell$$

By symmetry in $k\ell$:

$$f_{jk}f_{k\ell}f_j^{-1}f_k^{-1}+f_{jk}f_{j\ell}f_j^{-1}f_j^{-1} = f_{j\ell}f_{\ell k}f_j^{-1}f_\ell^{-1}+f_{j\ell}f_{jk}f_j^{-1}f_j^{-1}$$

That is, $f_{jk}f_k^{-1} = f_{j\ell}f_\ell^{-1}$. Thus

$$c_j = f_{jk}f_j^{-1}f_k^{-1} = f_{j\ell}f_j^{-1}f_\ell^{-1}$$ is independent of k or ℓ. So

$f_{jk} = c_j f_j f_k = c_k f_k f_j$ by symmetry of $f_{jk} = F_{jk}(0)$.

Putting $f_{jk} = cf_j f_k$, $b_k = cf_k$ and $v \to V$ we have

Theorem 1.

Canonical Form of the Characteristic Equations:

$$V_{jk} = \delta_{jk} + a_{jk}^{\lambda}V_\lambda + b_k V_j V_k$$

with the initial conditions $V_j(0) = 0$. □

Recall the relation $L_j = V_{\varepsilon j}(0)f_\varepsilon V_\varepsilon$. In the canonical form, $V_{jk}(0) = \delta_{jk}$ so we get $L_j = f_j V_j$. That is, up to scaling factors, V is grad L.

With L and V determined, the generating function $e^{a\cdot x - tL(a)}$ determines (implicitly) the polynomials $J_n(x,t)$. It is convenient to introduce $U = V^{-1}$ and $M(v) = L(U(v))$. U exists locally since V' is non-degenerate at zero. Then the generating function is

$$\exp(x\cdot U(v) - tM(v)) = \sum_n \frac{v^n}{n!} J_n(x,t)$$

Notice that the operator $V_j(D)$, $D = (\frac{\partial}{\partial x_1}, \ldots, \frac{\partial}{\partial x_N})$, simply multiplies by v_j. Thus we have

$$V_j J_n = n_j J_{n-e_j},$$

with $\{e_j\}$ = standard basis for $\mathbb{Z}^N$, $(e_j)_k = \delta_{jk}$. That is, on J_n, V acts as a gradient or "lowering operator," just as D acts on the powers x^n.

<u>Note</u>. For orthogonal polynomials in several variables, there is a generalized three-term recurrence formula. In our case we quote without proof the recurrence for the canonical case:

$$J_{n+e_k} = x_k J_n - \sum_{\ell,m} n_m a_{\ell k}^m J_{n+e_\ell - e_m} - b_k n_k(|n|-1+t)J_{n-e_k}$$

with $|n| = \sum n_j$ for a multi-index $(n_1,\ldots,n_N)$.

We proceed with the generic solutions. See [2] for "separable" solutions corresponding to processes with independent coordinates (up to linear transformations).

III. Solutions

Recall that L arises naturally as the logarithm of the characteristic function of p_t.

The relation $L_j = f_j V_j$ suggests the substitution

$$V_j = c_j y_j / y$$

Substituting into the canonical equations yields:

$$c_j(y\, y_{jk} - y_j y_k) = \delta_{jk} y^2 + a^{\lambda}_{jk} c_\lambda y_\lambda y + b_k c_j c_k y_j y_k$$

The quadratic terms can be eliminated by choosing

$$-c_j = b_k c_j c_k \quad \text{or} \quad c_k b_k = -1.$$

Dividing by y yields the linear system

$$c_j y_{jk} = \delta_{jk} y + a^{\lambda}_{jk} c_\lambda y_\lambda$$

Exponential solutions of the form $k\, e^{\xi \cdot z}$ require $\xi = (\xi_1, \ldots, \xi_N)$ to satisfy

$$c_j \xi_j \xi_k - a^{\lambda}_{jk}\, c_\lambda \xi_\lambda - \delta_{jk} = 0.$$

We will proceed now by starting with L as the log of a combination of exponentials, as this will allow us to see directly the solvability and consistency conditions required.

We thus assume L to be of the form

$$L = \log\left(\sum_{\ell} c_\ell (\exp(c_\ell^{-1} C_{\ell\sigma} z_\sigma) - 1) + 1\right) - \sum_{\ell, m} C_{\ell m} z_m$$

What is the probabilistic significance of this expression? Except for the coordinate transformation C and the centering terms to satisfy $L'(0) = 0$, this is the log of the characteristic function of a <u>multinomial distribution</u>. As this is the N-dimensional version of a Bernoulli random walk, this explains the generic terminology "Bernoulli systems." This is the "general solution" to the original question of what types of processes "naturally" give rise to a system of orthogonal polynomials.

We will proceed to determine conditions on the parameters to yield the solutions we have been seeking. Namely, normalized solutions that yield orthogonal systems.

<u>Notations</u>. Set: $c = (c_1, \ldots, c_N)$, $E_\ell = \exp(c_\ell^{-1} C_{\ell\sigma} z_\sigma)$, $u = (1,1,\ldots,1)$, $\bar{c} = 1 - u \cdot c = 1 - \sum_j c_j$ and $\Delta = c \cdot E + \bar{c}$ so that $L = \log\Delta$ + centering terms.

We put $L_j = \gamma_j V_j$, with γ_j's as generic scaling factors.

We compute:

$$L_j = \sum_{\ell} C_{\ell j} E_\ell / \Delta - \sum_{\ell} C_{\ell j}$$

<u>Notations</u>. (1) Underlining will be used when indicating components of vectors or matrices, e.g. $R_{j\lambda} x_\lambda = \underline{Rx}_j$. (2) The same symbol will be used for a vector or the corresponding diagonal matrix according to context, e.g. $c = (c_1, \ldots, c_N)$ may also denote the matrix with entries $(c)_{jk} = c_j \delta_{jk}$. (3) In the following calculations, B denotes C^{-1}.

We continue with $L_j = \Delta^{-1}\underline{EC_j} - \underline{uC_j}$. Multiply by B:

$\gamma VB = \Delta^{-1}E - u$. Applying to c yields

$$\gamma VBc = \Delta^{-1}E\cdot c - u\cdot c$$
$$= \Delta^{-1}(\Delta - \bar{c}) + \bar{c} - 1 = \bar{c}(1 - \Delta^{-1}).$$

We have:

(i) $\Delta^{-1} = 1 - \gamma VBc/\bar{c}$ and $E/\Delta = u + \gamma VB$

Continuing, differentiate L_j to get

$$\gamma_j V_{jk} = \Delta^{-1}\sum_\ell C_{\ell j}C_{\ell k}c_\ell^{-1}E_\ell - \Delta^{-2}\underline{EC_j}\,\underline{EC_k}$$
$$= \sum_\ell C_{\ell j}C_{\ell k}c_\ell^{-1}(1+\underline{\gamma VB_\ell}) - (\gamma_j V_j + \underline{uC_j})(\gamma_k V_k + \underline{uC_k})$$

The condition $V_{jk}(0) = \delta_{jk}$ gives us the <u>normalization</u> <u>condition</u>:

(ii) $$\gamma_j\delta_{jk} = \underline{C^t c^{-1}C}_{jk} - \underline{uC_j}\,\underline{uC_k}$$

Next we want to check orthogonality. Particularly, $\langle(e^{x\cdot U-tM})^2\rangle$ has to have no "cross-terms" as a function of v. Taking the expectation:

$$\exp(tL(2U(v)) - 2tL(U(v)))$$

has to be free of cross-terms. Notice that linear terms in L cancel out automatically. Since

$$L = \log\Delta + \text{centering}$$

we can work with

$$M = \log\Delta = -\log(1 - \gamma vBc/\bar{c})$$

by equation (i) above.

In computing L(2U) note that the substitution $z \to 2z$ simply replaces E by E^2 so that

$$L(2U(v)) = \log(\Delta^2\sum_\ell c_\ell(1 + \underline{\gamma vB_\ell})^2 + \bar{c})$$

Let us calculate $\exp(L(2U) - 2L(U))$:

$$(\Delta^2\sum c_\ell(1 + \underline{\gamma vB_\ell})^2 + \bar{c})(1 - \gamma vBc/\bar{c})^2$$
$$= \sum c_\ell(1+2\underline{\gamma vB_\ell}+(\underline{\gamma vB_\ell})^2) + \bar{c} - 2\gamma vBc + (\gamma vBc)^2/\bar{c}$$

(since the factor Δ^2 cancels for the first term)

$$= 1 + \sum_\ell c_\ell(\underline{\gamma vB_\ell})^2 + (\gamma vBc)^2/\bar{c}$$

We consider

$$\bar{c}\sum c_\ell(\underline{\gamma vB_\ell})^2 + (\gamma vBc)^2$$
$$= \bar{c}\sum_{i,j,\ell} c_\ell\gamma_i v_i B_{i\ell}\gamma_j v_j B_{j\ell} + \sum_{p,q,r,s} c_p\gamma_r v_r B_{rp}c_q\gamma_s v_s B_{sq}$$

Cancelling γv terms, we have the <u>orthogonality</u> <u>condition</u>:

$$\bar{c}\sum_\ell c_\ell B_{r\ell}B_{s\ell} + \sum_{p,q} c_p B_{rp}c_q B_{sq} = 0$$

or

(iii) $\quad \bar{c}\,\underline{BcB}^{t}_{rs} + \underline{Bc}_r\underline{Bc}_s = 0 \quad$ for $r \neq s$

Now set $\Gamma = C^t c^{-1} C$. We can express (iii) in the form

(iv) $\quad \Gamma^{-1}_{jk} = \eta_j \delta_{jk} - \bar{c}^{-1}\underline{Bc}_j\underline{Bc}_k$

for some constants η_j. And the companion normalization condition (ii) is:

(v) $\quad \Gamma_{jk} = \gamma_j \delta_{jk} + \underline{uC}_j\underline{uC}_k$

Lemma.

Let $A = a - tS$ where a is a diagonal matrix and S is a symmetric rank one matrix, i.e. S is of the form xx^t for some vector x. Then

(vi) $\quad A^{-1} = a^{-1} + (t/(1-rt))a^{-1}Sa^{-1}$

where $r = \sum_{\ell} a_{\ell}^{-1} x_{\ell}^{2}$.

(The proof is a direct calculation.)

Apply the Lemma to Γ, equation (v), to get

(vii) $\quad \Gamma^{-1}_{jk} = \gamma_j^{-1}\delta_{jk} - (1+g)^{-1}\gamma_j^{-1}\underline{uC}_j\underline{uC}_k\gamma_k^{-1}$

where $g = \sum_{\ell} \gamma_{\ell}^{-1}(\underline{uC}_{\ell})^2$.

Compare this with (iv). This suggests $\eta_j = \gamma_j^{-1}$ which leads to

(viii) $\quad (1+g)^{-1}\gamma_j^{-1}\underline{uC}_j\underline{uC}_k\gamma_k^{-1} = \bar{c}^{-1}\underline{Bc}_j\underline{Bc}_k$

Notice that

$$\sum_{\ell}\underline{uC}_{\ell}\underline{Bc}_{\ell} = u_{\rho}C_{\rho\lambda}B_{\lambda\sigma}c_{\sigma} = u\cdot c = 1 - \bar{c}$$

Thus, multiplying (viii) by $\underline{uC}_j\underline{uC}_k$ and summing yields

(ix) $\quad (1+g)^{-1}g^2 = (1-\bar{c})^2/\bar{c}$

with the solutions $\bar{c} = 1+g$ or $(1+g)^{-1}$. The first solution turns out to be "spurious." We continue with $\bar{c} = (1+g)^{-1}$. Going back to (viii), we have

$$\bar{c}^2\gamma_j^{-1}\underline{uC}_j\underline{uC}_k\gamma_k^{-1} = \underline{Bc}_j\underline{Bc}_k$$

or

(x) $\quad \bar{c}\,\underline{uC}_j = \gamma_j\underline{Bc}_j$

Note that this is the same as

(xi) $\quad B^t\gamma Bc = \bar{c}\,u$.

Now reconsider the normalization condition (ii)

$$\underline{C^tc^{-1}C}_{jk} = \gamma_j\delta_{jk} + \underline{uC}_j\,\underline{uC}_k$$

Multiply by B^t to get

$$c_j^{-1} C_{jk} = \underline{B^t\gamma}_{jk} + \underline{uC}_k$$

Multiply by B on the right:

$$c_j^{-1} \delta_{jk} = \underline{B^t\gamma B}_{jk} + 1$$

This is our key condition:

(xii) $\quad B^t \gamma B = c^{-1} - J$

where J is the "all-ones matrix" with $J_{jk} = 1$, for all j,k. Applying this relation to the vector c yields

$$B^t \gamma B c = \bar{c} u$$

which is just (xi). Thus we have checked the consistency of the normalization and orthogonality conditions.

<u>Theorem 2.</u>

General Bernoulli Generator

The generator

$$L = \log\Big(\sum_\ell c_\ell (\exp(c_\ell^{-1} C_{\ell\sigma} z_\sigma) - 1) + 1\Big) - \sum_{\ell,m} C_{\ell m} z_m$$

is a "general Bernoulli generator" with lowering operator $V = \gamma^{-1} L'$ provided the condition $B^t \gamma B = c^{-1} - J$ holds, where $B = C^{-1}$. □

Let us see how the characteristic equations come out. Just prior to equation (ii) we calculated

$$\gamma_j V_{jk} = \sum_\ell C_{\ell j} C_{\ell k} c_\ell^{-1} (1 + \underline{\gamma VB}_\ell) - (\gamma_j V_j + \underline{uC}_j)(\gamma_k V_k + \underline{uC}_k)$$

Multiplying out and dividing through by γ_j gives

$$V_{jk} = \delta_{jk} + a_{jk}^\lambda V_\lambda + b_k V_j V_k$$

with the identifications

$$b_k = -\gamma_k$$

$$\gamma_j a_{jk}^\ell = C_{\mu j} C_{\mu k} c_\mu^{-1} B_{\ell\mu} \gamma_\ell - \underline{uC}_j \gamma_k \delta_{k\ell} - \underline{uC}_k \gamma_j \delta_{j\ell}$$

Using the condition (xii) solved for γB, this first term may be expressed directly in terms of C as $C_{\mu j} C_{\mu k} c_\mu^{-1} (C_{\mu\ell} c_\ell^{-1} - u_\ell \underline{uC}_\mu)$.

We would like to find an expression for the generating function for the orthogonal polynomials in the general multinomial case. Recall relation (i):

$$E = \Delta(u + \gamma VB)$$

so $\quad c_\ell^{-1} C_{\ell\sigma} z_\sigma = \log(\Delta(1 + \underline{\gamma VB}_\ell))$

and $\quad z_\ell = B_{\ell\mu} c_\mu \log(\Delta(1 + \underline{\gamma VB}_\mu))$

i.e. $\quad U_\ell = \log \prod_m \Delta^{B_{\ell m}c_m}(1+\gamma V\underline{B}_m)^{B_{\ell m}c_m}$

We thus have

$$G = e^{x\cdot U - tM} = \prod_{\ell,m} (1+\gamma v\underline{B}_m)^{x_\ell B_{\ell m}c_m} \cdot \Delta^{x_\lambda B_{\lambda\mu}c_\mu} \cdot e^{-tM}$$

We have, including centering terms,

$$M = \log \Delta - u_\lambda C_{\lambda\mu} U_\mu$$

We find that

$$G = \prod_m (1+\gamma v\underline{B}_m)^{x\underline{B}_m c_m} \cdot \Delta^{xBc} \cdot \Delta^{-t} \cdot \Delta^{tu\cdot c} \cdot \prod_m (1+\gamma v\underline{B}_m)^{tc_m}$$

$$= \prod_m (1+\gamma v\underline{B}_m)^{c_m(x\underline{B}_m+t)} \Delta^{xBc-\bar{c}t}$$

where from (i) above,

$$\Delta^{-1} = 1 - \gamma VBc/\bar{c}$$

Using the key condition (xii) we can express

$$\Delta^{-1} = 1 - uCV$$

We have:

<u>Theorem 3.</u>

The generating function for the general Bernoulli polynomials is

$$G(x,t;v) = \prod_m (1+\gamma v\underline{B}_m)^{c_m(x\underline{B}_m+t)} \cdot (1-uCv)^{\bar{c}t-xBc} \qquad \square$$

This completes the basic discussion. In the next two sections we will present ways of formulating solvability conditions for our characteristic equations, i.e. conditions on the parameters a^ℓ_{jk} and b_k, that afford us insight into the structures at hand.

IV. Projective Formulation and Group-Theoretical Structure

The characteristic equations may be linearized by going to the projective coordinates on $\mathbb{C}^{N+1}$, namely we write $Y = \binom{V}{y}$ with our original V recovered as $V \leftarrow V/y$. The substitution $V \to V/y$ is similar to the logarithmic substitution in the previous section, but here we get a linear system in $N + 1$ coordinates.

Substitute $V_j \to V_j/y$ in $V_{jk} = \delta_{jk} + a^\lambda_{jk}V_\lambda + b_k V_j V_k$. We have

$yV_{jk} - V_j y_k = \delta_{jk}y^2 + a^\lambda_{jk}V_\lambda y + b_k V_j V_k$. We split this up to form the linear system

$$\frac{\partial}{\partial z_k} Y = X_k Y$$

where $X_k = \begin{pmatrix} A_k & e_k \\ -b_k e_k^t & -c_k \end{pmatrix}$

with A_k an $N \times N$ submatrix, e_k denoting the standard basis column vector, $-b_k e_k^t$ a $1 \times N$ row vector, and c_k a constant. We compare with the standard equations and check that

$$(A_k)_{j\ell} = a_{jk}^{\ell} - c_k \delta_{j\ell}$$

The c's may be chosen freely. It is convenient to choose c_k so that tr $X_k = 0$. (This is particularly useful in the case $N = 1$:

$X = \begin{pmatrix} \alpha & 1 \\ -\beta & -\alpha \end{pmatrix}$ is the standard form.) Since initially $V(0) = 0$, the solution sought is found by computing

$$Y(z) = \exp\Big(\sum_k z_k X_k\Big) Y_0$$

with $Y_0 = e_{N+1}$.

Then writing $Y(z) = \begin{pmatrix} V(z) \\ y(z) \end{pmatrix}$ we recover

$$V(z) \leftarrow V(z)/y(z)$$

For example, with $N = 1$, we readily compute

$$Y(z) = ((\cosh qz)I + q^{-1}(\sinh qz)X)Y_0$$

where $q^2 = \alpha^2 - \beta$. This gives the generic solution for $N = 1$

$$V(z) = \sinh qz/(q \cosh qz - \alpha \sinh qz).$$

In this form, "the $s\ell(N+1)$-formulation," the solvability conditions are simply expressed by the commutativity requirements $X_k X_\ell = X_\ell X_k$. We compute:

$$X_k X_\ell = \begin{pmatrix} A_k A_\ell - E_{k\ell} b_\ell & A_k e_\ell - c_\ell e_k \\ -b_k e_k^t A_\ell + e_\ell^t c_k b_\ell & c_k c_\ell \end{pmatrix}$$

where $E_{k\ell}$ denotes the elementary matrix with entries $(E_{k\ell})_{ij} = \delta_{ik}\delta_{j\ell}$.

Setting $\alpha_k = A_k + c_k I$, so that $(\alpha_k)_{j\ell} = a_{jk}^{\ell}$, we have:

<u>Theorem 4</u>.

The solvability relations for the characteristic equations:

(1) $[\alpha_k, \alpha_\ell] = E_{k\ell} b_\ell - E_{\ell k} b_k$

(2) $\alpha_k e_\ell = \alpha_\ell e_k$

(3) $b_k e_k^t \alpha_\ell = b_\ell e_\ell^t \alpha_k$

where [,] denotes the usual commutator. □

The second observation this formulation allows is to notice that the solutions transform under a group of linear transformations. Specifically, we can transform $Y \to RY$ with R an N + 1 by N + 1 matrix. To preserve the property $V(0) = 0$ we must have R in the block form

$$\begin{pmatrix} E & 0 \\ g^t & 1 \end{pmatrix}$$

where the (N+1, N+1) entry can be generically scaled to 1.

The action $Y \to RY$ becomes on the original V

$$RV = EV/(g\cdot V + 1)$$

a fractional linear transformation.

If RY satisfies, ∂ denoting a generic $\partial/\partial z$,

$$\partial(RY) = X(RY)$$

Then Y satisfies

$$\partial Y = (R^{-1}XR)Y.$$

Let us compute this action of R on X and determine how the parameters a and b transform. First check that

$$R^{-1} = \begin{pmatrix} F & 0 \\ -g^tF & 1 \end{pmatrix} \quad \text{with } FE = I.$$

Calculation of $R^{-1}XR$ yields, with $X = \begin{pmatrix} A & e \\ -b^t & -c \end{pmatrix}$,

$$\begin{pmatrix} FAE + Feg^t & Fe \\ -g^tFAE - g^tFeg^t - b^tE - cg^t - g^tFe - c \end{pmatrix}$$

Thus:

<u>Theorem 5.</u>

Under the linear action $R = \begin{pmatrix} E & 0 \\ g^t & 1 \end{pmatrix}$, the parameters of X transform as:

$$\begin{aligned} A &\to E^{-1}AE + E^{-1}eg^t \\ e &\to E^{-1}e \\ b^t &\to b^tE + g^t(E^{-1}AE + E^{-1}eg^t + c) \\ c &\to g^tE^{-1}e + c \end{aligned}$$ □

As k runs from 1 to N, the e's in X_k run through the standard basis on $\mathbb{R}^N$. If we want to consider transformations preserving the e's, which means they preserve the normalization $V_{jk}(0) = \delta_{jk}$, then E must be the identity. In this case the transformation simplifies to:

$$\begin{aligned} A &\to A + eg^t \\ b^t &\to b^t + g^t(A + eg^t + c) \\ c &\to c + g^te \end{aligned}$$

Remarks. (1) You also need conditions to guarantee that the new b^t has only one non-zero component in the correct place - the $k^{\underline{th}}$, for X_k.
(2) In the $N = 1$ case, the explicit representation of the group action induced on the J_n's can be computed. It shows that the J_n's are indeed a representation space for the affine group - represented by matrices of the form $\begin{pmatrix} E & 0 \\ g & 1 \end{pmatrix}$. From the above discussion we see that this statement holds for general $N > 1$ as well.

V. Differential-Geometric Formulation

We can interpret the characteristic equations as "the $j^{\underline{th}}$ component of" the system

$$\frac{\partial V}{\partial z_k} = g_k + A_k V + b_k(V,V)$$

where V denotes, as usual, the vector (V_j),

g_k is a vector with $(g_k)_j = \delta_{jk}$

A_k is a matrix with $(A_k)_j^\ell = a_{jk}^\ell$

b_k is a vector of quadratic forms with $(b_k)_j^{\ell m} = b_k \delta_{j\ell}\delta_{km}$

(notice this is not symmetrized in ℓm).

We write the system using differential forms:

$$dV = g + AV + b(V,V)$$

with $dV = \sum \frac{\partial V}{\partial z_k} dz_k$, $g = \sum g_k dz_k$, $A = \sum A_k dz_k$, $b = \sum b_k dz_k$.

It is natural now to assume that g, A and b are functions of z, not necessarily constants. We take the components of g to be g_{jk}. The solvability conditions for a system of the above form are

$$d^2 V = 0.$$

Let us see what this says explicitly.

Differentiating the characteristic equations with general coefficients

$$\begin{aligned} V_{jk\ell} &= g_{jk\ell} + a_{jk\ell}^\lambda V_\lambda + a_{jk}^\lambda V_{\lambda\ell} + b_k(V_{j\ell}V_k + V_j V_{k\ell}) + b_{k\ell} V_j V_k \\ &= g_{jk\ell} + a_{jk\ell}^\lambda V_\lambda + a_{jk}^\lambda (g_{\lambda\ell} + a_{\lambda\ell}^\mu V_\mu + b_\ell V_\lambda V_\ell) \\ &\quad + b_k(V_k(g_{j\ell} + a_{j\ell}^\mu V_\mu + b_\ell V_j V_\ell) + V_\ell(g_{k\ell} + a_{k\ell}^\mu V_\mu + b_\ell V_k V_\ell)) + b_{k\ell} V_j V_k \end{aligned}$$

Since V' is non-degenerate, we can think of V_j as a coordinate system and compare terms: $V_{jk\ell} = V_{j\ell k}$.

$0^{\underline{th}}$ order: $g_{jk\ell} + a_{jk}^\lambda g_{\lambda\ell} = g_{j\ell k} + a_{j\ell}^\lambda g_{\lambda k}$

Multiplying by $dz_\ell \wedge dz_k$, $\ell < k$, and summing yields

$$dg = A \wedge g$$

$1^{\underline{st}}$ order:

Taking the coefficient of V_m and comparing yields:

$$a^m_{jk\ell}-a^m_{j\ell k}+a^\lambda_{jk}a^m_{\lambda\ell}-a^\lambda_{j\ell}a^m_{\lambda k} = b^{\lambda m}_{\ell j}g_{\lambda k}+b^{m\mu}_{\ell j}g_{\mu k}-b^{\lambda m}_{kj}g_{\lambda\ell}-b^{m\mu}_{kj}g_{\mu\ell}$$

where $b^{rs}_{pq} = b_p\delta_{rq}\delta_{sp}$ (as defined above). Note that both indices ℓm of $b^{\ell m}_{kj}$ are contracted with g. We write, then,

$$dA - A \wedge A = b \times g$$

$2^{\underline{nd}}$ order:

Checking the terms of the form V_rV_s it is easy to see that terms not involving V_j cancel out. We find

$$b_k a^m_{k\ell} + b_{k\ell}\delta_{km} = b_\ell a^m_{\ell k} + b_{\ell k}\delta_{\ell m}$$

or

$$b_{k\ell}\delta_{km} - b_{\ell k}\delta_{\ell m} = b_\ell a^m_{\ell k} - b_k a^m_{k\ell}$$

Thinking of $b_\ell a^m_{\ell k}$, e.g., as $b^{n\mu}_{\ell j}\delta_{jn}\delta_{\ell\mu}a^m_{\mu k}$, with $j = n$, we write these relations in the form

$$db = b \wedge A$$

Summarizing:

Theorem 6.

The solvability relations for the characteristic equations may be expressed by the differential relations:

(1) $dg = A \wedge g$

(2) $b \times g = dA - A \wedge A$

(3) $db = b \wedge A$ □

These afford the following interpretation. Consider g as a generalized "metric form," A as a "connection form" or gauge potential and b as essentially "curvature" or a gauge field. Then (1) says that the covariant derivative of g is zero - Ricci's Lemma of classical differential geometry. (2) says that $b \times g$ is the curvature, or what is called in physics a gauge field. (3) is a "Bianchi identity."

The case $b \equiv 0$ yields Gaussian and Poisson solutions - these correspond to zero curvature. The general Bernoulli-type solutions, and in fact gamma distributions as well, correspond to non-trivial curvature. The field V may be termed a "superfield" as it generates by differentiation the "physical fields" g, a, and b.

VI. Remarks

1. The constructions involved are all local; everything depends only on a neighborhood of zero. Thus it seems likely that the theory can be

extended, say, to homogeneous spaces. It would be particularly interesting to find the special function theory that would arise.

2. Processes of Bernoulli type - multinomial, gamma, Poisson, Gaussian - can be described directly from the viewpoint of stochastic analysis as those processes for which the exponential martingale, appropriately modified for the cases of discrete time, is the generating function of the iterated stochastic integrals of the process such that these are functions of $w(t)$ and t. The classic example is, of course, Brownian motion with iterated integrals being the Hermite polynomials ([4]). It turns out that care must be taken to distinguish between discrete and continuous time, e.g. random walks vs. Brownian motion. The parameter b acts as a time-scaling factor. In the differential-geometric interpretation we may say that "curvature creates discreteness of time." Is this a physically exact analogy? A fascinating lead... .

3. The general system $V_{jk} = g_{jk} + a^{\lambda}_{jk} V_{\lambda} + b^{\lambda\mu}_{jk} V_{\lambda} V_{\mu}$ may provide an effective approach to the study of gauge fields. The equations with constant coefficients give us a probabilistic/quantum - theoretical structure. With variable coefficients we have a differential-geometric structure. Elucidation of this connection may lead to a definite quantum theory - relativity theory link. In any case, the mathematics of these systems appears to be a rich field for study.

References

1. L.D. Faddeev and A.A. Slavnov, *Gauge fields, introduction to quantum theory*, Benjamin/Cummings, 1980.
2. P. Feinsilver, " Moment Systems and Orthogonal Polynomials in Several Variables," J. Math. Anal. Appl., 85,2, 1982, 385-405.
3. P. Feinsilver, *Special Functions, Probability Semigroups, and Hamiltonian Flows*, Springer LNM 696, 1978.
4. H.P. McKean, *Stochastic Integrals*, Academic Press, 1969.
5. J. Meixner, "Orthogonale Polynomsysteme mit einem besonderen gestalt der erzeugenden Funktion," J. London Math. Soc., 9, 1934, 6 - 13.
6. G.-C. Rota, ed., *Finite Operator Calculus*, Academic Press, 1975.
7. I.M. Sheffer, "Some Properties of Polynomial Sets of Type Zero," Duke Math. J., 5, 1939, 590-622.
8. M. Spivak, *A Comprehensive Introduction to Differential Geometry*, Publish or Perish, 1979.

SELF-DECOMPOSABILITY ON $\mathbb{R}$ AND $\mathbb{Z}$

Gunnar Forst
Matematisk Institut, Universitetsparken 5
DK-2100 København Ø, Denmark.

Summary. The set $L(\mathbb{R})$ of self-decomposable probability measures on $\mathbb{R}$ is studied in terms of characteristic functions using a certain differential operator and its inverse. In particular a natural bijection onto $L(\mathbb{R})$, introduced by Wolfe, is interpreted via these operators.

In a similar way a bijection of certain sets of probability measures on $\mathbb{Z}$ is discussed, and this leads to a notion of discrete self-decomposability on $\mathbb{Z}$ which extends the notion of discrete self-decomposability on $\mathbb{Z}_+$ as defined by Steutel and van Harn.

Introduction

Motivated by a study of certain stochastic difference equations, Wolfe [8] considered stochastic integrals of infinitely divisible processes on $\mathbb{R}$, and obtained in particular the existence of a bijection between the set $L(\mathbb{R})$ of self-decomposable probabilities on $\mathbb{R}$ and a subset $I_{\log}(\mathbb{R})$ of the set of infinitely divisible probabilities on $\mathbb{R}$. A similar study for probabilities on Banach spaces has been made by Jurek and Vervaat [4].

The purpose of the present paper is to give, for the case of probabilities on $\mathbb{R}$, simple analytical descriptions of the above mentioned bijection $\Phi: I_{\log}(\mathbb{R}) \to L(\mathbb{R})$ and, in the case of probabilities on $\mathbb{Z}$, to give a discrete analogue Φ_d of Φ. This bijection Φ_d is then used to define a natural notion of self-decomposability on $\mathbb{Z}$ which extends the notion of discrete self-decomposability on $\mathbb{Z}_+$ introduced by Steutel and van Harn [5].

In §1 we study infinitely divisible probability measures on $\mathbb{R}$ in terms of their characteristic functions $\hat{\mu}$, or rather the functions $\psi = -\log \hat{\mu}$. It turns out that the operator V given by

$$V\psi(y) = \int_0^1 \psi(ty)\frac{1}{t}\,dt \ , \quad y \in \mathbb{R} \ ,$$

defined for $\psi = -\log \hat{\mu}$ where $\mu \in I(\mathbb{R})$ satisfies

$$\int_{-\infty}^{\infty} \log(1+|x|)d\mu(x) < \infty \ ,$$

(i.e. $\mu \in I_{\log}(\mathbb{R})$) induces the bijection Φ .

The operator V and its inverse, the differential operator S given by

$$S\psi(y) = y\psi'(y) \ , \quad y \in \mathbb{R} \ ,$$

which are intimately related to the family of multiplications used in the definition of self-decomposability, thus give simple analytical descriptions of Φ and Φ^{-1} .

While Φ is not weakly continuous, it is shown that Φ^{-1} is weakly continuous.

It is also pointed out that Φ^{-1} maps the set of generalized Γ-convolutions on $\mathbb{R}_+$ into the set of so-called generalized convolutions of mixtures of exponential distributions on $\mathbb{R}_+$.

In §2 we consider the set $I(\mathbb{Z})$ of infinitely divisible probabilities supported by $\mathbb{Z}$. Since Φ does not map $I(\mathbb{Z}) \cap I_{\log}(\mathbb{R})$ into $I(\mathbb{Z})$ we seek a discrete analogue of Φ which maps into $I(\mathbb{Z})$. The operator V from §1 corresponds to a simple transformation acting on Lévy measures, and a natural discrete analogue of this transformation acting on Lévy measures for elements of $I(\mathbb{Z})$ induces a bijective map Φ_d defined on $I(\mathbb{Z}) \cap I_{\log}(\mathbb{R})$.

To show that Φ_d is a discrete version of Φ we first consider the case of probabilities on $\mathbb{Z}_+$. The restriction of Φ_d to $I(\mathbb{Z}_+) \cap I_{\log}(\mathbb{R})$, where $I(\mathbb{Z}_+)$ is the set of infinitely divisible probabilities on $\mathbb{Z}_+$, is a bijection onto the set $L(\mathbb{Z}_+)$ of discrete self-decomposable probabilities on $\mathbb{Z}_+$ defined by Steutel and van Harn [5]. Then it is shown that the elements of $\Phi_d(I(\mathbb{Z}) \cap I_{\log}(\mathbb{R}))$ can be characterized by a kind of self-decomposability condition on $\mathbb{Z}$. Also the probabilities $\mu \in I(\mathbb{Z})$ for which $\Phi_d\mu$ is a convolution power of μ are identified with the "strictly stable" probabilities on $\mathbb{Z}$.

Finally the mapping Φ_d is expressed in terms of generating functions (the case $\mathbb{Z}_+$) and characteristic functions.

§1. Self-decomposability on $\mathbb{R}$

A probability measure μ on $\mathbb{R}$ is called self-decomposable if for every $c \in]0,1[$ there exists a probability μ_c on $\mathbb{R}$ such that

$$\mu = (T_c\mu) * \mu_c \ , \tag{1.1}$$

where $T_c\mu$ denotes the image measure of μ under the multiplication $T_c: x \mapsto cx$ of $\mathbb{R}$.

The set of self-decomposable probabilities on $\mathbb{R}$ is denoted $L(\mathbb{R})$, and it is well known that $L(\mathbb{R})$ is a subset of the set $I(\mathbb{R})$ of infinitely divisible probabilities. Also for every $c \in]0,1[$ the c-component of $\mu \in L(\mathbb{R})$, i.e. the unique μ_c such that (1.1) holds, belongs to $I(\mathbb{R})$.

The set $I(\mathbb{R})$ is determined by the Lévy-Khinchin representation in the following way: A probability μ on $\mathbb{R}$ belongs to $I(\mathbb{R})$ if and only if the characteristic function $\hat{\mu}$ of μ has the form $\hat{\mu} = \exp(-\psi)$, where

$$\psi(y) = ay^2 + iby + \int_{\mathbb{R}\setminus\{0\}} \left(1 - e^{-ixy} - \frac{ixy}{1+x^2}\right) d\sigma(x) , \qquad y \in \mathbb{R} , \tag{1.2}$$

for some (unique) triple (a,b,σ) (called the representing triple for μ or ψ) of numbers $a \geq 0$, $b \in \mathbb{R}$ and a non-negative measure σ on $\mathbb{R}\setminus\{0\}$ (in fact the Lévy measure for μ) satisfying the condition

$$\int_{\mathbb{R}\setminus\{0\}} \frac{x^2}{1+x^2}\, d\sigma(x) < \infty . \tag{1.3}$$

Let $J(\mathbb{R})$ denote the set of functions given by (1.2) with (a,b,σ) as specified.

For $\mu \in I(\mathbb{R})$ with $\psi = -\log \hat{\mu} \in J(\mathbb{R})$, it is easy to see that $\mu \in L(\mathbb{R})$ if and only if

$$\psi(\cdot) - \psi(c\cdot) \in J(\mathbb{R}) \quad \text{for all} \quad c \in]0,1[. \tag{1.4}$$

In terms of the representing triple (a,b,σ) for μ the condition for self-decomposability is that

$$\sigma = \frac{h(x)}{|x|}\, 1_{\mathbb{R}\setminus\{0\}}(x)dx , \tag{1.5}$$

where $h: \mathbb{R}\setminus\{0\} \to [0,\infty[$ is increasing on $]-\infty,0[$ and decreasing on $]0,\infty[$ (and (1.2) holds). (We may assume that h is left-continuous on $]0,\infty[$ and right-continuous on $]-\infty,0[$.)

The subset of $I(\mathbb{R})$ defined by

$$I_{\log}(\mathbb{R}) = \{\mu \in I(\mathbb{R}) \;\big|\; \int_{\mathbb{R}} \log(1+|x|)d\mu(x) < \infty\}$$

will be important for the sequel. It is easy to prove the following

Lemma 1.1. Let $\mu \in I(\mathbb{R})$ with $\psi = -\log \hat{\mu} \in J(\mathbb{R})$ and Lévy measure σ. Then the following are equivalent:

(i) $\mu \in I_{\log}(\mathbb{R})$,

(ii) $\int_0^1 |1-\hat{\mu}(y)| \frac{1}{y} \, dy < \infty$,

(iii) $\int_0^1 |\psi(y)| \frac{1}{y} \, dy < \infty$,

(iv) $\int_{|x|\geq 1} \log(1+|x|) d\sigma(x) < \infty$. □

We shall now discuss the bijection of $I_{\log}(\mathbb{R})$ onto $L(\mathbb{R})$ defined in Wolfe [8] by means of certain stochastic integrals. This bijection has a simple description in terms of the associated functions in $J(\mathbb{R})$.

For a continuous function $\psi: \mathbb{R} \to \mathbb{C}$ satisfying

$$\int_0^1 \frac{|\psi(ty)|}{t} \, dt < \infty \quad \text{for all} \quad y \in \mathbb{R} , \tag{1.6}$$

we define a function $V\psi: \mathbb{R} \to \mathbb{C}$ by

$$V\psi(y) = \int_0^1 \frac{\psi(ty)}{t} \, dt \quad \text{for} \quad y \in \mathbb{R} . \tag{1.7}$$

Theorem 1.2. The mapping V defined by (1.7) is a bijection of $J_{\log}(\mathbb{R}) = \{-\log \hat{\mu} \in J(\mathbb{R}) \mid \mu \in I_{\log}(\mathbb{R})\}$ onto the set $L(\mathbb{R}) = \{-\log \hat{\mu} \in J(\mathbb{R}) \mid \mu \in L(\mathbb{R})\}$.

Proof. Let $\psi \in J_{\log}(\mathbb{R})$. Then condition (1.6) is satisfied by Lemma 1.1 and the function $V\psi$ is clearly continuous. Moreover, since $J(\mathbb{R})$ is a convex cone of functions which is closed in the topology of local uniform convergence and stable under composition with the multiplications (T_c) , we see that $V\psi \in J(\mathbb{R})$. For $c \in]0,1[$ we find

$$V\psi(y) - V\psi(cy) = \int_c^1 \frac{\psi(ty)}{t} \, dt \ , \quad y \in \mathbb{R} ,$$

and by the same argument as for $V\psi$, this function belongs to $J(\mathbb{R})$, i.e. $V\psi$ satisfies (1.4), and therefore $V\psi \in L(\mathbb{R})$.

Let now conversely $\varphi \in L(\mathbb{R})$, say with representing triple (a,b,σ) where σ is given by (1.5) in terms of the function $h: \mathbb{R} \setminus \{0\} \to [0,\infty[$. Let $\tilde{\sigma}$ be the non-negative measure on $\mathbb{R} \setminus \{0\}$ defined by

$$h(x) = \begin{cases} \tilde{\sigma}([x,\infty[) & \text{for } x > 0 , \\ \tilde{\sigma}(]-\infty,x]) & \text{for } x < 0 . \end{cases} \tag{1.8}$$

By a) of Lemma 1.3 below, $\tilde{\sigma}$ is a Lévy measure and the function $\psi \in J(\mathbb{R})$ with representing triple $(\tilde{a},\tilde{b},\tilde{\sigma})$ where $\tilde{a} = 2a$ and

$$\tilde{b} = b - \int_{\mathbb{R}\setminus\{0\}} \left(\operatorname{Arctan} x - \frac{x}{1+x^2}\right) d\tilde{\sigma}(x) \ , \tag{1.9}$$

belongs to $J_{\log}(\mathbb{R})$ by b) of Lemma 1.3. Putting

$$K(x,y) = 1 - e^{-ixy} - \frac{ixy}{1+x^2} \quad \text{for} \quad x,y \in \mathbb{R}$$

we find for $y \in \mathbb{R}$, cf. Wolfe [8],

$$\begin{aligned}
V\psi(y) &= \int_0^1 \frac{1}{t}\left(\tilde{a}(ty)^2 + i\tilde{b}(ty) + \int_{\mathbb{R}\setminus\{0\}} K(x,ty)d\tilde{\sigma}(x)\right)dt \\
&= \frac{1}{2}\tilde{a}y^2 + i\tilde{b}y + \int_{\mathbb{R}\setminus\{0\}}\left(\int_0^1 \frac{1}{t}\left(K(tx,y) + \frac{ixty}{1+(tx)^2} - \frac{ixty}{1+x^2}\right)dt\right)d\tilde{\sigma}(x) \\
&= ay^2 + i\tilde{b}y + \int_{\mathbb{R}\setminus\{0\}} iy\left(\operatorname{Arctan} x - \frac{x}{1+x^2}\right)d\tilde{\sigma}(x) + \int_{\mathbb{R}\setminus\{0\}}\int_0^1 \frac{1}{t} K(tx,y)dtd\tilde{\sigma}(x) \\
&= ay^2 + iby + \int_0^\infty \left(\int_0^x \frac{1}{u} K(u,y)du\right)d\tilde{\sigma}(x) + \int_{-\infty}^0 \int_x^0 \frac{1}{|u|} K(u,y)dud\tilde{\sigma}(x) \\
&= ay^2 + iby + \int_{\mathbb{R}\setminus\{0\}} K(u,y)\,\frac{h(u)}{|u|}\,du = \varphi(y) \ .
\end{aligned}$$

It follows that V, which is clearly injective, is a bijection of $J_{\log}(\mathbb{R})$ onto $L(\mathbb{R})$. □

<u>Lemma 1.3</u>. Let $h: \,]0,\infty[\, \to [0,\infty[$ be a decreasing left-continuous function such that $\lim_{t\to\infty} h(t) = 0$, and let $\tilde{\sigma}$ be the non-negative measure on $]0,\infty[$ defined by

$$\tilde{\sigma}([t,\infty[) = h(t) \quad \text{for} \quad t > 0 \ .$$

Then we have

a) $$\int_0^1 t^2\,\frac{h(t)}{t}\,dt < \infty \iff \int_0^1 x^2\,d\tilde{\sigma}(x) < \infty$$

b) $$\int_1^\infty \frac{h(t)}{t}\,dt < \infty \iff \int_1^\infty \log(1+x)d\tilde{\sigma}(x) < \infty \ ,$$

c) $$\int_0^1 t\,\frac{h(t)}{t}\,dt < \infty \iff \int_0^1 x d\tilde{\sigma}(x) < \infty \ .$$

<u>Proof</u>. Let $\varepsilon \in \,]0,1[$ be a continuity point for h. A simple calculation gives

$$\int_0^\varepsilon th(t)dt = \frac{\varepsilon^2}{2}\,h(\varepsilon) + \int_0^\varepsilon \frac{x^2}{2}\,d\tilde{\sigma}(x) \ , \tag{1.10}$$

which shows a). Similar calculations show b) and c). $\square$

Let $\varphi\colon \mathbb{R} \to \mathbb{C}$ be a continuous function which is differentiable on $\mathbb{R}\setminus\{0\}$ and define the function $S\varphi\colon \mathbb{R} \to \mathbb{C}$ by

$$S\varphi(y) = \begin{cases} y\varphi'(y) & \text{for } y \neq 0 , \\ 0 & \text{for } y = 0 . \end{cases} \tag{1.11}$$

The operator S, which is "the inverse of V", can be used to give the following limit-version $(c \uparrow 1)$ of the conditions (1.4), cf. also Steutel and van Harn [5].

Corollary 1.4. Let $\mu \in I(\mathbb{R})$ with $\varphi = -\log \hat{\mu} \in J(\mathbb{R})$. Then $\mu \in L(\mathbb{R})$ if and only if φ is differentiable on $\mathbb{R}\setminus\{0\}$ and $S\varphi \in J(\mathbb{R})$ (then necessarily $S\varphi \in J_{\log}(\mathbb{R})$).

Proof. If $\mu \in L(\mathbb{R})$ then there exists $\psi \in J_{\log}(\mathbb{R})$ such that $\varphi = V\psi$, and it follows that φ is differentiable on $\mathbb{R}\setminus\{0\}$ and that $S\varphi = \psi \in J_{\log}(\mathbb{R}) \subseteq J(\mathbb{R})$.

Suppose now conversely that φ is differentiable on $\mathbb{R}\setminus\{0\}$ with $\psi = S\varphi \in J(\mathbb{R})$. Let us see that $\psi \in J_{\log}(\mathbb{R})$. The function $\operatorname{Re}\varphi$ is differentiable on $\mathbb{R}\setminus\{0\}$ with $S(\operatorname{Re}\varphi) = \operatorname{Re}\psi \in J(\mathbb{R})$, and for $y \in \mathbb{R}$ and $\varepsilon \in]0,1[$ we have

$$\int_\varepsilon^1 \frac{\operatorname{Re}\psi(ty)}{t}\,dt = \int_\varepsilon^1 \frac{ty(\operatorname{Re}\varphi)'(ty)}{t}\,dt = (\operatorname{Re}\varphi(y) - \operatorname{Re}\varphi(\varepsilon y)) ,$$

which has a finite limit as $\varepsilon \downarrow 0$, and it follows that $\operatorname{Re}\psi$ satisfies (1.6), hence that $\operatorname{Re}\psi$ and therefore also ψ belongs to $J_{\log}(\mathbb{R})$. Now it is easy to see that $V\psi = \varphi$, hence that $\varphi \in L(\mathbb{R})$ and $\mu \in L(\mathbb{R})$. $\square$

Remark. The operators V and S are closely connected with the family $(T_c)_{0<c<1}$ of multiplications used in the definition of self-decomposability. For $t \geq 0$ we consider the operator P_t on the space of continuous functions $f\colon \mathbb{R} \to \mathbb{C}$ given by $P_t f(x) = f(e^{-t}x)$, $x \in \mathbb{R}$. Clearly $P_s P_t = P_{s+t}$ for $s,t \geq 0$ and $P_0 = \mathrm{Id}$, so $(P_t)_{t\geq 0}$ is a semigroup. The formal infinitesimal generator of $(P_t)_{t\geq 0}$ is

$$\lim_{t\to 0} \tfrac{1}{t}(P_t f(x) - f(x)) = -\,Sf(x) , \quad x \in \mathbb{R} ,$$

where the pointwise limit exists for all $x \in \mathbb{R}$ if and only if f is differentiable on $\mathbb{R}\setminus\{0\}$. The formal potential operator for $(P_t)_{t\geq 0}$ is

$$\int_0^\infty P_t f(x)\,dt = \int_0^1 f(ux)\tfrac{1}{u}\,du = Vf(x) , \quad x \in \mathbb{R} ,$$

where the integral converges pointwise for all $x \in \mathbb{R}$ if and only if $f(0) = 0$ and $\int_0^1 \frac{1}{u}|f(u)|du < \infty$.

The bijection $V: J_{\log}(\mathbb{R}) \to L(\mathbb{R})$ from Theorem 1.2 induces a bijective mapping $\Phi: I_{\log}(\mathbb{R}) \to L(\mathbb{R})$, where $\Phi\mu$ for $\mu \in I_{\log}(\mathbb{R})$ is defined to be the element of $L(\mathbb{R})$ associated with $V(-\log \hat{\mu})$ i.e.

$$(\Phi\mu)^{\wedge} = \exp(-V(-\log \hat{\mu})) .$$

It follows from the proof of Theorem 1.2, that Φ is the bijection considered by Wolfe, cf. Theorem 2 and Corollary 3 of [8] (with $\gamma = 1$).

Equation (1.8) which relates the Lévy measure $\tilde{\sigma}$ for $\mu \in I_{\log}(\mathbb{R})$ with the Lévy measure σ for $\Phi\mu$ can also be expressed

$$\sigma = \int_0^1 \frac{1}{u} T_u \tilde{\sigma}\, du \qquad \text{(vague integral)} \tag{1.12}$$

i.e. formally $\sigma = V\tilde{\sigma}$, cf. (1.7), or $\tilde{\sigma} = S\sigma$ (derivative taken in Schwartz distribution sense).

The bijection Φ (and also Φ^{-1}) is clearly a <u>convolution</u> <u>homomorphism</u> i.e.

$$\Phi(\mu_1 * \mu_2) = (\Phi\mu_1) * (\Phi\mu_2) \quad \text{for} \quad \mu_1,\mu_2 \in I_{\log}(\mathbb{R}) ,$$

(note that $\mu_1 * \mu_2 \in I_{\log}(\mathbb{R})$) which commutes with <u>reflection</u> i.e.

$$\Phi(\check{\mu}) = (\Phi\mu)^{\vee} ,$$

where $\check{\mu}$ denotes the reflected measure of μ (the image of μ under the mapping $x \mapsto -x$).

The inverse bijection $\Phi^{-1}: L(\mathbb{R}) \to I_{\log}(\mathbb{R})$ can be described in the following "direct" way.

<u>Proposition 1.5</u>. Let $\nu \in L(\mathbb{R})$ and let $(\nu_c)_{0<c<1}$ be the family of c-components of ν . Then

$$\Phi^{-1}\nu = \lim_{c \to 1} (\nu_c)^{*\frac{1}{1-c}} \quad \text{weakly.}$$

<u>Proof</u>. With $\psi = -\log \hat{\nu} \in L(\mathbb{R})$ we have for $c \in]0,1[$

$$\left((\nu_c)^{*\frac{1}{1-c}}\right)^{\wedge}(y) = \exp\left(-\frac{1}{1-c}(\psi(y) - \psi(cy))\right), \quad y \in \mathbb{R} ,$$

which for $c \uparrow 1$ converges to $\exp(-S\psi(y))$ locally uniformly in y , and the assertion follows. □

To discuss continuity properties of Φ and Φ^{-1} we use the following well-known characterization of weak convergence in $I(\mathbb{R})$. Cf. e.g. Gnedenko and Kolmogorov [3].

Lemma 1.6. Let μ_n, $\mu \in I(\mathbb{R})$ have representing triples (a_n, b_n, σ_n) resp. (a,b,σ). Then $\mu_n \to \mu$ weakly if and only if

1°. $b_n \to b$,

2°. $\sigma_n \to \sigma$ weakly away from 0, in the sense that

$$\int f(x)d\sigma_n(x) \to \int f(x)d\sigma(x) ,$$

for every bounded continuous function $f\colon \mathbb{R}\setminus\{0\} \to \mathbb{C}$ which vanishes in some neighbourhood of 0,

3°.

$$\lim_{\varepsilon\to 0}\left(\liminf_{n\to\infty}\left(\int_{-\varepsilon}^{0} x^2 d\sigma_n(x) + a_n + \int_0^{\varepsilon} x^2 d\sigma_n(x)\right)\right)$$

$$= \lim_{\varepsilon\to 0}\left(\limsup_{n\to\infty}\left(\int_{-\varepsilon}^{0} x^2 d\sigma_n(x) + a_n + \int_0^{\varepsilon} x^2 d\sigma_n(x)\right)\right) = a . \quad □$$

The bijection Φ is not weakly continuous. Consider the sequence (μ_n) in $I_{\log}(\mathbb{R})$ where μ_n has representing triple $(0,0,\tilde{\sigma}_n)$ and

$$\tilde{\sigma}_n = \frac{1}{\log(n+1)}\, 1_{]n,n+1[}(x)dx , \quad n \in \mathbb{N} .$$

Since $(\tilde{\sigma}_n)$ converges to 0 weakly away from 0, we have that $\mu_n \to \varepsilon_0$ weakly. On the other hand, the Lévy measure σ_n for $\Phi\mu_n$ has a density which on $]0,n[$ equals $\frac{1}{t\log(n+1)}$ and it follows that (σ_n) does not converge to 0 weakly away from 0. Consequently $(\Phi\mu_n)$ does not converge weakly to $\Phi\varepsilon_0 = \varepsilon_0$.

Theorem 1.7. The mapping $\Phi^{-1}\colon L(\mathbb{R}) \to I_{\log}(\mathbb{R})$ is weakly continuous.

Proof. Let $\mu_n \in L(\mathbb{R})$ have representing triple (a_n, b_n, σ_n) and suppose that $\mu_n \to \mu$ weakly, where $\mu \in L(\mathbb{R})$ has representing triple (a,b,σ). Here σ_n and σ are given by (1.5) in terms of functions h_n and h.

The representing triple for $\Phi^{-1}\mu_n$ is $(\tilde{a}_n, \tilde{b}_n, \tilde{\sigma}_n)$ where $\tilde{a}_n = 2a_n$, $\tilde{\sigma}_n$ satisfies

$$h_n(t) = \begin{cases} \tilde{\sigma}_n([t,\infty[) & \text{for} \quad t > 0 , \\ \tilde{\sigma}_n(]-\infty,t]) & \text{for} \quad t < 0 , \end{cases}$$

and

$$\tilde{b}_n = b_n - \int_{\mathbb{R}\setminus\{0\}} \left(\text{Arctan } x - \frac{x}{1+x^2}\right) d\tilde{\sigma}_n(x) .$$

The representing triple $(\tilde{a},\tilde{b},\tilde{\sigma})$ for $\Phi^{-1}\mu$ is similarly defined in terms of a,b and h.

Note first that, since $\sigma_n \to \sigma$ weakly away from 0, we have that $h_n(x) \to h(x)$ at all continuity points x for h, and this implies that $\tilde{\sigma}_n \to \tilde{\sigma}$ weakly away from 0.

Let $\varepsilon > 0$ be such that ε and $-\varepsilon$ are continuity points for h. Then by (1.10)

$$\liminf_{n\to\infty}\left(\int_{-\varepsilon}^{0} x^2 d\tilde{\sigma}_n(x) + \tilde{a}_n + \int_0^{\varepsilon} x^2 d\tilde{\sigma}_n(x)\right)$$

$$= \liminf_{n\to\infty}\left(2\int_{-\varepsilon}^{0} x^2 \frac{h_n(x)}{|x|} dx + \varepsilon^2 h_n(-\varepsilon) + 2a_n + 2\int_0^{\varepsilon} x^2 \frac{h_n(x)}{x} dx + \varepsilon^2 h_n(\varepsilon)\right)$$

$$= \varepsilon^2(h(-\varepsilon) + h(\varepsilon)) + 2 \liminf_{n\to\infty}\left(\int_{-\varepsilon}^{0} x^2 \frac{h_n(x)}{|x|} dx + a_n + \int_0^{\varepsilon} x^2 \frac{h_n(x)}{x} dx\right) .$$

Letting $\varepsilon \to 0$ in (1.10) it follows by dominated convergence that

$$\lim_{\varepsilon\to 0} \varepsilon^2(h(-\varepsilon) + h(\varepsilon)) = 0$$

and consequently by Lemma 1.6.

$$\lim_{\varepsilon\to 0}\left(\liminf_{n\to\infty}\left(\int_{-\varepsilon}^{0} x^2 d\tilde{\sigma}_n(x) + \tilde{a}_n + \int_0^{\varepsilon} x^2 d\tilde{\sigma}_n(x)\right)\right) = 2a .$$

In the same way

$$\lim_{\varepsilon\to 0}\left(\limsup_{n\to\infty}\left(\int_{-\varepsilon}^{0} x^2 d\tilde{\sigma}_n(x) + \tilde{a}_n + \int_0^{\varepsilon} x^2 d\tilde{\sigma}_n(x)\right)\right) = 2a .$$

Finally

$$\tilde{b}_n = b_n - \int_{\mathbb{R}\setminus\{0\}} \left(\text{Arctan } x - \frac{x}{1+x^2}\right) d\tilde{\sigma}_n(x)$$

$$\to b - \int_{\mathbb{R}\setminus\{0\}} \left(\text{Arctan } x - \frac{x}{1+x^2}\right) d\tilde{\sigma}(x) = \tilde{b} \quad ,$$

since $\tilde{\sigma}_n \to \tilde{\sigma}$ weakly away from 0, and

$$\lim_{\varepsilon\to 0}\left(\int_{-\varepsilon}^{0} |x|^3 \, d\tilde{\sigma}_n(x) + \int_0^{\varepsilon} |x|^3 \, d\tilde{\sigma}_n(x)\right) = 0$$

uniformly in n . . □

Remark. The map Φ simplifies when restricted to the set $P(\mathbb{R}_+)$ of probabilities on $\mathbb{R}_+$ (where it perhaps is more natural to use Laplace transforms instead of characteristic functions). A probability $\mu \in P(\mathbb{R}_+)$ belongs to $I(\mathbb{R}_+) = I(\mathbb{R}) \cap P(\mathbb{R}_+)$ if and only if

$$-\log \hat{\mu}(y) = iby + \int_0^{\infty} (1-e^{-ixy})d\sigma(x) \ , \quad y \in \mathbb{R} \ , \tag{1.13}$$

with a unique couple (b,σ) of $b \geq 0$ and a non-negative measure (the Lévy measure for μ) σ on $]0,\infty[$ satisfying

$$\int_0^{\infty} \frac{x}{1+x} \, d\sigma(x) < \infty \ . \tag{1.14}$$

Going through the proof of Theorem 1.2 we see that Φ maps $I(\mathbb{R}_+) \cap I_{\log}(\mathbb{R})$ onto $I(\mathbb{R}_+) \cap L(\mathbb{R})$. In fact, μ represented by (b,σ) is mapped to ν represented by $(b,\tilde{\sigma})$ where $\tilde{\sigma}$ again satisfies (1.14) (Lemma 2.3 c)). Due to the absence of the modifying term $\frac{-ixy}{1+x^2}$ in (1.13) the "translation coefficient" b is unchanged by Φ.

The set $L(\mathbb{R}) \cap I(\mathbb{R}_+)$ has an interesting subset, the set $\Gamma(\mathbb{R}_+)$ of so-called <u>generalized Γ-convolutions</u>, introduced by Thorin [6]. Here $\mu \in I(\mathbb{R}_+)$ belongs to $\Gamma(\mathbb{R}_+)$ if and only if $-\log \hat{\mu}$ can be represented by (1.13) where σ has a density of the form $\frac{1}{x} h(x)$ with h <u>completely monotone</u> (and (1.14) holds). Another interesting subset of $I(\mathbb{R}_+)$ is the class $B(\mathbb{R}_+)$ introduced by Bondesson [2] of so-called <u>generalized convolutions of mixtures of exponential distributions</u>, defined to be the set of $\mu \in I(\mathbb{R}_+)$ where in the representation (1.13) the Lévy measure σ has a completely monotone density on $]0,\infty[$ (and (1.14) holds).

It is easy to see that Φ maps $B(\mathbb{R}_+) \cap I_{\log}(\mathbb{R})$ onto $\Gamma(\mathbb{R}_+)$. Considering the class $\Gamma(\mathbb{R})$ of <u>extended generalized Γ-convolutions</u> on $\mathbb{R}$, cf. Thorin [7], which is the smallest weakly closed subset of $I(\mathbb{R})$ containing $\Gamma(\mathbb{R}_+)$ and closed under reflection and arbitrary translations, and likewise the set $B(\mathbb{R})$ obtained from $B(\mathbb{R}_+)$ in the same way, we may state that Φ maps $B(\mathbb{R}) \cap I_{\log}(\mathbb{R})$ onto $\Gamma(\mathbb{R})$.

§2. Self-decomposability on $\mathbb{Z}$

Let $P(\mathbb{Z})$ be the set of probability measures on $\mathbb{Z}$ and let $I(\mathbb{Z})$ denote the infinitely divisible probabilities on $\mathbb{Z}$. The notion of divisibility is relative to $P(\mathbb{Z})$, i.e. $\mu \in P(\mathbb{Z})$ belongs to $I(\mathbb{Z})$ if and only if $\mu \in I(\mathbb{R})$ and the n'th convolution root $(n \in \mathbb{N})$ of μ belongs to $P(\mathbb{Z})$.

The Lévy measure of an element $\mu \in I(\mathbb{Z})$ is (identifying in the natural way a measure on $\mathbb{Z}$ with its coefficient sequence) a sequence $(b_n)_{n\in\mathbb{Z}^*}$, where $\mathbb{Z}^* = \mathbb{Z} \setminus \{0\}$, satisfying

$$b_n \geq 0 \quad \text{for} \quad n \in \mathbb{Z}^* \quad \text{and} \quad \sum_{n\in\mathbb{Z}^*} b_n < \infty , \tag{2.1}$$

and the map taking μ into its Lévy measure is a bijection of $I(\mathbb{Z})$ onto the set of sequences for which (2.1) hold.

By Lemma 1.1 a measure $\mu \in I(\mathbb{Z})$ belongs to the set

$$I_{\log}(\mathbb{Z}) = I(\mathbb{Z}) \cap I_{\log}(\mathbb{R})$$

if and only if its Lévy measure $(b_n)_{n\in\mathbb{Z}^*}$ satisfies

$$\sum_{n\in\mathbb{Z}^*} b_n \log(1+|n|) < \infty . \tag{2.2}$$

Lemma 2.1. Let $(b_n)_{n\geq 1}$ and $(c_n)_{n\geq 1}$ be sequences of non-negative numbers for which $\Sigma_{n=1}^{\infty} b_n < \infty$ and $c_n = \Sigma_{k=n}^{\infty} b_k$ for $n \geq 1$. Then

$$\sum_{n=1}^{\infty} \frac{1}{n} c_n < \infty \Leftrightarrow \sum_{n=1}^{\infty} b_n \log(1+n) < \infty .$$

This is clear, since $\Sigma_{j=1}^{n} \frac{1}{j} \sim \log n$ for $n \to \infty$.

Let $\underline{b} = (b_n)_{n\in\mathbb{Z}^*}$ be a sequence satisfying (2.1) and define a sequence $V_d\underline{b}$ by

$$(V_d\underline{b})_n = \begin{cases} \dfrac{1}{n} \displaystyle\sum_{j=n}^{\infty} b_j & \text{for } n \geq 1 \\[2ex] \dfrac{1}{|n|} \displaystyle\sum_{j=-\infty}^{n} b_j & \text{for } n \leq -1 . \end{cases} \tag{2.3}$$

By Lemma 2.1 the sequence $V_d\underline{b}$ satisfies (2.1) if (and only if) $\underline{b}$ satisfies (2.2), and V_d therefore induces a map Φ_d of $I_{\log}(\mathbb{Z})$ into $I(\mathbb{Z})$ in the following way. For $\mu \in I_{\log}(\mathbb{Z})$ with Lévy measure $\underline{b}$, $\Phi_d\mu$ is the unique element of $I(\mathbb{Z})$ with Lévy measure $V_d\underline{b}$.

The map V_d is the natural discrete analogue of the mapping considered in §1, cf. (1.12), of Lévy measures on $\mathbb{R} \setminus \{0\}$, and in the sequel we shall see that Φ_d is a discrete analogue of Φ. For this we consider first the restriction of Φ_d to probabilities on $\mathbb{Z}_+$ where the notion of discrete self-decomposability due to Steutel and van Harn [5] will be useful.

Let

$$P(\mathbb{Z}_+) = \left\{ \sum_{n=0}^{\infty} p_n \varepsilon_n \in P(\mathbb{Z}) \;\middle|\; p_0 > 0 \right\} .$$

Then $I(\mathbb{Z}_+) = I(\mathbb{Z}) \cap P(\mathbb{Z}_+)$ is the set of <u>infinitely divisible</u> elements of $P(\mathbb{Z}_+)$ i.e. for every $n \in \mathbb{N}$ the n'th convolution root exists as an element of $P(\mathbb{Z}_+)$. Also $\mu \in I(\mathbb{Z})$ belongs to $I(\mathbb{Z}_+)$ if and only if its Lévy measure is a measure on $\mathbb{Z}_+$.

A probability measure $\mu \in P(\mathbb{Z}_+)$ is called <u>discrete self-decomposable</u>, cf. [5], if for every $c \in]0,1[$ there exists $\mu_c \in P(\mathbb{Z}_+)$ such that

$$\mu = \tau_c(\mu) * \mu_c . \tag{2.4}$$

Here the family $(\tau_c)_{0<c<1}$ of "multiplications" on $P(\mathbb{Z}_+)$ is defined by

$$\tau_c\left(\sum_{k=0}^{\infty} p_k \varepsilon_0\right) = \sum_{k=0}^{\infty} p_k \left(\sum_{j=0}^{k} \binom{k}{j} (1-c)^{k-j} c^j \varepsilon_j\right) , \quad c \in]0,1[, \tag{2.5}$$

or in terms of <u>generating functions</u> $(\tilde{\mu}(z) = \sum_{k=0}^{\infty} p_k z^k)$

$$\widetilde{\tau_c(\mu)}(z) = \tilde{\mu}(1-c+cz) , \quad z \in [0,1] , \quad c \in]0,1[.$$

It was shown by Steutel and van Harn [5] that the set $L(\mathbb{Z}_+)$ of discrete self-decomposable probabilities on $\mathbb{Z}_+$ is a subset of $I(\mathbb{Z}_+)$, and also that $\mu \in I(\mathbb{Z}_+)$, with Lévy measure $(b_n)_{n\geq 1}$, belongs to $L(\mathbb{Z}_+)$ if and only if the sequence $(nb_n)_{n\geq 1}$ is decreasing.

<u>Theorem 2.2.</u> The restriction of Φ_d to the set

$$I_{\log}(\mathbb{Z}_+) = I(\mathbb{Z}_+) \cap I_{\log}(\mathbb{Z})$$

is a bijection onto the set $L(\mathbb{Z}_+)$.

<u>Proof.</u> Consider first $\mu \in I_{\log}(\mathbb{Z}_+)$ with Lévy measure $(a_n)_{n\geq 1}$. The Lévy measure for $\Phi_d\mu$ is then $(b_n)_{n\geq 1}$ given by $b_n = \frac{1}{n} \Sigma_{j=n}^{\infty} a_j$, and here $(nb_n)_{n\geq 1}$ is clearly decreasing, i.e. $\Phi_d\mu \in L(\mathbb{Z}_+)$.

Conversely, if $\nu \in L(\mathbb{R}_+)$ has Lévy measure $(b_n)_{n\geq 1}$ where $(nb_n)_{n\geq 1}$ is decreasing, then there exists a sequence $(a_n)_{n\geq 1}$ such that $nb_n = \Sigma_{j=n}^{\infty} a_j$. Clearly, $(a_n)_{n\geq 1}$ satisfies (2.1) and by Lemma 2.1 also (2.2), and $V_d\underline{a} = \underline{b}$, so the probability $\mu \in I_{\log}(\mathbb{Z}_+)$ with Lévy measure $\underline{a}$ fulfills $\Phi_d\mu = \nu$. □

The action of Φ_d on the whole set $I_{\log}(\mathbb{Z})$ will be described using the following "decomposition".

Lemma 2.3. Let $\mu \in I(\mathbb{Z})$. Then there exists a uniquely determined couple (μ^+,μ^-), $\mu^+,\mu^- \in I(\mathbb{Z}_+)$ such that

$$\mu = \mu^+ * (\mu^-)^\vee .$$

Proof. This is clear considering Lévy measures. □

Theorem 2.4. The map Φ_d is a bijection of $I_{\log}(\mathbb{Z})$ onto the set

$$L(\mathbb{Z}) = \{\mu^+ * (\mu^-)^\vee \in I(\mathbb{Z}) \mid \mu^+,\mu^- \in L(\mathbb{Z}_+)\} .$$

Proof. This is clear since Φ_d satisfies

$$\Phi_d(\mu*\nu) = (\Phi_d\mu)*(\Phi_d\nu) \quad \text{and} \quad \Phi_d(\overset{\vee}{\mu}) = (\Phi_d\mu)^\vee$$

for $\mu,\nu \in I_{\log}(\mathbb{Z})$ (note that $I_{\log}(\mathbb{Z})$ is stable under convolution and reflection). □

The set $L(\mathbb{Z})$, which is thus the discrete analogue of $L(\mathbb{R})$, can also be characterized by a self-decomposability condition similar to (2.4).

The "multiplications" $(\tau_c)_{0<c<1}$ of $P(\mathbb{Z}_+)$ are convolution homomorphisms of $P(\mathbb{Z}_+)$ into $P(\mathbb{Z}_+)$ such that $\tau_c(\varepsilon_0) = \varepsilon_0$ for all $c \in]0,1[$. It is not possible to extend these multiplications to convolution homomorphisms on $P(\mathbb{Z})$ (consider $\varepsilon_0 = \varepsilon_{-1}*\varepsilon_1$). But on the subset $I(\mathbb{Z})$ we may define the family $(\rho_c)_{0<c<1}$ of convolution homomorphisms extending $(\tau_c)_{0<c<1}$, using the decomposition from Lemma 2.3, by

$$\rho_c(\mu) = (\tau_c\mu^+) * (\tau_c\mu^-)^\vee$$

for $\mu \in I(\mathbb{Z})$ with $\mu = \mu^+*(\mu^-)^\vee$ and $\mu^+,\mu^- \in I(\mathbb{Z}_+)$.

Corollary 2.5. Let $\mu \in I(\mathbb{Z})$. Then $\mu \in L(\mathbb{Z})$ if and only if there exists for every $c \in]0,1[$ a probability measure $\mu_c \in I(\mathbb{Z})$ such that

$$\mu = (\rho_c\mu) * \mu_c .$$

Proof. The "only if" part is evident and the "if" part follows from the unicity assertion in Lemma 2.3. □

Remark. A simple discrete adaptation of the example given in §1 shows that $\Phi_d\colon I_{\log}(\mathbb{Z}) \to L(\mathbb{Z})$ is not weakly continuous. But, as in the case of $\mathbb{R}$, the inverse bijection $\Phi_d^{-1}: L(\mathbb{Z}) \to I_{\log}(\mathbb{Z})$ is weakly continuous.

To see this, consider a weakly convergent sequence (ν_n) in $L(\mathbb{Z})$ with limit $\nu \in L(\mathbb{Z})$. Denoting by $(c_k^n)_{k\in\mathbb{Z}^*}$ and $(c_k)_{k\in\mathbb{Z}^*}$ the corresponding Lévy measures, we have for $k \in \mathbb{Z}^*$ fixed

$$c_k^n \to c_k \quad \text{as} \quad n \to \infty .$$

The Lévy measures $(b_k^n)_{k\in\mathbb{Z}^*}$ and $(b_k)_{k\in\mathbb{Z}^*}$ for $\Phi_d^{-1}\nu_n$ and $\Phi_d^{-1}\nu$ satisfy

$$c_k^n = \begin{cases} \dfrac{1}{k} \displaystyle\sum_{j=k}^{\infty} b_j^n , & k \geq 1 \\ \dfrac{1}{|k|} \displaystyle\sum_{j=-\infty}^{k} b_j^n , & k \leq -1 , \end{cases} \qquad \text{for } n \in \mathbb{N}$$

and similarly for $(c_k)_{k\in\mathbb{Z}^*}$, and it follows that

$$b_k^n \to b_k \quad \text{and} \quad \sum_{k\in\mathbb{Z}^*} b_k^n \to \sum_{k\in\mathbb{Z}^*} b_k ,$$

i.e. that $\Phi_d^{-1}\nu_n \to \Phi_d^{-1}\nu$ weakly.

Let us now describe the action of Φ_d on $\mathbb{Z}_+$ in terms of generating functions.

It is well known that $\mu \in P(\mathbb{Z}_+)$ with generating function $\tilde{\mu}(z)$ belongs to $I(\mathbb{Z}_+)$ if and only if

$$\tilde{\mu}(z) = \exp\Big(\sum_{k=1}^{\infty} b_k(z^k-1)\Big) , \quad z \in [0,1] ,$$

where $(b_k)_{k\geq 1}$ satisfies (2.1) (in fact, $(b_k)_{k\geq 1}$ is the Lévy measure for μ).

A simple calculation shows that for $\mu \in I_{\log}(\mathbb{Z}_+)$

$$\log(\Phi_d\mu)^{\sim}(z) = \int_0^1 \frac{\log \tilde{\mu}(1-t+tz)}{t}\, dt , \quad z \in [0,1] , \tag{2.6}$$

and this equation is the discrete analogue (for the case of $\mathbb{Z}_+$) to (1.7). The multiplications $y \mapsto ty$ $(t \in]0,1[)$ on $\mathbb{R}$ have been replaced by the "multiplications" $z \mapsto 1-t+tz$ $(t \in]0,1[)$ on $[0,1]$.

It is easy to see that if $\nu \in L(\mathbb{Z}_+)$ then

$$z \mapsto \exp\Big((z-1)\frac{d}{dz}(\log \tilde{\nu}(z))\Big)$$

(= 0 for $z = 1$) is the generating function for a probability $\mu \in I_{\log}(\mathbb{Z}_+)$ and that $\Phi_d\mu = \nu$. The differential operator $\varphi \mapsto (\cdot - 1)\varphi'(\cdot)$ is thus the discrete analogue (for $\mathbb{Z}_+$) to the operator S on $\mathbb{R}$. See also [1].

Remark. Let Q_t for $t \geq 0$ be the operator on the space of continuous functions $f: [0,1] \to \mathbb{R}$ given by

$$Q_t f(x) = f(1-e^{-t} + e^{-t}x) \ , \quad x \in [0,1] \ .$$

Clearly $Q_t Q_s = Q_{t+s}$ for $s,t \geq 0$ and $Q_0 = \mathrm{Id}$ so $(Q_t)_{t\geq 0}$ is a semigroup with "infinitesimal generator"

$$\lim_{t\to 0} \frac{Q_t f(x) - f(x)}{t} = \begin{cases} -(1-x)f'(x) \ , & x \in [0,1[\\ 0 & x = 1 \end{cases}$$

where the pointwise limit exists for $x \in [0,1]$ if and only if f is differentiable on $[0,1[$.

The "potential operator" for $(Q_t)_{t\geq 0}$ is

$$\int_0^\infty Q_t f(x)dt = \int_0^\infty f(1-e^{-t} + e^{-t}x)dt = \int_0^1 \frac{f(1-u+ux)}{u} du$$

where the integral converges pointwise for $x \in [0,1]$ if and only if $f(1) = 0$ and $\int_0^1 \frac{|f(1-u)|}{u} du < \infty$.

The action of Φ_d on $I_{\log}(\mathbb{Z})$ can be described in terms of characteristic functions, however, not so explicitly as in the case of $\mathbb{Z}_+$. Let $\mu \in I_{\log}(\mathbb{Z})$ with Lévy measure $\underline{b} = (b_k)_{k\in\mathbb{Z}^*}$. The associated function $\psi = -\log\hat{\mu} \in L(\mathbb{R})$ can be written

$$\psi(y) = \psi^+(y) + \psi^-(y) \quad , \quad y \in \mathbb{R}$$

where

$$\psi^+(y) = \sum_{k=1}^{\infty} b_k(1-e^{-iky}), \quad \psi^-(y) = \sum_{k=-\infty}^{-1} b_k(1-e^{-iky}) \ .$$

A straightforward calculation now gives that the function associated with $\Phi_d\mu \in L(\mathbb{Z})$ is given by

$$-\log(\Phi_d\mu)^\wedge(y) = \sum_{k\in\mathbb{Z}^*}(1-e^{-iky})(V_d\underline{b})_k = \int_0^y i\,\frac{e^{-it}}{1-e^{-it}}\,\psi^+(t)dt + \int_0^y -i\,\frac{e^{it}}{1-e^{it}}\,\psi^-(t)dt \ ,$$

which is the discrete analogue (on $\mathbb{Z}$) of (1.7).

Remark. It was observed by Jurek and Vervaat [4] that $\mu \in I_{\log}(\mathbb{R})$ is strictly stable if and only if

$$\log(\Phi\mu)^{\wedge} = c \log \hat{\mu} \quad \text{for some} \quad c > 0 , \tag{2.7}$$

in which case $\alpha = \frac{1}{c} \in]0,2]$ is the stability exponent of μ. In terms of the map V the condition (2.7) is that $V(\log \hat{\mu}) = c \log \hat{\mu}$ for some $c > 0$.

Let now $\mu \in I_{\log}(\mathbb{Z})$ and suppose that

$$\log(\Phi_d\mu)^{\wedge} = c \log \hat{\mu} \quad \text{for some} \quad c > 0 .$$

Then the Lévy measure $\underline{b} = (b_n)_{n\in\mathbb{Z}^*}$ for μ satisfies $V_d\underline{b} = c\underline{b}$, and in particular, since $(V_d\underline{b})_1 = \Sigma_{j=1}^{\infty} b_j = cb_1$, we have $c \geq 1$. Let $\mu = \mu^+ * (\mu^-)$ be the decomposition of μ from Lemma 2.3. Then the function

$$f(z) = -\log \tilde{\mu}^+(z) = \sum_{n=1}^{\infty} b_n(z^n-1) , \quad z \in [0,1]$$

satisfies

$$cf'(z) = \sum_{n=1}^{\infty} cb_n n\cdot z^{n-1} = \sum_{n=1}^{\infty} z^{n-1} n \left(\frac{1}{n} \sum_{j=n}^{\infty} b_j\right)$$

$$= \sum_{j=1}^{\infty} b_j \sum_{n=1}^{j} z^{n-1} = \sum_{j=1}^{\infty} b_j \frac{1-z^j}{1-z} = \frac{1}{z-1} f(z) ,$$

hence $f(z) = \lambda(1-z)^{1/c}$ for some $\lambda \geq 0$, and it follows that μ^+, and by a similar argument also μ^-, is discrete stable as defined by Steutel and van Harn [5].

References.

[1] Berg, C. and G. Forst: Multiply self-decomposable probability measures on $\mathbb{R}_+$ and $\mathbb{Z}_+$. Z. Wahrscheinlichkeitstheorie verw. Gebiete 62, 147-163 (1983).

[2] Bondesson, L.: Classes of infinitely divisible distributions and densities. Z. Wahrscheinlichkeitstheorie verw. Gebiete 57, 39-71, (1981).

[3] Gnedenko, B. V. and A. N. Kolmogorov: Limit distributions for sums of independent random variables. Addison-Wesley, Reading Mass. 1954.

[4] Jurek, Z. J. and W. Vervaat: An integral representation for self-decomposable Banach space valued random variables. Z. Wahrscheinlichkeitstheorie verw. Gebiete 62, 247-262 (1983).

[5] Steutel, F. W. and K. van Harn: Discrete analogues of self-decomposability and stability. Ann. Probability 7, 893-899 (1979).

[6] Thorin, O.: On the infinite divisibility of the Pareto distribution. Scand. Actuarial J. 1977, 31-40.

[7] Thorin, O.: An extension of the notion of a generalized Γ-convolution. Scand. Actuarial J. 1978, 141-149.

[8] Wolfe, S. J.: On a continuous analogue of the stochastic difference equation $X_n = \rho X_{n-1} + B_n$. Stochastic Process. App. 12, 301-312 (1982).

LOIS DE PROBABILITE INFINIMENT DIVISIBLES

SUR LES HYPERGROUPES COMMUTATIFS, DISCRETS, DENOMBRABLES

Léonard GALLARDO
Faculté des Sciences
E.R.A. n° 839 du CNRS
B.P. n° 239
54506 VANDOEUVRE lès NANCY
FRANCE

et

Olivier GEBUHRER
Université Louis Pasteur
I.R.M.A.
7 Rue René Descartes
67084 STRASBOURG
FRANCE

Summary : This paper is a contribution to the study of infinitely divisible probability measures on hypergroups. In the case of a discrete, infinite countable hypergroup X without non trivial compact subhypergroups, we prove that the only infinitely divisible probability measures are of Poisson type as soon as (roughly speaking) we have a Levy continuity Theorem. More precisely : $\mu \in M_1(X)$ is infinitely divisible if and only if there exists a unique positive bounded Radon measure ν on X with $\nu(\{e\}) = 0$ and such that :

$$\hat{\mu}(\chi) = \exp\left[\sum_{x\in X}(\overline{\chi(x)}-1)\ \nu(\{x\})\right] \quad (\chi \in \hat{X}) ,$$

where $\hat{\mu}$ denotes the Fourier transform of μ (which is defined on the set $\hat{X}$ of hermitian characters of X) . This result is true in one of the two following cases :

C 1 : The support of the Plancherel measure on $\hat{X}$ contains the point $\mathbb{1}$ (= the character identically 1) .

C 2 : $\mathbb{1}$ is not an isolated point in $\hat{X}$ and there exists a neighbourhood V of $\mathbb{1}$ such that $\lim_{x\to\infty} \chi(x) = 0$ for every $\chi \in V - \{\mathbb{1}\}$.

These conditions are clearly only sufficient. They are illustrated by exemples in paragraph 6. In such cases we study also the problem of triangular arrays of probability measures on X .

1.- INTRODUCTION.

(1.1) La transformation de Fourier est l'un des outils les plus importants du calcul des probabilités. Cependant, dans les exposés classiques, l'essence profonde de certains résultats se trouve parfois cachée derrière la technicité des démonstrations. Un exemple typique est le théorème de continuité de Paul Levy (cf. par ex. [13], p. 49) dont Siebert (cf. [16]) a contribué (en 1978) à éclaircir la nature en le généralisant aux groupes.

L'étude des probabilités sur les groupes fournit ainsi très souvent, par l'introduction de nouvelles techniques, l'occasion de découvrir des phénomènes nouveaux et de mieux comprendre les situations classiques. Il en est de même, à notre avis, pour les probabilités sur les hypergroupes dont l'étude a débuté récemment (cf. [1] et [2]) quand on s'est aperçu, à la suite des Analystes (cf. [5], [8] et [17]) que de nombreuses propriétés démontrées dans le cas des groupes ne dépendaient pas de manière essentielle de la structure du groupe mais seulement de la structure de son algèbre de convolution des mesures bornées.

On peut placer dans ce cadre l'étude des probabilités sur un couple de Guelfand (cf. [3], [11] et sa très riche bibliographie) qui trouve son origine dans les propriétés particulièrement intéressantes qu'ont, dans un groupe semi-simple, les mesures biinvariantes par un sous-groupe compact.

(1.2) Parmi les définitions possibles d'un hypergroupe (Dunkl [5], Jewett [8], Spector [17]), c'est celle de Jewett qui semble à l'heure actuelle la plus utilisée (cf. les articles [1] et [2] de Bloom et Heyer qui nous serviront de référence) :

Un espace de Hausdorff (non vide) localement compact X, muni d'une involution notée $-$ (qui est un homéomorphisme $x \to x^-$ de X tel que $x^{--} = x$) est un hypergroupe si l'espace vectoriel $M(X)$ des mesures complexes bornées sur X est muni d'une opération $*$ (appelée convolution) telle que $(M(X),*,\sim)$ est une algèbre de Banach involutive $(\mu^\sim(A) = \overline{\mu(A^-)}$ pour $\mu \in M(X))$ vérifiant en plus les conditions suivantes :

1) L'application bilinéaire $(\mu,\nu) \mapsto \mu * \nu$ de $M(X) \times M(X)$ dans $M(X)$ est non négative (i. e. $\mu * \nu \geqslant 0$ si μ et $\nu \geqslant 0$) et sa restriction à $M^+(X) \times M^+(X)$ est continue lorsque $M^+(X)$ est muni de la topologie faible.

2) Quels que soient x et $y \in X$, $\delta_x * \delta_y$ est une probabilité et $\text{supp}(\delta_x * \delta_y)$ est compact.

3) L'application $(x,y) \mapsto \text{Supp}(\delta_x * \delta_y)$ de $X \times X$ dans l'espace des sous-ensembles compacts de X est continue pour la topologie particulière donnée en [8], § 2.5.

4) Il existe un unique élément $e \in X$ tel que $\delta_e * \delta_x = \delta_x * \delta_e = \delta_x$ pour tout $x \in X$ et le point e appartient au support de $\delta_x * \delta_y$ si et seulement si $x = y^-$.

Les axiomes de Spector sont analogues à la différence près que dans la condition 2) il n'est pas exigé que le support de $\delta_x * \delta_y$ soit compact et la condition 3) est remplacée par une condition plus faible de régularité des supports (cf. [17], p. 646).

(1.3) Quelques rappels utiles pour la lisibilité de l'article : Pour $f \in C(X)$ (= l'ensemble des fonctions continues bornées sur X), on définit les translatées à droite et à gauche (respectivement) par

$$f_x(y) = f^y(x) = \int_X f\, d(\delta_x * \delta_y) = \langle f, \delta_x * \delta_y \rangle$$

et on dit qu'une mesure positive σ non nécessairement bornée est une mesure de Haar à gauche (resp. à droite) si $\langle \sigma, f_x \rangle = \langle \sigma, f \rangle$ (resp. $\langle \sigma, f^y \rangle = \langle \sigma, f \rangle$)

pour tout $f \in C_k(X)$ et tout $x \in X$ (resp. tout $y \in X$) .

Un résultat fondamental de Spector (cf. [18]) est que tout hypergroupe commutatif (i. e. $(M(X),*)$ est une algèbre commutative) admet une mesure de Haar σ , unique à un coefficient multiplicatif près. De plus $\operatorname{supp} \sigma = X$.

Le dual d'un hypergroupe commutatif X est l'ensemble $\hat{X} = \{\chi \in C(X)\ ;\ \chi_x(y) = \chi(x)\ \chi(y)$ et $\chi(x^-) = \overline{\chi(x)}$ pour tous x et $y \in X\}$ qui est un espace de Hausdorff localement compact (pour la topologie de la convergence uniforme sur les compacts) qui s'identifie à la partie hermitienne du spectre de Guelfand de l'algèbre de Banach commutative $L^1_{\mathbb{C}}(X,\sigma)$.

La transformée de Fourier de $\mu \in M(X)$ (resp. de $f \in L^1(X,\sigma)$) est alors définie par :

$$\hat{\mu}(\chi) = \int_X \overline{\chi(x)}\ \mu(dx) \quad (\text{resp. } \hat{f}(\chi) = \int_X \overline{\chi(x)}\ f(x)\ d\sigma(x)) \quad (\chi \in \hat{X})$$

et si $\mu \in M(\hat{X})$, la transformée de Fourier réciproque est donnée par :

$$\check{\mu}(x) = \int_{\hat{X}} \chi(x)\ \mu(d\chi) \quad (x \in X) .$$

On a alors la formule de réciprocité (cf. [8], p. 73).

$$\int_X \check{\nu}\ d\mu^- = \int_{\hat{X}} \hat{\mu}\ d\nu \quad \text{si} \quad \nu \in M(\hat{X}) \quad \text{et} \quad \mu \in M(X) .$$

Un résultat essentiel (dû à Levitan cf. [8], p. 41) affirme alors l'existence d'une mesure de Plancherel $\hat{\sigma}$ sur $\hat{X}$ i. e. $\hat{\sigma}$ est une mesure de Radon positive telle que

$$\int_X |f(x)|^2\ \sigma(dx) = \int_{\hat{X}} |\hat{f}(\chi)|^2 \hat{\sigma}(d\chi) ,$$

pour toute $f \in L^1(X,\sigma) \cap L^2(X,\sigma)$, d'où une extension de la transformation de Fourier en une isométrie de $L^2(X,\sigma)$ sur $L^2(\hat{X},\hat{\sigma})$.

Il est à noter que les résultats qui précèdent sont encore valables avec les axiomes de Spector.

II.- OBJET DE L'ARTICLE ET PRESENTATION DES RESULTATS. Dans le cas où X est un groupe abélien, on sait aussi bien en analyse harmonique (cf. [15], p. 59) qu'en probabilités (cf. [14]) que les mesures positives $\mu \in M(X)$ telles que $\mu * \mu = \mu$ (idempotents) jouent un rôle très important. Par exemple (cf. [14], théorème 7.2, p. 106) toute loi de probabilité infiniment divisible sur X est le produit de convolution d'un idempotent (= la mesure de Haar normalisée d'un sous-groupe compact de X) et d'une probabilité infiniment divisible (sans facteur idempotent) dont la transformée de Fourier est donnée explicitement par la formule de Levy-

Khintchine (cf. [14], p. 103).

Dans le cas où X est un hypergroupe commutatif, Jewett ([8], p. 63) puis Bloom et Heyer ([1], p. 327) ont montré que tout idempotent de $M^+(X)$ est la mesure de Haar normalisée d'un sous-hypergroupe compact H de X (H est fermé, $H^- = H$ et $H * H \subset H$ où pour deux sous-ensembles A et B de X, on a posé $A * B = \bigcup \{\text{supp } \delta_x * \delta_y \; ; \; x \in A \, , \; y \in B\}$.

Le problème de la représentation de Levy-Khintchine d'une probabilité infiniment divisible sur X se trouve posé, mais faute en particulier d'une bidualité entre X et $\hat{X}$ qui fait déjà qu'on ne dispose pas toujours d'un "bon" théorème de continuité de Paul Levy, les difficultés semblent importantes. Il n'est pas inutile (pour le moment) de résoudre certains cas particuliers pour y voir un peu plus clair. C'est ce que nous faisons ici dans le cas où X est discret infini dénombrable.

Le résultat essentiel de cet article est le suivant :

Théorème 2 : Soit X un hypergroupe commutatif, discret infini dénombrable, sans sous-hypergroupe compact non trivial et vérifiant l'une des deux conditions suivantes :

C 1 : Le support de la mesure de Plancherel $\hat{\sigma}$ contient le point $\mathbb{I}$.

C 2 : $\mathbb{I}$ n'est pas isolé dans $\hat{X}$ et il existe un voisinage V de $\mathbb{I}$ tel que $\lim\limits_{x \to \infty} \chi(x) = 0$ pour tout $\chi \in V - \{\mathbb{I}\}$.

Alors $\mu \in M_1(X)$ est infiniment divisible si et seulement s'il existe une (unique) mesure positive ν sur X avec $\nu(\{e\}) = 0$ telle que

$$\hat{\mu}(\chi) = \exp\left[\sum_{x \in X} (\overline{\chi(x)} - 1) \, \nu(\{x\})\right] \quad (\forall \, \chi \in \hat{X}) \, .$$

On notera ici que comme X est discret, $\hat{X}$ est compact (cf. [17], p. 665).

Commentaires sur les hypothèses :

1) X est sans sous-hypergroupe compact non trivial si et seulement si $M^+(X)$ est sans idempotent. Cette condition est automatiquement vérifiée lorsque $\hat{X}$ est connexe, ce qui est le cas dans beaucoup d'exemples intéressants comme ceux où la structure d'hypergroupe provient de polynômes orthogonaux (cf. le § 6).

2) Dans le cas où l'hypothèse C 1 (resp. C 2) est vérifiée, on a un théorème de continuité de Paul Levy (cf. le § 3).

3) Il existe des hypergroupes X qui vérifient à la fois les deux conditions C 1 et C 2 (cf. le § 6).

4) Si $\hat{X}$ est un hypergroupe pour la "multiplication ponctuelle" (i. e. si pour tous χ_1 et χ_2 dans $\hat{X}$ il existe une probabilité $\delta_{\chi_1} * \delta_{\chi_2}$ telle que

$\chi_1(x)\,\chi_2(x) = (\delta_{\chi_1} * \delta_{\chi_2})^{\vee}(x)$ pour tout $x \in X$ et $\hat{X}$ est un hypergroupe pour cette convolution), alors la condition C 1 est vérifiée car $\operatorname{supp} \hat{\sigma} = \hat{X}$.

5) Le théorème 2 est encore valable avec les axiomes plus généraux de Spector, mais comme nous l'a fait remarquer H. Heyer, l'affirmation du théorème II.5.7 de [17] selon laquelle on a toujours $\operatorname{supp} \hat{\sigma} = \hat{X}$ est erronée.

III.- THEOREME(S) DE CONTINUITE DE PAUL LEVY. On ne suppose pas dans ce paragraphe que X est discret dénombrable.

(3.1) Théorème 1 : Soit X un hypergroupe commutatif à base dénombrable, vérifiant la condition suivante :

C 1 : Le support de la mesure de Plancherel $\hat{\sigma}$ contient le point $\mathbb{1}$ (= le caractère identiquement 1) .

Alors pour une suite (μ_n) de mesures de probabilités sur X, les assertions suivantes sont équivalentes :

i) La suite μ_n converge en loi vers une probabilité μ.

ii) La suite de fonctions continues bornées $\hat{\mu}_n$ converge $\hat{\sigma}$-presque partout sur $\hat{X}$ vers une fonction φ continue en $\mathbb{1}$.

Démonstration : Le seul point à démontrer est ii) $\Rightarrow$ i). Comme X est à base dénombrable, on peut extraire de (μ_n) une sous-suite (μ_{n_k}) faiblement convergente vers une mesure positive μ de masse $\leqslant 1$. Soit $\psi \in C_k(\hat{X})$. Alors par réciprocité de Fourier, on a

$$\int_{\hat{X}} \hat{\mu}_{n_k}.\psi \, d\hat{\sigma} = \int_X \check{\psi} \, d\bar{\mu}_{n_k} .$$

Or (cf. [1], p. 321) on a que $\check{\psi} \in C_o(X)$ de sorte que par limite faible, on a :

$$\lim_{k\to+\infty} \int_X \check{\psi}\, d\bar{\mu}_{n_k} = \int_X \check{\psi}\, d\bar{\mu} = \int_{\hat{X}} \hat{\mu}.\psi \, d\hat{\sigma} .$$

Le théorème de convergence dominée de Lebesgue donne d'autre part :

$$\lim_{k\to+\infty} \int_{\hat{X}} \hat{\mu}_{n_k}.\psi \, d\hat{\sigma} = \int_{\hat{X}} \varphi.\psi \, d\hat{\sigma} .$$

Donc $\int_{\hat{X}} \varphi.\psi \, d\hat{\sigma} = \int_{\hat{X}} \hat{\mu}.\psi \, d\hat{\sigma}$ pour tout $\psi \in C_k(\hat{X})$. Il en résulte que $\varphi = \hat{\mu}$ $\hat{\sigma}$-presque partout. Or φ est continue en $\mathbb{1}$ dont tout voisinage est chargé par $\hat{\sigma}$; on a donc $\hat{\mu}(\mathbb{1}) = \varphi(\mathbb{1}) = 1$ et μ est une probabilité. La conclusion en résulte par un argument de Helly-Bray.

(3.2) Remarque : La technique de démonstration précédente est celle de Siebert (cf. [16]).

Pour une version voisine du théorème 1 (mais qui ne correspond pas aux besoins de notre théorème 2), voir Bloom et Heyer [1].

(3.3) Théorème 1' : Soit X un hypergroupe commutatif à base dénombrable et vérifiant la condition suivante :

C 2 : $\mathbb{1}$ n'est pas isolé dans $\hat{X}$ et il existe un voisinage V de $\mathbb{1}$ tel que $\lim_{x\to\infty} \chi(x) = 0$ pour tout $\chi \in V - \{\mathbb{1}\}$.

Alors pour une suite (μ_n) de mesures de probabilité sur X, les deux assertions suivantes sont équivalentes :

i) μ_n converge en loi vers une probabilité μ.

ii) La suite de fonctions $\hat{\mu}_n$ converge simplement sur $\hat{X}$ vers une fonction φ continue au point $\mathbb{1}$.

Démonstration : En extrayant une sous-suite μ_{n_k} (comme dans (3.1)) convergente faiblement vers μ, on a ici immédiatement que $\hat{\mu}_{n_k}(\chi) \to \hat{\mu}(\chi)$ pour tout $\chi \in V - \{\mathbb{1}\}$ et donc $\varphi(\chi) = \hat{\mu}(\chi)$ pour tout $\chi \in V - \{\mathbb{1}\}$. La conclusion résulte alors de la continuité de φ en $\mathbb{1}$ et du fait que $\mathbb{1}$ n'est pas isolé.

(3.4) Remarque : Le théorème 1' est trivial mais nous montrerons au § 6 qu'il y a des situations intéressantes où il s'applique lorsque l'hypothèse C 1 du théorème 1 est en défaut.

IV.- DEMONSTRATION DU THEOREME 2.

(4.1) Définition : On dit qu'une probabilité μ sur X est infiniment divisible (I. D. en abrégé) si pour tout entier $n > 0$, il existe une probabilité μ_n sur X telle que $(\mu_n)^{*n} = \mu$ ou, ce qui est équivalent, $(\hat{\mu}_n)^n = \hat{\mu}$.

(4.2) Proposition : Soit μ une probabilité I. D. sur un hypergroupe X commutatif vérifiant les hypothèses du théorème 2 et soit μ_n une suite de probabilités telles que $(\mu_n)^{*n} = \mu$ pour tout $n > 0$. Alors μ_n converge en loi (i. e. étroitement) vers δ_e.

Démonstration : Soit $Z_\mu = \{\chi \in \hat{X} ; \hat{\mu}(\chi) = 0\}$. Si $\chi \in Z_\mu$, on a clairement $\lim_{n\to\infty} \hat{\mu}_n(\chi) = 0$ et si $\chi \notin Z_\mu$, on a $\lim_{n\to\infty} \hat{\mu}_n(\chi) = 1$, mais comme $\hat{\mu}(\chi)$ ne s'annule pas dans un voisinage de $\mathbb{1}$, la fonction $\varphi = \lim_{n\to\infty} \hat{\mu}_n$ est identiquement égale à 1 dans un voisinage de $\mathbb{1}$. Les théorèmes 1 ou 1' nous assurent alors que μ_n tend vers une probabilité ν. Or $\hat{\nu} = \varphi = 0$ ou 1, donc ν est un idempotent qui ne peut être que δ_e par hypothèse.

(4.3) Remarque : Dans la démonstration de la proposition (4.2), on n'a pas utilisé le fait que X est discret.

(4.4) Proposition (comportement de μ_n en e) : Soit μ une probabilité I. D. sur un hypergroupe X vérifiant les hypothèses du théorème 2 et soit μ_n comme dans (4.2). Alors il existe une constante $C \geqslant 0$ ne dépendant que de μ , telle que :

$$\mu_n(\{e\}) = 1 - \frac{C}{n} + o(\frac{1}{n}) .$$

De plus, $C > 0$ si et seulement si $\mu \neq \delta_e$.

Démonstration : Pour tout $z \in \mathbb{C} - \{0\}$, on pose

$$\operatorname{Log} z = \operatorname{Log} |z| + i \operatorname{Arg} z \quad (\text{avec } \operatorname{Arg} z \in [-\pi,+\pi[) .$$

La fonction $\operatorname{Log} \hat{\mu}$ est mesurable sur $\hat{X}$ et comme $Z_{\mu_n} = \emptyset$ pour tout n (car $|\hat{\mu}_n(\chi)| = |\hat{\mu}(\chi)|^{1/n}$ et $\hat{\mu}$ ne s'annule pas), on a

$$\operatorname{Log} \hat{\mu}_n(\chi) = \frac{1}{n} \operatorname{Log} |\hat{\mu}(\chi)| + \frac{i}{n} \operatorname{Arg} \hat{\mu}(\chi) \quad (\forall \chi \in \hat{X}) ,$$

donc $\hat{\mu}_n$ converge vers 1 uniformément sur $\hat{X}$. Il en résulte que pour n assez grand

$$\operatorname{Log} \hat{\mu}_n(\chi) = \sum_{p \geqslant 1} \frac{1}{p} [\hat{\mu}_n(\chi) - 1]^p \quad (\forall \chi \in \hat{X}) ,$$

où la série de droite converge uniformément sur $\hat{X}$. On a donc

(4.4.1) $$\operatorname{Log} \hat{\mu}(\chi) = n \operatorname{Log} \hat{\mu}_n(\chi) = n [\hat{\mu}_n(\chi) - 1] (1 + \alpha_n(\chi)) ,$$

où la suite de fonctions α_n converge uniformément vers zéro sur $\hat{X}$ quand n tend vers $+\infty$. On en déduit que

(4.4.2) $$\hat{\mu}_n(\chi) = 1 + \frac{1}{n} \operatorname{Log} \hat{\mu}(\chi) + \frac{1}{n} \varepsilon_n(\chi) \quad (\forall \chi \in \hat{X}) ,$$

où la suite de fonctions ε_n sur $\hat{X}$ converge uniformément vers 0 quand n tend vers $+\infty$.

Maintenant par inversion de Fourier et comme X est dénombrable, on a, pour toute $\nu \in M_1(X)$:

(4.4.3) $$\int_{\hat{X}} \hat{\nu}(\chi)\, \chi(x)\, d\hat{\sigma}(\chi) = \nu(\{x\}) \int_{\hat{X}} |\chi(x)|^2\, d\hat{\sigma}(\chi) = \frac{\nu(\{x\})}{\sigma(\{x\})} \quad (\forall x \in X) .$$

Si on suppose que $\hat{\sigma}$ est normalisée de telle sorte que $\hat{\sigma}(\hat{X}) = 1$ (car $\hat{X}$ est compact), on a $\sigma(\{e\}) = 1$. Utilisons alors (4.4.3) avec $x = e$ et $\nu = \mu_n$. Grâce à (4.4.2), on obtient alors

$$\mu_n(\{e\}) = 1 + \frac{1}{n}\int_{\hat{X}} \operatorname{Log}\hat{\mu}(\chi)\, d\hat{\sigma}(\chi) + \frac{1}{n}\int_{\hat{X}} \varepsilon_n(\chi)\, d\hat{\sigma}(\chi) ,$$

car $\operatorname{Log}\hat{\mu}$ est $\hat{\sigma}$-intégrable car $\hat{\mu}$ est une fonction continue et ne s'annulant pas sur le compact $\hat{X}$. Posons alors

$$C = -\int_{\hat{X}} \operatorname{Log}\hat{\mu}(\chi)\, d\hat{\sigma}(\chi) .$$

Il est clair que $C > 0$ si et seulement si $\mu \neq \delta_e$. Comme de plus $\varepsilon_n(\chi)$ converge uniformément vers 0 sur $\hat{X}$, la proposition est démontrée.

(4.5) <u>Proposition</u> : <u>Les hypothèses et les notations étant les mêmes qu'en</u> (4.4), <u>on pose</u> $\nu_n(\{e\}) = 0$ <u>et</u> $\nu_n(\{x\}) = n\mu_n(\{x\})$ <u>si</u> $x \neq e$. <u>Alors la suite de mesures positives</u> ν_n <u>converge vaguement quand</u> n <u>tend vers l'infini vers une mesure positive</u> ν <u>de masse</u> $\nu(X) \leqslant C$.

<u>Démonstration</u> : D'après (4.4.3) et (4.4.2), on a pour $x \neq e$:

(4.5.1) $$\mu_n(\{x\}) = \frac{1}{n}\sigma(\{x\})\Big(\int_{\hat{X}} \operatorname{Log}\hat{\mu}(\chi)\,\chi(x)\, d\hat{\sigma}(\chi) + \int_{\hat{X}} \varepsilon_n(\chi)\,\chi(x)\, d\hat{\sigma}(\chi)\Big)$$
$$= \frac{1}{n}(C(x) + \theta_n(x)) ,$$

où $\lim_{n\to+\infty} \theta_n(x) = 0$ et $C(x) \geqslant 0$ $(\forall x \in X - \{e\})$. Ainsi on voit immédiatement que la suite ν_n converge vaguement vers la mesure positive ν définie par $\nu(\{e\}) = 0$ et $\nu(\{x\}) = C(x)$ pour $x \neq e$. Mais compte tenu de (4.5.1) et de l'expression de $\mu_n(\{e\})$, quel que soit $n > 1$, on a :

$$\frac{1}{n} \sum_{x\in X-\{e\}} (C(x) + \theta_n(x)) + 1 - \frac{C}{n} + \frac{\theta_n(e)}{n} = 1 ,$$

il en résulte que :

$$\sum_{x\in X-\{e\}} (C(x) + \theta_n(x)) = C - \theta_n(e) ,$$

donc $\sum_{x\in X-\{e\}} C(x) = \nu(X) \leqslant C$ d'après le lemme de Fatou.

(4.6) <u>Proposition</u> <u>(condition nécessaire du théorème 2)</u> : <u>Avec les hypothèses de</u> (4.2) <u>et les notations de</u> (4.4), <u>si</u> μ <u>est</u> I. D., <u>on a</u> :

$$\hat{\mu}(\chi) = \exp\Big[\sum_{x\in X}(\overline{\chi(x)} - 1)\ \nu(\{x\})\Big] \ \Big(= \exp\int_X (\overline{\chi(x)} - 1)\ \nu(dx)\Big) .$$

<u>Démonstration</u> :

A) Supposons d'abord que X vérifie la condition C 1. Définissons alors une fonction g sur $\hat{X}$ de la façon suivante :

$$g(\chi) = -C + \sum_{x\in X-\{e\}} C(x)\ \overline{\chi(x)} .$$

Cette fonction est continue et bornée sur $\hat{X}$ et elle possède la propriété suivante :

$$\int_{\hat{X}} [\operatorname{Log} \hat{\mu}(\chi) - g(\chi)]\ \chi(x)\ d\hat{\sigma}(\chi) = 0 ,$$

pour tout $x \in X$. Il en résulte que $\operatorname{Log} \hat{\mu} = g$ $\hat{\sigma}$-presque partout (puisque $\operatorname{Log} \hat{\mu}$ est continue bornée sur $\hat{X}$) . L'hypothèse C 1 implique alors que $g(\mathbb{1}) = \operatorname{Log} \hat{\mu}(\mathbb{1}) = 0$ et donc que l'on a : $C = \sum_{x \neq e} C(x) = \nu(X)$. Soit maintenant I une partie finie de X qui contient e . On a :

$$\nu_n(I^c) = n(1 - \sum_{x \in I} \mu_n(\{x\}))$$

$$= C - (\sum_{x \in I - \{e\}} C(x) + \sum_{x \in I} \theta_n(x)) .$$

Ainsi $\overline{\lim_{n \to +\infty}}\ \nu_n(I^c) = C - \sum_{x \in I - \{e\}} C(x)$. Il en résulte donc d'après ce qui précède que $\lim_{\mathcal{F}} (\overline{\lim_{n \to \infty}}\ \nu_n(I^c)) = 0$, où $\mathcal{F}$ est le filtre naturel des sections de X . La suite ν_n converge donc étroitement vers ν . Or d'après (4.4.1), $\operatorname{Log} \hat{\mu}(\chi) = \lim_{n \to +\infty} n(\hat{\mu}_n(\chi) - 1) = \lim_{n \to \infty} \int_X (\overline{\chi(x)} - 1)\ \nu_n(dx) = \int_X (\overline{\chi(x)} - 1)\ \nu(dx)$ d'après ce qui précède, puisque la fonction $x \mapsto \overline{\chi(x)} - 1$ est bornée sur X .

B) Supposons maintenant la condition C 2 vérifiée. On a

$$n(\hat{\mu}_n(\chi) - 1) = n(\sum_{x \neq e} \overline{\chi(x)}\ \mu_n(\{x\}) + \mu_n(\{e\}) - 1)$$

$$= \int_X \overline{\chi(x)}\ \nu_n(dx) + (-C + \theta_n(e)) .$$

Ainsi pour $\chi \in V - \{\mathbb{1}\}$, on a immédiatement :

$$\operatorname{Log} \hat{\mu}(\chi) = \int_X \overline{\chi(x)}\ \nu(dx) - C ,$$

d'où

$$\hat{\mu}(\chi) = \exp(-C + \nu(X)) \exp \int_X (\overline{\chi(x)} - 1)\ \nu(dx) ,$$

pour tout $\chi \in V - \{\mathbb{1}\}$. Mais $\hat{\mu}$ est continue et $\hat{\mu}(\mathbb{1}) = 1$, donc $\nu(X) = C$ et ainsi les ν_n conservent la masse. La fin de la démonstration est alors la même qu'en A).

(4.7) <u>Fin de la démonstration du théorème 2</u> : Il reste à montrer la condition suffisante. Soit $a > 0$ et $x \in X$. Posons

$$h_{a,x}(\chi) = \exp [a(\overline{\chi(x)} - 1)] \quad (\chi \in \hat{X}) .$$

Alors $h_{a,x}$ est clairement la transformée de Fourier de la mesure de probabilité $e^{-a} \sum_{k=0}^{+\infty} \frac{a^k}{k!} \delta_x^{*k}$. Ainsi d'après le théorème de continuité de Paul Levy,

$\chi \mapsto \exp\left[\int_X (\overline{\chi(x)} - 1)\ \nu(dx)\right]$ est la transformée de Fourier d'une probabilité, pour toute mesure positive bornée ν. De plus, elle est clairement infiniment divisible et le théorème 2 est prouvé.

(4.8) Remarque : Pour un hypergroupe discret dénombrable commutatif (avec éventuellement des sous-hypergroupes compacts) et vérifiant C 1 ou C 2, tout ce qui précède montre que toute probabilité infiniment divisible μ sans facteur idempotent (i. e. $\lim \hat{\mu}_n \equiv 1$) admet la représentation de Levy-Khintchine du théorème 2. Dans le cas général où μ a un facteur idempotent ω (i. e. $\lim \hat{\mu}_n = \hat{\omega}$), nous n'avons pas encore réussi à factoriser ω, mais ceci semble faisable et on peut raisonnablement conjecturer qu'on a alors $\mu = \omega * \nu$ où ν est une probabilité I. D. sans facteur idempotent.

V.- LIMITES DES SYSTEMES TRIANGULAIRES DE PROBABILITES.

(5.1) Définition : Soit (n_i) une suite d'entiers tendant vers $+\infty$ avec i et soit (μ_{ij}), $1 \leq j \leq n_i$, un système triangulaire d'éléments de $M_1(X)$. Une probabilité μ sur X est dite limite du système (μ_{ij}) si on a :

a) $\hat{\mu}_i(\chi) = \prod_{j=1}^{n_i} \hat{\mu}_{ij}(\chi) \to \hat{\mu}(\chi)$ $(\forall \chi \in \hat{X})$ quand $i \to +\infty$.

b) $\sup_{1\leq j\leq n_i} |\hat{\mu}_{ij}(\chi) - 1| \to 0$ uniformément pour $\chi \in \hat{X}$ (condition infinitésimale cf. [14]).

(5.2) Remarque : Pour X discret dénombrable, la condition infinitésimale est équivalente à chacune des deux conditions suivantes :

b') $\sup_{1\leq j\leq n_i} \mu_{ij}(X - \{e\}) \to 0$ quand $i \to +\infty$.

b'') $\inf_{1\leq j\leq n_i} \mu_{ij}(\{e\}) \to 1$ quand $i \to +\infty$.

Ceci se vérifie facilement.

(5.3) Proposition : Si les μ_{ij} sont symétriques (i. e. $\mu_{ij} = \bar{\mu}_{ij}$) et si X est un hypergroupe vérifiant les conditions du théorème 2, alors la limite μ du système (μ_{ij}) est une probabilité I. D.

Démonstration : Les fonctions $\hat{\mu}_{ij}(\chi)$ sont alors réelles et il est facile de voir alors que grâce à la condition infinitésimale, on a :

$$\hat{\mu}(\chi) = \lim_{i\to+\infty} \exp\left[\sum_{j=1}^{n_i} \operatorname{Log} \hat{\mu}_{ij}(\chi)\right]$$

$$= \lim_{i\to+\infty} \exp\left[\sum_{j=1}^{n_i} (\hat{\mu}_{ij}(\chi) - 1)\right].$$

On a ainsi

$$\hat{\mu}(\chi) = \lim_{i\to+\infty} \prod_{j=1}^{n_i} \exp\left[\int_X (\overline{\chi(x)} - 1)\, d\mu_{ij}(x)\right]$$

$$= \lim_{i\to+\infty} \prod_{j=1}^{n_i} \varphi_{ij}(\chi) \;,$$

où φ_{ij} est la transformée de Fourier d'une loi de probabilité I. D. d'après le théorème 2. Ainsi μ est également I. D.

(5.4) Proposition : Soit (μ_{ij}) un système triangulaire sur un hypergroupe X vérifiant les hypothèses du théorème 2. Sous l'hypothèse additionnelle

$$\sup_i \sum_{j=1}^{n_i} |\hat{\mu}_{ij}(\chi) - 1| < +\infty \quad (\forall \chi \in \hat{X}) \;,$$

la limite μ du système (μ_{ij}) est une probabilité I. D.

Démonstration : On a

$$-\sum_{j=1}^{n_i} \mathrm{Log}\,\hat{\mu}_{ij}(\chi) = \sum_{j=1}^{n_i} \sum_{k=1}^{\infty} \frac{(1-\hat{\mu}_{ij}(\chi))^k}{k} \;,$$

pour i assez grand pour que pour tout $j \in [1,n_i]$ on ait $|1-\hat{\mu}_{ij}(\chi)| < 1$. L'expression précédente vaut alors :

$$\sum_{j=1}^{n_i}(1-\hat{\mu}_{ij}(\chi)) + \sum_{j=1}^{n_i} \sum_{k=2}^{\infty} \frac{1}{k}(1-\hat{\mu}_{ij}(\chi))^k \;.$$

Mais le module du 2ème terme peut être majoré par :

$$\frac{1}{2} \sum_{j=1}^{n_i} \sum_{k=2}^{+\infty} |1-\hat{\mu}_{ij}(\chi)|^k = \frac{1}{2} \sum_{j=1}^{n_i} \frac{|1-\hat{\mu}_{ij}(\chi)|^2}{1 - |1-\hat{\mu}_{ij}(\chi)|}$$

$$\leq \sup_{1\leq j\leq n_i} |1-\hat{\mu}_{ij}(\chi)| \sum_{j=1}^{n_i} |1-\hat{\mu}_{ij}(\chi)| \;.$$

Mais cette dernière somme est bornée par hypothèse et donc toute l'expression tend vers zéro quand $i \to +\infty$. Ainsi comme en (5.3), on a

$$\hat{\mu}(\chi) = \lim_{i\to+\infty} \exp\left[\sum_{j=1}^{n_i}(\hat{\mu}_{ij}(\chi) - 1)\right] \;,$$

et la fin de la démonstration est la même.

(5.5) Remarque : L'hypothèse additionnelle équivaut à $\sup_i \sum_{j=1}^{n_i}(1-\mu_{ij}(\{e\})) < +\infty$. Dans le cas où X est un groupe abélien, on peut se passer de cette condition car on peut montrer, grâce aux théorèmes de structure des groupes abéliens, qu'elle est toujours vérifiée (cf. [14], p. 91).

VI.- EXEMPLES ET REMARQUES.

(6.1) Si $X = \mathbb{Z}^n$ avec la convolution usuelle $(\delta_x * \delta_y = \delta_{x+y})$, notre théorème 2 indique le fait bien connu que les seules lois infiniment divisibles sont poissoniennes.

(6.2) Soit $(P_n(x))_{n\in\mathbb{N}}$ une suite de polynômes (avec $d^\circ P_n = n$ et normalisés par $P_n(1) = 1$) orthogonaux sur $[-1,+1]$ par rapport à une mesure positive $d\psi(x)$. Supposons que cette suite vérifie la propriété de linéarisation à coefficients $\geqslant 0$ i. e. pour tous m et n , on a

(6.2.0) $$P_m(x)\ P_n(x) = \sum_r C(m,n,r)\ P_r(x) ,$$

avec $C(m,n,r) \geqslant 0$ pour tout $r \in \mathbb{N}$. Alors on peut munir l'ensemble $\mathbb{N}$ des entiers naturels d'une structure d'hypergroupe discret commutatif d'élément neutre δ_o et d'involution l'identité par :

$$\delta_n * \delta_m = \sum_r C(m,n,r)\ \delta_r .$$

Comme $C(m,n,m+n) > 0$, un tel hypergroupe n'a pas de sous-hypergroupe compact non trivial. Par exemple :

(6.2.1) Les polynômes de Jacobi $P_n^{\alpha,\beta}(x)$ (normalisés) vérifient (6.2.0) si $\alpha \geqslant \beta > -1$ et $\alpha + \beta + 1 \geqslant 0$ (cf. [7]). Si $X = \mathbb{N}$ avec la structure d'hypergroupe associée, on a $\hat{X} = [-1,1]$ et pour $\chi \in \hat{X}$: $\chi(n) = P_n^{\alpha,\beta}(\chi)$ $(n \in X)$. La mesure de Plancherel est la mesure $d\hat{\sigma}(\chi) = (1-\chi)^\alpha(1+\chi)^\beta d\chi$ par rapport à laquelle les $P_n^{\alpha,\beta}$ sont orthogonaux sur $[-1,1]$. Dans ce cas la condition C 1 du théorème 2 est toujours vérifiée.

(6.2.2) Si $\alpha = \beta = {}^1/_2$, les polynômes de (6.2.1) sont les polynômes de Tchebychev de seconde espèce et il est bien connu que

$$P_n^{1/2,1/2}(\cos\theta) = \frac{\sin(n+1)\ \theta}{(n+1)\ \sin\theta} \quad (\theta \in [0,\pi]) .$$

Pour tout $\theta \in]0,\pi[$, on a donc $\lim_{n\to\infty} P_n^{1/2,1/2}(\cos\theta) = 0$ et la condition C 2 est donc aussi vérifiée dans le cas d'un hypergroupe associé à ces polynômes. On peut même montrer qu'il en va de même pour tous les polynômes de Gegenbauer-Jacobi avec $\alpha = \beta > -\frac{1}{2}$.

(6.2.3) Soit q un entier $\geqslant 2$. La suite P_n de polynômes (cf. [12]) définie par les relations de récurrence :

$$P_o(x) \equiv 1 \quad , \quad P_1(x) \equiv x$$

$$x\ P_n(x) = \frac{q}{q+1}\ P_{n+1}(x) + \frac{1}{q+1}\ P_{n-1}(x)$$

vérifie (6.2.0) et permet de munir $X = \mathbb{N}$ d'une structure d'hypergroupe pour laquelle $\hat{X} = [-1,+1]$ et

$$d\hat{\sigma}(\chi) = \frac{q+1}{2\pi}(1-\chi^2)^{-1}(\rho_q - \chi^2)\ 1_{[-\rho_q,\rho_q]}(\chi)\ d\chi\ ,$$

où $\rho_q = \frac{2\sqrt{q}}{q+1} < 1$ si $q \geq 2$. La condition C 1 n'est pas vérifiée ici mais la condition C 2 l'est, comme on peut le voir sur l'expression explicite de P_n donnée dans [12], p. 110.

(6.3) Soit G un groupe localement compact et K un sous-groupe compact. L'espace des double-classes $X = K\backslash G/K = \{KgK\,;\, g \in G\}$ est un hypergroupe (cf. [2]) avec $e = K$ et $(KgK)^{-} = Kg^{-1}K$ pour la convolution : $\delta_{KxK} * \delta_{KyK} = \int_K \delta_{KxkyK}\, dk$ (dk est la mesure de Haar normalisée de K) et la mesure invariante est $\sigma = \int_G \delta_{KgK}\, dg$. Lorsque (G,K) est un couple de Guelfand (i. e. $L^1(X,\sigma)$ est commutative), on peut voir que $\hat{X}$ s'identifie (cf. [3]) à l'ensemble des fonctions φ continues bornées sur G , hermitiennes (i. e. $\varphi(g^{-1}) = \overline{\varphi(g)}$) , biinvariantes par K , telles que $\int_K \varphi(xky)\, dk = \varphi(x)\, \varphi(y)$ et $\varphi(e) = 1$. Par exemple :

(6.3.1) G = le groupe des isométries d'un arbre homogène T d'ordre q (cf. [11]) et K le stabilisateur d'un point $s_o \in T$. Alors $X = K\backslash G/K \cong \mathbb{N}$ avec $\sigma(\{n\}) = q^{n-1}(q+1)$ et $\hat{X}$ est paramétré par $[-1,1]$, avec $\chi(n) = P_n(\chi)$ $(\chi \in \hat{X}, n \in \mathbb{N})$ où P_n est la suite de polynômes considérée en (6.2.3). Cet hypergroupe X est donc celui de l'exemple (6.2.3).

(6.4) Le dual de certains hypergroupes du type décrit en (6.3) peut parfois être muni d'une structure d'hypergroupe obtenue par multiplication ponctuelle (voir le § II). Par exemple :

(6.4.1) $G = SO(d)$, $K = SO(d-1)$, alors $X = K\backslash G/K \cong [-1,1]$ et $\hat{X}$ peut être paramétré par $\mathbb{N}$. Soit $n \in \hat{X}$, alors $n(x) = P_n^{\alpha,\alpha}(x)$ $(x \in [-1,1])$ avec $\alpha = \frac{d}{2} - 1$ et $P_n^{\alpha,\alpha}$ un polynôme de Jacobi. La structure d'hypergroupe sur $\hat{X}$ obtenue par multiplication ponctuelle est alors celle de (6.1) avec $\alpha = \beta = \frac{d}{2} - 1$.

(6.4.2) Si (G,K) est un espace symétrique de type compact, son dual qui est discret dénombrable est un hypergroupe commutatif vérifiant la condition C 1 (cf. [4]).

(6.5) Le dual d'un groupe de Lie compact G (i. e. l'ensemble $\hat{G}$ des classes de représentations unitaires irréductible de G) peut être muni d'une structure d'hypergroupe discret dénombrable commutatif par une convolution provenant des formules de Clebsch-Gordon pour le produit tensoriel des représentations (cf. [6]). Certains cas particuliers coïncident avec des hypergroupes décrits en (6.2) et (6.4). Ces hypergroupes vérifient la condition C 1.

(6.6) A Oberwolfach nous avons pris connaissance du travail de R. Lasser (cf. [10]) qui obtient une formule de Levy-Khintchine pour les semi-groupes de convolution sur un hypergroupe commutatif X dont le dual $\hat{X}$ est aussi un hypergroupe et satisfaisant à une certaine condition (notée F.) de régularité des caractères de $\hat{X}$. Sa méthode d'approche est tout à fait différente de la nôtre. Il y a des hypergroupes discrets dénombrables qui satisfont ses conditions.

(6.7) Dans le cas particulier décrit en (6.2.1) avec $\alpha = \beta$ (polynômes de Gegenbauer) H. Heyer nous a indiqué que notre formule de Levy-Khintchine pouvait se déduire du théorème 5.1 de [9] qui décrit les semi-groupes de convolution sur X.

BIBLIOGRAPHIE

[1] W. R. BLOOM et H. HEYER, The Fourier transform for probability measures on hypergroups. Rendiconti di Matematica (2), 1982, Vol. 2, p. 315-334.

[2] W. R. BLOOM et H. HEYER, Convergence of convolution products of probability measures on hypergroups. Rendiconti di Matematica (3), 1982, Vol. 2, p. 547-563.

[3] P. BOUGEROL, Un mini cours sur les couples de Guelfand. Publications du Laboratoire de Statistique et Probabilités, Université Paul Sabatier, Toulouse (1983).

[4] J. L. CLERC et B. ROYNETTE, Un théorème central limite. Analyse Harmonique sur les groupes de Lie II. Lecture Notes n° 739 (1979), p. 122-132.

[5] C. F. DUNKL, The measure algebra of a locally compact hypergroup. Trans. Amer. Math. Soc. 179 (1973), p. 331-348.

[6] Y. GUIVARC'H, M. KEAN et B. ROYNETTE, Marches aléatoires sur les groupes de Lie. Lecture Notes in Math. n° 624, (1977).

[7] E. HYLLERAS, Linearization of products of Jacobi polynomials. Math. Scand. n° 10 (1962), p. 189-200.

[8] R. I. JEWETT, Spaces with an abstract convolution of measures. Advances in Math. 18 (1975), p. 1-101.

[9] M. KENNEDY, A stochastic process associated with ultraspherical polynomials. Proc. Roy. Irish. Acad. 61 (1961), p. 89-100.

[10] R. LASSER, Orthogonal polynomials and hypergroups : Contributions to Analysis and Probability Theory. Preprint, Technische Universität München (1981).

[11] G. LETAC, Problèmes classiques de Probabilité sur un couple de Guelfand. Analytic Methods in Probability Theory. Lecture Notes in Math. 861 (1981), p. 93-127.

[12] G. LETAC, Dual Random walks and special functions on homogeneous trees. Publications de l'Institut Elie Cartan n° 7, Nancy (1983), p. 97-142.

[13] E. LUKACS, Characteristic functions. Second Edition, Griffin Editor, London (1970).

[14] K. R. PARTHASARATHY, Probability measures on metric spaces. Academic Press New-York-London (1967).

[15] W. RUDIN, Fourier analysis on groups. Interscience Publishers, Second Printing (1967).

[16] E. SIEBERT, A new proof of the generalized continuity theorem of Paul Levy. Math. Ann. 233 (1978), p. 257-259.

[17] R. SPECTOR, Aperçu de la théorie des hypergroupes. Analyse harmonique sur les groupes de Lie I. Lecture Notes in Math. n° 497, (1975), p. 643-673.

[18] R. SPECTOR, Mesures invariantes sur les hypergroupes. Trans. Amer. Math. Soc. 239 (1978), p. 147-165.

CONTINUED FRACTION METHODS FOR RANDOM WALKS ON $\mathbb{N}$ AND ON TREES

Peter Gerl

1. Introduction

During the last years many results for random walks on groups have been found. Especially the class of free groups has been studied in detail (see e.g. [4], [5], [10], [15]). Local limit theorems for random walks on free groups (with two or more generators) always have the form

$$p^{(2n)}(e,e) \underset{(n\to\infty)}{\sim} c \cdot \frac{r^n}{n^{3/2}} ,$$

where $r < 1$ depends on the number of generators but the denominator $n^{3/2}$ is always the same. More generally all local limit theorems for random walks on groups proved as yet have the form

$$p^{(n)}(e,e) \underset{(n\to\infty)}{\sim} c \cdot \frac{r^n}{n^{a/2}} ,$$

where $0 < r \le 1$ and a is a nonnegative integer (e.g. [2], [11]).

Now the Cayley-graph of a free group F_s with s generators is a homogeneous tree of degree s (s edges through each vertex) and so random walks on free groups can be considered as random walks on homogeneous trees. Therefore it seems natural to study random walks on more general trees (or even graphs); this will be done in this paper.

In a random walk on a graph a (one-step) transition occurs from one vertex v to an adjacent vertex with probability $\frac{1}{d(v)}$, where d(v) is the degree of v (uniform distribution at every vertex).

We will be mainly interested in the following quantities: Let 0 be a vertex and

$p_{00}^{(n)}$ = probability to reach 0 in n steps if starting in 0

$f_{00}^{(n)}$ = probability to reach 0 for the first time in n steps if starting in 0.

To get properties of $p_{oo}^{(n)}$ and $f_{oo}^{(n)}$, we relate the random walk on the tree to a random walk on $\mathbb{N}_o$, the non-negative integers. Now the generating functions for these sequences can be expressed via continued fractions; results from the theory of continued fractions then yield often information about the generating functions, their behaviour in the complex plane and their singularities. Then the method of Darboux (or something similar) furnishes information about the asymptotic behaviour of $p_{oo}^{(n)}$ and $f_{oo}^{(n)}$. Most results in this paper are just translations of theorems about continued fractions into the random walk language.

For general results about continued fractions see [7] or [9] or [14].

2. Random walks on $\mathbb{N}_o$ and continued fractions

Let $\mathbb{N}_o = \mathbb{N} \cup \{0\} = \{0,1,2,...\}$. We consider random walks (= discrete Markoff chains) on $\mathbb{N}_o$, where 0 is reflecting and one-step transitions can only occur to neighboring states, i.e. the one-step transition probabilities are

$$P(0 \to 1) = p_o = 1$$

$$P(i \to i+1) = p_i, \quad P(i \to i-1) = q_i$$

with $\quad p_i + q_i = 1, \; 0 < p_i < 1 \quad (i = 1,2,...).$

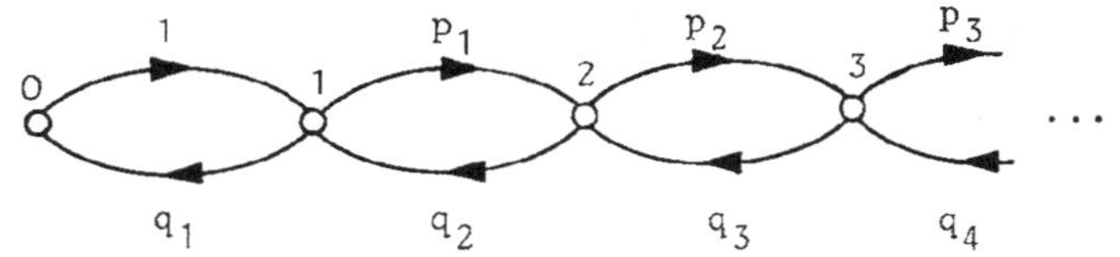

(2.1)

Let $\quad p_{ij}^{(n)}$ = probability to be in state j in n steps if starting in i

$f_{ij}^{(n)}$ = probability to be in state j for the first time in $n(\geq 1)$ steps if the first step is $i \to i+1$

and write for the corresponding generating functions

$$G_o(z) = \sum_{n=0}^{\infty} p_{oo}^{(2n)} z^n \qquad (p_{oo}^{(o)} = 1) \tag{2.2}$$

$$F_i(z) = \sum_{n=1}^{\infty} f_{ii}^{(2n)} z^n \qquad (i = 0,1,2,...). \tag{2.3}$$

The converge certainly for $|z| < 1$. Now we get easily the following relations (see e.g. [6] for this method):

$$G_o(z) = \frac{1}{1-F_o(z)}\ , \qquad (2.4)$$

$F_{i-1}(z)$:

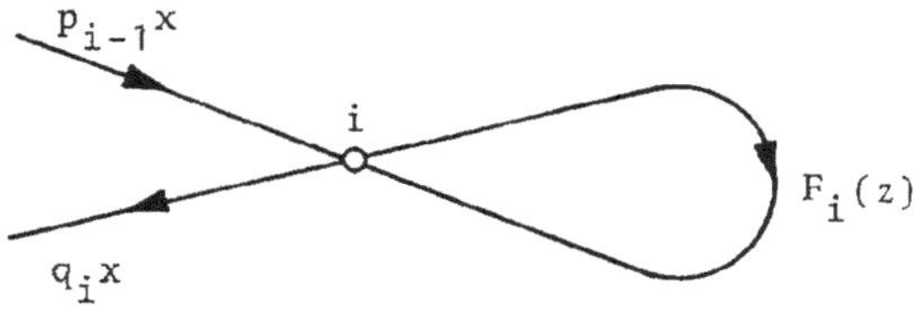

$$F_{i-1}(z) = \frac{p_{i-1}q_i\ z}{1-F_i(z)} \qquad (z = x^2),\ (i = 1,2,3,\ldots). \qquad (2.5)$$

Then (2.4) and (2.5) yield immediately the following continued fraction expansions

$$F_o(z) = \cfrac{q_1 z}{1 - \cfrac{p_1 q_2 z}{1 - \cfrac{p_2 q_3 z}{1 - \ldots}}} = \frac{q_1 z|}{|1} - \frac{p_1 q_2 z|}{|1} - \frac{p_2 q_3 z|}{|1} - \ldots \qquad (2.6)$$

$$G_o(z) = \cfrac{1}{1 - \cfrac{q_1 z}{1 - \cfrac{p_1 q_2 z}{1 - \ldots}}} = \frac{1|}{|1} - \frac{q_1 z|}{|1} - \frac{p_1\ q_2 z|}{|1} - \ldots \qquad (2.7)$$

These relations should be understood at first as formal equations but well known theorems about continued fractions ([7] : 4.58 and 4.60, [9], [14]) give:

Equality holds in (2.6) and (2.7) at least for $|z| < 1$. These continued fractions represent functions holomorphic in the cut z-plane $\mathbb{C}\setminus[1,\infty)$ and are the analytic continuation of F_o, G_o to this region. Therefore we get

Proposition 2.1: *The generating function* F_o *(and* G_o*) has on its circle of convergence only one singularity which is real (and* ≥ 1*).*

Unfortunately almost nothing is known about the type of this singularity; but in certain cases it can be seen to be a pole or a branch point (see below, 3.).

Now [14], Th. 11.1 shows that

$$F_o(1) = 1 - \frac{1}{1+\sum_{i=1}^{\infty} \frac{q_1 q_2 \cdots q_i}{p_1 p_2 \cdots p_i}} ; \tag{2.8}$$

from this we get the well-known result (e.g. [3]):

The random walk (2.1) *is recurrent (transient) iff* $F_o(1) = 1$ (<1) *iff* $\sum_{i=1}^{\infty} \frac{q_1 \cdots q_i}{p_1 \cdots p_i} = \infty$ $(< \infty)$.

We can be more precise: Call (2.1) the (p,q)-walk and the one obtained from this by interchanging p_i with q_i $(i = 1,2,3,\ldots)$ the (q,p)-walk. The generating function corresponding to (2.6) will be denoted by

$$F_o^{q,p}(z) = \frac{p_1 z |}{| 1} - \frac{q_1 p_2 z |}{| 1} - \frac{q_2 p_3 z |}{| 1} - \cdots . \tag{2.9}$$

Then we get from [14], formula (75.6)

$$F_o^{q,p}(z) = \frac{F_o(z) - z}{F_o(z) - 1} . \tag{2.10}$$

Since the (p,q)-walk is positive recurrent iff it is recurrent ($F_o(1) = 1$) and $F_o'(1) < \infty$ we get from (2.10):

If the (p,q)-walk is positive recurrent then by the rule of de l'Hospital:

$$F_o^{q,p}(1) = \lim_{z\to 1} \frac{F_o'(z)-1}{F_o'(z)} = 1 - \frac{1}{F_o'(1)} < 1 ,$$

i.e. the (q,p)-walk is transient. If the (p,q)-walk is transient, then the (q,p)-walk is recurrent; it cannot be null recurrent because this would give

$$F_o(1) = 1 - \frac{1}{\lim_{z\to 1} F_o^{q,p'}(z)} = 1 ,$$

and the (p,q)-walk would be recurrent. Alltogether we have therefore:

The (p,q)-walk is		The (q,p)-walk is
transient	iff	positive recurrent
null recurrent	iff	null recurrent
positive recurrent	iff	transient.

Combined with the result after (2.8) we obtain (e.g. [3]):

Proposition 2.2: *The random walk* (2.1) *is*

transient *iff* $\sum \frac{q_1\cdots q_i}{p_1\cdots p_i} < \infty$, $\left(\sum \frac{p_1\cdots p_i}{q_1\cdots q_i} = \infty\right)$

null recurrent *iff* $\sum \frac{q_1\cdots q_i}{p_1\cdots p_i} = \infty$, $\sum \frac{p_1\cdots p_i}{q_1\cdots q_i} = \infty$

positive recurrent *iff* $\left(\sum \frac{q_1\cdots q_i}{p_1\cdots p_i} = \infty\right)$, $\sum \frac{p_1\cdots p_i}{q_1\cdots q_i} < \infty$

(In the first and last case the series in parentheses diverge automatically).

Sufficient conditions for transience are given in

Proposition 2.3: *Each of the following conditions is sufficient for the transience of the random walk* (2.1):

(i) *for some* r *with* $0 < r \le \frac{1}{2}$:

$$q_1 \le 1-r, \; p_n q_{n+1} \le r(1-r) \quad (n = 1,2,\dots)$$

and at least one inequality is strict.

(ii) *one can write* $p_i = \frac{r_i}{s_i}$, $q_i = \frac{s_i - r_i}{s_i}$ *with* $0 < r_i < s_i$, *such that* $s_i \ge r_{i-1}(s_i - r_i) + 1 \quad (i = 1,2,3\dots)$

and at least one inequality is strict.

Proof: (i) Follows from [14], Th. 38.1 and 38.2.

(ii) Applying an equivalence transformation to the continued fraction (2.6) gives:

$$F_o(z) = \frac{(s_1 - r_1)z|}{|s_1} - \frac{r_1(s_2 - r_2)z|}{|s_2} - \dots$$

Taking $z = 1$ the result follows from [9], Satz 2.19.

Applying an equivalence transformation to (2.7) we can write

$$G_o(z) = \frac{1|}{|k_1} - \frac{z|}{|k_2} - \frac{z|}{|k_3} - \dots,$$

where $k_1 = 1$, $k_2 = \frac{1}{q_1}$, $k_3 = \frac{q_1}{p_1 q_2}$, $k_4 = \frac{p_1 q_2}{q_1 p_2 q_3}$, ...

Therefore [14], Ex. 17.2 gives the

Proposition 2.4: *For the random walk* (2.1) *we have*

$$(i)\quad \frac{p_{oo}^{(2n)}}{p_{oo}^{(2n-2)}} < \frac{p_{oo}^{(2n+2)}}{p_{oo}^{(2n)}} \qquad (n = 1,2,\ldots)$$

$$(ii)\quad \lim_{n\to\infty} \frac{p_{oo}^{(2n+2)}}{p_{oo}^{(2n)}} \quad \text{ex.} = R^{-1} < \infty .$$

If the random walk (2.1) is also recurrent Prop. 2.4 yields in combination with [8] immediately a strong ratio limit theorem.

3. Local limit theorems. Examples

We consider at first limit periodic continued fractions where the partial numerators tend geometrically fast to some limit.

Theorem 3.1: *Consider a random walk* (2.1) *such that*

$$|p_n q_{n+1} - \alpha| = \gamma_n Q^n, \qquad (n = 0,1,2,\ldots) \tag{3.1}$$

$$\alpha > 0, \quad 0 < Q < 1, \quad \sum \gamma_n < \infty.$$

Let R = *radius of convergence of* $G_o(z)$. *Then either*

$$(i)\quad 1 \le R < \frac{1}{4\alpha} \quad \text{and} \quad p_{oo}^{(2n)} \underset{(n\to\infty)}{\sim} c \cdot \frac{1}{R^n} \quad (c = \text{constant}) \tag{3.2}$$

or

$$(ii)\quad R = \frac{1}{4\alpha} \quad \text{and} \quad p_{oo}^{(2n)} \underset{(n\to\infty)}{\sim} c \cdot \frac{(4\alpha)^n}{n^{a/2}} \tag{3.3}$$

$$(c = \text{constant},\ a \in \mathbb{N}_o\setminus\{2\} = \{0,1,3,4,\ldots\})$$

Proof: Since the power series (2.2) of $G_o(z)$ has positive coefficients the point $z = R$ is a singularity of $G_o(z)$. It is evident that $R \ge 1$; now it is shown in [12], Thm. 4.2 that $z = \frac{1}{4\alpha}$ is a branch point of order 1 of the function represented by the continued fraction (2.7). Since this continued fraction is holomorphic in $\mathbb{C}\setminus[1,\infty)$ ([7], Cor. 4.60) and meromorphic in $\mathbb{C}\setminus[\frac{1}{4\alpha},\infty)$ ([7], Thm. 5.15) we get:

If $1 \le R < \frac{1}{4\alpha}$ the point $z = R$ is a pole of $G_o(z)$ and if $R = \frac{1}{4\alpha}$ the point $z = R$ is a branch point of order 1 of $G_o(z)$. We consider these two cases separately:

1) $1 \leq R < \frac{1}{4\alpha}$. For z close to R (= a pole) we may write

$$G_o(z) = \sum_{n=-k}^{\infty} c_n(R-z)^n, \qquad c_{-k} \neq 0, \quad k \geq 1.$$

The method of Darboux (see e.g. [1], Thm. 4) then gives

$$p_{oo}^{(2n)} \sim c \cdot \frac{1}{R^n n^{1-k}} \qquad (c = \text{constant}).$$

Since always $0 \leq R^n p_{oo}^{(2n)} < \infty$ (see e.g. [13]) the only possibility is $k = 1$.

2) $R = \frac{1}{4\alpha}$. We distinguish two subcases:

a) $G_o(\frac{1}{4\alpha}) = \infty$. Then by (2.4) we have $F_o(\frac{1}{4\alpha}) = 1$ and therefore $F_o(z)$ is bounded in $|z| \leq \frac{1}{4\alpha}$. F_o has also a branch point of order 1 at $z = \frac{1}{4\alpha}$ and so

$$F_o(z) = \sum_{n\geq 0} d_n(\frac{1}{4\alpha} - z)^{n/2} = 1 + (\frac{1}{4\alpha} - z)^{k/2}\, h(z),$$

where $k \geq 1$ and $h(\frac{1}{4\alpha}) \neq 0$. This gives

$$G_o(z) = -(\frac{1}{4\alpha} - z)^{-k/2}\, h_1(z)$$

and the method of Darboux yields

$$p_{oo}^{(2n)} \sim c \cdot \frac{(4\alpha)^n}{n^{1-k/2}} .$$

Since $k \geq 1$, the only possibilities (as in case 1) are $k = 1$ or 2.

b) $G_o(\frac{1}{4\alpha}) < \infty$. In this case we have

$$G_o(z) = \sum_{n\geq 0} e_n\, (\frac{1}{4\alpha} - z)^{n/2} = G_o(\frac{1}{4\alpha}) + e_s\, (\frac{1}{4\alpha} - z)^{s/2} + \dots ,$$

where $s \geq 1$ and $e_s \neq 0$. From this we derive

$$p_{oo}^{(2n)} \sim c \cdot \frac{(4\alpha)^n}{n^{1+s/2}} ,$$

which proves the theorem.

The case $\alpha = 0$, which was excluded in Theorem 3.1, can be handled similarly: If $\lim_{n\to\infty} p_n q_{n+1} = 0$, then $G_o(z)$ is meromorphic

([7], Th. 5.14). This in combination with Prop. 2.4 and the method of Darboux gives

Theorem 3.2: *If for the random walk* (2.1) *we have*

$$\lim_{n\to\infty} p_n q_{n+1} = 0,$$

then

$$p_{oo}^{(2n)} \underset{(n\to\infty)}{\sim} c \cdot \frac{1}{R^n}, \tag{3.4}$$

where $R \geq 1$. *(i.e. the random walk is* R-*positive in the terminology of* [13]).

Example 3.1: Let $0 < r \leq \frac{1}{2}$ and put

$$q_1 = r,\ q_n = \frac{r^n |}{| 1} - \frac{r^{n-1} |}{| 1} - \ldots - \frac{r^2 |}{| 1-r},\ p_n = 1 - q_n. \tag{3.5}$$

Then $p_{n-1}q_n = r^n$ $(n = 1,2,3,\ldots)$ and by (2.6)

$$F_o(z) = \frac{rz |}{| 1} - \frac{r^2 z |}{| 1} - \frac{r^3 z |}{| 1} - \ldots,$$

which is Ramanujan's continued fraction. From [9], § 24 and § 28 one gets

$$F_o(z) = 1 - \frac{H(-z)}{H(-rz)},$$

and further $H(z) = H(rz) + rzH(r^2 z)$, where $H(z)$ is an entire function. This last relation gives

$$F_o(z) = \frac{rz - F_o(rz)}{1 - F_o(rz)}. \tag{3.6}$$

Therefore $F_o(z) = 1$ is equivalent to $z = \frac{1}{r}$, so $G_o(z)$ has a pole at $z = \frac{1}{r}$ (≥ 2). Putting $G_o = \frac{1}{1-F_o}$ in (3.6) gives

$$G_o(z) = \frac{1}{1-rz} \quad \frac{1}{G_o(rz)}; \tag{3.7}$$

Since $0 < G_o(1) < \infty$, we see from (3.7) that $G_o(z)$ has a pole of order 1 at $z = \frac{1}{r}$. Darboux's method then yields

$$p_{oo}^{(2n)} \underset{(n\to\infty)}{\sim} c \cdot r^n$$

in agreement with Theorem 3.2. So the random walk given by (3.5) is $\frac{1}{r}$ - positive.

Example 3.2: Let $0 < a < c$ and put

$$q_{2n-1} = \frac{a+n-1}{c+2n-2}\ ,\quad q_{2n} = \frac{n}{c+2n-1}\ ,\quad p_n = 1 - q_n \quad (n = 1,2,\ldots) \tag{3.8}$$

Then (see [14], p. 340 and [9], § 28)

$$G_o(z) = F(a,1,c,z) = 1 + \frac{a}{c}z + \frac{a(a+1)}{c(c+1)}z^2 + \ldots,$$

the hypergeometric series of Gauss. Therefore

$$p_{oo}^{(2n)} = \frac{a(a+1)\ \ldots\ (a+n-1)}{c(c+1)\ \ldots\ (c+n-1)} \underset{(n\to\infty)}{\sim} \frac{\Gamma(c)}{\Gamma(a)} \cdot \frac{1}{n^{c-a}}$$

for the random walk (3.8). Although we have

$$\lim_{n\to\infty} p_n q_{n+1} = \frac{1}{4}\ ,$$

Theorem 3.1 is not applicable, since the convergence to the limit is not fast enough!

Example 3.3: Let $0 < r < 1$ and put

$$q_n = r,\ p_n = 1-r \quad (n = 1,2,\ldots) \tag{3.9}$$

Then a short calculation gives

$$G_o(z) = \frac{-1+2r+\sqrt{1-4r(1-r)z}}{2r(1-z)}\ .$$

Therefore (apply e.g. [1], Thm. 4) we have:
If $0 < r < \frac{1}{2}$:

$$p_{oo}^{(2n)} \sim \frac{1-r}{\sqrt{\pi}(1-2r)^2} \quad \frac{(4r(1-r))^n}{n^{3/2}}$$

if $r = \frac{1}{2}$:

$$p_{oo}^{(2n)} \sim \frac{1}{\sqrt{\pi}}\ \frac{1}{\sqrt{n}}$$

if $\frac{1}{2} < r < 1$:

$$p_{oo}^{(2n)} \sim 2 - \frac{1}{r}\ ;$$

this is of course well known.

<u>Example 3.3a:</u> Let $0 < r < 1$ and put

$$q_{2n-1} = 1-r, \quad q_{2n} = r, \quad p_n = 1 - q_n \quad (n = 1,2,\ldots). \tag{3.10}$$

This random walk can be obtained from the one in Example 3.3 by interchanging q_{2n-1} with $1-q_{2n-1}$. But then there is a formula for the corresponding generating functions resp. continued fractions ([14], (75.2)); namely if G_o is as in Example 3.3 and H_o the corresponding function for the random walk (3.10) then

$$H_o(z) = \frac{1}{1-z} \quad G_o\left(\frac{-z}{1-z}\right).$$

This gives

$$H_o(z) = \frac{-1+2r}{2r} + \frac{1}{2r}\sqrt{\frac{1-(1-2r)^2 z}{1-z}} .$$

Therefore $z = 1$ is a singularity of $H_o(z)$ and Darboux's method gives

$$p_{oo}^{(2n)} \underset{(n\to\infty)}{\sim} \frac{c}{\sqrt{n}} \qquad (c = \text{constant}),$$

so (3.10) is always a null-recurrent random walk.

4. Random walks on $\mathbb{N}$-trees

<u>Definition:</u> *An* $\mathbb{N}$-*tree is a tree with a root* 0 *such that all vertices at distance* n *from* 0 *have the same degree* d_n.

In a random walk on a graph we go from a vertex v to an adjacent one with probability $\frac{1}{d(v)}$ (where $d(v)$ = degree of v).

Now we see that a random walk on an $\mathbb{N}$-tree can be treated as a random walk on $\mathbb{N}_o$:

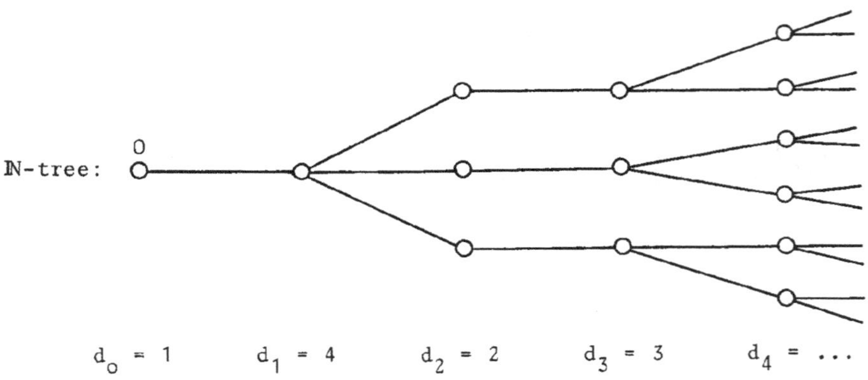

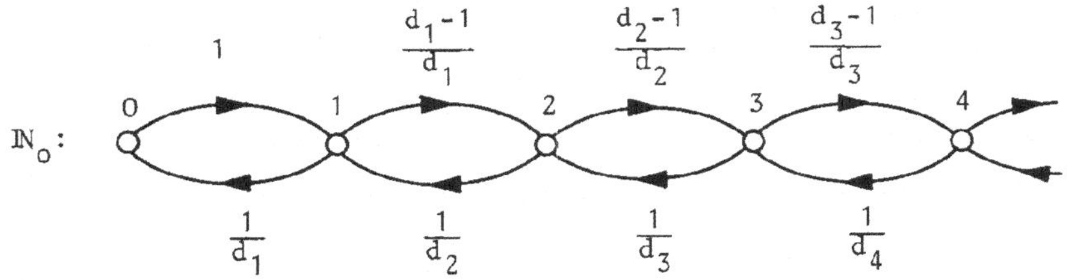

So if (d_n) is the degree sequence of an $\mathbb{N}$-tree then the random walk on it corresponds to the following random walk on $\mathbb{N}_o$:

$$q_o = 1, \quad q_n = \frac{1}{d_n}, \quad p_n = \frac{d_n-1}{d_n}, \qquad (n = 1,2,\ldots). \tag{4.1}$$

$(d_n = \text{integer} \geq 2)$.

Therefore all earlier results are applicable to random walks on $\mathbb{N}$-trees and we get in particular:

Proposition 4.1: *(i) No random walk on an $\mathbb{N}$-tree is positive recurrent.*
(ii) The random walk (4.1) *is recurrent iff* $\sum_{n=1}^{\infty} \frac{1}{(d_1-1)(d_2-1)\ldots(d_n-1)} = \infty$

Proof: (i) follows from Prop. (2.2), since

$$p_n = \frac{d_n-1}{d_n} \geq q_n = \frac{1}{d_n}.$$

(ii) follows from Prop. (2.2).

Example 4.1: Put $d_1 = d_2 = d_3 = \ldots = r (\in \mathbb{N})$, i.e. we consider a homogeneous tree of degree r:

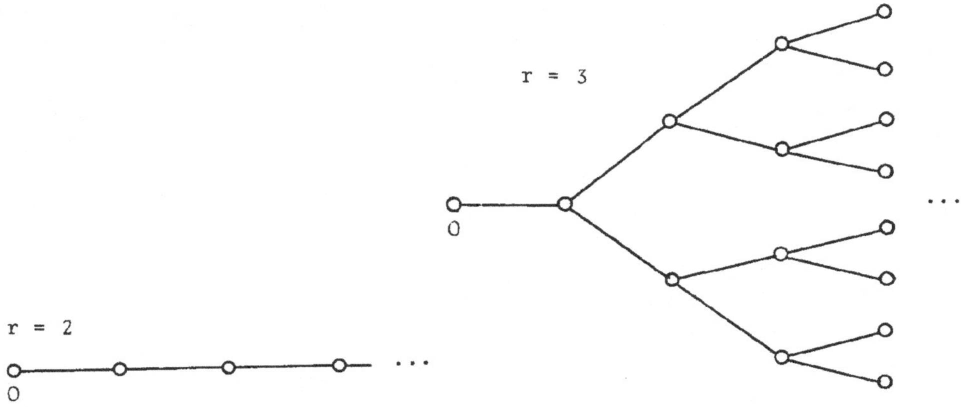

Then we obtain from Example (3.3) with r replaced by $\frac{1}{r}$:

$$p_{oo}^{(2n)} \sim \frac{1}{\sqrt{\pi}} \; \frac{1}{\sqrt{n}} \qquad \text{for } r = 2$$

$$p_{oo}^{(2n)} \sim \frac{1}{\sqrt{\pi}} \; \frac{r(r-1)}{(r-2)^2} \; \left(\frac{4(r-1)}{r^2}\right)^n \frac{1}{n^{3/2}} \qquad \text{for } r = 3,4,\ldots$$

(Since homogeneous trees of even degree represent the Cayley-graphs of free groups, compare these results with [4], [5])

More interesting is

Example 4.2: Put $d_1 = 2$, $d_2 = d_3 = \ldots = r \geq 3$, i.e. we consider a tree of the following kind:

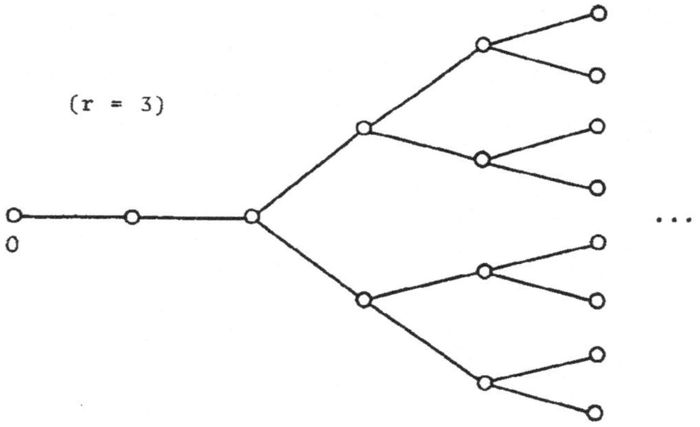

Some calculations give

$$G_o(z) = \frac{1|}{|1} - \frac{\frac{1}{2}z|}{|1} - \frac{\frac{1}{2r}z|}{|1} - \frac{\frac{r-1}{r^2}z|}{|1} - \frac{\frac{r-1}{r^2}z|}{|1} - \ldots$$

$$= \frac{4r - 3rz + 6z - 8 - z\sqrt{r^2 - 4(r-1)z}}{10z + 4r - 6rz - 8 + 2rz^2 - 2z^2} \, .$$

This function has radius of convergence z_o given by

$z_o = \frac{8}{9}$ for $r = 3$, $\quad z_o = \frac{r-1}{2r-4}$ for $r \geq 4$

and from this we find

$$p_{oo}^{(2n)} \sim c(\tfrac{8}{9})^n \frac{1}{n^{3/2}} \qquad \text{for} \quad r = 3 \qquad \text{(transient)}$$

$$p_{oo}^{(2n)} \sim c(\tfrac{3}{4})^n \frac{1}{n^{1/2}} \qquad \text{for} \quad r = 4 \qquad \text{(null-recurrent)}$$

$$p_{oo}^{(2n)} \sim c(\tfrac{r-1}{2r-4})^n \qquad \text{for} \quad r \geq 5 \qquad (\tfrac{2r-4}{r-1} \text{ - positive}).$$

Example 4.3: Put $d_1 = d_2 = 2$, $d_3 = d_4 = \ldots = r \geq 3$, i.e. we consider a tree of the following kind:

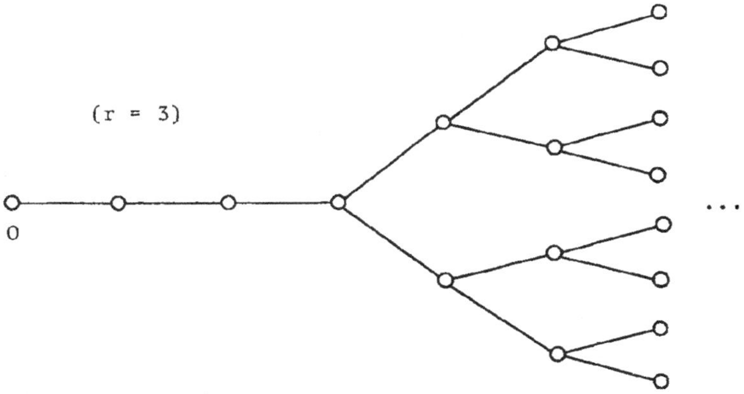

Then

$$G_o(z) = \frac{1|}{|1} - \frac{\frac{1}{2}\, z|}{|1} - \frac{\frac{1}{4}\, z|}{|1} - \frac{\frac{1}{2r}\, z|}{|1} - \frac{\frac{r-1}{r^2}\, z|}{|1} - \frac{\frac{r-1}{r^2}\, z|}{|1} - \ldots$$

$$= \frac{1}{1-F_o(z)}$$

and some calculations give

$$F_o(z) = \frac{2z(r-z+\sqrt{r^2-4(r-1)z}\,)}{4(r-z)-rz+(4-z)\sqrt{r^2-4(r-1)z}} \quad .$$

From this one finds that $F_o(z) = 1$ leads to

$$(z-1)(z^2+(6r-12)z+16-8r) = 0.$$

Since $F_o(1) = \frac{2r-3}{3r-5} < 1$ the only root z_o of this equation with $1 < z_o \le \frac{r^2}{4(r-1)}$ (this is by [12], Thm. 4.2 a singularity of $G_o(z)$) is given by

$$z_o = 6 - 3r + \sqrt{(9r-10)(r-2)} < \frac{r^2}{4(r-1)} .$$

Therefore $G_o(z)$ has a pole at z_o (because $F_o(z_o) = 1$ and $F_o(z)$ is holomorphic for $|z| < \frac{r^2}{4(r-1)}$) and we get from Thm. 3.1:

$$p_{oo}^{(2n)} \underset{(n\to\infty)}{\sim} c \cdot \frac{1}{z_o^n} ,$$

i.e. this random walk is z_o-positive.

If the degree sequence $d_n \to \infty$ $(n\to\infty)$ then results similar to this last example can be written down.

Example 4.4: Put $d_1 = 3$, $d_2 = d_3 = \ldots = 2$, i.e. consider the following tree:

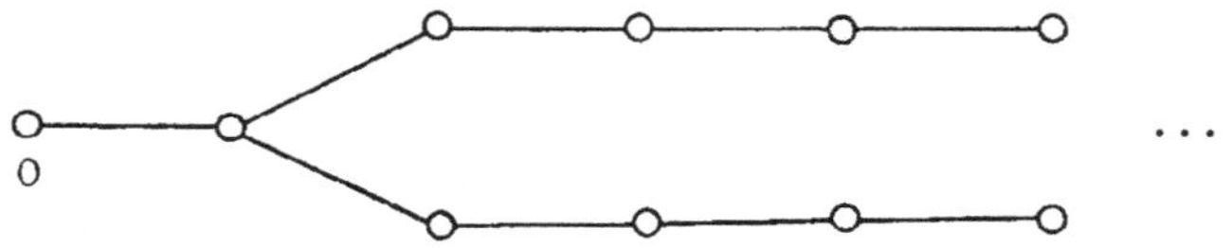

A simple calculation shows that

$$p_{oo}^{(2n)} \sim \frac{c}{n^{3/2}} .$$

The same random walk could be derived from the following graph:

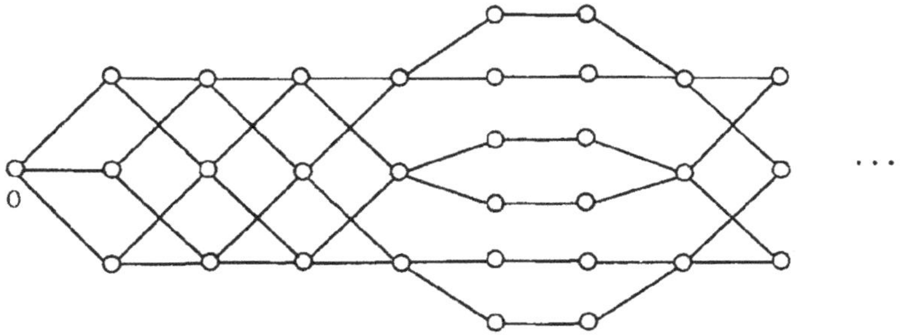

From this example one can see that random walks on different types of graphs can be treated as random walks on $\mathbb{N}_0$. Many variations of this kind are possible.

5. Outlook

Even the few results derived in this paper seem to indicate that it should be interesting to study random walks on trees and more generally on graphs. As the examples show one gets in most cases an R-transitive or an R-positive random walk. So let as pose as a

Problem: *Characterize R-positive trees (by some notions from graph theory).*

It seems so that a tree is R-transient if it is not too far away (what should this mean?) from a homogeneous tree. Can this be made precise?

Another question concerns local limit theorems. For example if $\alpha > 0$ is given does there exist a tree such that

$$p_{oo}^{(2n)} \sim c \frac{1}{R^n n^\alpha} \quad ?$$

Or what numbers α can appear in such a relation?

References

[1] Bender, E.A.: Asymptotic methods in enumeration. SIAM Review 16(1974), 485-515

[2] Bougerol, Ph.: Comportement asymptotique des puissances de convolution d'une probabilité sur un espace symétrique. Astérisque 74 (1980), 29-45

[3] Chung, K.L.: Markov chains, Springer-Verlag, Berlin 1967

[4] Gerl, P.: Irrfahrten auf F_2. Monatsh. Math. 84 (1977), 29-35

[5] Gerl, P.: Ein Gleichverteilungssatz auf F_2. In: Probability measures on groups, Lecture Notes in Math. 706 (1979), 126-130

[6] Howard, R.A.: Dynamic probabilistic systems. Vol.I. John Wiley & Sons, New York 1971

[7] Jones, W.B. - Thron, W.J.: Continued fractions. Addison-Wesley Publishing Comp.Reading, Mass. 1980

[8] Orey, S.: Strong ratio limit property. Bull. Am. Math. Soc. 67 (1961), 571-574

[9] Perron, O.: Die Lehre von den Kettenbrüchen. Bd. II. Teubner, Stuttgart 1957

[10] Sawyer, S.: Isotropic random walks in a tree. Z. Wahrscheinl. und Verw. Gebiete 42 (1978), 279-292

[11] Spitzer, F.: Principles of random walk. Van Nostrand, Princeton 1964

[12] Thron, W.J. - Waadeland, H.: Analytic continuation of functions defined by means of continued fractions. Math. Scand. 47 (1980), 72-90

[13] Vere-Jones, D.: Ergodic properties of nonnegative matrices I. Pac. J. of Math. 22 (1967), 361-386

[14] Wall, H.S.: Analytic theory of continued fractions. Van Nostrand, Toronto 1948.

[15] Woess, W.: A local limit theorem for random walks on certain discrete groups. In: Probability measures on groups. Lecture Notes in Mathematics 928 (1982), 467-477

Peter Gerl
Institut für Mathematik
Universität Salzburg
Petersbrunnstraße 19
A-5020 Salzburg/Austria

RESULTS IN SEMIGROUPS IN THE CONTEXT OF NON-HOMOGENEOUS MARKOV CHAINS: TAIL IDEMPOTENTS AND THEIR STRUCTURE FOR INFINITE-DIMENSIONAL NONNEGATIVE MATRICES

By

S.Gibert and A.Mukherjea

University of South Florida, Tampa, Florida 33620

1. This is an expanded version of the first part of a 3-part survey talk given by the second-named author in the conference. The second part of the talk is more or less contained in the second chapter of [7] and the third part in [4]. To avoid repetition, we present here more general versions of results that already appeared in print and were discussed during the talk.

Here we describe tail idempotents of a non-homogeneous Markov chain and their connections with the tail sigma-field of the chain. We also describe their structure and how they appear even for general nonnegative matrices. The results here have their origin in the following semigroup result of I. Csiszàr.

THEOREM 1 [2]. Let (x_n) be a sequence of elements in S, a compact Hausdorff first countable topological semigroup. Let us write for $n > k$, $x_{k,n}=x_{k+1}\cdots x_n$. Then, given any subsequence (n_t), there exists a further subsequence $(p_t) \subset (n_t)$ such that for each nonnegative integer k,

$$x_{k,p_t} \to y_k, \; y_{p_t} \to y_\infty = y_\infty y_\infty \text{ and } y_k y_\infty = y_k.$$

The assertion of the theorem remains valid also for non-compact first countable topological semigroups S, provided that S is a subspace (topologically) of a compact Hausdorff space S_0 and the limit points of the sequence (x_{k,n_t}), surely existing in S_0, also exist in S. □

Theorem 1 has been used effectively in the context of various problems in measures on semigroups and groups (see [3,7,9,10]. This result has been used in [9] to study certain aspects of non-homogeneous Markov chains. We now describe this briefly.

Let (X_n) be a non-homogeneous Markov chain with finite state space $E=\{1,2,\dots,m\}$. Write:

$$(P_n)_{ij}=Pr(X_n=j \mid X_{n-1}=i), \text{ when } Pr(X_{n-1}=i) > 0;$$
$$=a_{ij}(n), \text{ when } Pr(X_{n-1}=i)=0,$$

where $\sum_j a_{ij}(n) = 1$ and each $a_{ij}(n)$ is nonnegative. By Theorem 1, given any sequence (n_t) of positive integers, there exists $(p_t) \subset (n_t)$ such that for each nonnegative integer k,

$$P_{k,p_t} \to Q_k, \; Q_{p_t} \to Q=Q^2 \text{ and } Q_kQ=Q_k.$$

Here Q as well as each Q_k is an $m \times m$ stochastic matrix. The matrix Q is called the tail idempotent corresponding to (p_t). Though the tail idempotents can be different for different subsequences, they all have the same rank. Let p be this rank. Then it is well-known that there is a partition $\{T,C_1,C_2,\dots,C_p\}$ of E, called the basis of Q such that

$$Q_{ij}=0, \text{ whenever either } j \in T \text{ or } i \text{ and } j \text{ are in different C-classes };$$
$$=Q_{kj} \;(> 0), \text{ whenever } i,j \text{ and } k \text{ are in the same C-class.}$$

Let T_∞ be the tail sigma-field of (X_n). Then it is known ([1,9]) that for i,j in the same C-class of Q, the events $\{X_{p_t}=i \text{ i.o.}\}$ and $\{X_{p_t}=j \text{ i.o.}\}$ are equivalent events in T_∞, and each is an atom in T_∞ iff $\lim_{t\to\infty} Pr(X_{p_t}=i) > 0$. Moreover, the number of distinct atoms in T_∞ is precisely the number of such C-classes, and, these atoms generate T_∞; also,

$$Pr(X_{p_t}=j \text{ i.o.} \mid X_k=i) = [Q_k]_{ij}/Q_{jj},$$

whenever $Pr(X_k=i) > 0$ and $j \notin T$.

In this paper, we like to show that there is a more direct relationship between tail-idempotents and T_∞; more precisely, we will establish the existence of a continuous linear bijection from QB into $\mathcal{F}_\infty$, where $B=\{f:E \to [0,1]\} \subset L_\infty(E,\beta)$ with weak*-topology, $(Qf)(i)= \sum_j Q_{ij}f(j)$ for $f \in B$, $\beta(i)=\lim_{t\to\infty} \Pr(X_{p_t}=i)$, and $\mathcal{F}_\infty$ is the set of [0,1]-valued $\mathcal{T}_\infty$- measurable random variables with topology derived from the weak*-topology of L_∞- random variables. This problem becomes more interesting in the context when E is at most countable, possibly infinite. While we will do this in section 4, in sections 2 and 3 we tackle another two problems that come up quite naturally. We describe these here.

First, the conditions under which tail-idempotents (which are also stochastic matrices) exist are not at all clear from Theorem 1 in the case when E is infinite. In section 2, we describe reasonable conditions that ensure the existence of non-zero tail idempotents in the more general context of infinite-dimensional nonnegative matrices. In section 3, we will describe the structure of infinite nonnegative idempotent matrices. While this structure is known in the finite case ([8]), the finite case arguments do not carry over in the infinite case.

2. Tail-idempotents for infinite-dimensional nonnegative matrices:

Let (P_n) be a sequence of nonnegative matrices with state space the positive integers satisfying the following conditions:

(a) All products $P_{k,n}=P_{k+1}\cdots P_n$ are well-defined.

(b) There exists $M > 0$ such that $(P_{k,n})_{ij} < M$ for all i,j,k,n.

(c) There exists a subsequence (n_t) of positive integers such that for each nonnegative k, $P_{k,n_t} \to Q_k$ (poitwise) and for every $i \geq 1$, $\sum_{j=1}^{\infty} (Q_k)_{ij} < \infty$ and

$$\lim_{t\to\infty} \sum_{j=1}^{\infty} \left|(P_{k,n_t})_{ij} - (Q_k)_{ij}\right| = 0.$$

DEFINITION. We say that a sequence of nonnegative matrices A_n star-converges to the matrix A if $\lim_{n\to\infty} \sum_{j=1}^{\infty} |(A_n)_{ij}-A_{ij}| = 0$ and

$\sum_{j=1}^{\infty} A_{ij} < \infty$. □

All matrices in this section are infinite-dimensional and nonnegative.

LEMMA 1. Suppose that the sequence B_n star-converges to B and the sequence A_n converges to A pointwise such that for some $M > 0$, $(A_n)_{ij} < M$ for all i,j and n. Then the sequence B_nA_n converges pointwise to BA. □

LEMMA 2. Suppose that the sequence A_n star-converges to A and the sequence B_n converges to B pointwise such that for some $M > 0$, $(B_n)_{ij} < M$ for all i,j and n. Then the sequence B_nA_n converges to BA pointwise. □

LEMMA 3. Consider the sequence P_n satisfying the conditions (a), (b) and (c). Let Q' be a pointwise limit of the sequence (Q_{n_t}). Then for each nonnegative k,

$$Q_kQ'=Q_k. \qquad (1)$$ □

LEMMA 4. Consider (1). Suppose that there exist k,i and s such that $(Q_k)_{is} > 0$. Then, $\sum_{j=1}^{\infty} (Q')_{sj} < \infty$. □

LEMMA 5. If the j-th column of Q' (as in Lemma 3) is not all zeros, then the j-th row of Q' has a finite sum. □

THEOREM 2. Let the sequence P_n be as in Lemma 3. Let Q' be a pointwise limit of the Q_{n_t}'s. Let $T=\{j: Q'_{ij}=0$ for each i$\}$. If Q_k is not the zero matrix for some k, then Q' is also so and E-T is non-empty and Q', restricted to E-T, is a nonnegative idempotent matrix with no zero columns. If Q'' is another pointwise limit of the Q_{n_t}'s, then Q'' has the same ''T'' set as Q', and Q'' restricted to E-T is also idempotent with no zero columns; moreover,

$$Q''Q'=Q''. \qquad (2)$$ □

Proof. Suppose that Q_k is not the zero matrix for some k. By (1), Q' is non-zero and therefore, E-T is non-empty. Now for $i \notin T$

and $j \not\in T$, by Lemma 5, $\sum_{k=1}^{\infty} Q'_{ik} < \infty$, and therefore,

$$(Q'^2)_{ij} = \sum_{k=1}^{\infty} Q'_{ik}Q'_{kj} \leq M.\sum_{k=1}^{\infty} Q'_{ik} < \infty.$$

It is clear that Q'^2 is well-defined and $(Q'|_{T^c})\cdot(Q'|_{T^c}) = Q'^2|_{T^c}$.

Also, for $k < s$, we have:

$$Q_k = P_{k,s}Q_s. \qquad (3)$$

If $j \not\in T$, then there exists i such that $Q'_{ij} > 0$. Choose N such that $(Q_N)_{ij} > 0$. Choose s_o such that for $s \equiv n_t > s_o > N$,

$$(P_{N,s})_{ij} > d > 0. \qquad (4)$$

Given $\varepsilon > 0$. Choose K such that

$$\sum_{u=K+1}^{\infty} (Q_N)_{iu} < d\varepsilon. \qquad (5)$$

By (3), for s in (4) , we have (using (5)):

$$d\varepsilon > \sum_{u=K+1}^{\infty}(Q_N)_{iu} = \sum_{u=K+1}^{\infty} \sum_{v=1}^{\infty} (P_{N,s})_{iv}(Q_s)_{vu}$$

$$\geq (P_{N,s})_{ij}\cdot \sum_{u=K+1}^{\infty} (Q_s)_{ju}.$$

It follows from (4) that for sufficiently large t,

$$\sum_{u=K+1}^{\infty}(Q_{n_t})_{ju} < \varepsilon. \qquad (6)$$

Thus, we have proven that if $j \not\in T$ and $\lim_{t\to\infty} Q_{p_t} = Q'$ (pointwise), then

$$\lim_{t\to\infty} \sum_{k=1}^{\infty} \left| (Q_{p_t})_{jk} - Q'_{jk} \right| = 0. \qquad (7)$$

Now we establish that Q', restricted to $E-T$, is idempotent. To prove this, let $i \not\in T$, $j \not\in T$. Let $\varepsilon > 0$. Choose K so that

$$\sum_{u=K+1}^{\infty} Q'_{iu} < \varepsilon/2M \quad\text{and}\quad \sum_{u=K+1}^{\infty} (Q_s)_{iu} < \varepsilon/2M$$

for $s \equiv n_t$ larger than some s_o. Then,

$$\left|Q'_{ij} - (Q'^2)_{ij}\right| = \lim_{t\to\infty} \left|(Q_{p_t}Q')_{ij} - (Q'^2)_{ij}\right|$$

$$\leq \lim_{t\to\infty} \sum_{u=1}^{K} \left|(Q_{p_t})_{iu} - Q'_{iu}\right| \cdot Q'_{uj} + \lim_{t\to\infty} \sum_{u=K+1}^{\infty}(Q_{p_t})_{iu}Q'_{uj} +$$

$$\sum_{u=K+1}^{\infty} Q'_{iu}Q'_{uj} < 3\varepsilon. \qquad \square$$

3. Structure of infinite-dimensional nonnegative idempotent matrices:

Let $Q=Q^2$ be a non-zero nonnegative matrix with state space E= the set of positive integers. Define the set S as

$$S=\{j \in E: \text{ either the } j\text{-th row of } Q \text{ is all zeros or the } j\text{-th column of } Q \text{ is all zeros }\}.$$

Since $Q_{ij}=\sum_{k\in E} Q_{ik}Q_{kj}$, it is clear that $E-S$ is non-empty, Q being non-zero. Let P be the restriction of Q on $E-S$. Then, $P=P^2$. Also, P has no zero rows or columns. (Notice that if $P_{ij}=0$ for some $i \notin S$ and each $j \in E-S$, then there is a k in S such that $Q_{ik}>0$; but, this is impossible since $Q_{ik}=\sum_{s\notin S} Q_{is}Q_{sk}$. Similarly, P cannot have a zero column.) First, let us describe the structure of P. We claim that $E-S$ can be partitioned into disjoint classes $\{C_1,C_2,\ldots\}$ such that

(i) $P_{ij}>0$ iff i and j are in the same C-class;

(ii) P restricted to any single C-class has rank one with one as the sum of its diagonal entries.

We carry out the proof of this claim in several steps.

STEP I. For each $i \in E-S$, $P_{ii}>0$. (8)

Proof of Step I. Suppose, if possible, that $P_{ii}=0$ for some $i \notin S$. With no loss of generality, we can assume that $1 \notin S$ and $P_{11}=0$. Define the set $A=\{j \in E-S: P_{1j}>0\}$. Since P has no zero row or columns, A is non-empty. Also, S^c-A contains 1. For $j \in S^c-A$,

$$0=P_{1j}=\sum_{k\in A} P_{1k}P_{kj},$$

and therefore, $P_{kj}=0$ for $k \in A$ and $j \in S^c-A$. This means that P and P^2 are of the forms

P:

	S^c-A	A
S^c-A	P_1	P_2
A	0	P_3

P^2:

	S^c-A	A
S^c-A	P_1^2	$P_1P_2+P_2P_3$
A	0	P_3^2

Since $P=P^2$, we have: $P_1=P_1^2$, $P_2=P_1P_2+P_2P_3$ and $P_3=P_3^2$. It follows that $P_1P_2=P_1(P_1P_2+P_2P_3) = P_1P_2+P_1P_2P_3$ or $P_1P_2P_3 = 0$. Notice that P_2P_3 is a $(S^c-A)\times A$ matrix; if there are $j \in S^c-A$ and $s \in A$ such that $(P_2P_3)_{js} > 0$, then choosing k such that $(P_1)_{kj} > 0$ (such k exists since P_1 has no zero columns), we see that

$$(P_1P_2P_3)_{ks} \geq (P_1)_{kj}(P_2P_3)_{js} > 0,$$

a contradiction. Thus, $P_2P_3=0$. Similarly, noting that P_3 has no zero rows since P has none, it follows that $P_2=0$. Now recall that $1 \in S^c-A$. By the definition of A, $P_{1j}=0$ for each j in S^c-A; this means that the first row of P is a zero row, a contradiction. This proves (8).

STEP II. For $i \in S^c$, $j \in S^c$,

$$P_{ij}=0 \text{ iff } P_{ji}=0. \qquad (9)$$

Proof of Step II. Suppose that $P_{1j}=0$ for some $j \in S^c$. Define the set $B=\{k \in S^c: P_{1k} > 0\}$. Then, $1 \in B$ (by step I) and $j \in S^c-B$. For $k \in S^c-B$, $0 = P_{1k}= \sum_{s\in B} P_{1s}P_{sk}$ so that $P_{sk}=0$ if $s \in B$ and $k \in S^c-B$. Thus, P and P^2 are of the forms

P:

	B	S^c-B
B	P_4	0
S^c-B	P_5	P_6

P^2:

	B	S^c-B
B	P_4^2	0
S^c-B	$P_5P_4+P_6P_5$	P_6^2

Since $P=P^2$, we have:

$$P_4=P_4^2,\ P_6=P_6^2 \text{ and } P_5=P_5P_4+P_6P_5.$$

Hence, $P_5P_4=(P_5P_4+P_6P_5)P_4=P_5P_4+P_6P_5P_4$ so that $P_6P_5P_4=0$. This means that $P_5P_4=0$, since if $s \in S^c-B$ and $t \in B$, choosing k such that $(P_6)_{ks} > 0$, we have : $(P_6)_{ks}(P_5P_4)_{st} \leq (P_6P_5P_4)_{kt} = 0$. By a similar argument, $P_5P_4=0$ leads to $P_5=0$. It is now clear from the form of P that $P_{j1}=0$. This establishes (9).

STEP III. Let $i \notin S$ and $C_i=\{j: P_{ij} > 0\}$. Then the $C_i \times C_i$ block of P is a positive idempotent matrix.

Proof of Step III. The proof is simple since for s and t in C_i, $P_{si} > 0$ (by step II) and $P_{st}=(P^2)_{st} \geq P_{si}P_{it} > 0$.

STEP IV. A positive idempotent matrix D has rank one.

Proof of Step IV. Write: $D_{ir}=a(i,r,k)D_{kr}$. Since for each r, $D_{kr} \geq D_{ki}D_{ir}$, it follows that

$$a(i,r,k)=[D_{ir}/D_{kr}] \leq 1/D_{ki}.$$

This means that

$$\sup_r a(i,r,k) = \beta(i,k) \leq 1/D_{ki} < \infty. \tag{10}$$

Also, notice that for any t,k and r,

$$D_{tr}/D_{kr} \geq D_{tk} > 0. \tag{11}$$

We now have:

$$\sum_t [\beta(i,k)-a(i,t,k)]D_{kt}D_{tr}$$
$$= \beta(i,k)D_{kr} - D_{ir} = [\beta(i,k)-a(i,r,k)]D_{kr}.$$

Therefore, for any t,

$$0 \leq [\beta(i,k)-a(i,t,k)]D_{kt}\cdot[D_{tr}/D_{kr}] \leq \beta(i,k)-a(i,r,k).$$

It follows by (11) that

$$0 \leq [\beta(i,k)-a(i,t,k)]D_{kt}D_{tk} \leq \inf_r\{\beta(i,k)-a(i,r,k)\} = 0.$$

Hence, $\beta(i,k)=a(i,t,k)$ for each t. Thus, rank (D) = 1.

It is clear from the above four steps that we can now partition E-S into disjoint equivalent classes $\{C_1,C_2,\ldots.\}$, where the equivalence relation ''r'' can be defined by i (r) j iff $P_{ij} > 0$, such that $P|_{C_i}$, for each i, is a positive rank one matrix and $P_{ij}=0$ iff i and j are in different C-classes.

For the matrix Q, that we started with, we can now state without proof the following assertions:

(a) if the i-th column of Q is all zeros, then for j and k in the same C-class of E-S,

$$Q_{ij}/Q_{jj} = Q_{ik}/Q_{jk} \ ;$$

(b) if the i-th row of Q is all zeros, then for j and k in the same C-class of E-S,

$$Q_{ji}/Q_{jj} = Q_{ki}/Q_{kj}.$$

4. Tail-idempotents and the tail sigma-field of a non-homogeneous Markov chain:

Let (X_n) be a non-homogeneous Markov chain with state space E, which is countable, possibly infinite. Let (P_n) be a sequence of stochastic matrices, associated with (X_n), as described in section 1. We make the following basic assumption:

''There exists a subsequence (n_t) such that for each nonnegative integer k, $P_{k,n_t} \to Q_k$ (pointwise) as $t \to \infty$, where each Q_k is a stochastic matrix.''

Then, by results of sections 2 and 3, it follows that if Q is a pointwise limit point of the Q_{n_t}'s, then $Q \neq 0$ and E can be partitioned into disjoint classes $\{T, C_1, C_2, \ldots\}$ such that

(i) $T = \{j \in E: Q_{ij} = 0$ for each $i \in E\}$;

(ii) E-T is nonempty;

(iii) Q, restricted to E-T, is a stochastic idempotent matrix with no zero columns and basis $\{C_1, C_2, \ldots\}$.

The matrix Q is called a tail-idempotent of (X_n). Notice that β defined by

$$\beta(i) = \lim_{t\to\infty} \Pr(X_{n_t} = i) = \sum_j \Pr(X_0 = j).\ (Q_0)_{ji}$$

is a probability measure on E. Consider the set

$$B = \{f: E \to [0,1]\} \subset L_\infty(E, \beta) \quad \text{(with weak*-topology)}.$$

We now make the following observation:

For f and g in $L_\infty(E,\beta)$, $f=g$ a.e.(β) => $Qf=Qg$ a.e.(β) and for $k \geq 0$, $Q_kf=Q_kg$ a.e.(β_k), where $\beta_k(i)=Pr(X_k=i)$. (12)

Proof of (12). Let $\beta(i) > 0$. Suppose that $f=g$ a.e.(β). It is easily verified that $Q_{ik} > 0 \Rightarrow \beta(k) > 0$. It follows immediately that $(Qf)(i)=(Qg)(i)$. For the second assertion, notice that when $\beta(j) = 0$ and $\beta_k(i) > 0$, then $(Q_k)_{ij}= \lim_{t\to\infty} Pr(X_{n_t}=j \mid X_k=i) = 0$; and, when $\beta(j) > 0$, then $f(j)=g(j)$. Thus, $Q_kf=Q_kg$ a.e.(β_k). □

Now we are going to use some of Kingman's ideas from [6] to establish that QB as well as $\mathcal{F}_\infty$ is a ''best target'' of a projective system of compact convex subsets and thus, they are ''isomorphic''.

Let $E_k=\{i \in E: Pr(X_k=i) > 0\}$. Consider the set

$$\mathcal{F}_{kk}=\{f:E_k\to [0,1]\} \subset L_\infty(E_k,\beta_k)$$

with weak*-topology. Define $P_{mn}: \mathcal{F}_{nn} \to \mathcal{F}_{mm}$ by

$$(P_{mn}f)(i)= \sum_j (P_{m,n})_{ij}f(j).$$

Write: $\mathcal{F}_{mn}=P_{mn}(\mathcal{F}_{nn}) \subset \mathcal{F}_{mm}$. Notice that $n < s \Rightarrow \mathcal{F}_{ms} \subset \mathcal{F}_{mn}$. The mappings P_{mn} are continuous, and therefore, the sets $\mathcal{F}_{mn}$, $n > m$, form a decreasing sequence of compact sets so that the set

$$\mathcal{F}_m= \cap\{\mathcal{F}_{mn}: n > m\}$$

is nonempty.

LEMMA 6 [6]. $P_{mn}(\mathcal{F}_n)=\mathcal{F}_m$. □

LEMMA 7. Q_k, as a map from $QB \subset L_\infty(E,\beta)$ with weak*-topology to $\mathcal{F}_k \subset L_\infty(E_k,\beta_k)$ with weak*-topology, is a continuous surjection. □

Proof. Let us prove only that $Q_k(QB)=Q_kB=\mathcal{F}_k$. Let $g \in Q_kB$. Then there is $h \in B$ such that $h(i)=0$ whenever $\beta(i)=0$ and $g=Q_kh$. For each $n_t > k$, define: $h_{n_t}(i)=h(i)$ if $\beta_{n_t}(i) > 0$, $=0$ otherwise. Then, $h_{n_t} \in \mathcal{F}_{n_tn_t}$ and $h_{n_t}(i) \to h(i)$, whenever $\beta(i) > 0$. Hence,

$$g=Q_kh= \lim_{t\to\infty} P_{k,n_t}h_{n_t} \in \mathcal{F}_{kn_t} \subset \mathcal{F}_{kn_{t-1}}.$$

It follows that $g \in \mathcal{F}_k$. Conversely, if $g \in \mathcal{F}_k$, there exist h_{n_t} in $\mathcal{F}_{n_t n_t}$ such that $g=P_{k,n_t} h_{n_t}$; if h is a pointwise limit point of (h_{n_t}), then $h \in B$ and $g=Q_k h$. □

Kingman defined in [6] the map $P_{m,\infty}: \mathcal{F}_\infty \to \mathcal{F}_m$ by

$$P_{m,\infty} Z = E(Z \mid X_m).$$

Notice that $P_{m,\infty}Z=P_{m,n}P_{n,\infty}Z \in P_{mn}\mathcal{F}_{nn} = \mathcal{F}_{mn}$ for each $n > m$, and thus, $P_{m,\infty}Z \in \mathcal{F}_m$. It is also shown in [6] that $P_{m,\infty}$ is a continuous surjection.

The system $(P_{mn}, \mathcal{F}_n)$ form a projective system of compact convex sets and the following triangle

$$\mathcal{F}_m \xleftarrow{P_{mn}} \mathcal{F}_n, \quad \mathcal{F}_s \xrightarrow{P_{ms}} \mathcal{F}_m, \quad \mathcal{F}_s \xrightarrow{P_{ns}} \mathcal{F}_n \qquad m < n < s, \quad P_{ms}=P_{mn}P_{ns}$$

commutes. The set QB is a ''target'' of this system in the sense that the following diagram

$$\mathcal{F}_m \xleftarrow{P_{mn}} \mathcal{F}_n, \quad QB \xrightarrow{Q_m} \mathcal{F}_m, \quad QB \xrightarrow{Q_n} \mathcal{F}_n \qquad Q_m=P_{mn}Q_n$$

commutes. In this sense, $\mathcal{F}_\infty$ is also a target of this system as $P_{m,\infty}=P_{mn}P_{n,\infty}$. Kingman [6] has shown that the set $\mathcal{F}_\infty$ is actually a ''best target'' or the projective limit of the system in the sense that for any other ''target'' U of the system, there is a map $\pi: U \to \mathcal{F}_\infty$ such that the diagram

$$\mathcal{F}_m \xleftarrow{\pi_m} U, \quad U \xrightarrow{\pi} \mathcal{F}_\infty, \quad \mathcal{F}_\infty \xrightarrow{P_{m,\infty}} \mathcal{F}_m \qquad P_{m,\infty}\pi=\pi_m$$

commutes. Our aim here is to show that the set QB is also a ''best target'' of the projective system in the sense that QB and $\mathcal{F}_\infty$ are isomorphic.

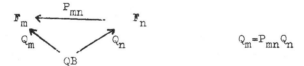

Consider the map G: QB $\rightarrow F_\infty$ defined by

$$G(Qf) = \lim_{t\to\infty} (Q_t f)(X_t). \tag{13}$$

Notice that G is well-defined since because of (12), for each t, $(Q_t f)(X_t) = (Q_t g)(X_t)$ almost surely whenever f=g a.e.(β), and moreover, the limit in (13) exists by a martingale convergence theorem.

THEOREM 3. The mapping G is a continuous bijection. □

Proof. First, we observe that given $\varepsilon > 0$, there exists a subsequence $(p_t) \subset (n_t)$ such that

$$\Pr(D_\varepsilon) > 1-\varepsilon, \tag{14}$$

where $D_\varepsilon = \{X_{p_t} \in E-K \text{ for each } t\}$ and $K = \{i \in E: \beta(i)=0\}$.

Suppose that $Qf_n \rightarrow Qf$ in $B \subset L_\infty(E,\beta)$ with weak*-topology. Then if $\beta(i) > 0$, $(Qf_n)(i) \rightarrow (Qf)(i)$ as $n \rightarrow \infty$. It can be easily verified that for $\beta(i) > 0$,

$$\lim_{n\to\infty} [\lim_{t\to\infty} Q_{p_t}(f_n)(i)] = \lim_{t\to\infty} Q_{p_t}(f)(i). \tag{15}$$

(The reason for (15) is that when $i \notin T$, the i-th row of Q is a probability vector and $Q_{p_t} \rightarrow Q$ pointwise.)

Now let y be any L_1- random variable. Given $\varepsilon > 0$, there exists $h > 0$ such that

$$\int_{D_h^c} |y|\, dP < \varepsilon. \tag{16}$$

Notice that by (15) and the dominated convergence theorem,

$$\lim_{n\to\infty} \{ \int_{D_h} [\lim_{t\to\infty} (Q_{p_t} f_n)(X_t)]\, ydP - \int_{D_h} [\lim_{t\to\infty} (Q_{p_t} f)(X_t)] ydP\}$$

is zero. The continuity of G follows easily.

To prove that G is one-one, let f,g $\in$ B such that G(Qf)=G(Qg) almost surely. Let $\beta(i) > 0$. Then, $\Pr(A) > 0$, where $A = \{X_{n_t} = i \text{ i.o.}\}$. This means that there exists $w \in A$ such that

$$\lim_{t\to\infty} (Q_{n_t} f)(X_{n_t}(w)) = \lim_{t\to\infty} (Q_{n_t} g)(X_{n_t}(w)).$$

Thus, $(Qf)(i)=(Qg)(i)$ whenever $\beta(i) > 0$.

To prove that G is a surjection, let $Z \in \mathcal{F}_\infty$. Then, $P_{t,\infty}Z = E(Z \mid X_t)$. Hence, by a martingale convergence theorem,

$$\lim_{t\to\infty} (P_{t,\infty}Z)(X_t) = Z \text{ a.s.}$$

By the definition of $P_{t,\infty}$ and Lemma 7, for each t, $P_{t,\infty}Z \in Q_tB$ and therefore, there exist $\theta_t \in B$ such that $Q_t\theta_t = P_{t,\infty}Z$. Now let $\varepsilon > 0$. Because of (14), there exists a subsequence $(q_t) \subset (n_t)$ such that

(i) $\theta_{q_t} \to \theta$ pointwise;

(ii) $\Pr(D_\varepsilon) > 1-\varepsilon$, where $D_\varepsilon = \{X_{q_t} \in$ the support of β for each t $\}$.

Then, on D_ε,

$$G(Q\theta) = \lim_{t\to\infty} Q_{q_t}(\theta)(X_{q_t})$$

$$= \lim_{t\to\infty} Q_{q_t}(\theta_{q_t})(X_{q_t})$$

$$= \lim_{t\to\infty} (P_{q_t,\infty}Z)(X_{q_t}) = Z.$$

Since $\varepsilon > 0$ is arbitrary, the surjection property follows. □

It is relevant to point out here that one easy consequence of Theorem 3 is that the tail-sigma field T_∞ is atomic whenever there exists a tail-idempotent (non-zero) and T_∞ is generated by the atoms $\{X_{n_t}=j \text{ i.o.}\}$, where $\beta(j) > 0$. Also, the events $\{X_{n_t}=j \text{ i.o.}\}$ and $\{X_{n_t}=k \text{ i.o.}\}$ are equivalent atoms in $\mathcal{T}_\infty$, whenever j and k belong to the same C-class of Q and $\beta(j) > 0$.

REFERENCES

1. Cohn, H. (1982). Math. Proc. Cambridge Philo. Soc., 528-534.
2. Csiszar, I. (1966). Z. Wahrscheinlichkeitstheorie v. Geb. 5, 279-295.
3. Hofmann, K.H. and A. Mukherjea. (1981). Math. Ann. 256, 535-548.
4. Högnäs, G. and A. Mukherjea. (1983). Springer-Verlag Lecture Notes in Math., this volume.

5. Iosifescu, M. (1972). Z. Wahrscheinlichkeitstheorie v. Geb.24, 159-166.
6. Kingman, J.F.C. (1975). Math. Proc. Cambridge Philo. Soc. 77, 171-183.
7. Mukherjea, A. (1979). Probabilistic analysis and related topics, Vol. 2, 143-203.
8. Mukherjea, A. (1980). Trans. Amer. Math. Soc. 262, No. 2, 505-520.
9. Mukherjea, A. (1983). Math. Zeitschrift 183, 293-309.
10. Ruzsa, Imre Z. (1981). Preprint of the Math. Inst. of the Hungarian Academy of Sciences, 1-17.

EXPOSANTS CARACTERISTIQUES DES PRODUITS DE MATRICES ALEATOIRES EN DEPENDANCE MARKOVIENNE

Y. CUIVARC'H

On s'intéresse ici à la relation de récurrence dans $\mathbb{R}^d$: $Y_{n+1} = X_{n+1} Y_n$ où X_n est une suite de matrices aléatoires inversibles gouvernée par une chaîne de Markov et on précise en particulier les modes de croissance de la longueur de Y_n suivant la condition initiale Y_o. Plus précisément on considère l'espace vectoriel réel $V = \mathbb{R}^d$ muni du produit scalaire canonique, une chaîne de Markov d'espaces d'états X et de noyau de transition P, une application borélienne f de X dans le groupe linéaire $G\ell(V)$ et la suite X_n définie par $X_n = f(x_n)$ où (x_n) est une trajectoire de la chaîne.

Si l'on considère une probabilité π sur X invariante par la chaîne de noyau P ($\pi P = \pi$) et la mesure P_π invariante par translation correspondante sur l'espace des trajectoires $\Omega = X^Z$, le théorème ergodique multiplicatif [16],[17]fournit l'existence p.p. d'exposants caractéristiques (dits de Liapunoff) $\gamma_1 \leq \gamma_2 \leq \ldots \leq \gamma_d$ caractérisant le mode de croissance de $||Y_n||$ suivant le choix de la condition initiale $Y_o = v$ dans un certain sous-espace $V_i(\omega)$, dès que $\text{Log } ||f(x)||$ est intégrable. Ecrivant $S_n = X_n X_{n-1} \ldots X_1$ et $Y_n = S_n v$, le plus grand exposant γ_d est par exemple donné par $\gamma_d = \lim_n \frac{1}{n} \text{Log } ||S_n||$ et une suite aléatoire de sous-espaces emboîtés $V_i(\omega)$ est définie par la formule $V_i = \{v \in V \text{ ; } \lim_n \frac{1}{n} \text{Log } ||S_n(v)|| \leq \gamma_i\}$

On pourra se reporter à [13] pour un exposé d'ensemble de ces notions et de leurs applications. Au point de vue des applications (cf. par exemple [3] et [7]) l'aspect aléatoire de X_n peut être rattaché à un phénomène de propagation en milieu inhomogène ou bien à un bruit [1]. Le résultat principal de ce travail dit que, si la chaîne de Markov est "suffisamment ergodique" et si la fonction f prend "suffisamment de valeurs", les γ_i sont tous distincts et donc les V_i sont de dimension i. La situation peut donc être comparée à celle d'une relation de récurrence à coefficients constants $Y_{n+1} = X Y_n$ où la matrice X serait diagonale à coefficients distincts en module.

Si les matrices X_n sont de déterminant 1 et si $d \geq 2$, on voit en raison de la relation $\sum_{i=1}^{d} \gamma_i = \lim_n \frac{1}{n} \operatorname{Log}|\det S_n|$ que $\gamma_d > 0$, ce qui s'interprète comme une croissance exponentielle de la norme de S_n. Ce fait a été justifié dans le cas indépendant en [4] et dans le cas Markovien en [9],[19],[22]. On voit donc que des propriétés nouvelles apparaissent en dimension supérieure à 1 et cet aspect multidimensionnel jouera ici un rôle essentiel. La méthode ici développée utilise la chaîne auxiliaire sur $X \times V$ de trajectoires $(x_n, S_n v)$; du fait de la linéarité elle admet une projection naturelle sur $X \times \mathcal{V}$ où $\mathcal{V}$ est l'espace des directions de V et $||S_n v||$ apparaît comme une fonctionnelle multiplicative de cette chaîne dont l'étude préliminaire est donc essentielle ; ce travail préliminaire fait l'objet de la partie I et s'appuie sur les notions développées en [10], en particulier les propriétés de convergence du produit S_n^{-1} considéré comme application projective de $\mathcal{V}$, ce qui fournit une construction a priori des $V_i(\omega)$. Dans le cas indépendant des résultats analogues avaient été obtenus en [18] et [21] sous l'hypothèse que la loi de X_1 admette une densité.
On pourra trouver en [10] une version condensée des résultats ici obtenus et pour des développements s'appuyant sur ce type de résultat on pourra consulter [11],[12],[14],[21].

I - CONVERGENCES EN DIRECTION -

Rappelons d'abord quelques notions développées en [5] et introduisons diverses notations. Soit $\mathcal{V}$ l'espace projectif associé à V, c'est-à-dire l'ensemble des directions de V. Chaque élément u de $G\ell(V)$ définit par passage au quotient une application projective de $\mathcal{V}$ notée encore u, l'image de $x \in \mathcal{V}$ étant notée u.x.

L'intérêt de $\mathcal{V}$ réside ici dans ses propriétés de compacité ; il est utile de considérer pour un endomorphisme u de V, l'application "quasi-projective" définie par passage au quotient en dehors du sous-espace des directions contenues dans le noyau de u.

En effet on peut alors, de toute suite u_n d'applications projectives, extraire une sous-suite convergeant simplement, en dehors d'un sous-espace projectif, vers une application quasi-projective. On dira qu'une mesure m sur $\mathcal{V}$ est propre si elle ne charge pas de sous-espace projectif. Considérons un semi-groupe $T \subset G\ell(V)$ et disons que T est totalement irréductible s'il ne laisse invariante aucune réunion finie de sous-espaces.

Disons aussi que T est proximal sur $\mathcal{V}$ [5] si pour tout couple x,y de $\mathcal{V}$ il existe une suite $t_n \in T$ telle que $\lim_n t_n.x = \lim_n t_n.y$. Cette condition est par exemple réalisée (et on peut voir que c'est le cas essentiel), si T contient une matrice diagonale dont les coefficients sont de modules distincts et si de plus T ne laisse pas de sous-espace invariant. Il en est clairement ainsi si l'image de T dans le groupe projectif est topologiquement dense. Mais ces propriétés sont vraies de manière nettement plus générale : par exemple si T est un sous-groupe discret de covolume fini du groupe $S\ell(V)$ des matrices unimodulaires il en est bien ainsi d'après la propriété de Selberg et le théorème de densité de Borel [5] .

L'utilisation des applications quasi-projectives permet de montrer [6] que si T est proximal sur $\mathcal{V}$ et si m est une mesure de probabilité, il existe une suite t_n de T telle que $t_n.m$ converge (vaguement) vers une mesure de Dirac.

On supposera le noyau de transition P de la chaîne sur X défini par une fonction k : $P(x,dy) = k(x,y)\ \pi(dy)$ où π est invariante sous P et k strictement positive $\pi \times \pi$ p.p. ; en particulier, on a $\int k(x,y)\ d\pi(x) = 1$

On supposera de plus que $\operatorname{Sup}_x k(x,y)$ est π-intégrable et l'ensemble de ces conditions sera noté A dans la suite. Il est clair que A implique la constance des fonctions P-invariantes. On note S_π le support de $f(\pi)$ et T_π le semi-groupe fermé engendré par S_π. Dans la suite de ce paragraphe interviennent des noyaux de transition $x \to \lambda_x$ de X dans $\mathcal{V}$ vérifiant l'équation

(1) $\lambda_x = \int f(y)\ \lambda_y\ P(x,dy).$

On peut obtenir un tel noyau en considérant une probabilité λ sur $X \times \mathcal{V}$ de projection π, invariante sous le noyau P_1^* défini par

$P_1^*\ \Phi(x,v) = \int \Phi\,[y, f(x)\cdot v]\ \ k(y,x)\,dy$ et en désintégrant λ sous la forme

$\lambda = \int \delta_y \times \lambda_y \quad d\,\pi(y).$

Si l'on munit l'ensemble des probabilités sur $X \times \mathcal{V}$ de projection π de la topologie de la convergence faible sur les fonctions $\Phi(x,v)$ continues en v, telles que $\operatorname{Sup}_{v \in \mathcal{V}} |\Phi(x,v)|$ soit dans $\mathbb{L}^1(\pi)$, on obtient un convexe compact invariant sous P_1^* et ceci fournit l'existence de λ d'après le théorème de Markov-Kakutani.

On a alors le :

Théorème 1.

Supposons que T_π *opère de manière totalement irréductible et proximale sur* $\mathcal{V}$ *et que* P *vérifie la condition* A. *Soit* $x \to \lambda_x$ *un noyau vérifiant l'équation* 1 . *Alors la suite de probabilités* $X_1 X_2 \ldots X_n \lambda_{X_n}$ *est une martingale convergeant p.p. vers une mesure de Dirac* $\delta_{Z(\omega)}$.

Le noyau λ *solution de* (1) *est unique et si* m *est une mesure propre sur* $\mathcal{V}$, *la suite* $X_1 X_2 .. X_n m$ *converge aussi vers* $\delta_{Z(\omega)}$.

On établit d'abord 3 lemmes.

Lemme 1.

Si le noyau P *vérifie la condition* A , *il existe une probabilité* ν *sur* $\mathcal{V}$ *et une constante* C *telles que* (π - *p.p.*) *les mesures* λ_x *soient équivalentes à* ν *et que, de plus* $\lambda_x \leq C\nu$.

Preuve.

Posons $K(y) = \sup_x k(x,y)$ et $C = \int K(y)\, d\,\pi(y)$.

Alors $\lambda_x = \int f(y)\, \lambda_y\, k(x,y)\, d\,\pi(y) \leq \int f(y)\, \lambda_y\, K(y)\, d\,\pi(y)$.

La probabilité ν est alors définie par $\nu = \frac{1}{C} \int f(y)\, \lambda_y\, K(y)\, d\,\pi(y)$ et l'on a :

$\lambda_x \leq C\nu$.

Pour voir l'équivalence de λ_x et ν il suffit d'observer que la condition

$0 = \lambda_x(A) = \int f(y)\, \lambda_y(A)\, k(x,y)\, d\,\pi(y)$ implique (p.p.) $f(y)\, \lambda_y(A) = 0$ puisque $k>0$

et donc : $\nu(A) = \frac{1}{C} \int f(y)\, \lambda_y(A)\, K(y)\, d\pi(y) = 0$.

Lemme 2.

La mesure ν *définie au lemme 1 est propre dès que* T_π *est totalement irréductible.*

Preuve :

Soit μ une probabilité sur $\mathcal{V}$ et considérons la famille des sous-espace H de dimension minimum tels que $\mu(H) > 0$. Observons que si $H \neq H'$ on a $\dim H \cap H' < \dim H$ et donc $\mu(H \cap H') = 0$. Il en découle que la famille des $\mu(H)$ est sommable et donc que, pour tout δ, l'ensemble des H tels que $\mu(H)>\delta$ est fini. On en déduit finalement l'existence de $H°$ tel que $\mu(H°)$ soit la borne supérieure des $\mu(H)$. On désignera par $L°$ la réunion finie des sous-espaces $H°$ où $\mu(H)$ est maximum.

Notons $H°_x$ et $L°_x$ les éléments associés à λ_x et $H°$, $L°$ ceux associés à ν.

Puisque λ_x est équivalente à ν, il est clair que $\dim H°_x = \dim H°$. Considérons l'équation $\lambda_x(H°_x) = \int f(y)\,\lambda_y\,(H°_x)\,P(x,dy)$ et notons que $f(y)\,\lambda_y\,(H°_x) \leq \lambda_y(H°_y)$.

La fonction $\psi(x) = \lambda_x(H°_x)$ vérifie donc $\psi(x) \leq \int \psi(y)\,P(x,dy)$.

D'après la condition A on en déduit :

$\psi(x) = cte = f(y)\,\lambda_y(H°_x) = \lambda_y(H°_y)$.

On a donc $f(y)^{-1}\,H°_x \subset L°_y$ et $f(y)^{-1}\,L°_x \subset L°_y$ $(\pi \times \pi$ p.p.).

On déduit de cette relation que le nombre de sous-espaces constituant $L°_x$ est constant, puis que $L°_x = f(y)\,L°_y$.

Comme les deux membres sont indépendants, on en conclut $L°_x = L°$

$f(y)\,L°_y = f(y)\,L° = L°$.

Si T_π est totalement irréductible, la dernière relation implique $L° = \mathcal{V}$

Lemme 3.

Soit g_n *une suite d'applications projectives convergeant en dehors d'un sous-espace projectif* H *vers une application quasi-projective* ζ, μ_n *une suite de probabilités convergeant vers une probabilité* μ *telle que* $\mu(H) = 0$. *Alors* $g_n\,\mu_n$ *converge vers* $\zeta\mu$.

Preuve.

Soit u continue comprise entre 0 et 1, égale à 1 sur un voisinage de H et nulle en dehors d'un voisinage plus grand. Alors si ψ est continue, on a, en posant $v = 1 - u$:

$$g_n\,\mu_n(\Phi) - \zeta\mu(\Phi) = \mu_n\,[\Phi \circ g_n - \Phi \circ \zeta] + (\mu_n - \mu)(\Phi \circ \zeta)$$

$$|g_n\,\mu_n(\Phi) - \zeta\mu(\Phi)| \leq ||\Phi||_\infty\,\mu_n(u) + ||(\Phi \circ g_n - \Phi \circ \zeta)v||_\infty + |(\mu_n - \mu)(\Phi \circ \zeta v)|$$

Le choix de u permet de rendre $\mu(u)$ petit donc $\mu_n(u)$ aussi pour n grand.

Comme $(\Phi \circ \zeta)\,v$ est continue, pour n grand le dernier terme peut être rendu petit. Enfin la petitesse du terme central résulte de la convergence uniforme de g_n vers ζ en dehors d'un voisinage de H.

Preuve du théorème.

La suite $X_1 \dots X_n \lambda_{x_n}$ est bien une martingale car $E_\pi(X_1 \dots X_{n-1} X_n \lambda_{x_n} / X_k \; ; \; k \leq n-1)$ n'est autre que $\int X_1 \dots X_{n-1} f(y) \lambda_y P(x_{n-1}, dy)$ par définition de P_π. En tenant compte de l'équation d'invariance, cette intégrale devient $X_1 \dots X_{n-1} \lambda_{x_{n-1}}$.

Si alors d est une distance sur le compact des mesures de probabilité on a, en raison de la convergence de cette martingale vers la probabilité α_ω :

$$\text{p.p.} \quad \lim_{n\to\infty} \sup_{p\geq 0} d\,[X_1 \dots X_n \lambda_{x_n}, X_1 \dots X_n \dots X_{n+p} \lambda_{x_{n+p}}] = 0.$$

Il est commode d'utiliser ici l'espace $\Omega = X^Z$ des trajectoires bilatères. On obtient par stationnarité, une sous-suite m d'entiers positifs telle que :

$$\lim_{m\to\infty} \sup_{p\geq 0} d\,[X_{-m} \dots X_0 \lambda_{x_0}, X_{-m} \dots X_0 \, X_1 \dots X_p \lambda_{x_p}] = 0$$

Si alors les parties $A_1 \dots A_{p-1}, A_p$ de X vérifient $\pi(A_i) > 0$, la condition $k > 0$ implique π-p.p. $P_{x_0}\{x_1 \in A_1, \dots, x_p \in A_p\} > 0$ et ceci montre p.p. l'existence de $y_1, \dots, y_p$ dans $A_1, \dots, A_p$ avec, en posant $f(y_i) = g_i$:

$$\lim_{m\to\infty} d\,[X_{-m} \dots X_0 \lambda_{x_0}, X_{-m} \dots X_0 g_1 \dots g_{p-1} \, g_p \, \lambda_{y_p}] = 0.$$

Notons que les λ_y varient dans un compact formé de mesures propres d'après les lemmes 1 et 2.
La relation précédente est d'ailleurs vraie pour tout p et lorsque chacun des A_i décrit un ensemble dénombrable. En particulier fixons $A_p = A$ de façon que l'adhérence C de l'ensemble des mesures de la forme $f(a)\lambda_a$ $(a \in A)$ ne contienne que des mesures propres, ce qui est possible d'après les lemmes 1 et 2, et prenons les A_i de la forme $f^{-1}(U)$ où U décrit une base dénombrable d'ouverts du support S_π de $f(\pi)$.

Si alors τ est une application quasi-projective adhérente à la suite $X_{-m} \dots X_0$ on obtient d'après le lemme 3 pour $\gamma \in (S_\pi)^{p-1}$ un η voisin de γ et un λ' de C avec $\tau\eta\lambda' = \tau\lambda_{x_0}$ donc aussi :

un λ'' de C avec $\tau\lambda_{x_0} = \tau\gamma\lambda''$ où τ et γ sont indépendants. Les propriétés de proximalité et d'irréductibilité de $T_\pi = \bigcup_{p\geq 0} (S_\pi)^p$ permettent de faire converger $\gamma\lambda''$ vers une mesure de Dirac δ_z telle que $\tau(z)$ soit défini. Ceci montre que $\tau\lambda_{x_0}$ est une mesure de Dirac et donc, puisque λ_{x_0} est propre, que τ est constante. La suite $X_{-m} \dots X_0 \lambda_{x_0}$ n'a donc comme valeurs d'adhérence que des mesures de Dirac ; on a aussi pour toute mesure propre λ' : $\lim d\,[X_{-m} \dots X_0 \lambda_{x_0}, X_{-m} \dots X_0 \lambda'] = 0.$

Par stationnarité on obtient une nouvelle sous-suite notée encore m telle que λ' variant dans un ensemble dénombrable dense formé de mesures propres :

$\lim_m d[X_1\ldots X_m\lambda_{x_m}, X_1\ldots X_m\lambda'] = 0$. Ceci donne $\lim_m X_1\cdots X_n\lambda_{x_m} = \alpha_\omega = \lim_m X_1\ldots X_m\lambda'$ et l'arbitraire de λ' permet de conclure que α_ω est une mesure de Dirac.

Etant données deux solutions λ et μ de l'équation (1), écrivons :

$$\alpha_\omega = \lim_n X_1\ldots X_n \lambda_{x_n}$$

$$\beta_\omega = \lim_n X_1\ldots X_n \mu_{x_n}$$

$$\zeta = \lim_k X_1\ldots X_{n_k}$$

Les lemmes 1 et 2 disent que λ_{x_n} et μ_{x_n} varient dans un compact C' formé de mesure propres et le lemme 3 donne donc $\alpha_\omega = \zeta\mu'$ $\beta_\omega = \zeta\mu''$ avec $\mu', \mu'' \in C'$.

Comme $\alpha_\omega = \delta_{Z(\omega)}$ et que μ' est propre, ζ est constante d'image $Z(\omega)$. On a donc aussi $\beta_\omega = \delta^Z_{(\omega)}$ et $\mu_{x_o} = E_\pi(\beta_\omega|X_k\ ;\ k\leq 0) = E_\pi(\alpha_\omega|\ X_k\ ;\ k\leq 0) = \lambda_{x_o}$

D'où $\lambda = \mu$.

On vient de voir que les valeurs d'adhérence de la suite $X_1\ldots X_n$ ont toutes pour image $Z(\omega)$. Ceci donne si m est propre $\lim_k X_1\ldots X_{n_k} m = \zeta m = \delta_{Z(\omega)}$ d'où la dernière assertion du théorème.

II - EXPOSANTS CARACTERISTIQUES -

On reprend les notations du début et on note B l'espace des drapeaux sur $\mathcal{V}$, c'est-à-dire l'espace des suites de $d-1$ sous-espaces projectifs strictement emboîtés et distincts de $\mathcal{V}$

Deux quotients de B jouent un rôle important, l'espace des hyperplans de $\mathcal{V}$ qui s'identifie à l'espace projectif $\mathcal{V}^*$ associé au dual de V et l'espace B_2 des éléments de contact, c'est-à-dire des couples formé d'un point de $\mathcal{V}$ et d'une droite passant par ce point. On prolonge de manière naturelle les applications linéaires de V aux puissances extérieures et en particulier pour $g\in G\ell(V)$ et $v'\in V^*$, on pose $gv' = {}^tg^{-1}(v')$.

Pour un élément de contact $\xi = (\overline{v}, \overline{v\wedge w})$ défini par le vecteur v et le bivecteur $v\wedge w$, on pose :

$$\sigma(g,\xi) = \frac{||gv \wedge gw||}{||gv||^2}$$ où $g \in G\ell(V)$ et l'on observe la relation de cocycle

$\sigma(gh,\xi) = \sigma(g,h\cdot\xi)\ \sigma(h,\xi)$.

Le théorème justifié en I donnera des informations sur $\sigma(S_n,\xi)$ en raison de la

Proposition.

Soit m *une mesure de probabilité propre sur* $\mathcal{V}^*$ *et* u_n *une suite d'applications projectives telles que* $u_n^{-1} m$ *converge vers une mesure de Dirac* δ_s. *Alors si l'origine de l'élément de contact* ξ *n'est pas dans l'hyperplan* s *on a*

$$\lim_n \sigma(u_n,\xi) = 0.$$

Preuve.

En changeant éventuellement de produit ecalaire, on peut supposer v orthogonal à s et on peut choisir une base orthonormée $(e_1,e_2,\ldots,e_d)$ avec $e_1 = v$, $e_1^* = s$ $v \wedge w = e_1 \wedge e_2$.

Ecrivons $g \in G\ell(V)$ sous la forme polaire $g = k\,a\,k'$ où k et k' sont orthogonale et a diagonale de coefficients vérifiant : $a_1 \geq a_2 \geq \ldots \geq a_d > 0$.
Si alors $u_n = k^n a^n k'^n$ on a $u_n^{-1} m = (k'^n)^{-1}(a^n)^{-1}(k^n)^{-1} m$.

Supposant que k^n et k'^n convergent le long de la sous-suite n_i vers k et k', on a aussi :

$$\lim_n k'^{-1}(a^n)^{-1} k^{-1} m = \delta_{e_1^*}$$

Comme $k^{-1}m$ est propre, ceci implique $a_j = o(a_1)$ $(j > 1)$

et $k'^{-1} e_1^* = e_1^*$

Puisque k'^n est orthogonale, on a donc $\lim_n k'^n e_1 = e_1$.

On a d'autre part la majoration :

$$\sigma(u_n,\xi) = \frac{||a^n k'^n e_1 \wedge a^n k'^n e_2||}{||a_n k'^n e_1||^2} \leq \frac{||a^n k'^n e_2||}{||a^n k'^n e_1||}$$

Or $||a^n k'^n e_2|| \leq a_1^n <k'^n e_2, e_1> + a_2^n = o(a_1^n)$

et $||a^n k'^n e_1|| \geq a_1^n |<k'^n e_1, e_1>| \sim a_1^n$. D'où $\lim_n \sigma(u_n,\xi) = 0$.

Les divers résultats vont se déduire du théorème fondamental suivant qui résulte du théorème 1 de I appliqué à la fonction f^{-1}.

THEOREME 2.

Supposons que le noyau P *vérifie la condition* A *et que le semi-groupe* T_{π}^{-1} *opère de manière proximale et totalement irréductible sur* $\mathcal{V}^*$. *Alors, pour tout élément de contact* ξ *à* $\mathcal{V}$, *la suite* $\frac{1}{n}\,\mathrm{Log}\,\sigma(S_n(\omega),\xi)$ *converge p.p. vers une fonction strictement négative et ne prenant qu'un nombre fini de valeurs.*

On a d'abord le lemme suivant, bien connu en théorie ergodique [2] :

Lemme 1.

Soit (E,T,η) *un système dynamique où* η *est une mesure invariante finie,* h *une fonction* η*-intégrable telle que* $\sum_{0}^{n-1} h \circ T^k$ *converge p.p. vers* $-\infty$ *alors* $\int_E h\,d\eta < 0$.

Une construction classique permet de se ramener au cas T *inversible.*

Lemme 2.

Soient X_k *des variables aléatoires stationnaires à valeurs dans* $G\ell(V)$ *telles que* $\mathrm{Log}\,||X_1(\omega)||$ *soit intégrable,* γ *le plus grand exposant du produit* $S_n = X_n \ldots X_1$ *et* Σ *une partie génératrice de* V *qui en projection dans* $\mathcal{V}$ *est non dénombrable. Alors il existe* $v \in \Sigma$ *tel que* $\frac{1}{n}\,\mathrm{Log}||S_n(\omega)v||$ *converge p.p. vers* γ.

Preuve.

Soit $W(\omega)$ le sous-espace de dimension $\leq d-1$ défini par

$$W(\omega) = \{ v \in V \; ; \; \lim_n \frac{1}{n}\,\mathrm{Log}\,||S_n(\omega)v|| < \gamma \} \quad \text{et} \quad A_v = \{ \omega \; ; \; v \in W(\omega) \} .$$

Si $v_1, v_2, \ldots, v_d$ forment une base de V, l'intersection des A_{v_i} est vide.

Soit $\Sigma_1 \subset \Sigma$ une partie non dénombrable de Σ telle que les partie à d éléments de Σ_1 soient des bases de V, ce qui est possible d'après les hypothèses sur Σ. Alors la famille des mesure $|A_v|$ des parties A_v $(v \in \Sigma_1)$ est sommable de somme majorée par d ce qui implique l'existence de $v \in \Sigma_1$ avec $|A_v| = 0$ et donc $v \notin W(\omega)$ p.p.

Introduisons le noyau P_2 sur $X \times B_2$ défini par la formule :

$$P_2\psi(x,\xi) = \int \psi[\, y, f(x)\xi\,]\; P(x,dy)$$

et considérons l'ensemble C des mesures λ^2 sur $X \times B_2$ de projection π qui sont P_2-invariantes.

Soit P_1 le noyau projection de P_2 sur $X \times \mathcal{V}$

Lemme 3.

Pour tout $v \in V$ *la suite* $\frac{1}{n} \operatorname{Log} \|S_n(\omega)\, v\|$ *converge p.p. vers le plus grand exposant de Liapunoff.*

Preuve.

D'après le lemme 2, il existe $v \in V$ tel que $\frac{1}{n} \operatorname{Log}\|S_n(\omega)\, v\|$ converge p.p. vers le plus grand exposant γ. Comme [16] cette convergence a lieu dans $\mathbb{L}^1(P_\pi)$, on a

$$\gamma = \lim_n \frac{1}{n} \int \operatorname{Log} \|S_n(\omega)\, v\|\, dP_\pi(\omega).$$

D'autre part on peut écrire :

$$\operatorname{Log} \|S_n(\omega)v\| = \sum_{k=0}^{n-1} F(x_{k+1}, S_k \cdot v) \quad \text{avec} \quad F(x,w) = \operatorname{Log} \|f(x)w\|$$ et en intégrant :

$$\frac{1}{n} \int \operatorname{Log} \|S_n(\omega)v\|\, dP_\pi(\omega) = \frac{1}{n} \sum_{k=0}^{n-1} < P_1^k\, F, \pi \times \delta_v >$$

Avec la topologie introduite en I, l'ensemble des probabilités sur $X \times \mathcal{V}$ de projection π est compact. On peut donc extraire de la suite

$\frac{1}{n} \sum_{k=0}^{n-1} (\pi \times \delta_v)(P_1)^k$ une sous-suite convergente vers une probabilité λ^1 de projection π vérifiant $\lambda^1 P_1 = \lambda^1$. A la limite on a donc

$$\gamma = \lim_n \frac{1}{n} \int \operatorname{Log}\|S_n(\omega)v\|\, dP_\pi(\omega) = \lambda^1(F).$$

Si maintenant on remplace v par un vecteur w quelconque, le même raisonnement fournit, d'après l'unicité de la mesure invariante λ^1 résultant du théorème 1 de A :

$$\lim_n \frac{1}{n} \int \operatorname{Log} \|S_n(\omega)\, w\|\, dP_\pi(\omega) = \lambda^1(F).$$

Comme pour tout w on a p.p. $\lim_n \frac{1}{n} \operatorname{Log}\|S_n(\omega)w\| \leq \gamma$ on en conclut que sauf sur un ensemble négligeable, on a $\gamma = \lim_n \frac{1}{n} \operatorname{Log}\|S_n(\omega)w\|$

Lemme 4.

Supposons que pour tout λ^2 *de* C, *l'intégrale* $\int \mathrm{Log}\,\sigma[f(x),\xi]\, d\lambda^2(x,\xi)$ *soit négative. Alors pour tout* ξ , *la suite* $\frac{1}{n}\mathrm{Log}\,\sigma[S_n(\omega),\xi]$ *converge p.p. vers une fonction strictement négative et ne prenant qu'un nombre fini de valeurs.*

Preuve.

D'après la stationnarité de la suite X_k, on sait déjà d'après [16] que P_π p.p la suite $\frac{1}{n}\mathrm{Log}\,\sigma[S_n(\omega),\xi]$ converge pour tout ξ vers $C(\xi,\omega)$ dont les valeurs en nombre fini sont des différences d'exposants de Liapunoff.
D'autre part, utilisant la chaine de noyau P_2 et le raisonnement du lemme 3, on obtient, avec une mesure λ^2 qui est P_2-invariante :

$$\lim_n \frac{1}{n}\int \mathrm{Log}\,\sigma[S_n(\omega),\xi]\, dP_\pi(\omega) = \lambda^2(G) \quad \text{où} \quad G(x,\xi) = \mathrm{Log}\,\sigma[f(x),\xi]$$

Choisissons $\xi = (\overline{x},\overline{x\wedge y})$ de façon que $\frac{1}{n}\mathrm{Log}\,\|S_n(\omega)x \wedge S_n(\omega)y\|$ converge p.p vers le plus grand exposant dans $\Lambda^2 V$, ce qui est possible d'après le lemme 2.

On a alors :

$$\gamma_{d-1} + \gamma_d = \lim_n \frac{1}{n}\int \mathrm{Log}\,\|S_n(\omega)x \wedge S_n(\omega)y\|\, dP\pi(\omega)$$

et $\gamma = \gamma_d = \lim_n \frac{1}{n}\int \mathrm{Log}\,\|S_n(\omega)x\|\, dP_\pi(\omega)$ en vertu du lemme 3.

On en déduit $\lambda^2(G) = -\gamma_d + \gamma_{d-1} = \int \mathrm{Log}\,\sigma[f(x),\xi]\, d\lambda^2(x,\xi)$ et donc vu l'hypothèse du lemme, $\gamma_d > \gamma_{d-1}$.

Enfin si $\xi = (\overline{x},\overline{x\wedge y})$ est quelconque, on a

$$\lim_n \frac{1}{n}\mathrm{Log}\,\sigma[S_n(\omega),\xi] \leq \gamma_d + \gamma_{d-1} - 2\lim_n \frac{1}{n}\mathrm{Log}\,\|S_n(\omega)x\| = \gamma_{d-1} - \gamma_d < 0.$$

Preuve du théorème.

Pour démontrer l'inégalité

$$\int \operatorname{Log} \sigma[f(x),\xi]\ d\lambda^2(x,\xi) < 0$$

à laquelle on est amené par le lemme 4, on considère l'espace $E = X^{\mathbb{N}} \times B^2$ muni de la mesure $\eta = \int P_x \times \lambda_x^2\, d\pi(x)$ et la transformation T définie par

$$T(\omega,\xi) = (\theta\omega, f(x_o)\xi)$$

où $x_o(\omega)$ est la première coordonnée de ω et θ la translation sur $X^{\mathbb{N}}$

Montrons l'invariance de la mesure η :

On a d'abord

$$\eta = \int P_z \theta \times f(z)\lambda_z^2\, d\pi(z) = \int P_t \times f(z)\lambda_z^2\, P(z,dt)d\pi(z)$$

Mais l'équation $\lambda^2 P_2 = \lambda^2$ permet d'écrire

$$\int \delta_z \times \lambda_z^2\, d\pi(z) = \int \delta_t \times f(z)\lambda_z^2\, P(z,dt)d\pi(z)$$

et donc $\eta T = \int P_z \times \lambda_z^2\, d\pi(z) = \eta$

D'autre part si l'on pose

$h(\omega,\xi) = \operatorname{Log} \sigma[f(x_o), \xi]$ on a

$$\int \operatorname{Log} \sigma[f(x),\xi]\ d\lambda^2(x,\xi) = \int h(\omega,\xi)\ dP_x(\omega)\ d\lambda_x^2(\xi)\ d\pi(x) = \int h\ d\eta$$

Ayant observé que $\sum_0^{n-1} h \circ T^k(\omega,\xi) = \operatorname{Log} \sigma[S_n(\omega),\xi]$ il suffit donc de voir, en raison du lemme 1, que $\lim_n \sigma[S_n(\omega),\xi] = -\infty$ η p.p.

Les hypothèses permettent d'appliquer le théorème 1 de I à la suite de probabilités sur $\mathcal{V}^*$ $\alpha_n = X_1^{-1} \ldots X_n^{-1}\ \nu'$ où ν' est propre et celui-ci donne : $\lim_n \alpha_n = \delta_{Z(\omega)}$ avec $Z(\omega) \in \mathcal{V}^*$.

Il suffit donc de voir, en raison de la proposition, que pour ω fixé, l'ensemble des ξ dont l'origine appartient à l'hyperplan $Z(\omega)$ de $\mathcal{V}$ est $\lambda_{x_o}^2$-négligeable, ce qui signifie que la projection $\lambda_{x_o}^1$ de $\lambda_{x_o}^2$ sur $\mathcal{V}$ ne charge pas l'hyperplan $Z(\omega)$. Mais l'invariance de λ^2 sous P_2 donne l'invariance de $\lambda^1 = \int \delta_z \times \lambda_z^1\ d\pi(z)$ sous P_1 et la totale irréductibilité de T_π^{-1} sur $\mathcal{V}^*$

donne la même propriété sur $\mathcal{V}$.

D'après le lemme 2 de I λ_z^1 est propre π-p.p. ce qui donne bien

$\lambda_{x_0}^1\ [Z(\omega)] = 0$.

REMARQUES

$C(\omega,\xi)$ est une constante dès que λ^2 est unique, ce qui est assuré, d'après la proposition de I dès que T_π agit de manière proximale et totalement irréductible sur les espaces projectifs de V et $\Lambda^2 V$.

Dans le cas général, en raison de l'égalité des limites de $\frac{1}{n} \operatorname{Log}||S_n(\omega)v||$ et $\frac{1}{n} \operatorname{Log}||S_n(\omega)\ w||$ on a :

$$\operatorname{Lim}_n \frac{1}{n} \int \operatorname{Log} \frac{||S_n v \wedge S_n w||}{||S_n v||\ ||S_n w||}\ dP_\pi(\omega) \leq C < 0$$

ce qui s'interprète géométriquement comme une décroissance exponentielle vers zéro de l'angle de $S_n v$ et $S_n w$. En particulier, dans le cas indépendant $[f = Id\ , \pi = p]$, ceci permet de montrer [14] que l'opérateur sur $\mathcal{V}$ défini par convolution avec p se comporte essentiellement comme un barycentre de contractions, situation que l'on rencontre dans l'étude de modèles d'apprentissage.

Considérons l'espace B des drapeaux sur $\mathcal{V}$, le noyau $\tilde{P}$ sur $X \times B$ défini par : $\tilde{P}\ \psi(x,b) = \int \psi\ [y,f(x)b]\ P(x,dy)$ et les mesures $\tilde{\lambda}$ sur $X \times B$ de projection π qui sont $\tilde{P}$-invariantes. Décomposons $g \in G\ell(V)$ en produit d'une matrice orthogonale $k \in K$ et d'une matrice triangulaire supérieure t et désignons par $a_i(g,k)$ les coefficients diagonaux de la partie triangulaire de gk. En fait $a_i(g,k)$ ne dépend que de g et de l'image canonique de k dans B et il sera donc également noté $a_i(g,b)$. On vérifie également la relation de cocycle

$$a_i(gg',b) = a_i(g,g'.b)\ a_i(g',b).$$

Rappelons [16] que, P_π-pp, les suites $\frac{1}{n} \operatorname{Log}||S_n(\omega)v||$ convergent pour tout $v \in V$ et que les limites possibles prennent d valeurs au plus, appelées exposants de Liapunoff du produit de matrices aléatoires $S_n(\omega)$. On peut calculer [16], ces exposants en introduisant la transformation $\tilde{\theta}$ sur $\Omega \times B$ définie par $\tilde{\theta}(\omega,b) = [\theta\omega, X_0(\omega)\cdot b]$ où $\Omega = X^Z$ et $X_0(\omega) = f\ [x_0(\omega)]$ et en choisissant une mesure $\tilde{\eta}$ de projection P_π sur Ω qui soit $\tilde{\theta}$-invariante ; ces exposants avec leurs multiplicités sont alors donnés par les intégrales :

$\int\int \text{Log } a_i[X_0(\omega),b] d\tilde{\eta}(\omega,b)$. Dans le cas présent, celles-ci se réduisent à

$$\gamma_i = \iint_{X\times B} \text{Log } a_i\,[f(x),b]\, d\,\tilde{\lambda}(x,b)$$

On posera enfin $\sigma_i(g,b) = \dfrac{a_{i+1}(g,b)}{a_i(g,b)}$.

Théorème 3.

Supposons que le noyau P *vérifie la condition* A , *que le semi-groupe* T_π^{-1} *opère de manière proximale et totalement irréductible sur les espaces projectifs associés aux puissances extérieures de* V *et soit* $\tilde{\lambda}$ *une mesure* $\tilde{P}$*-invariante sur* $X \times B$ *de projection* π . *Alors, pour tout* $b \in B$, *la suite* $\frac{1}{n} \text{Log } \sigma_i[S_n(\omega),b]$ *converge* P_π*-pp vers le nombre strictement négatif* $\iint \text{Log } \sigma_i[f(x),b]\, d\tilde{\lambda}(x,b)$. *En particulier, les exposants de Liapunoff de* $S_n(\omega)$ *prennent* $d = \dim V$ *valeurs distinctes égales aux intégrales* $\iint \text{Log } a_i\,[f(x),b]\, d\tilde{\lambda}(x,b)$.

Preuve.

Si k est une rotation transformant le drapeau défini par la base canonique (e_i) en b on a $(a_1\ldots a_i)(g,b) = \|gk(e_1\wedge\ldots\wedge e_i)\|$ et ceci fournit, par les mêmes considérations qu'au théorème 2, et en raison de la remarque suivant ce théorème, que, pour tout $b \in B$, la suite considérée ici converge bien vers une constante égale à l'intégrale $\iint \text{Log } \sigma_i\,[f(x),b]\, d\tilde{\lambda}(x,b)$. Pour voir que cette intégrale est négative, il suffit donc de voir que $\lim_n \frac{1}{n} \text{Log } \sigma_i[S_n(\omega),b] < 0$.

Or $\sigma_i(g,b) = \dfrac{a_{i+1}}{a_i} = \|gE_i\wedge gF_i\|\,\|gE_i\|^{-2}$ où $E_i = e_1\wedge\ldots\wedge e_i$ et $F_i = e_1\wedge\ldots\wedge e_{i-1}$.

Il suffit donc d'appliquer le théorème 2 en remplaçant V par $\Lambda^i V$, ξ par $(E_i, E_i\wedge F_i)$, puisque le dual de $\Lambda^i V$ n'est autre que $\Lambda^{d-i}V$.

Les intégrales $\gamma_i = \iint_{X\times B} \text{Log } a_i[f(x),b]\, d\tilde{\lambda}(x,b)$ vérifient donc $\gamma_i > \gamma_{i+1}$, ce qui justifie l'assertion relative aux exposants de Liapunoff.

Le théorème précédent admet une extension lorsqu'on remplace le groupe linéaire, ou plutôt son sous-groupe unimodulaire $Sl(d,\mathbb{R})$ par un groupe semi-simple sans facteur compact à centre fini. Précisons d'abord quelques notations empruntées à [15] . Si G est un tel groupe d'algèbre de Lie $\mathfrak{g}$, on peut le faire opérer par l'action adjointe sur $\mathfrak{g}$ ce qui plonge G dans $Sl(\mathfrak{g})$. Ayant fixé un sous-groupe compact maximal K de G et un produit scalaire sur $\mathfrak{g}$ qui soit K-invariant, on désigne par A un sous-groupe connexe abélien formé d'éléments auto-adjoints et qui soit maximal. Un tel sous-groupe est isomorphe à $\mathbb{R}^t$ et formé de matrices diagonales. Si $\mathfrak{A}$ est l'algèbre de Lie de A, on désigne par $\alpha_1, \alpha_2, \dots, \alpha_r$ un système fondamental de racines simples de $\mathfrak{A}$ dans $\mathfrak{g}$ et toute racine α est alors combinaison linéaire des α_i à coefficients entiers de même signe, ce qui précise en particulier la notion de racine positive.

Posons alors : $\mathfrak{g}_\alpha = \{x \in \mathfrak{g} \ ; \ [a,x] = \alpha(a)x \quad \forall\, a \in \mathfrak{A}\}$

$$\mathfrak{N} = \bigoplus_{\alpha>0} \mathfrak{g}_\alpha \ , \quad \tilde{\mathfrak{N}} = \bigoplus_{\alpha>0} \mathfrak{g}_{-\alpha}$$

et désignons par N, $\tilde{N}$ les groupes (nilpotents) correspondants. Si M est le centralisateur de A dans K, la frontière de Furstenberg de G [5] est alors $B = G/MAN$. Plus généralement, on prolonge au cas présent les notations déjà introduites précédemment, le rôle de $Sl(d,\mathbb{R})$ étant joué par $G \subset Gl(\mathfrak{g})$. En particulier, B correspond à l'espace des drapeaux de V déjà rencontré et l'équation (1) concerne ici un noyau $x \to \tilde{\lambda}_x$ de X dans B. On notera G_π le sous-groupe fermé de G engendré par T_π. Soit a(g) la composante sur A de g dans la décomposition de G sous forme d'Iwasawa G = KAN et pour une forme linéaire α sur $\mathfrak{A}$, posons $\sigma_\alpha(g,k) = e^\alpha[a(gk)]$ pour $g \in G$, $k \in K$, expression qui ne dépend de k que par son image $\bar{k}$ dans B.

Les propriétés d'irréductibilité seront ici renforcées : une hypothèse commode (mais trop forte) est la densité de G_π au sens algébrique dans $G \subset Sl(\mathfrak{g})$. Ceci signifie que tous les polynômes nuls sur G_π s'annullent sur G.

Théorème 4.

Supposons que le noyau P *vérifie la condition* A , *que* G_π *soit algébriquement dense dans* G , *que* T_π^{-1} *opère de manière proximale sur* B *et soit* $\tilde{\lambda}$ *une solution de* (1) .

Si ρ *est une représentation irréductible de* G , *le plus grand exposant caractéristique du produit de matrices* $\rho(S_n)$ *est simple.*

Si α est une forme linéaire négative sur $\mathcal{A}$, alors pour tout $b \in B$ *, la suite* $\frac{1}{n}$ Log σ_α $[S_n(\omega,b)]$ *converge* P_π*-p.p vers le nombre strictement négatif* $\iint \operatorname{Log} \sigma_\alpha[f(x),b]\, d\tilde{\lambda}(x,b)$.

Enfin $\tilde{\lambda}$ *solution de* (1) *est unique.*

Preuve.

Soit V l'espace de la représentation irréductible considérée, $\mathcal{V}$ l'espace projectif associé. D'après le théorème de Lie, il existe un vecteur propre dans V pour le sous-groupe résoluble AN ; ce vecteur est unique à un scalaire près en raison de l'irréductibilité de G donc de G_π; ceci implique que MAN fixe le point v de correspondant et on obtient donc une application continue $g \to g.v$ de $B = G/MAN$ dans V.

La proximalité de G_π sur B implique la proximalité de G_π sur l'image de B dans V ; mais cette image engendre V tout entier par irréductibilité et ceci implique la proximalité de G_π sur V.

La densité de G_π dans G implique d'autre part que G_π est totalement irréductible sur V, comme l'est G.

Par dualité, la représentation de G dans V^* définie par $\rho^*(g)[v'] = [\rho(g^{-1})]^t(v')$ est irréductible et les propriétés vraies pour V le sont aussi pour V^*. On peut donc appliquer le théorème 2 qui donne pour $\xi = (\bar{x}, \overline{x \wedge y})$ la convergence de $\frac{1}{n}$ Log $\sigma[\rho[S_n(\omega)], \xi]$, vers un nombre négatif $C(\omega,\xi)$.

Comme il a été vu au cours de la démonstration de ce théorème, ceci revient à dire que le plus grand exposant caractéristique est simple.

Remarquons, pour la suite, que d'après le lemme 3 $\lim_n \frac{1}{n}$ Log $\rho[S_n(\omega)x]$ est une constante ; si ρ n'est pas irréductible, on peut la décomposer en somme directe de représentation irréductible et écrire pour $x = \sum_i x_i$: $\|\rho[S_n(\omega)]x\|^2 = \sum_i \|\rho[S_n(\omega)]x_i\|^2$ ce qui fournit la constance en x de la limite considérée et donc le fait que $C(\omega,\xi)$ dépend de $x \wedge y$ seulement. On voit de plus que la dépendance en $x \wedge y$ provient de l'annulation de certaines composantes irréductibles de $x \wedge y$; comme cette annulation ne dépend que de l'orbite de $x \wedge y$ sous G, il en est de même de $C(\omega,\xi)$.

Pour obtenir l'assertion relative à σ_α on considère une représentation irréductible de G telle que le sous-groupe MAN soit le stabilisateur d'un point v de $\mathcal{V}$, ce qui identifie B à une sous-variété compacte de $\mathcal{V}$. Pour un élément de contact ξ tangent en v à B, on a alors $\sigma(g,\xi) = \sigma(kt,\xi) = \sigma(t,\xi)$ où $k \in K$, $t \in AN$; d'autre part, $\sigma(g,\xi) = \sigma(t,\xi)$ n'est autre que le coefficient de multiplication des distances dans la direction de ξ sous l'action de t laquelle conserve v et donc l'espace tangent en v à B. Mais l'action de $t \in MAN$ sur cet espace tangent s'identifie à celle de Adt sur $\widetilde{\mathfrak{n}}$, action définie par passage au quotient suivant l'algèbre de Lie de MAN. Or, on a pour $\xi_i \in \mathfrak{g}_{-\alpha_i}$ et $t \in MAN$:

$$\mathrm{Adt}(\xi_i) = e^{-\alpha_i}[a(t)]\ \xi_i \mod \mathfrak{m}+\mathfrak{a}+\mathfrak{n}$$

ce qui donne $\sigma(g, \xi_i) = e^{-\alpha_i}[a(g)] = \sigma_{-\alpha_i}(g,e)$

et $\sigma(g,k\xi_i) = \sigma_{-\alpha_i}(g,k)$.

On en déduit d'après ce qui précède, la convergence de $\frac{1}{n}$ Log $\sigma[S_n(\omega),k\xi_i]$, vers un nombre qui est indépendant de ω et k et égal à $\lim_n \frac{1}{n}$ Log $\sigma_{-\alpha_i}[S_n(\omega),k]$

Or d'après l'invariance de $\tilde{\lambda}$ et la propriété de multiplicateur de $\sigma_{-\alpha_i}$, on a

$$\frac{1}{n}\int\int \mathrm{Log}\ \sigma_{-\alpha_i}[S_n(\omega,k)]\ dP_x(\omega)d\tilde{\lambda}(x,\bar{k}) = \int\int \mathrm{Log}\ \sigma_{-\alpha_i}[f(x),k]\ d\tilde{\lambda}(x,\bar{k}).$$

Ceci fournit le résultat voulu dans le cas $\alpha = -\alpha_i$. Dans le cas général, α est combinaison linéaire à coefficients négatifs des α_i et le résultat s'étend donc à ce cas. L'unicité de $\tilde{\lambda}$ découle directement de I.

Considérons maintenant une représentation quelconque ρ de G dans un espace vectoriel V et fixons une base de V dans laquelle les $\rho(t)$ pour $t \in AN$ sont triangulaires supérieures de coefficients diagonaux a_i. Notons $a_i(g,k)$ les coefficients diagonaux de la partie triangulaire de $\rho(gk)$ $(g \in G,\ k \in K)$ et observons que la dépendance en k se réduit à celle de $\bar{k}$ image de k dans B. Avec ces notations on a le complément suivant au théorème 4.

Corollaire.

Avec les notations du théorème 4, soit ρ une représentation quelconque de G. Les exposants caractéristiques du produit $\rho[S_n(\omega)]$ sont donnés avec leurs multiplicités par les intégrales $\int\int_{X\times B} \mathrm{Log}\ a_i[f(x),b]\ d\tilde{\lambda}(x,b)$.

Preuve.

Posons comme en [15] pour $k \in K$, $\omega \in \Omega$ $\hat{S}_n(\omega,k) = \rho[T_n(\omega,k)]$ où T_n est défini par la décomposition d'Iwasawa $S_n(\omega)k = K_n T_n$ avec $K_n \in K$, $T_n \in AN$. Si l'on considère la transformation $\hat{\theta}$ définie par $\theta(\omega,k) = [\hat{\theta}\omega, X_o(\omega).k]$ où $X_o(\omega).k$ est la partie orthogonale du produit $X_o(\omega)k$ et si l'on munit $\Omega \times K$ d'une mesure invariante se projetant sur Ω suivant la mesure naturelle, $\hat{S}_n(\omega)k$ apparait comme un nouveau produit de matrices aléatoires triangulaires dont les exposants caractéristiques sont ceux de $\rho[S_n(\omega)]$. Ces derniers sont donc donnés avec leurs multiplicités [16] par les intégrales des termes diagonaux, c'est-à-dire par les nombres $\iint_{X\times B} \operatorname{Log} a_i\,[f(x),b]\,d\tilde{\lambda}(x,b)$ en raison de l'unicité de la mesure invariante $\tilde{\lambda}$.

Exemples.

1) groupe symplectique réel $G = Sp(2n)$

Ce groupe intervient en particulier dans les problèmes de propagation des ondes dans un tube [12] . Il s'agit du groupe des matrices laissant invariante la forme bilinéaire alternée sur $\mathbb{R}^{2n}$: $\langle x,y\rangle = \sum_{i+j=2n} (x_i y_j - x_j y_i)$. La frontière de Furstenberg B est ici formée des drapeaux isotropes, c'est-à-dire des suites de n sous-espaces distincts emboîtés contenus dans un sous-espace isotrope maximal (de dimension n). La dualité symplectique permet de compléter naturellement ce drapeau en un drapeau d'ordre 2n. Les coefficients d'une matrice diagonale symplectique sont deux à deux inverses et on notera $\lambda_i(g)$ $[i = 1,2,\dots,n]$ les n premiers coefficients de la partie diagonale de $g \in G$ dans la décomposition d'Iwasawa $G = KAN$.

Les autres coefficients sont alors les λ_i^{-1} dans l'ordre inverse. Sous les conditions d'application du corollaire, prenant ici pour ρ la représentation naturelle, les exposants caractéristiques du produit de matrices symplectiques $S_n(\omega)$ sont donc donnés par les intégrales $\pm \int \operatorname{Log} \lambda_i[f(x)b]\, d\tilde{\lambda}(x,b)$.

Les racines de A dans $\mathfrak{g}$ sont ici les $\lambda_i^{\pm 1}$ $\lambda_j^{\pm 1}$ avec $i \leq j$ et les racines positives sont donc les logarithmes de $\lambda_i\ \lambda_j^{-1}$ $(i<j)$ et λ_i^2.

Le théorème 4 dit alors que $S_n(\omega)$ possède n exposants positifs distincts égaux aux intégrales des Log λ_i et n exposants négatifs opposés des précédents. En particulier, il y a 2n exposants distincts.

2) groupe orthogonal.

Il s'agit du groupe des matrices d'ordre n concervant la forme quadratique

$$\langle x,y\rangle = \sum_{i=1}^{q} x_i\, x_{n+1-i} + \sum_{j=q+1}^{n-q} x_j^2 .$$

La frontière de Furstenberg est encore formée des drapeaux isotropes de q sous-espaces emboîtés contenus dans un sous-espace isotrope (de dimension q) maximal. La dualité permet de compléter un tel drapeau en une suite de 2q sous-espaces emboîtés dont les dimensions successives diffèrent de 1 ou n-2q (pour les termes centraux). Considérons une décomposition d'Iwasawa $G = KAN$ et notons $\lambda_1(g),\dots,\lambda_q(g)$ les q premiers termes de la partie diagonale de g ; les n-2q suivants sont alors égaux à 1 et les q derniers à $\lambda_q^{-1}(g),\dots,\lambda_1^{-1}(g)$. Les racines de A dans $\mathfrak{g}$ sont données par 0, $(\lambda_i\ \lambda_j^{\pm 1})^{\pm 1}$ $(i<j)$

et λ_i si $n \neq 2q$. En particulier, les racines positives sont $\lambda_i \lambda_j^{\pm 1}$ (i>j) et λ_i si $n \neq 2q$. Donc si $n \neq 2q$ il y a q exposants positifs distinct, q exposants négatifs opposés et n-2q exposants nuls. Si n = 2q il y a q ou q-1 exposants positifs distincts auxquels correspondent autant d'exposants négatifs ; il y a 0 ou 2 exposants nuls.

BIBLIOGRAPHIE

[1] L. Arnold et W. Kliemann : "Lyapunov exponents of linear stochastic systems" Preprint Université de Brême 1983.

[2] G. Atkinson : "Recurrence of co-cycles and random walks". Journal of the London Math. society - Vol.13 part 3 (1976) p.486-488.

[3] Cohen : "Ergodic theorems in demography " B.A.M.S. juin 1979.

[4] Furstenberg : "Non commuting random products" A.M.S. 108, 1963, p.337-428.

[5] Furstenberg : "Boundary theory and stochastic processes on homogeneous spaces" Proc. Symp. Pure Math. Vol.26, (1972) p.139-229.

[6] Furstenberg : "Séminaire Bourbaki", (1979-1980), n°559.

[7] Gol'dseid , Molchanov, Pastur : " A pure point spectrum of the stochastic one-dimensional Schrödinger operator". Funct. Anal. Appl. 11-1 (1977) p.1-10.

[8] Y. Guivarc'h : "Quelques propriétés asymptotiques des produits de matrices aléatoires". Lecture Notes in Math. 774 (1980) p.176-250.

[9] Y. Guivarc'h : "Marches aléatoires à pas markovien". C.R.A.S. Paris t.289 (1979).

[10] Y. Guivarc'h : "Exposants de Lyapunoff des produits de matrice aléatoires en dépendance markovienne". C.R.A.S. t292 (2 fév. 1981) p.327-329.

[11] Y. Guivarc'h et A. Raugi : "Frontières de Furstenberg, Propriétés de contraction et théorèmes de convergence". Séminaires de l'Université de Rennes I (1980), A paraitre dans Z. Wahrsch.

[12] J. LAcroix : "Localisation pour l'opérateur de Schrödinger à potentiel aléatoire dans un ruban". A paraitre.

[13] F. Ledrappier : "Quelques propriétés des exposants caractéristiques". Cours d'Eté de Saint-Flour (1983) . A paraitre dans Springer Lecture Notes.

[14] E. LEPAGE : "Théorèmes limites pour les produits de matrices aléatoires". C.R.A.S. t.292 (9 fév. 1981) p.379-382.

[15] Mostow : "Strong rigidity of locally symetric spaces". Annals of Math. Studies 78 (1973).

[16] Osedelcts : Trans Moscow Math Soc 19, (1968) p.197-231.

[17] Ragunathan : " A proof of Osedelet's multiplicative ergodic theorem". Israel J. of Math. Vol.32 n°4 (1979) p.356-362.

[18] A. Raugi : "Fonctions harmoniques et théorèmes limites sur les groupes de Lie". Bul. S.M.F. mémoire 54, (1977).

[19] Royer : "Croissance exponentielle de produits markoviens de matrices aléatoires". Université P. et M. Curie, (mai 1979).

[20] Tutubalin : " Some theorems of the type of laws of large numbers". Theory of Proba. 1969) p.313-319.

[21] Tutubalin : "The central limit theorem for products of matrices". Symposias math. (1977) p.101-116.

[22] Virtser : Theory of Proba 24 B (1979) p.361-370.

IRMAR
Université de Rennes 1
Campus de Beaulieu
35042 Rennes Cédex
France

Remarks on [semi-] stable probabilities.

W. Hazod, Dortmund

We continue the investigations on stable probabilities on locally compact groups [6]. There we obtained the following result: let A be the generating distribution of a semigroup of probability measures on a locally compact group G, let $(\tau_t)_{t>0} \subseteq \mathrm{Aut}(G)$ be a group of automorphisms,

such that (i) $\tau_t \tau_s = \tau_{ts}$, $t,s>0$, (ii) $t \longrightarrow \tau_t$ is continuous

and (iii) $t \longrightarrow \tau_t$ is contracting on G, i.e. $\lim_{t\to 0} \tau_t(x) = e$

for every $x \in G$. If A is stable w.r.t. (τ_t),

i.e. if $\tau_t(A) = tA + X(t)$, $X(t)$ a primitive term, and if G is a simply connected nilpotent Lie group, then there exists a uniquely determined generating distribution $\mathring{A}$ on $\mathcal{G}$, the Lie algebra of G, such that $\mathring{A}$ is operator-stable in the sense of M. Sharpe [13]. Conversely, if $\mathring{A}$ is operator stable w.r.t. $(\mathring{\tau}_t)_{t>0}$ on the vector-space $\mathcal{G}$, and if in addition the (vector space-) automorphisms are automorphisms of the Lie algebra, then there exists a uniquely determined generating distribution A on G, which is stable w.r.t. some group (τ_t).

Now we want to find similar results for non-nilpotent Lie groups. Therefore the automorphisms are not supposed to act contracting on the whole group G. Furthermore we consider a more general class of distributions called semistable.
It turns out (§ 2 theorem 2.1 and 2.8) that for Lie groups we have the following situation: If there exists a [semi-] stable distribution A, then there exists a nilpotent, connected, simply connected analytic subgroup $H \subseteq G$ supporting a shift $A_1 = A - Y, Y \in \mathcal{G}$, of A. Again A_1 may be identified with an operator-[semi-] stable generating distribution $\mathring{A}_1$ on the Lie algebra $\mathcal{J}$ of H. The structure of operator semi-stable distributions on the vector space $\mathcal{J}$ is completely known via the Lēvy-Hinčin-representation, s. [9], [11] .

There are some results concerning the structure of [semi-] stable distributions on general locally compact groups, but examples in § 4 show, that the general situation is much more complicated than the case of a Lie group. (Locally compact groups are always assumed to be metrizable.)

§ 1. Preliminaries.

Let G be a locally compact group, Aut(G) the corresponding group of automorphisms. A one-parameter multiplicative group of automorphisms is a map $\mathbb{R}_+^* \ni t \mapsto \tau_t \in \mathrm{Aut}(G)$, such that $\tau_t \tau_s = \tau_{ts}$, $t,s > 0$.

(τ_t) is called strongly [weakly] continuous, if $t \mapsto \tau_t$ $[t \mapsto \tau_t(x)]$ is a continuous map $\mathbb{R}_+^* \to \mathrm{Aut}(G)$ $[\mathbb{R}_+^* \to G$ for every $x \in G]$.
(τ_t) is contracting in $x \in G$ if $\tau_t(x) \xrightarrow[t \to 0]{} e$.

For any $\tau \in \mathrm{Aut}(G)$ we define: τ acts contracting in $x \in G$ if $\tau^k(x) \xrightarrow[k \to \infty]{} e$. (τ_t) resp. τ acts contracting on a set $E \subseteq G$, if (τ_t) resp. τ is contracting in x for any $x \in E$.

$\mathfrak{G}$ denotes the Lie algebra of G.

As usual $\mathcal{D}(G)$ denotes the space of testfunctions with compact support and $\mathcal{M}\mathcal{D}(G) \subseteq \mathcal{D}(G)'$ the cone of generating distributions of semigroups of probabilities on G. Let $A \in \mathcal{M}\mathcal{D}(G)$ be the generating distribution of the continuous convolution semigroup $(\mu_t)_{t \geq 0}$. Then we write $\mu_t = \mathcal{E}xp(tA)$.
[For details see [8], [5]].

In the following definitions we do not suppose that (τ_t) is continuous or contracting:

1.1 Definition A continuous convolution semigroup $(\mu_t = \mathcal{E}xp(tA))_{t \geq 0}$ is called <u>stable in the strict sense</u> w.r.t. a group of automorphisms $(\tau_t) \subseteq \mathrm{Aut}(G)$, if

(1.1.a) $\tau_t(\mu_s) = \mu_{ts}$, $t,s > 0$, equivalently if

(1.1.b) $\tau_t(A) = t \cdot A$, $t > 0$ (see e.g. [6]).

(μ_t) resp. A is called <u>stable in the wide sense</u>, if for $t > 0$ there exists a primitive distribution $X(t) \in \mathfrak{G}$, such that

(1.1.c) $\tau_t(A) = tA + X(t)$, $t > 0$.

1.2 Definition $(\mu_t)_{t \geq 0}$ resp. $A \in \mathcal{M}\mathcal{D}(G)$ is called semistable w.r.t. $\tau \in \mathrm{Aut}(G)$ and $c \in (0,1)$, if

(1.2.a) $\tau(A) = c.A$ (semistable in the strict sense).
(1.2.b) $\tau(A) = cA + X$, $X \in \mathcal{G}$ (semistable in the wide sense).
A different generalisation of stability is

1.3 Definition $(\mu_t)_{t \geq 0}$ resp. $A \in \mathcal{M}(G)$ is called self-decomposable w.r.t $(\tau_t)_{t>0}$, if for $t \in (0,1)$ there exist $B(t) \in \mathcal{M}(G)$, such that

$$A = \tau_t(A) + B(t).$$

Obviously any stable generating distribution A is self-decomposable: If $\tau_t(A) = tA + X(t)$, $t>0$, then for $t \in (0,1)$ $A = \tau_t(A) - X(t) + (1-t)A$. The structure of (operator-)-semistable and - self-decomposable probabilities on vector spaces is well known (s.e.g. [9], [11], and the literature cited in [6]). The intersection of the classes of semistable and of self-decomposable measures contains the stable measures, but the two classes are not comparable (s. [10]). Therefore the following simultaneous generalisation is sometimes useful:

1.4. Definition $(\mu_t)_{t \geq 0}$ resp. $A \in \mathcal{M}(G)$ is called semi-selfdecomposable w.r.t. $\tau \in \mathrm{Aut}(G)$ if there exists $B \in \mathcal{M}(G)$, such that

(1.4) $$A = \tau(A) + B.$$

Remark Any semistable distribution is semi-self-decomposable: Assume $\tau(A) = cA + X$ for some $c \in (0,1)$, $X \in \mathcal{G}$. Then $A = \tau(A) + (1-c)A - X = \tau(A) + B$, with $B := (1-c)A - X \in \mathcal{M}(G)$.

The following simple lemma plays an important role in the sequel:

1.5. Lemma Let $A \in \mathcal{M}(G)$ be semi-self-decomposable w.r.t. $\tau \in \mathrm{Aut}(G)$, i.e. $A = \tau(A) + B$, $B \in \mathcal{M}(G)$. Denote by Q resp. Q_B the Gaussian part of A resp. B and by η resp. η_B the corresponding Lévy measures. Then

(1.5.a) $Q = \tau(Q) + Q_B$ and (1.5.b) $\eta = \tau(\eta) + \eta_B$.

Therefore: If η is bounded, then we have $\tau(\eta) = \eta$ (and $\eta_B = 0$).

Obviously $\tau(Q)$ resp. $\tau(\eta)$ are the Gaussian part resp. the Lévy-measure of $\tau(A)$.

Therefore $A = \tau(A) + B$ implies

$Q = \tau(Q) + Q_B$ and $\eta = \tau(\eta) + \eta_B$.

Assume that the positive measure η is bounded. Then η_B is bounded too and by $\|\eta\| = \|\tau(\eta)\|$ and $\|\eta\| = \|\tau(\eta)\| + \|\eta_B\|$

we obtain $\eta_B = o$.

Therefore $\eta = \tau(\eta)$. ⌋

Now we obtain some corollaries:

<u>1.6 Corollary</u> Let A be semistable w.r.t. (τ, c). Denote by Q resp. η the Gaussian part resp. the Lévy-measure, then

(1.6.a) $\tau(Q) = cQ$

(1.6.b) $\tau(\eta) = c\,\eta$ and $\eta = o$ or η is unbounded.

⌈ We have (see the remark to definition 1.4):

$A = \tau(A) + B$, $B = (1-c)A - X$, $X \in \mathcal{G}$, $c \in (o,1)$.

Therefore $Q_B = (1-c)Q$

$Q = \tau(Q) + Q_B = \tau(Q) + (1-c)Q$, i.e. (1.6.a) holds.

Furthermore we have $\eta_B = (1-c)\eta$, therefore $\eta = \tau(\eta) + (1-c)\eta$, i.e. (1.6.b) holds.

If $\eta \neq o$ we conclude $\tau(\eta) \neq \eta$, then by 1.5 η must be unbounded. ⌋

<u>1.7 Corollary</u> Let $\Gamma \in G$ be a τ - invariant neighbourhood of e. Let A be semi-self-decomposable w.r.t. τ.

Let η resp. ν be the Lévy-measures of A resp. B and denote by η_o resp. ν_o the restrictions to $\complement\Gamma$.

Then $\tau(\eta_o) = \eta_o$ and $\nu_o = o$.

If A is semistable w.r.t. (τ, c), then $\eta_o = 0$.

⌈ As Γ and $\complement\Gamma$ are τ - invariant we obtain

$\eta_o = \tau(\eta_o) + \nu_o$ (resp. $\tau(\eta_o) = c\,\eta_o$ if A is semistable).

On the other hand the restriction of a Lévy-measure to the complement of a neighbourhood of e is bounded, therefore

$\eta_o, \tau(\eta_o)$ and ν_o are bounded measures.

Now by 1.5 resp. 1.6 we conclude $\tau(\eta_0) = \eta_0$ resp. $\eta_0 = 0$. ‖

1.8 Corollary Let $G_1 \subseteq G$ be an open subgroup. Then the group H generated by $\bigcup_{k\in\mathbb{Z}} \tau^k(G_1)$ is an open τ- invariant subgroup $\supseteq G_1$.
If A is semi-self-decomposable w.r.t. τ, i.e. $A = \tau(A) + B$ and if $\eta_0 := \eta|_{\complement H}$, then $\tau(\eta_0) = \eta_0$ and $\eta_B|_{\complement H} = 0$.
If A is semistable w.r.t. (τ, c) for some $c \in (0,1)$, then $\eta|_{\complement H} = 0$.

1.9 Corollary Let G be a Lie group and let G_0 be the connected component of e. Let A be a generating distribution with Lévy- measure η. Denote $\eta_1 := \eta|_{G_0}$, $\eta_0 := \eta|_{\complement G_0}$.

Then we have: If A is semi-self-decomposable, then

$$\tau(\eta_0) = \eta_0, \quad \eta_1 = \tau(\eta_1) + \eta_B, \quad \eta_B|_{\complement G_0} = 0.$$

If A is semistable w.r.t. (τ, c), then $\eta_0 = 0$. We may therefore suppose w.l.o.g. that $G = G_0$.

1.10 Corollary Let G be a Lie group w.l.o.g. connected and let $\mathfrak{g}$ be the Lie algebra of G. Denote by $\Gamma := \exp(\mathfrak{g})$ the range of the exponential map. For a generating distribution A with Lévy measure η we denote $\nu_1 := \eta|_\Gamma$, $\nu_0 := \eta|_{\complement\Gamma}$. Then we have:
If A is semi-self-decomposable, then

$$\tau(\nu_0) = \nu_0, \quad \nu_1 = \tau(\nu_1) + \eta_B.$$

If A is semistable w.r.t. (τ, c), then $\nu_0 = 0$, i.e. A is concentrated on the range of the exponential map.

Some simple observations:

1.11 Lemma Assume $\tau \in \mathrm{Aut}(G)$ resp. $(\tau_t) \subseteq \mathrm{Aut}(G)$, and $A \in \mathcal{M}(G)$ with Lévy-measure η and Gaussian part Q.
Then we have:

(1.11a) $A = \tau(A) + B \Longleftrightarrow Q = \tau(Q) + Q_B$ and $\eta \geq \tau(\eta)$

(1.11b) $A = \tau_t(A) + B(t), t>0 \Longleftrightarrow Q = \tau_t(Q) + Q_{B(t)}$ and $\eta \geq \tau_t(\eta)$, $t > 0$.

(1.11c) $\tau(A) = cA + X \Longleftrightarrow \tau(Q) = cQ$ and $\tau(\eta) = c\eta$

(1.11d) $\quad \tau_t(A) = tA + X(t), t>o \iff \tau_t(Q) = tQ$

and $\tau_t(\eta) = t\eta,\ t > o.$

The implication "$\Rightarrow$" of (1.11a) resp.(1.11.c) is already proved in 1.5 resp. 1.6. The same arguments hold for (1.11b) and (1.11d).

Conversely: Assume $Q = \tau(Q) + Q_B$ for some Gaussian part Q_B and $\eta \geq \tau(\eta)$.

Define $\eta_B := \eta - \tau(\eta)$. Then we obtain immediately that η_B is a Lévy-measure.

Fix a Lévy map $\Gamma : \mathcal{D}(G) \longrightarrow \mathcal{D}(G)$ (s.e.g. [8]), and define

$$B_1 : f \longmapsto Q_B(f) + \int_{G\setminus\{e\}} (f(x) - f(e) + \Gamma f(x))d\eta_B(x).$$

Then $B_1 \in \mathcal{M}(G)$ and, if P is the primitive part of A,

$$A(f) - \tau(A)(f) = P(f) - \tau(P)(f) + Q(f) - \tau(Q)(f)$$

$$+ \int_{G\setminus\{e\}} [f(x) - f(e) + \Gamma f(x)]d\eta(x) - \int_{G\setminus\{e\}} [f(\tau(x)-f(e)+\Gamma(f\circ\tau)(x)]d\eta$$

$$= \overbrace{(P - \tau(P))(f)}^{I(f)} + Q_B(f) -$$

$$\underbrace{- \int_{G\setminus\{e\}} [f(x) - f(e) + \Gamma f(x)]d(\eta - \tau(\eta))(x) + \int_{G\setminus\{e\}} [(\Gamma f)(\tau(x)) - (\Gamma(f\circ\tau))(x)]d\eta(x)}_{II(f)} =$$

$$(I(f) + II(f)) + Q_B(f) + \int_{G\setminus\{e\}} [\ \dots\ ]d\eta_B(x) = I(f) + II(f) + B_1(f) =: B(f)$$

As $f \longmapsto I(f)$ and $f \longmapsto II(f)$ are obviously primitive we obtain the desired result, namely $B \in \mathcal{M}(G)$ and $A = \tau(A) + B$.

The other implications are proved in a similar way. $\square$

<u>1.12 Corollary</u> If $A \in \mathcal{M}(G)$ is semi-self-decomposable/self-decomposable/semistable/stable w.r.t. τ resp. (τ_t), then the adjoint distribution $\tilde{A}$ and the symmetrized distribution $A_s := \frac{1}{2}(A + \tilde{A})$ have the same property.

If $A = \tilde{A}$ is [semi-] stable, then it is [semi-] stable in the strict sense.

Follows immediately: Assume e.g. $A = \tau(A) + B$. Then $Q = \tau(Q) + Q_B$ and $\eta = \tau(\eta) + \eta_B$. Therefore $\tilde{\eta} = \tau(\tilde{\eta}) + \tilde{\eta}_B$, and by the preceding lemma $\tilde{A} = \tau(\tilde{A}) + \tilde{B}$. If $A = \tilde{A}$ and if $\tau(A) = cA + X$, then we obtain in the same way $\tau(A) = \tau(\tilde{A}) = c\tilde{A} - X = cA + X$. So we conclude $X = 0$.
The other assertions are proved in a similar way.

If $A \in \mathcal{M}^1(G)$ is semi-self-decomposable, $A = \tau(A) + B$, $B \in \mathcal{M}^1(G)$, then there exists a sequence $\{B(k)\}_{k\in\mathbb{N}} \subseteq \mathcal{M}^1(G)$, s.t. $A = \tau^k(A) + B(k)$, $k \in \mathbb{N}$ (and $B(1) = B$).

[We see immediately that $B(k) = \sum_{l=0}^{k-1} \tau^l(B)$.]

We put $C(k) := \tau^{-k}B(k)$, $k \in \mathbb{N}$. Then the defining equation is equivalent to

(1.13a) $\tau^{-k}(A) = A + C(k), \quad k \in \mathbb{N}$,

Therefore for any $k, l \in \mathbb{N}$

(1.13b) $C(k+l) = C(k) + \tau^{-k} C(l)$.

For self-decomposable distributions we obtain an analogous result:

$A = \tau_t(A) + B(t)$, $0 < t < 1$ iff

(1.13c) $\tau_s(A) = A + C(s), \quad s > 1,$ where

(1.13d) $C(s) := \tau_s(B(1/s)), \quad s > 1.$

For $s,t > 1$ we obtain $C(s\cdot t) = C(s) + \tau_s(C(t))$.

To describe the [semi-] stable distributions in a similar manner, we introduce "space - time" - transformations:

<u>1.13 Definition</u> Fix $\tau \in \mathrm{Aut}(G)$, $c \in (0,1)$ resp. $(\tau_t)_{t>0} \subseteq \mathrm{Aut}(G)$. We define $\xi : \mathcal{D}'(G) \longrightarrow \mathcal{D}'(G)$ resp. $\xi_t : \mathcal{D}'(G) \longrightarrow \mathcal{D}'(G)$ via $\xi(F) := c^{-1}\tau(F)$ resp. $\xi_t(F) := t^{-1}\tau_t(F)$.

If A is semistable w.r.t. (τ, c), i.e. $\tau(A) = cA + X$, then

(1.14a) $$\tau^k(A) = c^kA + X(k), \quad k \in \mathbb{Z}, \; X(k) \in \mathcal{G}.$$

(Obviously $X(1) = X, X(k) = \sum_{o}^{k-1} c^{j} \tau^{k-1-j}(X)$, $k \in \mathbb{N}$,

$X(o) = 0$, $X(-k) = -\sum_{o}^{k-1} c^{-j} \tau^{j-k+1}(X)$, $k \in \mathbb{N}$).

If we put $Y(k) := c^{-k} X(k)$, $k \in \mathbb{Z}$, we obtain

(1.14.b) $\xi^{k}(A) = A + Y(k)$, $k \in \mathbb{Z}$,

where $Y(\cdot)$ is a solution of the cocycle equation

$$Y(k+l) = Y(l) + \xi^{l}(Y(k)), \quad k,l \in \mathbb{Z}.$$

If A is stable w.r.t. (τ_t), i.e. $\tau_t(A) = tA + X(t)$, we obtain

(1.14.c) $X(ts) = tX(s) + \tau_s(X(t))$, $t,s > o$.

If we put again $Y(t) := t^{-1} X(s)$, we obtain

(1.14.d) $\xi_t(A) = A + Y(t)$, where

(1.14.e) $Y(ts) = Y(s) + \xi_s(Y(t))$, $t,s > 0$.

We summarize the results in the following proposition:

1.14 Proposition

a) A is semi-self-decomposable, $A = \tau(A) + B$, iff there exist $C(k) \in \mathcal{M}(G)$, $k \in \mathbb{N}$, such that

$$\tau^{-k}(A) = A + C(k), \quad \text{where}$$

$$C(k+l) = C(k) + \tau^{-k} C(l), \quad k,l \in \mathbb{N}.$$

b) A is self-decomposable, i.e. $\tau_t(A) = A + B(t)$, $t \in (o,1)$ iff there exist $C(s) \in \mathcal{M}(G)$, $s > 1$, such that

$$\tau_s(A) = A + C(s), \quad s > 1,$$

where $C(st) = C(s) + \tau_s\, C(t)$, $s,t > 1$.

c) A is semistable, i.e. $\tau(A) = cA + X$,

iff there exist $Y(k) \in \mathcal{G}$, $k \in \mathbb{Z}$, such that

$\xi^{k}(A) = A + Y(k)$, $k \in \mathbb{Z}$,

where $Y(k+l) = Y(l) + \xi^{l}(Y(k))$, $k,l \in \mathbb{Z}$.

d) A is stable, i.e. $\tau_t(A) = tA + X(t)$, $t > o$,

iff there exist $Y(t) \in \mathfrak{g}$, $t > 0$,
such that

$$\xi_t(A) = A + Y(t), \text{ where}$$
$$Y(st) = Y(s) + \xi_s(Y(t)),\ s,t > 0.$$

§ 2 The Lévy-Hinčin formula for semistable distributions on Lie groups

We start with the following simple observation (s. [6] for the case of stable distributions):

Assume that G is a Lie group, connected, simply connected and nilpotent with Lie algebra $\mathfrak{g} \cong \mathbb{R}^d$. Then $\exp : \mathfrak{g} \longrightarrow G$ is a C^∞-isomorphism. Therefore, we define for $f \in C_o(G)$ $[\mathcal{D}(G)]$ $\mathring{f} := f \circ \exp \in C_o(\mathfrak{g})$ $[\mathcal{D}(\mathfrak{g})]$ and similar for $F \in M^b(G)$ $[\mathcal{D}(G)']$ we define

$\mathring{F} \in M^b(\mathfrak{g})$ $[\mathcal{D}(\mathfrak{g})']$ via $< F,f > = < \mathring{F},\mathring{f} >$.

We obtain $C_o(G) \cong C_o(\mathfrak{g})$, $M^b(G) \cong M^b(\mathfrak{g})$, $\mathcal{D}(G) \cong \mathcal{D}(\mathfrak{g})$, $\mathcal{D}(G)' \cong \mathcal{D}(\mathfrak{g})'$.

Furthermore, if we put for $\tau \in \mathrm{Aut}(G)$ $\mathring{\tau} := d\tau$ (the differential of τ) $\in \mathrm{Aut}(\mathfrak{g})$, then $\mathrm{Aut}(G) \cong \mathrm{Aut}(\mathfrak{g})$. So we can identify measures distributions and automorphisms of the group G with the corresponding objects on the vector space $\mathfrak{g} \cong \mathbb{R}^d$.

2.1 Theorem If $A \in \mathcal{M}\mathcal{D}(G)$ is stable [self-decomposable] w.r.t. $(\tau_t)_{t>0}$ resp. semistable [semi-self-decomposable] w.r.t. $\tau \in \mathrm{Aut}(G)$ and $c \in (0,1)$, then $\mathring{A} \in \mathcal{M}\mathcal{D}(\mathfrak{g})$ is operator-stable [operator-self-decomposable] resp. operator-semistable [operator-semi-self-decomposable] w.r.t. $(d\tau_t = \mathring{\tau}_t)_{t>0}$ resp. $(d\tau = \mathring{\tau}, c)$ and vice versa.

[The proof follows immediately with similar arguments as in [6] thm. 3.6.]

Now we want to obtain similar results for stable distributions on general Lie groups: It is shown, that a semistable distribution is always supported by a connected, simply connected nilpotent analytic subgroup, therefore theorem 2.1 may be applied to the general case.

The first results are valid for any locally compact group:

2.2 Proposition Let G be a locally compact group. Assume that there exists $\tau \in \mathrm{Aut}(G)$, $c \in (0,1)$ and $A \in \mathcal{M}(G)$, such that A is semistable w.r.t. (τ, c).

Then the Lévy-measure of A is concentrated on a set $U \subseteq G$, on which τ acts contracting.

Proof: We consider a function $f \in C(G)$, $0 \le f \le 1$ such that $f \equiv 0$ in a neighbourhood U of e and $f \equiv 1$ outside of U^2. Then we have for the Lévy-measure η :

$\int | f \circ \tau^k | d\eta = \int f \circ \tau^k d\eta = \int fd\, \tau^k(\eta) = c^k \int fd\eta$.

Therefore $\Sigma \int | f \circ \tau^k | d\eta < \infty$, hence $f \circ \tau^k \to 0$ η-a.e.
We conclude, that $\tau^k(x) \to e$ for η a.a. $x \in G \setminus \{e\}$.

2.3 Proposition Let G be a locally compact group. Assume that $A \in \mathcal{M}(G)$ is strictly semistable w.r.t. (τ, c). Then there exists a subset $V \subseteq G$, such that τ acts contracting on V and such that any measure $\mu_t = \mathcal{E}xp(tA)$ is concentrated on V.

Proof: We have $\tau^k(\mu_t) = \mu_{c^k t}$, $k \in \mathbb{N}$, $t \in \mathbb{R}_+^*$. Assume $\alpha \in (0,1)$ and let ξ be the Lévy-measure of the one-sided stable distribution on $\mathbb{R}_+$ with Index α. Then $\eta_1 := \int_0^\infty \mu_t \, d\xi(t)$ is the Lévy-measure of a convolution semigroup on G (see e.g. [5] I § 5 and II § 2). A simple calculation yields $\tau(\eta_1) = c^\alpha \cdot \eta_1$.
So proposition 2.2 is applicable to η_1: η_1 is concentrated on a subset V on which τ acts contracting. The representation

$\eta_1 = \int_0^\infty \mu_t \, d\xi(t)$ shows that the measures μ_t, $t > 0$ are concentrated on V too.

2.4 Corollary Let A be semistable w.r.t. (τ, c). Let η be the Lévy-measure of A and let Q be the (symmetric) Gaussian part of A, and put $(\nu_t := \mathcal{E}xp(tQ))_{t \ge 0}$. Then we have:

There are subsets $U \subseteq G$, $V \subseteq G_0$, such that η is concentrated on U, ν_t is concentrated on V, and such that τ acts contracting on $U \cup V$.

[[This follows immediately from 2.2 and 2.3:
The existence of U is guaranteed by Prop. 2.2.
By Corollary 1.6 Q is semistable w.r.t. (τ, c) in the strict sense, therefore the existence of V is guaranteed by Prop. 2.3.]]

2.5 Proposition Assume that G is a connected finite dimensional group with Lie algebra $\mathfrak{g}$. Let A be semistable w.r.t. (τ, c) and assume that A is symmetric Gaussian. Then there exists a Gaussian distribution $\mathring{A}$ on the vector space $\mathfrak{g}$, such that $\langle \mathring{A}, f \rangle = \langle A, f \circ \log \rangle$ and we have:

If $\mathfrak{g}_1$ is the subspace on which $\mathring{A}$ is strict (i.e. on which the corresponding quadratic form in non-degenerate), then $d\tau$ acts contracting on $\mathfrak{g}_1$.

Proof: We know already (s. [15])that Gaussian distributions A on finite dimensional groups G correspond in a 1 - 1 - manner to Gaussian distributions on the (finite dimensional) vector space $\mathfrak{g}$. Obviously

$\tau(A) = cA$ iff $d\tau(\mathring{A}) = c\mathring{A}$.

Now $\mathring{A}$ is a symmetric Gaussian distribution on a finite dimensional vector space, which is semistable w.r.t. $(d\tau, c)$.

There exists a basis $X_1, \ldots, X_d$ of $\mathfrak{g}$, such that $A = \sum_1^m X_i^2$, $m \le d$.

Let $\mathfrak{g}_1$ be the subspace generated by $X_1, \ldots, X_m$.

$\mathring{A}$ is a full semistable distribution on $\mathfrak{g}_1$. Therefore by the Lévy-Hinčin formula (s. [9]) $d\tau$ acts contracting on $\mathfrak{g}_1$.

2.5 Lemma For any locally compact group, for any $\tau \in \mathrm{Aut}(G)$ we define

$C(\tau) := \{ x \in G \mid \tau^k(x) \underset{k \to \infty}{\longrightarrow} e \}$.

Then we have: $C(\tau)$ is a τ- invariant subgroup of G.

[Obvious].

2.6 Lemma For a Lie algebra $\mathfrak{g}$ of dimension d and $\mathring{\tau} \in \mathrm{Aut}(\mathfrak{g})$ we define $\mathcal{C}(\mathring{\tau}) := \{ X \in \mathfrak{g} \mid \mathring{\tau}^k(X) \longrightarrow 0 \}$. Then we have: $\mathcal{C}(\mathring{\tau})$ is a $\mathring{\tau}$ - invariant subalgebra of $\mathfrak{g}$.

[Obvious].

2.7 Proposition. Assume that G is a Lie group with Lie algebra $\mathfrak{g}$ and $\tau \in \mathrm{Aut}(G)$. Define as in 2.5 resp. 2.6 $C(\tau)$ resp. $\mathcal{C}(d\tau)$.

Then we obtain: (i) $C(\tau) \subseteq G_0$ and $C(\tau)$ is the (connected) analytic subgroup generated by $\mathcal{C}(d\tau)$; $C(\tau)$ is σ-compact, hence a Borel set of G.

(ii) $C(\tau)$ is simply connected, nilpotent, τ - invariant.

Proof: G_o is a neighbourhood of e and τ- invariant, therefore $\tau^k(x) \underset{k\to\infty}{\longrightarrow} e$ implies $x \in G_o$.

If we put $\Gamma := \exp(\mathfrak{g})$ the range of the exponential map, then we obtain in a similar manner $C(\tau) \subseteq \Gamma$.

Put $\Lambda := \exp \mathcal{C}(d\tau)$ then for $x = \exp X$, $X \in \mathcal{C}(d\tau)$, we obtain via $\tau^k(x) = \exp((d\tau)^k(X))$, that $x \in C(\tau)$. On the other hand assume $x \in C(\tau)$. Assume $V \subseteq \Gamma$ is a neighbourhood of e, such that the restriction $\exp : \log(V) \longrightarrow V$ is 1 - 1. Then there exists $k_o \in \mathbb{N}$, such that for any $k \in \mathbb{N}$ $\tau^{k_o+k}(x) \in V$. Put $X := \log \tau^{k_o}(x)$, then we have $\tau^{k_o+k}(x) = \exp(d\tau)^k(X)$ resp. $\log(\tau^{k_o+k}(x)) = (d\tau)^k(X) \longrightarrow 0$. Therefore $X \in \mathcal{C}(d\tau)$, hence $x \in \Lambda$. We obtain: $\Lambda = C(\tau) = \exp(\mathcal{C}(d\tau))$.

$C(\tau)$ is τ - invariant, $\mathcal{C}(d\tau)$ is $d\tau$ - invariant, hence the restrictions of τ resp. $d\tau$ to $C(\tau)$ resp. $\mathcal{C}(d\tau)$ are automorphisms of the group resp. the Lie algebra, which act contracting on the whole group $C(\tau)$ resp. algebra $\mathcal{C}(d\tau)$. Therefore (see e.g. [12]) $\mathcal{C}(d\tau)$ must be nilpotent and moreover $C(\tau)$ simply connected.

Now we are ready to prove the main result of this paper:

2.8 Theorem Let G be a Lie group with Lie algebra $\mathfrak{g}$. Assume $\tau \in \mathrm{Aut}(G)$ and $c \in (o,1)$, and assume further that $A \in \mathcal{MD}(G)$ is a generating distribution of a semigroup of probability measures $(\mu_t)_{t\geq o}$. If A is semistable w.r.t. (τ, c), i.e. $\tau(A) = cA + X$, $X \in \mathfrak{g}$, then there exist

(i) a primitive distribution $Y \in \mathfrak{g}$,

(ii) a connected, simply connected, τ- invariant, nilpotent, analytic subgroup $H \subseteq G$ with Lie algebra $\mathfrak{h} \subseteq \mathfrak{g}$,

such that $(\nu_t := \mathcal{E}xp(t(A-Y))$ is concentrated on H for $t \geq o$.

Therefore via theorem 2.1 to A-Y there corresponds uniquely a generating distribution $A_1 \in \mathcal{MD}(\mathfrak{h})$, which is operator-semistable in the sense of R. Jajte [9] w.r.t. $(d\tau, c)$.

Remarks 1. A - Y is semistable iff A is semistable.

2. ν_t need not to be a translate of μ_t. The relation between the semigroups (ν_t) and (μ_t) is given by the Trotter-Kato formula [5]

$$\mu_t = \lim_{n\to\infty} [\nu_{t/n} * \varepsilon_{\exp(\frac{t}{n} Y)}]^n ,$$

$$\nu_t = \lim_{n\to\infty} [\mu_{t/n} * \varepsilon_{\exp(-\frac{t}{n} Y)}]^n .$$

3. The Lévy-Hinčin formula for operator-semistable distributions ([9], [11]) on $\mathbb{R}^d$ gives a complete description of all possible semistable distributions on Lie groups G.

Proof of theorem 2.8: Put $H := C(\tau)$, $\mathcal{f} := \mathcal{L}(d\tau)$.
Then we know by 2.7 that H is connected, simply connected, nilpotent with Lie algebra $\mathcal{f}$. Fix a Lévy - map $\Gamma : \mathcal{D}(G) \longrightarrow \mathcal{D}(G)$ (see [8] for details) according to a basis $X_1, \ldots, X_d$ of $\mathcal{g}$, such that $X_1, \ldots, X_s$ is a basis of $\mathcal{f}$.

Now we consider the Lévy-Hinčin-Hunt-representation of A ([8])
$A = P + Q + \int_{G\setminus\{e\}} (\ldots\ldots) d\eta$. Then we know already by 2.2 - 2.6, that η is concentrated on $H = C(\tau)$ and that the semigroup $(\gamma_t := \mathcal{E}xp(t\,Q))$ generated by the Gaussian part Q is concentrated on H too.

By the choice of the basis $X_1, \ldots, X_d$ of $\mathcal{g}$ we have

$$\int_{G\setminus\{e\}} (f(x) - f(e) - \sum_{i=1}^{d} X_i(f)\ \xi_i(x))\, d\eta(x) =$$

$$= \int_{H\setminus\{e\}} (f(x) - f(e) - \sum_{i=1}^{s} X_i(f)\ \xi_i(x))\, d\eta(x), \quad f \in \mathcal{D}(G),$$

therefore the integral term $\int(\ldots)d\eta$ generates a semigroup $(\pi_t)_{t\geq 0}$, which is concentrated on H too.

If we put $Y := P$, we have $\nu_t := \mathcal{E}xp(t(A - P)) = \lim_{n\to\infty}(\gamma_{t/n} * \pi_{t/n})^n$.

On the other side (γ_t), (π_t), (ν_t) are semigroups on the Lie group H (with the natural topology), and the canonical embedding $H \longrightarrow G$ is continuous, therefore $(\nu_t)_{t\geq 0}$ is concentrated on $H \subseteq G$.

§ 3 Stable generating distributions.

As any stable distribution A is semistable the results of the preceding § 2 are applicable to stable distributions. Therefore theorem 2.8 is a direct generalization of [6] theorem 3.6. But we want to give some further information on stable distributions.
We begin with some considerations concerning semistable distributions on general locally compact groups G.

3.1 Proposition Let G be a locally compact group.

a) Assume that $A \in \mathcal{M}(G)$ is semistable w.r.t. (τ, c). Then there exists an open, σ - compact, τ - invariant subgroup G_1 supporting A.

b) Assume that $A \in \mathcal{M}(G)$ is stable w.r.t. $(\tau_t)_{t>o}$ and assume further that (τ_t) is weakly continuous. Then there exists an open, σ - compact subgroup G_1 supporting A, which is τ_t - invariant for any $t>o$.

Proof: a) Fix some open σ - compact subgroup H which contains the support of the Lévy measure of A. (As H is open, $H \supseteq G_o$, therefore the semigroup generated by A ($\mathcal{E}xp(tA) =: \mu_t$) is supported by H). To obtain a larger σ - compact, τ - invariant subgroup we put

$$G_1 := \langle \bigcup_{k \in \mathbb{Z}} \tau^k(H) \rangle := \bigcup_{n \in \mathbb{N}} [\bigcup_{k \in \mathbb{Z}} \tau^k(H)]^n .$$

b) Fix $t_o > 1$ and construct G_1 as in a).

G_1 is σ - compact and $\tau_{t_o^k}(G_1) = G_1 \quad \forall k \in \mathbb{Z}$.

The continuity assumption implies that for any $k \in \mathbb{Z}$ and t, such that $t_o^k \leq t$, and any $x \in G_1$

$$\tau_t(x) \in \tau_{t_o^k}(x) \cdot G_o \subseteq G_1 .$$

Therefore $\tau_t(G_1) = G_1$ for $t > o$.

3.2 Lemma For $\tau \in \mathrm{Aut}(G)$ denote by $\bar{\tau}$ the corresponding automorphism on G/G_o. $\tau \longmapsto \bar{\tau}$ is a homomorphism $\mathrm{Aut}(G) \longrightarrow \mathrm{Aut}(G/G_o)$.

If $(\tau_t)_{t>o}$ is weakly continuous, then $\bar{\tau}_t = \mathrm{id}|_{G/G_o}$ for $t > o$.

⟦ Obvious as G/G_o is totally disconnected and G_o is a characteristic subgroup of G. ⟧

3.3 Proposition a) Assume that $A \in \mathcal{M}(G)$ is strictly semistable w.r.t. (τ, c). If $(\mu_t := \mathcal{E}xp(tA))_{t\geq o}$, if $\pi : G \longrightarrow G/G_o$ is the canonical homomorphism and $\bar{\mu}_t := \pi(\mu_t)$, then we have: The generating distribution of $(\bar{\mu}_t)$ $\bar{A} \in \mathcal{M}(G/G_o)$ is strictly semistable w.r.t. $(\bar{\tau}, c)$. ($\bar{\tau}$ is defined as in 3.2).

b) If A is stable w.r.t. $(\tau_t)_{t>o}$ and if (τ_t) is weakly continuous, then A - and therefore $(\mu_t)_{t\geq o}$ - is supported by the connected component G_o.

Proof: We have $\tau(\mu_t) = \mu_{ct}$, $t > o$. Therefore

$$\bar{\mu}_{ct} = \pi(\mu_{ct}) = \pi(\tau(\mu_t)) = \bar{\tau}(\bar{\mu}_t) .$$

So we obtain a).

To prove b) we use Lemma 3.2.: Assume first that A is strictly stable. On the one hand we have by a) $\bar{\tau}_t(\bar{\mu}_s) = \bar{\mu}_{ts}$ for $t,s > o$, on the other hand $\bar{\tau}_t = id_{G/G_o}$, hence $\bar{\mu}_t = \varepsilon_{\bar{e}}$ for any $t \geq o$. So we have proved b) for strictly stable distributions A.
If A is stable w.r.t. (τ_t) then $A_s = \frac{1}{2}(A + \tilde{A})$ is strictly stable and the considerations above yield that A_s is concentrated on G_o. Now the support of the Lévy - measure of A is contained in the support of the Lévy measure of A_s and therefore in G_o, therefore A and hence (μ_t) is supported by G_o.

3.4 Theorem Let $A \in \mathcal{M}(G)$ be a generating distribution on a locally compact group, let $(\tau_t)_{t>o}$ be a multiplicative group of automorphisms of G, such that for $t > 0$

$$\tau_t(A) = tA + X(t) \quad \text{for some } X(t) \in \mathfrak{g} .$$

Assume that there exists a τ_t - invariant Borel set $B \subseteq G$ on which the Lévy-measure η of A is concentrated.

(a) If $t \longmapsto \tau_t(x)$ is continuous for $x \in B$, then there exists a τ_t - invariant subset $B^* \subseteq B$, such that $B^* \subseteq G_o$, $\eta(\complement B^*) = 0$, and such that e is accumulation point of any orbit $\{\tau_t(x) \mid t > o\}$, $x \in B^*$.

(b) If $(x,t) \longmapsto \tau_t(x)$ is jointly continuous $(B, \mathbb{R}_+^*) \longrightarrow B$, then there exists a τ_t - invariant subset $B^* \subseteq B$, such that τ_t acts contracting on B^*.

Proof: (a) In Prop. 3.3 we showed, that we may assume $B \subseteq G_o$. Fix $t_o \in (0,1)$. Then A is semistable w.r.t. $(\tau := \tau_{t_o}, c := t_o)$. Therefore there exists $B^* \subseteq G$, such that $\eta(\complement B^*) = 0$, and such that $\tau_{t_o^n}(x) \xrightarrow[n\to\infty]{} e$ for $x \in B^*$.

The proof of Prop. 2.2 shows that we may assume $B^* \subseteq B$. If τ_{t_o} acts contracting on x, then obviously τ_{t_o} acts contracting on the orbit $\{\tau_t(x) \mid t > o\}$. Therefore we may assume that

$$B^* = \bigcup_{x \in B^*} \{\tau_t(x) \mid t > o\}, \quad \text{i.e.} \quad \tau_t(B^*) = B^*.$$

[B^* depends on t_o. If we repeat the construction for all rational $t_o \in (0,1)$ we obtain a subset B^*, for which $\tau_{t^n}(x) \longrightarrow e$ for $x \in B^*$ and $t \in \mathbb{Q} \cap (0,1)$.]

(b) If the action of τ_t is jointly continuous we are able to show that $\tau_t(x) \xrightarrow[t\to o]{} e$ for $x \in B^*$: Let $(s_n) \subseteq \mathbb{R}_+^*$ be a sequence with $s_n \longrightarrow 0$. Fix $t_o \in (0,1)$ and define $m(n) \in \mathbb{N}$ via $t_o^{m(n)+1} < s_n \leq t_o^{m(n)}$. Put $r_n := s_n / t_o^{m(n)+1} \in (1, 1/t_o]$. W.l.o.g we may assume that r_n is convergent, $r_n \longrightarrow r_o \in [1, 1/t_o]$ say.

Therefore $\tau_{s_n}(x) = \tau_{r_n}(\tau_{t_o^{m(n)+1}}(x)) \xrightarrow[n\to\infty]{} \tau_{r_o}(e) = e$.

So we have proved: For any sequence $s_n \longrightarrow o$ and any $x \in B^*$ e is accumulation point of $\{\tau_{s_n}(x)\}$, therefore $\tau_t(x) \xrightarrow[t\to o]{} e$.

3.5 Remarks 1. There exist examples of groups and measures which fit into the framework of thm. 3.4 (s. 4.2.a).

2. The assumption of joint continuity of the action $(t,x) \longrightarrow \tau_t(x)$ ist not very restrictive as separate continuity implies joint continuity under very general assumptions.

3. If G is a Lie group it is well known, that weak continuity of (τ_t) implies strong continuity of (τ_t) (and therefore joint continuity of $(t,x) \longmapsto \tau_t(x)$).

4. The general assumption in this paper, namely that any locally compact group is assumed to be metrizable, was used in 2.2. and 3.4. The restriction is not very severe as it is always possible to approximate σ - compact groups by metrizable factor groups G/K, where K is τ_t- invariant.

§ 4 Illustrations.

If G is not a Lie group then semistable distributions need not to be concentrated on G_o. (Therefore the assumptions in 1.9, 2.8, 3.3b and 3.4 are necessary). By 3.3a it is sufficient to study measures on totally disconnected groups.

4.1.a Example There exists a totally disconnected locally compact group G, an automorphism τ, contracting on G, $c \in (0,1)$, a semigroup of probabilities (μ_t) with generating distribution A, semistable w.r.t. (τ, c), such that $\mathrm{supp}(\mu_t) = G$, $t > 0$:

Assume that there exists a sequence $\{H_n\}_{n \in \mathbb{Z}}$ of compact open normal subgroups, such that

(i) $H_n \subsetneq H_{n+1}$, $n \in \mathbb{Z}$, (ii) $\bigcup H_n = G$,

(iii) $\bigcap H_n = \{e\}$ and (iv) $\tau(H_n) = H_{n-1}$, $n \in \mathbb{Z}$.

We put $A_n := \omega_{H_n} - \varepsilon_e$ (ω_{H_n} is the Haar-measure on H_n). Then obviously $\tau(A_n) = A_{n-1}$.

Choose $c \in (0,1)$. Then for $N \in \mathbb{Z}$ $\nu_N := \sum_{k=N}^{\infty} c^n \omega_{H_n}$ is a bounded measure.

For $f \in C^b(G)$, $f \equiv 0$ in a neighbourhood U of e, the sum $\sum_N^{\infty} c^n \omega_{H_n}(f)$ converges to $\sum_{-\infty}^{\infty} c^n \omega_{H_n}(f)$.

(Indeed there exists $K \in \mathbb{N}$, such that $H_n \subset U$ for $n \leq -K$, therefore $\sum_{-\infty}^{\infty} c^n \omega_{H_n}(f) = \sum_{-K}^{\infty} c^n \omega_{H_n}(f)$).

Put $B : \mathcal{D}(G) \ni f \longmapsto \sum_{-\infty}^{\infty} c^n A_n(f) = \sum_{-\infty}^{\infty} c^n \omega_{H_n}(f - f(e))$,

then $B \in \mathcal{M}(G)$ with Lévy measure

$$\eta = \eta_B = \sum_{-\infty}^{\infty} c^n \omega_{H_n}.$$

Obviously $\tau(B) = c\, B$, $\tau(\eta) = c \cdot \eta$,

i.e. B is semistable w.r.t. (τ, c).

Remark $G = \lim_{n \in \mathbb{N}} G/H_{-n}$, therefore $\mu_t := \mathcal{E}xp(t\,B)$ is a limit of Poisson measures. But the automorphism τ is not representable as limit of automorphisms of the factor groups G/H_{-n}.
It remains to show that such a group G

exists: Put $K := \mathbb{Z}_2$, $H := \mathbb{Z}_2^{\mathbb{Z}} = \{f : \mathbb{Z} \to \mathbb{Z}_2\}$.
τ is defined by $\tau(f)(k) := f(k+1)$.
Now we put $H_o := \{f \in H \mid f(k) = 0, \ k \geq 0\}$,

$$\tau^n(H_o) =: H_n = \{f \in H \mid f(k) = 0, \ k \geq n\}, \quad n \in \mathbb{Z}.$$

So $\tau(H_n) = H_{n-1}$, $n \in \mathbb{Z}$.
Finally we put $G := \bigcup_{n \in \mathbb{Z}} H_n$.

Each H_n is a compact open subgroup of H_k for $k \geq n$ (w.r.t. product topology). We define a new topology on G, such that the induced topology on each H_n is the topology induced by H, and such that H_n is open in G.
Now the conditions (i) - (iv) of 4.1.a are fulfilled.

4.1.b Example Fix a prime number p and let $G := \Omega_p$ be the (locally compact abelian) group of p-adic numbers. Let $\Delta = \Delta_p$ be the (open compact) subgroup of p-adic integers. (See e.g. [7] chap. II § 10 for details.) Ω_p is a topological field containg the rationals $\mathbb{Q}$ as a dense subset.

For any $q \in \Omega_p \setminus \{o\}$ the map $\tau_q : x \mapsto q \cdot x$ is an automorphism of Ω_p.
The set $\Delta_k := \tau_{p^k}(\Delta)$, $k \in \mathbb{Z}$, is an open compact subgroup for any $k \in \mathbb{Z}$. We have

$$\tau_{p^k}(\Delta_1) = \Delta_{k+1}, \quad \Delta_0 = \Delta, \quad \text{and} \quad \Delta_k \subseteq \Delta_{k+1}, \quad k \in \mathbb{Z}.$$

Therefore Ω_p fulfills the hypothesis of 4.1.a with $G = \Omega_p$, $H_n = \Delta_n$, $\tau = \tau_p$.

4.2.a Example There exist compact, connected, finite dimensional groups and stable semigroups of probabilities (μ_t) with $\text{supp}(\mu_t) = G$, $t > 0$. The corresponding group of automorphims $(\tau_t)_{t>0}$ is contracting on a subset (the range of the exponential map), but not contracting on G. $t \mapsto \tau_t$ is not continuous in this example, and G is not second countable.
Choose $S := \mathbb{R}_d$, the real line with the discrete topology, and let G be the solenoidal group $G = \hat{S}$ (= $\beta(\mathbb{R})$, the Bohr compactification of $\mathbb{R}$).

Then $i : S = \mathbb{R}_d \longrightarrow \mathbb{R}$, $i(x) := x$, is a continuous 1 - 1 homomorphism, therefore the dual homomorphism $\hat{i} : \hat{\mathbb{R}} = \mathbb{R} \longrightarrow \hat{S} = G$ is continuous, injective and has dense range.
(Indeed G is one dimensional and $\hat{i} : \mathbb{R} \longrightarrow G$ is just the exponential map.)

Now let $(\nu_t)_{t \geq 0}$ be strictly stable on $\mathbb{R}$ with index α, define $\check{\tau}_t : x \longmapsto t^\alpha x$, $t > 0$, $x \in \mathbb{R}$. Then $\check{\tau}_t(\nu_s) = \nu_{st}$. $\check{\tau}_t$ can be regarded as automorphism of S, therefore the dual map $\tau_t : G \longrightarrow G$ is an automorphism of G.
$y \in \mathbb{R}_d$ is identified with a character γ_y of G, such that $\langle \tau_t(g), \gamma_x \rangle = \langle g, \gamma_{\check{\tau}_t(x)} \rangle$, $t > 0$, $x \in \mathbb{R}_d$, $g \in G$.

Define $\mu_t := \hat{i}(\nu_t)$, $t \geq 0$. Obviously

$$\tau_t(\mu_s) = \tau_t(\hat{i}(\nu_s)) = \hat{i}(\check{\tau}_t(\nu_s)) = \hat{i}(\nu_{ts}) = \mu_{ts}, \quad t,s > 0.$$

So $(\mu_s)_{s \geq 0}$ is stable w.r.t. $(\tau_t)_{t > 0}$.
The group (τ_t) is not contracting on G as G is compact, but on the range of the exponential map $\hat{i}(\mathbb{R})$ τ_t acts contracting:
for $x \in \mathbb{R}$ $\tau_t(\hat{i}(x)) = \hat{i}(t^\alpha x) \xrightarrow[t \to 0]{} \hat{i}(0) = e$.
On the other hand $\mu_t = \hat{i}(\nu_t)$ is concentrated on $\hat{i}(\mathbb{R})$.
We note that $t \longrightarrow \tau_t$ is not continuous: There exist elements $g \in G$ which are non - continuous characters on $\mathbb{R}$.

<u>4.2.b Example</u> A similar construction yields the existence of a compact connected group G with the following properties: There exists a c.c.s. $(\mu_t)_{t>0}$, a multiplicative group

$(\tau_t)_{t \in \mathbb{Q}_+^*} \subseteq \mathrm{Aut}(G)$ such that

$$\tau_s(\mu_t) = \mu_{s^{1/\alpha} t} \quad \text{for } s \in \mathbb{Q}_+^*,$$

i.e. $(\mu_t)_{t \geq 0}$ is semistable w.r.t. $(\tau_s, s^{1/\alpha})$ for any $s \in \mathbb{Q}_+^*$.

(τ_t) cannot be extended to a group $(\tau_t)_{t \in \mathbb{R}_+^*}$.

As in 4.2.a τ_t acts contracting on the (dense) range of the exponential map.

Put $S := \mathbb{Q}$ (with the discrete topology), $G = \hat{S}$, $i : \mathbb{Q} \to \mathbb{R}$, $\hat{i}: \mathbb{R} \to \hat{S} = G$. If $(\nu_t)_{t \geq 0}$ is defined as in 4.2.a), $\mu_t = \hat{i}(\nu_t)$, $t \geq 0$, and if we define now

$$\check{\tau}_t : \mathbb{Q} \to \mathbb{Q}\ , \ \tau_t(x) = tx,$$

we have $\check{\tau}_t \in \mathrm{Aut}(\mathbb{Q})$ for $t \in \mathbb{Q}_+^*$.

Then from $\check{\tau}_t(\nu_s) = \nu_{t^\beta s}$, $t,s \in \mathbb{R}_+^*$, we get

$$\tau_t(\mu_s) = \mu_{st^\beta} \quad \text{for} \quad s \in \mathbb{R}_+^*, \quad t \in Q_+^* .$$

(Here again $\tau_t : \hat{\mathbb{Q}} \to \hat{\mathbb{Q}}$ is the dual automorphism and $\beta := 1/\alpha$).

<u>4.3 Example</u> The important assertion 1.6., namly that the Lévy-measure of a semistable distribution is zero or unbounded, is not valid for [semi-] self-decomposable distributions.

Precisely: For any locally compact group G and any $\tau \in \mathrm{Aut}(G)$ there exist nontrivial bounded semi-self-decomposable distributions.

Similar: For any continuous group $(\tau_t) \subseteq \mathrm{Aut}(G)$ there exist bounded self-decomposable distributions.

Fix a bounded Poisson distribution

$D = \alpha(\lambda - \varepsilon_e)$, $\lambda \in M^1(G)$, $\alpha > 0$ and fix $c \in (0,1)$. Then we put

$A := \sum_{k=0}^{\infty} c^k \tau^k(D)$. A is bounded and $\tau^{-1}(A) = cA + \tau^{-1}(D)$.

Therefore $A = \tau(A) + [D + (1-c)\tau(A)] = \tau(A) + B$

with $B = D + (1-c)\tau(A)$, i.e. A is semi-self-decomposable w.r.t. τ.

A similar construction yields the existence of bounded self-decomposable distributions:

We put $A := \int_0^\infty e^t \tau_{e^{-t}}(D)\, dt$ and obtain for $s \in (0,1)$

$$A = \tau_s(A) + (1-s)A + s \cdot \int_0^{-\log s} e^{-t} \tau_{e^t}(D)\, dt.$$

Final remarks 1. It is natural to ask for the connection between the concept of semi-stability and limit laws for measures on groups. This will be worked out in a forthcoming paper.

2. The following generalization is quite natural: Let $K \subseteq G$ be a compact subgroup and consider a continuous convolution semigroup $(\mu_t, t \geq 0, \mu_0 = \omega_K)$ with nontrivial idempotent ω_K - the Haar measure on K. The generating distribution is defined as a distribution on $\mathcal{D}_K(G) := \{f \in \mathcal{D}(G) | f$ invariant under $K\}$. The examples (s. [1], [3]) lead to the following

Conjecture: If $A \in \mathcal{D}'_K(G)$ is a [semi-] stable generating distribution w.r.t. $(\tau_t) \subseteq \mathrm{Aut}(G)$ $[(\tau, c)]$ on a Lie group G, then there exist (i) a nilpotent, connected, simply - connected Lie group N, (ii) a homomorphism $\varphi : K \longrightarrow \mathrm{Aut}(N)$ (iii) an injection of the semidirect product $K \circledS_\varphi N \longrightarrow G$ and (iv) a [semi-] stable distribution A on N, which is (v) invariant under $\varphi(K)$, such that the canonical embedding of B into $\mathcal{D}_K(G)'$ via $N \longrightarrow K \circledS_\varphi N \longrightarrow G$ yields, up to a shift, the distribution A.

3. The representation of a [semi-] stable distribution A by the corresponding distribution $\mathring{A}$ on the vector space does not give informations about the generated measures $\mathcal{E}xp(tA)$ on G. In the case of the Heisenberg group stable semigroups w.r.t. a special class of automorphisms (τ_t) are studied in [4]. In the paper [16] results on (strictly) semistable measures on general groups are collected.

4. In the case of the Heisenberg group the list of stable generating distributions is completely known, see [2]. In the forthcoming paper [3] similar results are obtained for groups which are compact extensions of the Heisenberg group. There the authors consider the general situation of stable semigroups with non trivial idempotent factors.

Literature

1. P. BALDI: "Lois stables sur les déplacements de $\mathbb{R}^d$". Probability measures on groups. Proceedings. Lecture Notes in Math. 706 (1979) 1-9.

2. TH. DRISCH, L. GALLARDO: "Stable laws on the Heisenberg group". In: Probability measures on groups. Prodeedings. Oberwolfach 1983 (This volume)

3. TH. DRISCH, L. GALLARDO: "Stable laws on the diamond group". (In preparation).

4. P. GŁOWACKI: "Stable semigroups of measures on the Heisenberg group".(Preprint 1983).

5. W. HAZOD: "Stetige Halbgruppen von Wahrscheinlichkeitsmaßen und erzeugende Distributionen". Lecture Notes Math. Vol. 595 (1977).

6. W. HAZOD: "Stable probabilities on locally compact groups". In: Probability measures on groups. Proceedings, Lecture Notes Math. 928. Springer: Berlin Heidelberg, New York (1982).

7. E. HEWITT, K.E. ROSS: "Abstract Harmonic Analysis I". Berlin-Göttingen-Heidelberg: Springer 1963.

8. H. HEYER: "Probability measures on locally compact groups". Ergebnisse der Math. Berlin-Heidelberg-New York, Springer 1977.

9. R. JAJTE: " Semistable probability measures on $\mathbb{R}^N$". Studia Math.61 (1977) 29-39.

10. R. JAJTE, E. HENSZ: "On a class of limit laws". Theory Prob. Appl. 23 (1978) 206-211.

11. A. LUCZAK: "Operator semi-stable probability measures on $\mathbb{R}^N$". Coll. Math. 45 (1981) 287-299.

12. P.R. MÜLLER - RÖMER: "Kontrahierbare Erweiterungen kontrahierbarer Gruppen". J. reine angew. Math. 283/284 (1976) 238-264.

13. M. SHARPE: "Operator stable probability measures on vector groups". Trans. Amer. Math. Soc. 136 (1969) 51-65.

14. K. SCHMIDT: "Stable probability measures on $\mathbb{R}^n$". Z. Wahrscheinlichkeitstheorie verw. Geb. 33 (1975) 19-31.

15. E. SIEBERT: "Absolut-Stetigkeit und Träger von Gauß-Verteilungen auf lokalkompakten Gruppen". Math. Ann. 210, 129-147 (1974).

16. E. SIEBERT: "Semistable convolution semigroups on measurable and topological groups". (Preprint 1982). To appear in: Ann. Inst. Henri Poincaré .

Wilfried Hazod
Universität Dortmund
Abteilung Mathematik
Postfach 500 500

D-4600 Dortmund 50

ON THE LIMIT OF THE AVERAGE OF THE VALUES OF A FUNCTION AT RANDOM POINTS

By Göran Högnäs*and Arunava Mukherjea

1. In this paper, we study the asymptotic behavior of the sequence $(1/N) \sum_{n=1}^{N} f(S_n)$ as $N \to \infty$, where f is a bounded Borel measurable real-valued function and (S_n) is the sequence of partial sums of an i.i.d. sequence (X_i) of random variables. Here we consider the case of a lattice first, then the reals and finally Z_2 and R_2. Our results and methods in this paper are inspired by the brilliant paper of Meilijson [1]. In [1], complete results have been obtained in the case when $0 < E(X_i) < \infty$. There are also some results in [1] for the case $E(X_i)=0$, which are, however, not complete. The main purpose of our paper is to exploit the methods in [1] a little further to learn more about the case $E(X_i)=0$. Let us also point out that in the context of probability measures on locally compact groups, the problem of finding the limiting behavior of $(1/N) \sum_{n=1}^{N} \beta^{(n)}(A)$, where A is non-compact (resp. infinite) when the group is non-discrete (resp. discrete), is natural, but completely unexplored. [The corresponding problem when A is compact (resp. finite) is completely solved (in groups and even in the case of discrete and some other classes of semigroups) and the solution is well-known.] The results in [1] and this paper provide some answers in this context. Our main results in one dimension are the following:

<u>Theorem 1.</u> Let (X_i) be a i.i.d. sequence of integer-valued random variables with common probability distribution induced by the probability measure F. Suppose that the support of F generates Z, the

---

*This author's research is supported by a grant from the Academy of Finland.

integers as a group. Let f be a bounded function on Z such that for some real numbers a and b the following inequalities hold:

$$a \leq \underline{\lim}_{N\to\infty} \inf_{M\in Z} (1/N) \sum_{n=M+1}^{M+N} f(n) \leq \overline{\lim}_{N\to\infty} \sup_{M\in Z} (1/N) \sum_{n=M+1}^{M+N} f(n) \leq b. \quad (1)$$

Write $S_n = X_1+X_2+...+X_n$. Then we have:

$$a \leq \underline{\lim}_{N\to\infty} \inf_{\substack{M\in Z \\ j\geq 0}} (1/N) \sum_{n=1}^{N} E(f(S_{n+j}+M)) \leq \overline{\lim}_{N\to\infty} \sup_{\substack{M\in Z \\ j\geq 0}} (1/N) \sum_{n=1}^{N} E(f(S_{n+j}+M)) \leq b. \quad (2)$$

We remark that the result above also holds for any lattice instead of Z. The same is true for the next theorem.

Theorem 2. Under the hypotheses of Theorem 1, we have almost surely, for each M in Z,

$$a \leq \underline{\lim}_{N\to\infty} (1/N) \sum_{n=1}^{N} f(S_n+M) \leq \overline{\lim}_{N\to\infty} (1/N) \sum_{n=1}^{N} f(S_n+M) \leq b.$$

Our result in the non-lattice case is the following:

Theorem 3. Let F be a probability measure on the Borel sets of the reals such that the support of F is not contained in any lattice. Let (X_i) be a i.i.d. sequence of real random variables with common distribution induced by F and U be a uniformly distributed random variable on [0,1], which is independent of the X_i's. Let f be a bounded Borel measurable real function such that for some real numbers a and b, the following inequalities hold:

$$a \leq \underline{\lim}_{N\to\infty} (1/N) \inf_{M\varepsilon R} \int_M^{M+N} f(x)\, dx \tag{3}$$

$$\leq \overline{\lim}_{N\to\infty} (1/N) \sup_{M\varepsilon R} \int_M^{M+N} f(x)\, dx \leq b.$$

Then the following result holds:

$$a \quad \underline{\lim}_{N\to\infty} (1/N) \inf_{M\varepsilon Z^+} \sum_{n=M+1}^{M+N} E(fU+S_n)) \tag{4}$$

$$\leq \overline{\lim}_{N\to\infty} (1/N) \sup_{M\varepsilon Z^+} \sum_{n=M+1}^{M+N} E(f(U+S_n)) \leq b.$$

<u>Theorem 4.</u> Under the same hypotheses as in Theorem 3, the following hold:

(i) For almost all x (with respect to the Lebesgue measure),

$$a \leq \underline{\lim}_{N\to\infty} (1/N) \sum_{n=1}^{N} f(S_n+x) \leq \overline{\lim}_{N\to\infty} (1/N) \sum_{n=1}^{N} f(S_n+x) \leq b \quad \text{a.s.}$$

(ii) If the distribution of X_1 is not singular with respect to the Lebesgue measure, then the statement in (i) holds for x=0.

Though we used $\underline{\lim}$ and $\overline{\lim}$ above to maintain the form in which the results appeared in [1], it is relevant to point out that for any bounded sequence (x_n) of real numbers, the limits $\lim_{N\to\infty} \sup_{M\varepsilon Z} \frac{1}{N} \sum_{n=M+1}^{M+N} x_n$ and $\lim_{N\to\infty} \inf_{M\varepsilon Z} \frac{1}{N} \sum_{n=M+1}^{M+N} x_n$ both exist and define sublinear functionals on the Banach space ℓ_∞ of bounded sequences of real numbers. In fact, the maximal value of Banach limits on the sequence (x_n) is the above "lim sup" expression and the corresponding minimal value is the "lim inf" expression. [Note that Banach limits are linear functionals Φ on ℓ_∞ such that if $(x_n) \in \ell_\infty$, then (i) $\Phi[(x_n)] \geq 0$ if $x_n \geq 0$ for $0 < n < \infty$ (ii) $\Phi[(x_{n+1})] = \Phi[(x_n)]$ and (iii) $\Phi[(1)] = 1$, $(1) = (1,1,1,....)$.] It is also relevant to point out that in the continuous case, for an almost periodic function f on the reals,

$$\lim_{N\to\infty} \frac{1}{N} \int_M^{M+N} f(x)\, dx$$

exists uniformly with respect to all reals M and is the same for all real numbers M.

In section 2, we will discuss two key arguments which are essential for the proofs of the above theorems given in section 3. In section 4, we discuss the two-dimensional problem.

2. We present in this section two key arguments (in our context). For completeness, we include some details though these tools have been used earlier by other authors including Meilijson.

I. The Ornstein coupling.

Let F be a non-degenerate probability measure on the lattice $L_d = \{nd: n \in Z\}$ such that L_d is the smallest lattice containing its support. Let (X_i) be a sequence of i.i.d. L_d-valued random variables with distribution induced by F. We assume that $F(0) > 0$. Let J be a sufficiently large positive integer such that the measure F_J defined by

$$F_J(B) = F(B \cap [-J,J])/F([-J,J])$$

is non-degenerate and its support generates L_d. Let (Y_i) be a sequence of i.i.d. L_d-valued random variables with common distribution induced by F_J such that each Y_i is also independent of the X_j's.
Let us define for each positive integer n,

$$X'_n = X_n \text{ if } X_n > J;$$

$$= Y_n \text{ if } X_n \leq J.$$

Then, X'_n has also distribution induced by F. Write $Z_n = X_n - X'_n$ so that Z_ns are bounded and have symmetric distribution. Notice that since $F(0) > 0$, the support of the distribution of Z_n generates also the lattice L_d. Hence, the walk $Z_{on} = Z_1+Z_2+\ldots+Z_n$ is recurrent and the random variable T defined by

$$T = \inf\{n: Z_1+Z_2+\ldots+Z_n = d\}$$

is almost surely finite and a stopping time. Now we define the random variables (W_i) by

$$W_i = X'_i \text{ if } i \leq T;$$

$$= X_i \text{ if } i > T.$$

Then, the sequence (W_i) is an i.i.d. sequence with F inducing their common distribution. [Perhaps this needs a little proof. First, notice that the event $\{T=j\}$, $j < i$, depends only on $\{X_k, X'_k: k \leq j\}$ and therefore, is independent of the events $\{X_i=n\}$ as well as $\{X'_i=n\}$. Thus,

$$P(W_i=n) = \sum_{j=1}^{\infty} P(W_i=n,\ T=j)$$

$$= \sum_{j<i} P(X_i=n,T=j) + \sum_{j\geq i} P(X'_i=n,T=j)$$

$$= \sum_{j<i} P(X_i=n)P(T=j) + \sum_{j\geq i} P(X'_i=n,T=j)$$

$$= \sum_{j<i} P(X'_i=n,T=j) + \sum_{j\geq i} P(X'_i=n,T=j) = P(X'_i=n) = F(n).$$

The independence of the W_i's can be similarly established.]

Let $S_n^{(2)} = W_1+W_2+\ldots+W_n$. Notice that $n \geq T \Rightarrow S_n = S_n^{(2)} + d$.

We remark that by an exactly similar argument as above, we can find for any given N, N copies of (S_n), $(S_n^{(j)})$, $j=2,3,\ldots,N+1$, and a stopping time T such that

$$n \geq T \Rightarrow S_n^{(j+1)} = S_n^{(j)} + d,\ j=1,2,\ldots,N, \qquad S_n^{(1)} \equiv S_n.$$

Let us also remark that the coupling argument can be easily carried over to the reals, 2-dimensional lattices and R^2 with obvious modifications.

II. <u>The retardation argument.</u>

For simplicity, we consider here Z instead of the lattice L_d. Let F be a non-degenerate probability measure as in (I) and (X_i) be an i.i.d. sequence of integer-valued random variables with common distribution induced by F. Let (H_i) be also an i.i.d. sequence (and also independent of the X_i's) such that

$$P(H_i=0)=p \text{ and } P(H_i=1)=1-p, \text{ where } 0 < p < 1.$$

We define the i.i.d. sequence (X_i') by

$$X_i' = X_i \text{ if } H_i = 1;$$

$$= 0 \text{ if } H_i = 0.$$

Then, the distribution of X_i' is induced by the measure

$$F_1(x)=p+(1-p)F(x) \text{ if } x \geq 0;$$

$$= (1-p)F(x) \text{ if } x < 0.$$

Notice that $F_1(0) > 0$. Let $S_n = \sum_{i=1}^{n} X_i$ and $S_n' = \sum_{i=1}^{n} X_i'$. Suppose that f is a bounded function on Z such that for some real numbers a and b, the inequalities (2) hold for the random walk (S_n'). Then we claim that the same inequalities will also hold for the random walk (S_n).

To prove this, let $\varepsilon > 0$. Then there exists a positive integer N such that $n \geq N$ implies that for any positive integer j and any integer M,

$$a-\varepsilon < (1/n) \sum_{i=j+1}^{j+n} E(f(S_i'+M)) < b+\varepsilon. \tag{5}$$

Let us define the stopping times $T_1, T_2, \ldots$ successively by

$$T_{i+1} = \inf(\ n > T_i : H_n = 1).$$

Observe that for any positive integer j and any integer M,

$$E(f(S_{T_i}' + M))$$

$$= E[f(S_{T_i}' + M) . \sum_{n_1 < n_2 < \ldots < n_i} I_{\{T_1=n_1,\ldots,T_i=n_i\}}]$$

$$= \sum_{n_1 < n_2 < \ldots < n_i} E(f(X_{n_1}+\ldots+X_{n_i} + M)).P(T_1=n_1,\ldots,T_i=n_i)$$

$$= E(f(X_{j+1}+\ldots+X_{j+i}+M)).$$

In view of this, it is sufficient to show that given $\varepsilon > 0$, there is a positive integer n_o such that $n \geq n_o$ implies that for any positive integer j and any integer M,

$$a-\varepsilon < (1/n) \sum_{i=1}^{n} E(f(S'_{T_{j+i}} + M)) < b+\varepsilon. \tag{6}$$

To prove (6), we use (5). From (5), we have a.s. for $n \geq N$,

$$a-\varepsilon < [1/T_{j+n+1}-T_{j+1})]. \sum_{i=T_{j+1}}^{T_{j+n+1}} E(f(S'_i+M)) < b+\varepsilon. \tag{7}$$

This means that almost surely we have:

$$a-\varepsilon < \sum_{i=1}^{n} E(f(S'_{T_{j+i}} + M)).[\frac{T_{j+i+1}-T_{j+i}}{T_{j+n+1}-T_{j+1}}] < b+\varepsilon. \tag{8}$$

Since $E[(T_{j+i+1}-T_{j+i})/(T_{j+n+1}-T_{j+1})] = 1/n$, we have by taking expectations in (8),

$$a-\varepsilon < (1/n) \sum_{i=1}^{n} E(f(S'_{T_{j+i}} + M)) < b+\varepsilon.$$

This proves our claim.

Let us remark that the above argument carries over to the non-lattice case.

3. In this section, we present the proofs of the theorems stated in section 1.

Proof of Theorem 1. Because of the retardation argument in section 2, we assume with no loss of generality that $F(0) > 0$. This means that now we can use the Ornstein coupling to find desired copies of the random walk (S_n). It follows from condition (1) that given $\varepsilon > 0$, there is a positive integer n_o such that $n \geq n_o$ implies that for any integer m, we have:

$$a-(\varepsilon/2) < (1/n) \sum_{i=m+1}^{m+n} f(i) < b+(\varepsilon/2). \tag{9}$$

By the coupling method in section 2, there are n_o many random walks $(S_n^{(1)})$, $(S_n^{(2)})$,.....,$(S_n^{(n_0)})$, each having the same distribution as that of (S_n), and a stopping time T such that

$$n \geq T \Rightarrow S_n^{(i+1)} = S_n^{(i)} + 1 \quad \text{(almost surely)}$$

for $0 \leq i < n_o$. Here, $S_n^{(o)} = S_n$. Choose a positive integer A such that $P(T > A) < \varepsilon/2$. Let $L > 4A/\varepsilon$ and M be any integer. Then, for any positive integer j, on the event $\{T \leq A\}$, the expression

$$\sum_{1 \leq i \leq L} \sum_{1 \leq k \leq n_o} f(S_{i+j}^{(k)} + M)$$

is at most $An_o.||f|| + (b+\varepsilon/2)n_o(L-A)$, but at least $-An_o||f|| + (a-\varepsilon/2)n_o(L-A)$; on the event $\{T > A\}$, the above sum is at most $Ln_o||f||$ and at least $-Ln_o||f||$. Hence, taking expected values, the theorem follows easily.

Proof of Theorem 2. We will use Theorem 1 here. By this theorem, given $\varepsilon > 0$, there exists a positive integer N_o such that $N \geq N_o$ implies that for any integer m,

$$a-\varepsilon < \sum_{i=1}^{N} (1/N)E(f(S_i+m)) < b+\varepsilon. \tag{10}$$

Let M be any integer and N be an integer greater than N_o. We define:

$$Y_n = (1/N) \sum_{i=(n-1)N+1}^{nN} f(S_i+M) \quad ; \quad Y_o = 0.$$

For $n \geq 1$, let $Z_n = Y_n - E(Y_n | \Lambda_{(n-1)N})$, where Λ_t is the smallest sigma-algebra determined by $\{X_s : s \leq t\}$. Then the process $M_r = \sum_{n=1}^{r} Z_n$ is a martingale with mean zero and uniformly bounded increments so that $(1/r)M_r$ tends to zero almost surely. For $n > 1$,

$$\begin{aligned}
&E(Y_n | \Lambda_{(n-1)N}) \\
&= E((1/N) \sum_{i=(n-1)N+1}^{nN} f(S_i+M) | \Lambda_{(n-1)N}) \\
&= \sum_k E((1/N) \sum_{i=(n-1)N+1}^{nN} f(S_i+M) | S_{(n-1)N}=k).I_{\{S_{(n-1)N}=k\}} \\
&= \sum_k E((1/N) \sum_{i=1}^{N} f(X_{(n-1)N+1}+\ldots+X_{(n-1)N+i}+k+M) | S_{(n-1)N}=k) I_{\{S_{(n-1)N}=k\}} \\
&= \sum_k E((1/N) \sum_{i=1}^{N} f(S_i+k+M)).I_{\{S_{(n-1)N}=k\}},
\end{aligned}$$

and this belongs to $(a-\varepsilon, b+\varepsilon)$, by (10).

Now notice that

$$(1/r)M_r = (1/Nr)\sum_{i=1}^{Nr} f(S_i+M) - (1/r)\sum_{n=1}^{r} E(Y_n|\Lambda_{(n-1)N})$$

goes to zero as r tends to infinity, almost surely. The theorem follows.

Proof of Theorem 3. By condition (3), given $\varepsilon > 0$, there exists a positive integer N_o such that for $N \geq N_o$, for all real numbers M,

$$a-\varepsilon < (1/N)\int_M^{M+N} f(x)\,dx < b+\varepsilon. \tag{11}$$

Fix some $N > N_o$. By the coupling argument, there are then N many random walks $(S_n^{(1)}),\ldots,(S_n^{(N)})$, each having the same distribution as that of the walk (S_n), and a stopping time T such that

$$n \geq T \Rightarrow S_n^{(k+1)}-S_n^{(k)} \in (1,1+\varepsilon),\ 0 \leq j < N.$$

Notice that for any positive integers M and L,

$$\sum_{i=M+1}^{M+L} E(f(U+S_i))$$

$$= \sum_{i=M+1}^{M+L} \int f(x)\Pr(x-1 < S_i \leq x)\,dx$$

$$= \int f(x)E\{\sum_{i=M+1}^{M+L} I_{[S_i,S_i+1)}(x)\}\,dx$$

$$= E[\sum_{i=M+1}^{M+L} \int_{S_i}^{S_i+1} f(x)\,dx\ ].$$

Choose A so that $\Pr(T > A) < \varepsilon/2.||f||$. Notice that when $T \leq A$,

$$i \geq A \Rightarrow S_i^{(k+1)}-S_i^{(k)} \in (1,1+\varepsilon),\ 0 \leq k < N.$$

Choose L so large that

$$\varepsilon(L-A)/L + ||f||.(A/L) < 3\varepsilon/2.$$

Now since

$$(1/NL)\sum_{i=M+A}^{M+L}\ \sum_{k=1}^{N} \int_{S_i^{(k)}}^{S_i^{(k+1)}} f(x)\,dx$$

$$\leq (1/NL)(L-A)[\sup_m \int_m^{m+N} f(x)\,dx + ||f||2N\varepsilon],$$

it follows easily by taking expectations that

$$(1/L)\sum_{i=M+1}^{M+L} E(f(U+S_i)) < b+2\varepsilon \quad \text{(for sufficiently large L).}$$

The other side of (4) also follows similarly.

Proof of Theorem 4 (i). Suppose that there exists a Lebesgue measurable subset D with m(D) positive such that for each x in D, there is a set of positive probability where we have:

$$\overline{\lim_{N\to\infty}} \ (1/N) \sum_{n=1}^{N} f(S_n+x) \quad > \quad b.$$

Then, by the Hewitt-Savage zero-one law, this inequality holds almost surely. Hence,

$$\begin{aligned} 0 &> \int_D [\int \{b-\lim_{N\to\infty} (1/N) \sum_{n=1}^{N} f(S_n+x)\ \}dP]dx \\ &= \int_D E[b-\lim_{N\to\infty} (1/N) \sum_{n=1}^{N} f(S_n+U) \mid U+x]dx \end{aligned} \qquad (12)$$

But it can also be verified, exactly in the same way as in Theorem 2, that almost surely,

$$a \le \underline{\lim_{N\to\infty}} (1/N) \sum_{n=1}^{N} f(S_n+U) \le \overline{\lim_{N\to\infty}} (1/N) \sum_{n=1}^{N} f(S_n+U) \le b,$$

which, however, contradicts (12).

Proof of Theorem 4 (ii). We omit this part of the proof since it can be given following the same idea as given in page 198[1].

4. In this section, we consider the two-dimensional analogue of the problem considered in section 3. Let $((X_i,Y_i))_{i=1}^{\infty}$ be a sequence of i.i.d. random vectors on the lattice $Z_2=\{(i,j):i,j\in Z\}$. We assume that the distribution of (X_1,Y_1) is non-degenerate and that its support generates Z_2 as a group. It is easily observed that analogs of Theorems 1 and 2 hold in two dimensions with obvious modifications, but without any restrictions on the expectations of X_1 and Y_1.

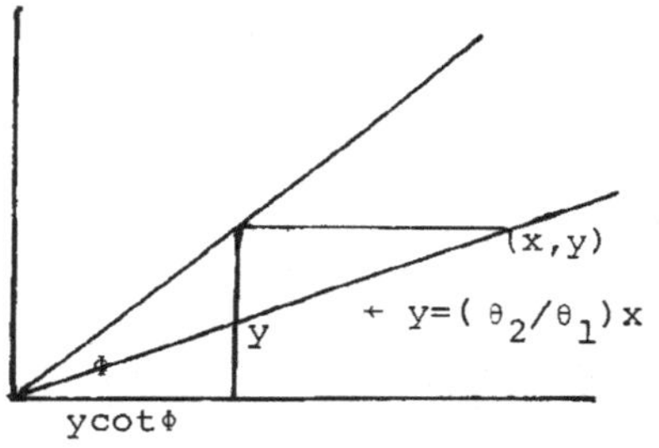

$(x,y) \in A_\Phi$

iff $0 < y\cot\Phi < x < \infty.$

Let us now assume that $E(X_1)=\theta_1 > 0$ and $E(Y_1)=\theta_2 > 0$. Let us define the cone A_Φ as the cone starting at the origin and having the line $y=(\theta_2/\theta_1)x$ as its bisector. Let f be a bounded function on Z_2 such that for some real numbers a and b, the following inequalities hold:

$$a \le \varliminf_{N\to\infty} \inf_{\substack{(M_1,M_2)\,\varepsilon\, A_\Phi \\ (M_1,M_2+N)\,\varepsilon\, A_\Phi}} (1/N^2) \sum_{i=M_1+1}^{M_1+N} \sum_{\substack{j=M_2+1 \\ (i,j)\,\varepsilon\, A_\Phi}}^{M_2+N} f(i,j) \tag{13}$$

$$\le \varlimsup_{N\to\infty} \sup_{\substack{(M_1,M_2)\,\varepsilon\, A_\Phi \\ (M_1,M_2+N)\,\varepsilon\, A_\Phi}} (1/N^2) \sum_{i=M_1+1}^{M_1+N} \sum_{\substack{j=M_2+1 \\ (i,j)\,\varepsilon A_\Phi}}^{M_2+N} f(i,j) \le b.$$

[Notice that if $f=I_{A_\Phi}$, then we can take above a=b=1.]
The set A_Φ is a semigroup under addition; that is, $(x_1,y_1)\,\varepsilon\, A_\Phi$ and $(x_2,y_2)\,\varepsilon\, A_\Phi$ imply that $(x_1+x_2,y_1+y_2)\,\varepsilon\, A_\Phi$. Let us write:

$S_n=X_1+X_2+\ldots+X_n$ and $T_n=Y_1+Y_2+\ldots+Y_n$.

Then we have the following results:

Theorem 5. If f satisfies (13), then the following inequalities hold:

$$a \le \varliminf_{N\to\infty} (1/N) \inf_{(M_1,M_2)\,\varepsilon A_\Phi} \sum_{n=1}^{N} E[f((S_n,T_n)+(M_1,M_2))] \tag{14}$$

$$\le \varlimsup_{N\to\infty} (1/N) \sup_{(M_1,M_2)\,\varepsilon\, A_\Phi} \sum_{n=1}^{N} E[f((S_n,T_n)+(M_1,M_2))] \le b.$$

Proof. First, notice that by the strong law of large numbers, almost surely, (S_n,T_n) will be eventually in A_Φ. By (13), given $\varepsilon > 0$, there exists a positive integer N_o such that $N \ge N_o$ implies that for any ordered pair (M_1,M_2) in A_Φ, we have:

$$a-\varepsilon < (1/N^2) \sum_{i=M_1+1}^{M_1+N} \sum_{j=M_2+1}^{M_2+N} f(i,j) < b+\varepsilon,\ (i,j)\,\varepsilon A_\Phi. \tag{15}$$

By a similar coupling procedure as is given in section 2, we can find N_o^2 new copies of our original walk (S_n,T_n), the random walks

$$\{(S_n^{(j)}, T_n^{(j)}): 1 \le j \le N_o^2\},$$

each having the same distribution as (S_n, T_n), and a stopping time T_o such that

$$i > T_o \Rightarrow (S_i^{(j+1)}, T_i^{(j+1)}) - (S_i^{(j)}, T_i^{(j)}) = (p+1, k)$$

where $j = kN_o + p$, $0 \le k \le N_o$ and $0 \le p < N_o$. [In other words, the stopping time T_o, and the new walks are constructed in such a way that for $i > T_o$, the set of points $\{(S_i^{(j)}, T_i^{(j)}): 1 \le j \le N_o^2\}$ fill up a square of side N_o and actually can be identified with the set $\{(S_i, T_i) + (p,k): 1 \le p,k \le N_o\}$.] Let us now define the random variable $c(\omega)$ by

$$C(\omega) = \sup\{n: (S_n^{(j)}(\omega), T_n^{(j)}(\omega)) \notin A_\Phi \text{ for some } j, 1 \le j \le N_o^2\}$$

Then, by the strong law of large numbers, $C(\omega)$ is finite almost surely. Let A be a number so large that

$$\Pr(C > A) < \varepsilon/2 \text{ and } \Pr(T_o > A) < \varepsilon/2.$$

Then, $\Pr(T_o \le A, C \le A) > 1-\varepsilon$. Let $(M_1, M_2) \in A_\Phi$ and L be a sufficiently large number. Consider the expression

$$(1/L) \sum_{i=1}^{L} \sum_{j=1}^{N_o^2} f((S_i^{(j)}, T_i^{(j)}) + (M_1, M_2)).$$

On the set $\{T_o \le A, C \le A\}$, this expression is almost surely at most

$$(1/L) A N_o^2 . ||f|| + (1/L)(L-A) N_o^2 (b+\varepsilon)$$

and at least

$$-(1/L)AN_o^2.||f|| + (1/L)(L-A)N_o^2(a-\varepsilon).$$

The theorem now follows easily by taking expectations.

Theorem 6. If f satisfies (13), then almost surely the following inequalities hold: For every (M_1,M_2) in A_Φ,

$$a \le \underline{\lim}_{N\to\infty} (1/N) \sum_{i=1}^{N} f((S_i,T_i) + (M_1,M_2))$$

$$\le \overline{\lim}_{N\to\infty} (1/N) \sum_{i=1}^{N} f((S_i,T_i) + (M_1,M_2)) \le b.$$

Proof. We sketch the proof briefly since the idea is the same as before. As in the one-dimensional case, we define

$$\text{(i)} \quad Y_n = (1/N) \sum_{i=(n-1)N+1}^{nN} f((S_i,T_i) + (M_1,M_2)); \; Y_o=0$$

$$\text{(ii)} \quad Z_n = Y_n - E(Y_n|F_{(n-1)N}), F_t \equiv \tau\{(X_s,T_s): s \le t\}.$$

Let $M_r = \sum_{n=1}^{r} Z_n$. We then have:

$$E(Y_n|F_{(n-1)N})$$

$$= \sum_{(j,k)} E((1/N) \sum_{i=1}^{N} f((S_i,T_i)+(j,k)+(M_1,M_2))I_{\{(S_{(n-1)N},T_{(n-1)N})=(j,k)\}}.$$

Note that those terms inside the summation for which $(j,k) \in A_\Phi$ will clearly belong to $(a-\varepsilon,b+\varepsilon)$; for the other terms, we use the strong law of large numbers to observe that since A_Φ is a semigroup and (S_i,T_i) is almost surely eventually in A_Φ, we will have almost surely

$$(1/r) \sum_{n=1}^{r} E(Y_n|F_{(n-1)N}) \in (a-\varepsilon,b+\varepsilon)$$

for sufficiently large r. The theorem now follows as in the one-dimensional case.

Note that the above results hold for any cone A_Φ about the line $y = \frac{\theta_2}{\theta_1} x$, not just the cone shown in the figure. Let us also mention that in (13), we could replace A_Φ by the truncated cone $A_{\Phi(M)}$, defined by

$$A_{\Phi(M)} = \{(x,y) \in A_\Phi : x,y \geq M\}$$

where M is a given positive integer. In the case when $\theta_2 = 0$ and $\theta_1 \neq 0$, the same results hold. However, there are obvious difficulties when $\theta_1 = \theta_2 = 0$ and we cannot use these cones at all.

REFERENCES

1. Isaac Meilijson (1973): The average of the values of a function at random points, Israel J. Math. 15, 193-203.

2. D. Ornstein (1969): Random walks I, Trans. Amer. Math. Soc. 138, 1-44.

G. Högnäs
Åbo Akademi
Åbo, Finland

A. Mukherjea
University of South Florida
Tampa, Florida 33620

NON-COMMUTATIVE SUBADDITIVE ERGODIC THEOREMS FOR SEMIFINITE VON NEUMANN ALGEBRAS

by

R. JAJTE

0. Introduction and preliminaries. The Kingman subadditive ergodic theorem for sequences can be formulated as follows:

THEOREM A (|8| |9|). *Let* (Ω,B,p) *be a probability space, and let* Θ *be a measure preserving transformation on* Ω. *Let* $\{f_n\}$ *be a sequence of integrable functions on* (Ω,B,p), *such that* $f_{n+k} \leq f_n + f_k \circ \Theta^n$ *for all integers* n,k *and* $\inf_n \int_\Omega f_n/n\, dp > -\infty$. *Then the sequence* $\frac{1}{n} f_n$ *converges in* L_1 *and almost everywhere.*

Our main purpose is to generalize this theorem to the von Neumann algebra context. Let us begin with some notation and definitions. Let A be a semifinite von Neumann algebra (acting in a Hilbert space H) with a semifinite faithful normal trace τ. $L_1(A,\tau)$ stands for the space of measurable operators integrable with respect to τ (see for example [13],[15]).

We shall identify L_1 with the predual A_* of A and the dual of L_1 with A ([4],[15]).

The symbol $L_1^{(h)}$ denotes the hermitian part of L_1. Proj A will stand for the lattice of all orthogonal projectors from A. Let $T = \{0,1,2,\ldots\}$ or $T = [0,\infty)$.

A map $\xi : T \to L_1^{(h)}(A,\tau)$ is said to be subadditive if there is a group $(u^t, t \in T)$ of τ-preserving $*$-automorphisms of A such that, for each $s,t \in T$, we have

1° $\xi_{s+t} \leq \xi_s + u^s \xi_t$

2° $\inf\limits_{t\in T} \frac{\tau(\xi_t)}{t} > -\infty$

(of course, u^s in 1° should be treated as the (unique) extension of u^s to L_1).

A net ξ_α is said to converge almost uniformly to $\hat{\xi}$ if, for each $\varepsilon > 0$, there is a projection $p_\varepsilon \in \mathrm{Proj}\, A$ such that $\tau(1 - p_\varepsilon) < \varepsilon$ and $\|p_\varepsilon(\xi_\alpha - \hat{\xi})p_\varepsilon\|_\infty \underset{\alpha}{\to} 0$. Here and elsewhere $\|.\|_\infty$ denotes the norm in A and $\|.\|_1$ denotes the norm in L_1, i.e. $\|x\|_1 = \tau(|x|)$, where $|x| = (x^*x)^{\frac{1}{2}}$.

A non-commutative individual ergodic theorem (of Birkhoff type) was first obtained by E.C.Lance [12], and then F.I.Yeadon proved the following

THEOREM B ([18][19]). *Let* $\alpha : L_1 \to L_1$ *be a positive linear map such that* $0 \le \alpha(x) \le I$, $\tau(\alpha(x)) \le \tau(x)$, *whenever* $x \in L^1 \cap A$ *and* $0 \le x \le I$. *Then, for any* $x \in L_1$, *the averages*

$$s_n(x) = n^{-1}\sum_{k=0}^{n-1} \alpha^k(x)$$

converge almost uniformly and in L_1.

In section 1 we shall prove the following subadditive ergodic theorem in the von Neumann algebra set-up.

THEOREM 1. If $(\xi_n, n = 0,1,2,\ldots)$ *is a subadditive sequence, then* $n^{-1}\xi_n$ *converges in* L_1 *and almost uniformly to a* u_**-invariant element* $\hat{\xi} \in L_1$.

To prove a similar result for the continuous parameter t, we need more restrictions on ξ. Let us remark that, as it was noticed by Kingman [8], some additional assumptions concerning ξ_t are necessary even in the classical case (when A is a commutative algebra $L^\infty(\Omega,\mu)$ over a probability space (Ω,μ)).

Let $e_{d\lambda}\{|\xi|\}$ denote the spectral measure of $|\xi|$, that is,

$|\xi| = \int_0^\infty \lambda e_{d\lambda}\{|\xi|\}$. A subadditive map ξ is said to be separable if there exists a sequence $0 \le s_k \le 1$ such that, for all $\lambda > 0$, we have

$$(1) \qquad \bigvee_{1\le k\le N} e_{(\lambda,\infty)}\{|\xi_{s_k}|\} \uparrow \bigvee_{0\le t\le 1} e_{(\lambda,\infty)}\{|\xi_t|\}$$

as $N\to\infty$ (in the strong operator topology).

For $\lambda > 0$ let us put

$$(2) \qquad p(\lambda) = \bigvee_{0\le t\le 1} e_{(\lambda,\infty)}\{|\xi_t|\}$$

and

$$(3) \qquad q(\lambda) = \bigvee_{0\le s,t\le 1} e_{(\lambda,\infty)}\{|u^s\xi_t|\}.$$

In section 2 the following theorem will be proved.

THEOREM 2. Let $\xi : [0,\infty) \to L_1^{(h)}(A,\tau)$ *be a separable subadditive map such that the (semi) group* $(u^t, t \in T)$ *of* $*$*-automorphisms is weakly continuous. If*

$$(4) \qquad \int_0^\infty \tau(p(\lambda))d\lambda < \infty$$

holds, then the limit $\lim_{t\to\infty} \frac{\xi_t}{t} = \hat{\xi}$ *exists in* L_1. *If*

$$(5) \qquad \int_0^\infty \tau(q(\lambda))d\lambda < \infty$$

holds, then $\frac{\xi_t}{t} \to \hat{\xi}$ (as $t\to\infty$) *almost uniformly. Moreover (in both cases), the limit* $\hat{\xi}$ *is* u_*^t*-invariant for all* t *and belongs to* L_1.

The two theorems just formulated generalize to the von Neumann algebra set up the Kingman ergodic theorems for subadditive stochastic processes. In the proof of Theorem 2 we shall follow the general idea of Kingman [8] although the technicalities are different and more complicated in the case of a von Neumann algebra. Namely, we shall reduce the proof to Theorem 1.

Of course, in the commutative case (i.e. when $A = L_\infty$), Theorem 1 reduces to Kingman's result (Theorem A). In the case of additivity in 1° our theorem gives the Yeadon result Theorem B for α being a *-automorphism of A. In particular, Theorem 1 extends the classical Birkoff theorem. Let us mention here a few other results concerning the non-commutative pointwise ergodic theorems (generalizing the classical results of Birkhoff, Akcoglu and others ([5][10]). Namely, we would like to mention, besides the papers of Lance and Yeadon, the works of I.Conze and N.Dang Ngoc [2], B.Kümmerer [11], H.Goldstein [7] and D.Petz [14]. It would be very interesting to prove the subadditive versions of the results just mentioned, in particular, the subadditive ergodic theorems for states and weights (also in the style of [1]).

1. PROOF of THEOREM 1. We divide the proof into a few steeps.

PROPOSITION 1. Let $\{\xi_n\}$ *be as in Theorem 1. Put*

$$(6) \qquad y_m = \frac{1}{m} \sum_{k=1}^{m} (\xi_k - u\xi_{k-1})$$

for each $m \geq 1$. *Then, for each* n, $1 \leq n < m$, *there is an element* $z_n \in L_1(A, \tau)$ *such that*

$$(7) \qquad \sum_{i=0}^{n-1} u^i y_m \leq \xi_n + \frac{1}{m} z_n .$$

Moreover, z_n *does not depend on* m, *and* $\sup_m \|y_m\|_1 < \infty$.

PROOF. [9],[3]. We have

$$my_m = (I - u)\Big(\sum_{k=1}^{m-1} \xi_k\Big) + \xi_m.$$

Consequently,

$$m \sum_{i=0}^{n-1} u^i y_m = (I - u^n)\Big(\sum_{k=1}^{m-1} \xi_k\Big) + \sum_{i=0}^{n-1} u^i \xi_m$$

$$= \sum_{k=1}^{n} \xi_k + \sum_{k=1}^{m-n-1} (\xi_{k+n} - u^n \xi_k) + \sum_{i=0}^{n-1} u^i (\xi_m - u^{n-i} \xi_{m-n+i}).$$

By the subadditivity of $\{\xi_n\}$, we get

$$\xi_{n+k} - u^n \xi_k \le \xi_n \quad \text{and} \quad \xi_m - u^{n-i} \xi_{m-n+i} \le \xi_{n-i};$$

therefore,

$$\sum_{i=0}^{n-1} u^i y_m \le \frac{1}{m}\Big(\sum_{k=1}^{n} \xi_k + (m-n)\xi_n + \sum_{i=0}^{n-1} u^i \xi_{n-i}\Big),$$

and we can put

$$z_n = \sum_{k=1}^{n} \xi_k - n\xi_n + \sum_{i=0}^{n-1} u^i \xi_{n-i}$$

(z_n does not depend on m).

Moreover, since $y_m \le \xi_1$, we have

$$\|y_m\|_1 = \|\xi_1 - y_m\|_1 + \|\xi_1\|_1$$

$$= \|\xi_1\|_1 + \tau(\xi_1 - y_m) = \|\xi_1\|_1 + \tau(\xi_1) - \tau\Big(\frac{\xi_m}{m}\Big)$$

$$\le \|\xi_1\|_1 + \tau(\xi_1) - \gamma = \text{const.}$$

PROPOSITION 2. If $\{x_n\}$ *is a sequence as in Theorem 1, then there exists an element* $v_0 \in L_1$ *such that* $\tau(v_0) = \gamma$, *and*

$$(8) \qquad \xi_n \geq \sum_{i=0}^{n-1} u^i v_0 \quad \textit{for every} \quad n.$$

PROOF. (comp. [8],[9]). Identifying L_1 with the predual A_*, we can treat the operators y_m as the elements of the second dual $L_1^{**} = A^*$, i.e. as the continuous linear functionals on A. The images in $L^{**} = A^*$ of the elements $\xi, y, \ldots$ are denoted by $\tilde{\xi}, \tilde{y}, \ldots$ In particular, $\tilde{y}_m(a) = \tau(y_m a)$ for $a \in A$. The sequence $\{y_m\}$ is bounded in norm (by Proposition 1), therefore it is compact in the weak * topology on A^* and has a limit point, say $u \in A^*$. Rewriting formula (4) for the functionals and passing to the limit with m, we obtain

$$(9) \qquad \sum_{i=0}^{n-1} \tilde{u}^i u_0 \leq \tilde{\xi}_n \qquad \text{for } n = 1,2,\ldots ,$$

where $(\tilde{u}\sigma)(a) = \sigma(T^{-1}a)$ for $\sigma \in A^*$ and $a \in A$. In particular, $\tilde{u}\tilde{y} = \widetilde{(uy)}$ for $y \in L_1$.

Each functional $\alpha \in A^*$ can be written in the form: $\alpha = \alpha_1 + \alpha_2$ where $\alpha_1 \in A_*$ and α_2 is the singular part of α ([16] , p.127). Formula (9), for $n = 1$, gives $u_0 \leq \tilde{\xi}_1$. Thus

$$\tilde{\xi}_1 - u_0 \in A_+^* \quad \text{with} \quad \tilde{\xi}_1 \in A_*, \quad \text{and}$$
$$\tilde{\xi}_1 - u_0 = \alpha + \beta \quad \text{where} \quad \alpha \quad \text{and} \quad \beta \quad \text{are}$$

normal and singular parts of $\tilde{\xi}_1 - u_0$, respectively. Finally we can write

$$(10) \qquad u_0 = \nu - \sigma \quad \text{where} \quad \nu \in A_*, \quad \text{and} \quad \sigma$$

is a linear functional which is positive and singular. Thus we have

$$(11) \qquad \sum_{i=0}^{n-1} u^i \nu \leq \tilde{\xi}_n + s_n$$

where $s_n = \sum_{i=0}^{n-1} \tilde{u}^i \sigma$.

Let us remark that $\tilde{u}\sigma$ is also singular. Indeed, supposing that there is a positive normal functional $\sigma_1 \leq \tilde{u}\sigma$, we would have $\tilde{u}^{-1}\sigma \leq \sigma$. Consequently, $\tilde{u}^{-1}\sigma_1 = 0$ and $\sigma_1 = 0$. This implies that the functional $s_n = \sum_{i=0}^{n-1} \tilde{u}^i\sigma$ is singular. Let $(p_t, t \in T)$ be a maximal orthogonal family of non-zero projections from A such that $s_n(p_t) = 0$ for $t \in T$. By the singularity of s_n, we have $\sum_T p_t = I$ (see [16], p.134). Let F be the directed set of all finite subsets of T. For $\alpha \in F$, we put $q_\alpha = \sum_{t \in \alpha} p_t$. Then $\sum_{i=0}^{n-1} u^i \nu(q_\alpha) \leq \xi_n(q_\alpha)$. Passing to the limit with α, we obtain, by the normality of ν and $\tilde{\xi}_n$,

$$n\nu(I) = \sum_{i=0}^{n-1} \tilde{u}^i \nu(I) \leq \tilde{\xi}_n(I) = \tau(\xi_n),$$

consequently, $\nu(I) \leq \tau(\frac{\xi_n}{n}) \to \gamma = \inf_n \tau(\frac{\xi_n}{n})$, so $\nu(I) \leq \gamma$.
On the other hand, $u_o(I) = \gamma$. Indeed

$$u_o(I) = \lim_m \tau(y_m) = \lim_m \tau(\frac{\xi_m}{m}) = \inf_m \tau(\frac{\xi_m}{m}) = \gamma.$$

Thus we have

$$\nu(I) \leq \gamma = u_o(I) = \nu(I) - \sigma(I), \quad \text{which}$$

implies $\sigma(I) = 0$ and, consequently, $u_o = \nu \in A_*$. By the Yeadon theorem [17], there is some $v_o \in L_1$ such that $u_o(a) = \tau(v_o a)$ for $a \in A$; of course, $\tau(v_o) = \gamma$ and (9) translates into (8). It means that v_o is the operator we are looking for. The proof is completed.

PROPOSITION 3. Let g_n, h_n *and* $\alpha_n(m)$ *be the elements of* $L_1(A,\tau)$ *satisfying the inequalities*

(12) $\quad 0 \leq g_n \leq h_m + \alpha_n(m)$ *for each* $n > m$; $\quad n,m \in N$.

Let us suppose that $\tau(h_m) \to 0$ *as* $m \to \infty$ *and* $\alpha_n(m) \to 0$ *almost uniformly as* $n \to \infty$ *for each* $m \in N$. *Then* $g_n \to 0$ *almost uniformly.*

PROOF. Let $\varepsilon > 0$ be given. Choose a subsequence $\{m_s\}$ of positive integers in such a way that $\tau(h_{m_s}) = \varepsilon^2_{m_s}$ with $\sum_{s=1}^{\infty} \varepsilon_{m_s} < \frac{\varepsilon}{2}$. Then, putting $E_s = e_{[0,\varepsilon_{m_s}]}(h_{m_s})$, we obtain $\tau(I - E_s) < \varepsilon_{m_s}$. Consequently, for $E = \bigwedge_{s=1}^{\infty} E_s$, we have $\tau(I - E) < \frac{\varepsilon}{2}$.

Let $\{p_m\}$ be a sequence of projectors from A, such that $\tau(I - p_m) < \frac{\varepsilon}{2^{m+1}}$ and $\|p_m \alpha_n(m) p_m\| \to 0$ as $n \to \infty$. Put $p = \bigwedge_{m=1}^{\infty} p_m$. Then, for each $m \in N$, $\|p\alpha_n(m)p\| \to 0$. In particular, $\|p\alpha_n(m)p\| < \varepsilon_m$ for $n > \beta(m)$. Putting $F = E \wedge P$, we have $\tau(I - F) < \varepsilon$ and $\|Fg_nF\| \leq \|Fh_{m_s}\| + \|F\alpha_n\ m\ F\| < 2\varepsilon_{m_s}$ for $n > \max(\beta(m_s), m_s)$, which means that $g_n \to 0$ almost uniformly.

Now, we are in a position to prove Theorem 1. For (ξ_n) as in the theorem, we can find $v_0 \in L_1$ such that (8) holds. Putting

$$\bar{\xi}_n = \xi_n - \sum_{i=0}^{n-1} u^i v_0 \qquad (n = 1,2,\ldots),$$

we obtain a non-negative subadditive sequence $\{\bar{\xi}_n\}$ with

$\inf_n \tau(\frac{\overline{\xi}_n}{n}) = 0$. By Theorem B, the averages $n^{-1} \sum_{i=0}^{n-1} u^i v_0$ converge almost uniformly and in L_1, so it is clear now that it suffices to prove our theorem for the subadditive sequences which are non--negative and satisfy the equality

$$\inf_n \tau(\frac{\xi_n}{n}) = 0.$$

In this case, the convergence (to zero) of $\frac{\xi_n}{n}$ in L_1 is obvious. It remains to show the almost uniform convergence of $\frac{\xi_n}{n}$ to zero. Fix $m \in N$. For $n > m$, we can write $n = mk + r$ with $0 \le r \le m-1$ and, by the subadditivity of $\{\xi_n\}$, we have

$$0 \le \xi_n \le \xi_{mk} + u^{mk} x_r \le \sum_{i=0}^{k-1} u^{im} \xi_m + u^{mk} \xi_r$$

$$\le \sum_{i=0}^{k-1} u^{im} \xi_m + u^{mk} (\xi_1 + \dots + \xi_{m-1}).$$

The averages $\frac{1}{k} \sum_{i=0}^{k-1} u^{im} \xi_m$ converge (in L_1 and almost uniformly) to a T-invariant element $\hat{\xi}_m$. Putting $\xi_1 + \dots + \xi_{m-1} = z$ and

$$\alpha_n(m) = \frac{1}{k} \sum_{i=0}^{k-1} u^{im} \frac{\hat{\xi}_m - \xi_m}{m} + \frac{1}{k} u^{mk} z,$$

we obtain

$$(13) \qquad \frac{\xi_n}{n} \le \frac{\hat{\xi}_m}{m} + \alpha_n(m)$$

(of course, for a fixed m, k depends on n). We shall show that

$$(14) \qquad \frac{1}{k} u^{km} z \to 0 \quad \text{almost uniformly as } k \to \infty \quad \text{for each } m \in N.$$

Indeed, let $\varepsilon > 0$ be given. Putting $u^m = S$, $z = \int_0^\infty \lambda e(d\lambda)$, (spectral resolution of z), we have $\int_0^\infty \lambda\tau(e(d\lambda)) < \infty$; $S^k e(d\lambda)$ is a spectral measure of $S^k z$ and $\tau(e(d\lambda)) = \tau(S^k e(d\lambda))$. Thus, having taken $0 < \sigma_n - 0$, we can write

$$\sum_{k=1}^{\infty} \tau(e_{(\sigma_n,\infty)}\{\tfrac{1}{k} S^k z\} = \sum_{k=1}^{\infty} \tau(e_{(k\sigma_n,\infty)}\{z\}) < \infty \quad (\text{because} \quad z \in L_1)$$

Choose N_n in such a way that

$$\sum_{k=N_n}^{\infty} \tau(e_{(k\sigma_n,\infty)}\{z\}) < \frac{\varepsilon}{2^n}$$

holds for $n = 1,2,\ldots$.

Putting

$$P_{kn} = e_{[0,\sigma_n]}\{\tfrac{1}{k} S^k z\}$$

and

$$P = \bigwedge_{n=1}^{\infty} \bigwedge_{k=N_n}^{\infty} P_{kn},$$

we have $\tau(I-P) \leq \sum_{k,n} \tau(I-P_{kn}) < \sum_{n=1}^{\infty} \frac{\varepsilon}{2^n} = \varepsilon.$

On the other hand,

$$\|P \tfrac{1}{k} S^k z P\| < \sigma_n \quad \text{for} \quad k > N_n,$$

which means that (14) holds.

Applying Proposition 3 to inequality (13), we get that $\frac{\xi_n}{n} \to 0$ almost uniformly. The proof of our theorem is completed.

2. *PROOF of THEOREM 2.* Taking $t = 0,1,2,\dots,$ we obtain the subadditive sequence $(\xi_0,\xi_1,\dots)$, and applying Theorem 1, we have that the limit $\lim_{n\to\infty}\frac{\xi_n}{n} = \hat{\xi}$ exists (in L_1 and almost uniformly). We shall show that also $\lim_{t\to\infty}\frac{\xi_t}{t} = \hat{\xi}$. To this end, let us denote by $n = n(t)$ the integer part of t, i.e. always $n \le t < n + 1$. Then, by the subadditivity of ξ, we have that $\xi_{n+1} \le \xi_t + u^t\xi_{n+1-t}$ and $\xi_t \le \xi_n + u^n\xi_{t-n}$ and, consequently,

$$(15) \qquad \frac{\xi_{n+1}}{n+1} - \frac{u^t\xi_{n+1-t}}{n+1} \le \frac{\xi_t}{t} \le \frac{\xi_n}{n} + \frac{u^n\xi_{t-n}}{n}.$$

To prove that $\frac{\xi_t}{t} \to \hat{\xi}$ in L_1 it is enough to show that

$$(16) \qquad \frac{u^t\xi_{n+1-t}}{n} \to 0$$

$$(17) \qquad \frac{u^n\xi_{t-n}}{n} \to 0$$

in L_1. But (16) and (17) easily follow from (4). Indeed, for example,

$$\tau(|u^t\xi_{n+1-t}|) = \tau(u^t|\xi_{n+1-t}|) = \tau(|\xi_{n+1-t}|)$$

$$= \int_0^\infty \tau(e_{(\lambda,\infty)}\{|\xi_{n+1-t}|\})d\lambda \le \int_0^\infty \tau(p(\lambda))d\lambda < \infty.$$

Let us now remark that if $\alpha_t \le \eta_t \le \beta_t$ for some $\alpha_t, \beta_t, \eta_t \in L_1 (t \in T)$, and $\alpha_t \to \zeta$ almost uniformly and $\beta_t \to \zeta$ almost uniformly, then $\eta_t \to \zeta$ almost uniformly. To prove this fact, we can assume that $\zeta = 0$. Then, for each $\varepsilon > 0$, there is some $p \in \mathrm{Proj}\, A$ with $\tau(1 - p) < \varepsilon$, such that $\|p\beta_t p\|_\infty \to 0$ and $\|p\alpha_t p\|_\infty \to 0$ as $t \to \infty$. But $p\alpha_t p \le p\eta_t p \le p\beta_t p$, and

$|(p\eta_t px,x)| \le \max\{|(p\alpha_t px,x)|, |(p\beta_t px,x)|\}$. Consequently, $\|p\eta_t p\|_\infty \le \max\{\|p\alpha_t p\|_\infty, \|p\beta_t p\|_\infty\} \to 0$, which means that $\eta_t \to 0$ almost uniformly. Thus, to complete the proof, it remains to show that (16) and (17) hold in the sense of almost everywhere convergence. To this end, let us put

$$(18) \qquad Q_m(\lambda) = u^m q(\lambda) \quad \text{for} \quad \lambda > 0 \quad \text{and} \quad m = 1,2,\ldots .$$

We shall prove that then

$$(19) \qquad Q_m(\lambda) = \bigvee_{0\le s,t\le 1} e_{(\lambda,\infty)}\{|u^{s+m}\xi_t|\}$$

holds. We need a few lemmas.

LEMMA 1. For a $*$*-automorphism* $u \in A$ *and* $q_1, q_2, \ldots, q_N \in \text{Proj } A$, *we have that*

$$(20) \qquad u\Big(\bigvee_{1\le s\le N} q_s\Big) = \bigvee_{1\le s\le N} uq_s .$$

PROOF. It is well known that, for two projectors $p, q \in A$, we have $p \wedge q = \lim_{n\to\infty} (pq)^n$ in the strong operator topology. Since u is ultrastrongly continuous, it is strongly continuous on the unit ball in A. Thus

$$(21) \qquad u(p \wedge q) = \lim_{n\to\infty} (up\cdot uq)^n = up \wedge uq$$

and then (20) follows easily.

LEMMA 2. Let $q \in \text{Proj } A$, *and let* $(u^t, t \ge 0)$ *be a strongly continuous (semi)group of* $*$*-automorphisms of* A. *Put*

$$(22) \qquad P_n = \bigvee_{0\le k\le 2^n} u^{k/2^n} q \qquad (n = 1,2,\ldots).$$

Then $P_n \uparrow \bigvee_{0\le t\le 1} u^t q$ *(in the strong operator topology).*

PROOF. Obviously, $\bigvee_{0\le t\le 1} u^t q \ge P_n$ for all n. If there were a $Q \in \mathrm{Proj}\, A$ such that $P_n \le Q < \bigvee_{0\le t\le 1} u^t q$, then there would exist an $x_0 \in H$ such that $u^{t_0} q x_0 \ne 0$ for some $t_0 \in [0,1]$ and $P_n x_0 = 0$ for all n. But it is impossible because we can take $0 \le k_s \le 2^{n_s}$ with $k_s / 2^{n_s} \to t_0$ and then, by the continuity of $(u^t,\ t \in T)$, we would have $0 \equiv \| u^{k_s/2^{n_s}} q x_0 \| \to \| u^{t_0} q x_0 \| \ne 0$.

LEMMA 3. If $q(\lambda)$ *is defined by* (5), *then*

$$(23) \qquad q(\lambda) = \bigvee_{0\le s\le 1} u^s \bigvee_{0\le t\le 1} e_{(\lambda,\infty)}\{|\xi_t|\}.$$

PROOF. Since, for the automorphism u^s, we have $u^s e_{(\lambda,\infty)}\{|\xi_t|\} = e_{(\lambda,\infty)}\{|u^s \xi_t|\}$, it is enough to show that

$$(24) \qquad u^s \bigvee_{0\le t\le 1} e_{(\lambda,\infty)}\{|\xi_t|\} = \bigvee_{0\le t\le 1} u^s e_{(\lambda,\infty)}\{|\xi_t|\}$$

for each fixed $s \in [0,1]$. By the separability of ξ, we have (1). Thus

$$(25) \qquad u^s \bigvee_{0\le t\le 1} e_{(\lambda,\infty)}\{|\xi_t|\} = u^s \lim_{n\to\infty} \bigvee_{1\le k\le n} e_{(\lambda,\infty)}\{|\xi_{s_k}|\}$$

$$= \lim_{n\to\infty} \bigvee_{1\le k\le n} u^s e_{(\lambda,\infty)}\{|\xi_{s_k}|\} = \bigvee_{0\le t\le 1} u^s e_{(\lambda,\infty)}\{|\xi_t|\}.$$

The last equality follows from the fact that the order structure of the sequence of projectors $\pi_k = e_{(\lambda,\infty)}\{|\xi_{s_k}|\}$ $(k = 1,2,\ldots)$ is the same as the structure of the sequence $k^s \pi_k$ (s-fixed, $k = 1,2,\ldots$), and that , for π_k's, formula (1) holds (see for example [4]). From lemmas 1,2 and 3 and the weak (hence strong) continuity of $(u^t,\ t \in T)$ formula (19) follows immediately.

Let us now remark that, for each $\delta > 0$.

(26) $$\sum_{m=1}^{\infty} \tau(Q_m(m\delta)) < \infty$$

holds. Indeed, $\sum_{m=1}^{\infty} \tau(Q_m(m\delta)) = \sum_{m=1}^{\infty} \tau(q(m\delta)) < \infty$ (because $\int_0^{\infty} \tau(q(\lambda))d\lambda < \infty$).

Let us fix $\varepsilon > 0$.

Take now $\varepsilon_k \to 0$ and choose N_k in such a way that

(27) $$\sum_{m=N_k}^{\infty} \tau(Q_m(m\varepsilon_k)) < \frac{\varepsilon}{2^k} \quad \text{for} \quad k = 1,2,\ldots .$$

Put

(28) $$P_{m,k} = I - Q_m(m\varepsilon_k)$$

and

(29) $$P = \bigwedge_{1\le k<\infty} \bigwedge_{N_k \le m < \infty} P_{m,k}.$$

We then have

(30) $$\tau(I - P) \le \sum_{k=1}^{\infty} \sum_{m=N_k}^{\infty} \tau(Q_m(m\varepsilon_k) < \varepsilon.$$

Moreover, for $n > N_k$, we have that

(31) $$\left\|\frac{u^n\xi_{t-n}}{n} P\right\| \le \left\|\frac{u^n\xi_{t-n}}{n} (I - Q_n(n\varepsilon_k))\right\|$$

$$\le \left\|\frac{u^n\xi_{t-n}}{n} \bigwedge_{0\le s\le 1} u^n e_{[0,\varepsilon_k]}\{|\frac{\xi_s}{n}|\}\right\|$$

$$\leq \left\| \left|\frac{u^n \xi_{t-n}}{n}\right| e_{[0,\varepsilon_k]}\left\{\left|\frac{u^n \xi_{t-n}}{n}\right|\right\} \right\| < \varepsilon_k$$

Similarly,

$$(32) \qquad \left\| \frac{u^t \xi_{n+1-t}}{n} \cdot P \right\|$$

$$\leq \left\| \frac{u^n(u^{t-n}\xi_{n+1-t})}{n} \bigwedge_{0\leq s,t\leq 1} e_{[0,\varepsilon_k]}\{|u^{n+s}\xi_t|\} \right\|$$

$$\leq \left\| \left|\frac{u^n}{n}\right| |u^{t-n}\xi_{n+1-t}| \, e_{[0,\varepsilon_k]}\{|u^{t-n}\xi_{n+1-t}|\} \right\| < \varepsilon_k$$

for $n > N_k$.

Formulae (30), (31) and (32) mean that (16) and (17) hold in the sense of almost uniform convergence. This completes the proof of our theorem.

REFERENCES

[1] M.A. Akcoglu and L. Sucheston, *A Ratio Ergodic Theorem for Superadditive Processes*, Z. für Wahrschein. verw. Geb. 44 (1978), 269-278.

[2] J.P. Conze and N. Dang Ngoc, *Ergodic theorems for non-commutative dynamical systems*, Invent. Math. 46 (1978), 1-15.

[3] Y. Derrienic, *Sur le theoreme ergodique sous-additif*, C.R. Acad. Sci. Paris Ser. A, vol. 281 (1975), 985-988.

[4] J. Dixmier, *Les algebres d'opérateurs dans l'éspace hilbertien (algebres de von Neumann)*, Paris 1957.

[5] A. Garsia, *Topics in Almost Everywhere Convergence*, Chicago 1970.

[6] R. Jajte, *Non-commutative subadditive ergodic theorem*, Bull. Acad. Pol. Sci., to appear.

[7] М.Ш. Голдштейн, Теоремы сходимости почти всюду в алгебрах фон Неймана, J. Operator Theory 6 (1981), 233-311.

[8] J.F.C. Kingman, *Subadditive ergodic theory*, Ann. Prob. 1 (1973), 883-909.

[9] J.F.C. Kingman, *Subadditive processes*, Ecole d'Eté de Probabilités de Saint Flour V-1975, Lecture Notes 539 (1976), 167-223.

[10] U. Krengel, *Recent progress on ergodic theorems*, Asterisque 50 (1977), 151-192.

[11] B. Kümmerer, *A non-commutative individual ergodic theorem*, Invent. Math. 46 (1978), 139-145.

[12] C. Lance, *Ergodic theorems for convex sets and operator algebras*, Invent. Math. 37 (1976), 201-214.

[13] E. Nelson, *Notes on non-commutative integration*, I. Funct. Analysis 15 (1974), 103-116.

[14] D. Petz, *Ergodic theorems in von Neumann algebras*, Acta Sci. Math., to appear.

[15] I.E. Segal, *A non-commutative extension of abstract integration*, Ann. of Math. 57 (1953), 401-457.

[16] M. Takesaki, *Theory of operator algebras I*, Berlin-New York 1979.

[17] F.J. Yeadon, *Non-commutative* L^p*-spaces*, Proc. Camb. Philos. Soc. 77 (1975), 91-102.

[18] F.J. Yeadon, *Ergodic theorems for semifinite von Neumann algebras I*, J. London Math. Soc. (2) 16 (1977) 326-332.

[19] F.J. Yeadon, *Ergodic theorems for semifinite von Neumann algebras II*, Math. Proc. Camb. Phil. Soc. 88 (1980), 135-147.

Institute of Mathematics
Łódź University
ul.Banacha 22
90-238 Łódź, Poland.

$*$-REGULARITY OF LOCALLY COMPACT GROUPS

Eberhard Kaniuth
Fachbereich Mathematik / Informatik
der Universität-Gesamthochschule Paderborn
D-4790 Paderborn

Let A be a Banach $*$-algebra and $\mathrm{Prim}_* A$ the set of all primitive ideals of A, i.e. of all kernels of topologically irreducible $*$-representations of A. $\mathrm{Prim}_* A$ is endowed with the hull-kernel-topology: the closure of $E \subseteq \mathrm{Prim}_* A$ is given by $\bar{E} = h(k(E))$, where $k(E) = \cap \{P; P \in E\}$ and $h(I) = \{P \in \mathrm{Prim}_* A;\ I \subseteq P\}$ for $I \subseteq A$. The ideal theory of A is based on this structure space rather than on the space Prim A of algebraically simple A-modules. Every representation π of A extends uniquely to a representation $\tilde{\pi}$ of the enveloping C^*- algebra $C^*(A)$ of A. Thus there is a continuous mapping

$$\phi : \mathrm{Prim}\ C^*(A) \to \mathrm{Prim}_* A,\ P \to P \cap A$$

from $\mathrm{Prim}\ C^*(A) = \mathrm{Prim}_* C^*(A)$ onto $\mathrm{Prim}_* A$. A is called <u>$*$-regular</u>, if ϕ is a homeomorphism. If A is commutative, then this means that the algebra of Gelfand transforms of A is a regular function algebra on the hermitian part of the spectrum of A.

A locally compact group G is called $*$-regular if its L^1-algebra $L^1(G)$ is $*$-regular. For a unitary representation π of G, we also denote by π the corresponding $*$-representation of $L^1(G)$ and then by $\tilde{\pi}$ the extension to $C^*(G) = C^*(L^1(G))$. The dual space $\hat{G}$ of G is the set of equivalence classes of irreducible unitary representations of G, equipped with the inverse image of the hk-topology on $\mathrm{Prim}\ C^*(G)$ under the mapping $\hat{G} \to \mathrm{Prim}\ C^*(G), \pi \to \ker \tilde{\pi}$. Evidently, $*$-regularity of G is then equivalent ot the following: given a closed subset E of $\hat{G}$ and $\pi \notin E$, there exists $f \in L^1(G)$ such that $\pi(f) \neq 0$ and $\rho(f) = 0$ for all $\rho \in E$. If G is abelian, then $\hat{G}$ can be identified with the dual group of G and this regularity condition is well known to hold, i.e. G is $*$-regular. The investigation of $*$-regularity of locally compact groups has been started in [2]. As a first step, the authors verify the following

<u>Lemma 1.</u> The following conditions on G are equivalent:

(i) ϕ is a homeomorphism;

(ii) $\ker \pi \subseteq \ker \rho \Rightarrow \ker \tilde{\pi} \subseteq \ker \tilde{\rho}$ for all representations π and ρ of G;

(iii) $\ker \pi \subseteq \ker \rho \Rightarrow \|\rho(f)\| \leq \|\pi(f)\|$ for all $f \in L^1(G)$ and all representations π and ρ of G.

The main result of [2] is

Theorem 1. (i) If G is $*$-regular, then G has to be amenable;
(ii) If G has polynomial growth, then G is $*$-regular.

(i) follows from the above lemma and the fact that G is amenable if the left regular representation of $L^1(G)$ extends faithfully to $C^*(G)$. Before indicating the proof of (ii), we recall the definition of a polynomially growing group: G has polynomial growth if for every compact subset K of G there is a polynomial p_K such that the Haar measure of K^n is bounded by $p_K(n)$ for all $n \in \mathbb{N}$. For instance, compact extensions of nilpotent locally compact groups are polynomially growing.

Suppose that π and ρ are unitary representations of G such that $\ker \pi \subseteq \ker \rho$ and $\|\pi(f)\| < \|\rho(f)\|$ for some $f^* = f \in C_c(G)$. Now an important functional calculus due to Dixmier [6] can be applied. Take $\varphi \in C_c^\infty(\mathbb{R})$ satisfying $\varphi(0) = 0$. Then, using the above growth condition for the support of f, Dixmier has shown that the integral

$$\varphi\{f\} = \int_{-\infty}^{\infty} \exp(i\lambda f)\hat{\varphi}(\lambda)d\lambda,$$

where $\hat{\varphi}$ denotes the Fourier transform of f, converges in $L^1(G)$. Moreover, for every unitary representation π of G, the equation

$$\pi(\varphi\{f\}) = \varphi(\pi(f))$$

holds, where the right hand side is defined by the usual functional calculus on the hermitian operator $\pi(f)$ in the Hilbert space of π. By the way, Dixmier's functional calculus turned out to be a very good tool in studying ideal theory of $L^1(G)$. Now, choose φ such that $\varphi(t) = 0$ for $t \leq \|\pi(f)\|$ and $\varphi(\|\rho(f)\|) = 1$. Then it follows that $\|\pi(\varphi\{f\})\| = 0$, but $\|\rho(\varphi\{f\})\| \geq 1$, i.e. $\varphi\{f\} \in \ker \pi$, but $\varphi\{f\} \notin \ker \rho$, a contradiction

In view of Theorem 1, the following problems arose:

(i) Do there exist amenable groups which fail to be $*$-regular and $*$-regular groups which are not polynomially growing?

(ii) Find, at least for special classes of locally compact groups, if and only if conditions for $*$-regularity.

(iii) Find further classes of $*$-regular groups.

Of course, the candidates to look at are the solvable groups. The answer to (i) is yes. The ax+b-group is $*$-regular, and the group G consisting of all matrices

$$\begin{pmatrix} 1 & x & z \\ 0 & a & y \\ 0 & 0 & 1 \end{pmatrix}, \quad x, y, z \in \mathbb{R}, \ a > 0,$$

turned out to be non-$*$-regular. In fact, G is the smallest dimensional connected solvable Lie group which is not $*$-regular. These assertions can be checked by applying a very deep and powerful criterion, due to Boidol [4], for $*$-regularity of connected groups.

Suppose that G is a connected group and $\pi \in \hat{G}$. Then there exists a unique closed normal subgroup N_π of G containing the commutator subgroup, such that π is weakly equivalent to the induced representation $U^{\pi|N_\pi}$ ($\pi \sim U^{\pi|N_\pi}$), i.e. $\ker \tilde{\pi} = \ker \widetilde{U^{\pi|N_\pi}}$. Let $K_\pi = \{x \in G;\ \pi(x) = I\}$. Then G is said to have a polynomially induced dual if N_π/K_π has polynomial growth for all $\pi \in \hat{G}$.

Theorem 2 [4]. A connected locally compact is $*$-regular if and only if it has a polynomially induced dual.

For exponential Lie groups this criterion coincides with the one in [3]. According to the above example, solvable groups of length 3 may fail to be $*$-regular. On the other hand, as a consequence of Theorem 2 (see [5, Theorem 2] and Lemma 2 below) one obtains

Corollary. Every connected metabelian group is $*$-regular.

There are two more positive results for metabelian groups. The first one has under a somewhat restrictive assumption on the action of A on $\hat{N}$ already been proved in [2].

Theorem 3 [5, Theorem 3]. Suppose that $G = A \ltimes N$ is a semidirect of abelian groups A and N. Then G is $*$-regular.
We are now going to show that metabelian discrete groups are $*$-regular. First we show that the main result of [9] implies that second countable metabelian groups are "weakly monomial". If N is a normal subgroup of G, λ a representation of N and $x \in G$, then λ^x denotes the representation of N defined by $\lambda^x(n) = \lambda(x^{-1}nx)$. Moreover, we write Z(G) for the center of G.

Lemma 2. Let G be a metabelian second countable locally compact group and N the closure of the commutator subgroup of G. For $\lambda \in \hat{N}$ set

$$N_\lambda = \{x \in N;\ \lambda(x) = 1\},\ G_\lambda = \{x \in G;\ \lambda^x = \lambda\} \text{ and}$$
$$H_\lambda = \{x \in G;\ xN_\lambda \in Z(G_\lambda/N_\lambda)\}.$$

Then, given $\pi \in \hat{G}$, there exist a $\lambda \in \hat{N}$ and a $\chi \in \widehat{H_\lambda}$ such that $\chi|N = \lambda$ and $\pi \sim U^\chi$.

Proof. By [9, Theorem 4.3] there are $\lambda \in \hat{N}$ and a homogeneous representations σ of G_λ satisfying $\sigma|N \sim \lambda$ and $\pi \sim U^\sigma$. σ is in fact a representation of G_λ/N_λ, and G_λ/N_λ is nilpotent of class 2. Now the kernel of a homogeneous representation of a separable C^*-algebra is a primitive ideal [7, (3.9.1) and (5.7.6)]. Thus

$\pi \sim U^\tau$ for some $\tau \in \widehat{G_\lambda/N_\lambda}$ such that $\tau|N \sim \lambda$. The assertion then follows from [11, Lemma 2].

Lemma 3. Let G be a second countable metabelian group. Then the mapping Prim $C^*(G) \to \mathrm{Prim}_* L^1(G)$ is injective.

Proof. Suppose that $\pi_1, \pi_2 \in \hat{G}$ such that ker π_1 = ker π_2, and choose λ_j and χ_j, $j = 1,2$, according to Lemma 2. For $f \in L^1(N)$, denote by μ_f the corresponding Radon measure on G. If $f \in \ker \pi_1|N$, then

$$g * \mu_f \in \ker U^{\pi_1|N} \subseteq \ker \pi_1 = \ker \pi_2,$$

and hence $\pi_2(g)\pi_2|N(f) = \pi_2(g * \mu_f) = 0$ for all $g \in C_c(G)$. This shows that $\ker \pi_1|N = \ker \pi_2|N$. Since N is abelian and $\pi_j|N$ is weakly equivalent to the G-orbit $G(\lambda_j)$ [8, Theorem 5.3], we conclude that $\overline{G(\lambda_1)} = \overline{G(\lambda_2)}$. G/N being abelian, we obtain $G_{\lambda_1} = G_{\lambda_2}$. Using the notations of Lemma 2, we have

$$G_{\lambda^x} = G_\lambda,\ H_{\lambda^x} = H_\lambda \text{ and } H_\lambda/\bigcap_{x\in G} N_{\lambda^x} = Z(G_\lambda/\bigcap_{x\in G} N_{\lambda^x}).$$

Moreover, $\overline{G(\lambda_1)} = \overline{G(\lambda_2)}$ implies

$$\bigcap_{x\in G} N_{\lambda_1^x} = \bigcap_{x\in G} N_{\lambda_2^x}.$$

Therefore, by the usual reduction, we can assume that $\bigcap_{x\in G} N_{\lambda_j^x} = \{e\}$. Then $H_{\lambda_1} = H_{\lambda_2}$ as $G_{\lambda_1} = G_{\lambda_2}$, and $\pi_j \sim U^{\chi_j}$ for some $\chi_j \in \hat{H}$, where $H = H_{\lambda_j}$, $j = 1,2$. Finally, the above argument applied to H shows that $\overline{G(\chi_1)} = \overline{G(\chi_2)}$ and hence $\pi_1 \sim \pi_2$.

Theorem 4. Every metabelian discrete group G is $*$-regular.

Proof. Notice first that by [5, Theorem 1] or [10, Lemma 1.1] we can assume that G is finitely generated. In view of Lemma 3 it remains to show that the mapping Prim $C^*(G) \to \mathrm{Prim}_* L^1(G)$ is closed. Suppose that $\pi \in \hat{G}$ and that $(\pi_\iota)_{\iota\in I}$ is a net in $\hat{G}$ such that $\pi_\iota \to \pi$. Let N denote the commutator subgroup of G, and for every π_ι choose λ_ι and χ_ι according to Lemma 2:

$$\lambda_\iota \in \hat{N},\ N_\iota = \{x \in N;\ \lambda_\iota(x) = 1\},\ G_\iota = \{x \in G;\ \lambda_\iota{}^x = \lambda_\iota\},$$
$$H_\iota = \{x \in G;\ xN_\iota \in Z(G_\iota/N_\iota)\},\ \text{and}\ \chi_\iota \in \hat{H}_\iota\ \text{satisfying}$$
$$\pi_\iota \sim U^{\chi_\iota}\ \text{and}\ \chi_\iota|N = \lambda_\iota.$$

Let S(G) denote the set of all subgroups of G. Fell has introduced a topology on S(G), a subbasis of which is given by the sets $U(C,V) = \{H \in S(G);\ H \cap C = \emptyset,\ H \cap V \neq \emptyset\}$, where C, $V \subseteq G$ and C is finite (see[8, p. 427]). S(G) is a compact (Hausdorff) space. Therefore we can assume that $H_\iota \to H$ in S(G). i.e. $x \in H$ iff $x \notin H_\iota$ for all $\iota \geq \iota(x)$. Now $H_\iota \supseteq N$ and G/N is finitely generated and abelian. Thus H/N is finitely generated, and we can assume that $H = \bigcap_{\iota \in I} H_\iota$. Setting $K = \bigcap_{\iota \in I} \bigcap_{x \in G} N_\iota^x$, we have $\pi_\iota \in \widehat{G/K}$ for all ι, so that we can assume $K = \{\varepsilon\}$. But then is H abelian since $H/\bigcap_{x \in G} N_\iota^x$ is abelian for all ι.

Now, $\ker \pi_\iota \to \ker \pi$ implies $\ker \pi_\iota|H \to \ker \pi|H$. Moreover, $\pi_\iota|H \sim G(\chi_\iota|H)$ and $\pi|H \sim G(\chi)$ for some $\chi \in \hat{H}$. H being abelian, we can assume that there exists a net $(x_\iota)_{\iota \subseteq I}$ in G such that

$(\chi_\iota|H)^{x_\iota} \to \chi$. Obviously, then

$$(H_\iota, \chi_\iota^{x_\iota}) \to (H,\chi)$$

in Fell's subgroup representation topology [8, § 2]. Since inducing is continuous in this topology [8, Theorem 4.2], it follows that

$$\pi_\iota \sim U^{\chi_\iota^{x_\iota}} \longrightarrow U^\chi.$$

Finally, π is weakly contained in $U^{\pi|H}$ and $U^{\pi|H} \sim U^\chi$. This shows that Φ is closed.

We conclude with some

<u>Remarks</u>. a) It is, of course, expected that every metabelian locally compact group is $*$-regular.

b) Poguntke [12] recently proved the remarkable result that an exponential Lie group is $*$-regular iff $L^1(G)$ is a symmetric Banach $*$-algebra.

c) It has been shown in [10] that, if G is a locally compact group with relatively compact conjugacy classes and ω a symmetric weight function on G with rate of growth 1, then the Beurling algebra $L^1_\omega(G)$ is $*$-regular iff ω is non-quasianalytic.
d) Barnes [1] has defined the interesting concept of a locally regular Banach $*$-algebra and shown that (i) a locally regular Banach $*$-algebra is $*$-regular, and (ii) $L^1(G)$ is locally regular if G has polynomial growth (compare Theorem 1).

References

1. Barnes, B.A.: Ideal and representation theory of the L^1-algebra of a group with polynomial growth.
   Colloq. Math. 45, 301-315 (1981)
2. Boidol, J., Leptin, H., Schürmann, J., Vahle, D.: Räume primitiver Ideale von Gruppenalgebren.
   Math. Ann. 236, 1-13 (1978)
3. Boidol, J.: *-regularity of exponential Lie groups.
   Invent. Math. 56, 231-238 (1980)
4. Boidol, J.: Connected groups with polynomially induced dual.
   J. Reine Angew. Math. 331, 32-46 (1982)
5. Boidol, J.: *-regularity of some classes of solvable groups.
   Math. Ann. 261, 477-481 (1982)
6. Dixmier, J.: Opérateurs de rang fini dans les représentations unitaires.
   Publ. Math. Inst. Hautes Etudes Sci. 6, 305-317 (1960)
7. Dixmier, J.: Les C^*-algèbres et leurs représentations.
   Paris: Gauthier-Villars 1964.
8. Fell, J.M.G.: Weak containment and induced representations of groups II.
   Trans. Amer. Math. Soc. 110, 424-447 (1964)
9. Gootman, E., Rosenberg, J.: The structure of crossed product C^*-algebras: a proof of the generalized Effros-Hahn conjecture.
   Invent. Math. 52, 283-298 (1979)
10. Hauenschild, W., Kaniuth, E., Kumar, A.: Ideal structure of Beurling algebras on $[FC]^-$ groups.
    J. Functional Analysis 51, 213-228 (1983)
11. Kaniuth, E.: On primary ideals in group algebras.
    Monatsh. Math. 93, 293-302 (1982)
12. Poguntke, D.: Algebraically irreducble representations of L^1-algebras of exponential Lie-groups.
    preprint

ASYMPTOTIC EQUIDISTRIBUTION ON LOCALLY COMPACT SEMIGROUPS

Franz Kinzl

§ 1. Introduction.

In the classical equidistribution theory a sequence (x_n) in a compact group G is called asymptotic equidistributed, if for every continuous function f

$$\frac{1}{N}\sum_{n=1}^{N} f(x_n) \to \lambda(f)$$

(where λ is the normed Haar measure on G) (cf. [11]). The usuage of the Haar measure can be avoided by the following simple reformulation: Let $\mu_N = \frac{1}{N}\sum_{i=1}^{n} \delta_{x_n}$ then for every $x \in G$ the measure $\mu_N - \delta_x * \mu_N$ tends to 0 in the weak-* topology as $N \to \infty$.

In order to get extensions of the theory the special sequence (μ_N) can be replaced by an arbitrary sequence and the weak-* topology by another topology on M(G) (M(G) denotes the set of all regualr measures on G). KERSTAN and MATTHES ([5], see also [4]) chose the norm topology and studied the asymptotic equidistribution of sequences of probability measures on locally compact abelian groups. GERL ([2]) and later MAXONES and RINDLER ([12],[13],[14])considered arbitrary nets of probability measures which converge to left invariance with respeet to certain topologies.

Now GERL's definitions can be carried over for locally compact semigroups and many results of this paper are also valid in the case of semigroups. But the proofs are different from the case of groups since in generally in semigroups there are no inverse elements, the multiplication is not homogenuous (i.e. the subset a.U ($a \in S$) need not be an open subset of S if U is open in S). Furthermore there is no invariant measure on a locally compact semigroups like the Haar measure on groups. These facts make great difficulties.

§ 2. Notations.

Let S be a locally compact semigroup.

M(S) denotes the set of all regular complex valued bounded Borel measures on S

$M(S)^r$, $M(S)^+$, $M(S)^-$ are the subsets of all real valued resp. nonnegativ resp. nonpositiv measures in M(S)

C(S) denotes the set of all complex valued bounded continuous functions on S

$C_o(S) = \{f \in C(S) : f$ vanishes at infinity$\}$

$C_{oo}(S) = \{f \in C(S) : f$ has compact support$\}$

Let $x \in S$, $f \in C(S)$, the function ${}_xf$ resp. f_x is defined by ${}_xf(s) = f(xs)$ resp. $f_x(s) = f(sx)$.

A function $f \in C(S)$ is called uniformly left (right) continuous, if the mapping $x \to {}_xf$ $(x \to f_x)$ from $S \to C(S)$ is continuous with respect to the norm topology on C(S).

LUC(S) resp. RUC(S) denotes the set of all uniformly left resp. right continuous functions on S

It is easy to see that LUC(S) and RUC(S) are norm-closed, translation invariant linear subspaces of C(S), which contain the constant functions.

Ist is wellknown that $M(S) = C_o(S)^*$ is a Banach algebra with convolution $*$ as multiplication:

$$\xi * \eta(f) = \int\int f(xy)\, d\xi(x)\ \ d\eta(y)\ , \quad \xi,\eta \in M(S),\ f \in C_o(S)$$

([3]).

P(S) denotes the set of all probability measures on S.

A measure μ on S is called left (right) absolutely continuous, if the mapping $S \to M(S)$, $x \to \delta_x * \mu$ $(x \to \mu * \delta_x)$ is continuous with respect to the norm of topology on M(S) (For the theory of absolutely continuous measures on semigroups see [7], [9], [17]).

$M_a(S)$ denotes the set of all left absolutely continuous measures on S.

$P_a(S) := P(S) \cap M_a(S)$.

The support Supp(ν) of a measure ν is defined by

$$\mathrm{Supp}(\nu) = \{x \in S : |\nu|(U) > 0 \text{ for every neighbourhood } U \text{ of } x\}$$

In order to get more informations of structure theory of semigroups or (probability) measures on semigroups consult [1], [15], [16],[17]

In the next chapter we shall consider semigroups which satisfy one of the following conditions:

(A) $M_a(S) * M_a(S) = M_a(S)$

(B) $M_a(S)$ has a left approximate identity (i.e. there exists a net (τ_β) in $P_a(S)$ such that for all $\sigma \in M_a(S)$: $||\tau_\beta * \sigma - \sigma|| \to 0$).

(C) S has a left unit e and for all compact neighbourhoods U of e there is a measure λ_U in $P_a(S)$: $\lambda_U(U) = 1$.

Examples:

1. Let $d^1(S) := \{x \in S:$ for every neighbourhood U of x there is a $\lambda \in M_a(S)^+$ such $\lambda(U) \neq 0\}$

   Let $S = d^1(S)$ and S may have a left unit. Then S satisfies (A), (B), (C). Such semigroups are called foundation semigroups (see [17], in which there are a lot of concrete examples). Every completely simple semigroup, $S = X \times G \times Y$ (X discrete), or specially the right simple semigroups which have an idempotent element, $S = G \times Y$, are foundation semigroups.

2. Let $S = [0,1]$ with usual multiplication. Then $M_a(S) = \{a . \delta_0 : a \in \mathbb{C}\}$. Therefore S satisfies (A), (B), (C).

3. Let $S = ([\frac{1}{2}, \frac{3}{4}], o)$ where $x \circ y = \min \frac{1}{2}, x . y)$

   Then $M_a(S)$ contains every measure whose support is contained in $[\frac{1}{2}, \frac{2}{3}]$ and measures which are absolutely continuous with respect to the Lebesgue measure on the intervall. None of the conditions is satisfied.

§ 3. Asymptotic equidistribution of nets.

Definition. Let (μ_α) be a net in P(S).

1. (μ_α) is called asymptotic equidistributed (= as.equ.) if for every $\nu \in P(S)$:

$$||\nu * \mu_\alpha - \mu_\alpha|| \to 0$$

2. (μ_α) is called weakly asymptotic equidistributed (= weakly as.equ.) if for every $\nu \in P(S)$ and $\sigma \in P_a(S)$:

$$||\nu * \sigma * \mu_\alpha - \sigma * \mu_\alpha|| \to 0$$

3. (μ_α) is called uniform asymptotic equidistributed (= un.as.equ.) if for every $\nu \in P(S)$ and $f \in LUC(S)$:

$$\nu * \mu_\alpha (f) - \mu_\alpha(f) \to 0.$$

As in the group case we have the following

Theorem 1. There exists an as.euq. net resp. un.as.equ. net if and only if there is a left invariant mean on M(S)* resp. LUC(S).

Proof. The first assertion results of [18]. To prove the second assertion first consider a measure ξ on S as a functional $\overline{\xi}$ on $X = LUC(S)$: $\overline{\xi}(f) = \int f\, d\xi$. Therefore every $\overline{\mu}$ ($\mu \in P(S)$) is a mean on X($M \subset X^*$ is called a mean, if $M(f) \geq 0$ for every $f \geq 0$ and $M(1) = 1$). Now the set $\mathfrak{M}$ of all means on X is weak-*compact in X and the weak-* closure of the set $\mathcal{P} = \{\overline{\mu} : \mu \in P(S)\}$ is equal to $\mathfrak{M}$. Therefore a weak-* accumulation point of an un.as.equ. net (μ_α) defines a mean M on X and for all $x \in S$ we have

$$0 \leftarrow \delta_x * \mu_\alpha(f) - \mu_\alpha(f) = \overline{\mu}_\alpha({}_xf) - \overline{\mu}_\alpha(f) \to M({}_xf) - M(f)$$

This means: M is a left invariant mean.

On the other hand if there exists a left invariant mean M on X then there is a net (μ_α) in P(S) such that $\overline{\mu}_\alpha(f) \to M(f)$ for every $f \in LUC(S)$. Therefore we have for every $x \in S$:

$$\delta_x * \mu_\alpha(f) - \mu_\alpha(f) \to 0$$

The proof will be complete by the following Theorem 3.

Proposition 2.

(a) Every as.equ. net is weakly as.equ.

(b) (μ_α) is as.equ. $\Leftrightarrow ||\sigma * \mu_\alpha - \mu_\alpha|| \to 0$ for every $\sigma \in P_a(S)$.

(c) (μ_α) is weakly as.equ. $\Leftrightarrow ||\sigma * \tau * \mu_\alpha - \tau * \mu_\alpha|| \to 0$ for every $\sigma, \tau \in P_a(S)$.

Proof.
(a) follows from

$$||\nu * \sigma * \mu_\alpha - \sigma * \mu_\alpha|| \leq || (\nu * \sigma) * \mu_\alpha - \mu_\alpha|| + ||\mu_\alpha - \sigma * \mu_\alpha|| \to 0$$

(b): => : trivial

<= : We use the fact that $P_a(S)$ is a twoside ideal in P(S). Therefore for every $\nu \in P(S)$ $\sigma \in P_a(S)$ we get

$$||\nu * \mu_\alpha - \mu_\alpha|| \leq ||\nu * \mu_\alpha - \nu * \sigma * \mu_\alpha|| + ||\nu * \sigma * \mu_\alpha - \mu_\alpha|| \leq$$
$$\leq ||\mu_\alpha - \sigma * \mu_\alpha|| + ||(\nu * \sigma) * \mu_\alpha - \mu_\alpha|| \to 0$$

(c): Similar to (b).

Theorem 3. The following conditions are equivalent:

(a) (μ_α) is un.as.equ.

(b) For every $f \in LUC(S)$ and $x \in S$: $\delta_x * \mu_\alpha(f) - \mu_\alpha(f) \to 0$

(c) For every $f \in LUC(S)$ and compact $K \subset S$:
$$\sup\{|\delta_x * \mu_\alpha(f) - \mu_\alpha(f)| : x \in K\} \to 0$$

(d) For every $f \in LUC(S)$ and compact $K \subset S$:
$$\sup\{|\nu * \mu_\alpha(f) - \mu_\alpha(f)| : \nu \in P(S),\ Supp(\nu) \subset K\} \to 0$$

Proof.
(a) =>(b): trivial

(b) =>(c):

Let $f \in LUC(S)$, K a compact subset of S, $\varepsilon > 0$ arbitrary. For every $s \in S$ there is an open set $U(s)$ such that $s \in U(s)$ and

$$||_s f - {}_t f|| < \frac{\varepsilon}{2}$$

for every $t \in U(s)$.

Now the family $\{U(s) : s \in K\}$ is an open cover of K and by the compactness of K there are elements $s_1, \ldots, s_n \in K$ such that K is covered by $\{U(s_i) : 1 \leq i \leq n\}$. Now there is an index α_o such that

$$|\delta_{s_i} * \mu_\alpha(f) - \mu_\alpha(f)| < \frac{\varepsilon}{2}$$

for every $\alpha \geq \alpha_o$, $1 \leq i \leq n$.

Let $x \in K$. Since x is an element of some $U(s_i)$ we have for $\alpha \geq \alpha_o$:

$$|\delta_x * \mu_\alpha(f) - \mu_\alpha(f)| \leq |\delta_x * \mu_\alpha(f) - \delta_{s_i} * \mu_\alpha(f)| + |\delta_{s_i} * \mu_\alpha(f) - \mu_\alpha(f)|$$

$$< ||\mu_\alpha|| \cdot ||_x f - {}_{s_i} f|| + \frac{\varepsilon}{2} < \varepsilon$$

(c) => (d):

Let K be a compact subset of S, let $\nu \in P(S)$ such that $\mathrm{Supp}(\nu) \subset K$. Then

$$|\nu * \mu_\alpha(f) - \mu_\alpha(f)| = |\int_K \int_S f(xy)\, d\nu(x)\, d\mu_\alpha(y) - \mu_\alpha(f)| =$$

$$= |\int_K [\delta_x * \mu_\alpha(f) - \mu_\alpha(f)]\, d\nu(x)| \leq$$

$$\leq \sup \{|\delta_x * \mu_\alpha(f) - \mu_\alpha(f)| : x \in K\}.$$

(d) => (a):

Let $f \in LUC(S)$, $\nu \in P(S)$, $\varepsilon > 0$ arbitrary. Then there exists a compact subset K of S such that $\nu(K) > 1 - \varepsilon$. Let $a = \nu(K)$, $b = 1 - a$. Let us define two probability measures on S by

$$\nu_1 := \frac{1}{a}\, \nu(. \cap K) \quad \text{and} \quad \nu_2 := \begin{cases} \nu & \text{if } a = 1 \\ \frac{1}{b}(\nu - a \cdot \nu_1) & \end{cases}$$

By assumption there exists an index α_o such that for every $\alpha \geq \alpha_o$: $|\nu_1 * \mu_\alpha(f) - \mu_\alpha(f)| < \varepsilon$. Therefore we get

$$|\nu * \mu_\alpha(f) - \mu_\alpha(f)| = |(a\nu_1 + b\nu_2) * \mu_\alpha(f) - (a+b)\mu_\alpha(f)| \leq$$

$$\leq a|\nu_1 * \mu_\alpha(f) - \mu_\alpha(f)| + b|(\nu_2 * \mu_\alpha - \mu_\alpha)(f)| <$$

$$< \varepsilon + b \,||\nu_2 * \mu_\alpha - \mu_\alpha|| \cdot ||f|| \leq$$

$$\leq \varepsilon\,(1 + 2||f||).$$

A completely simple locally compact semigroup S is isomorphic to the semigroup X x G x Y (thc so-called Rees product. Let e be an idempotent element of S then G = eSe is a locally compact group and X = E(Se) (Y = E(eS)) is a locally compact left (right)-zero-semigroup, consisting of all idempotents of Se (eS). The multiplication is given by $(x,g,y) \cdot (u,h,v) = (x,gyuh),v)$. S = X x G x Y has minimal right ideals, which are of the form $\{x\} \times G \times Y$, x X.

Proposition 4. If there exists an un.as.equ. net on a completely simple semigroup S, then S has exactly one minimal right ideal.

Proof. We have to prove that the first semigroup factor X of S = X x G x Y has exactly one element. Suppose X has at least two elements $\bar{x}_1$, $\bar{x}_2$. Then by Urysohn's lemma there is a continuous function $g: X \to [0,1]$ such that $g(\bar{x}_1) = 1$ and $g(\bar{x}_2) = 0$. The real valued function f on S defined by $f(x,g,y) = g(x)$ is continuous and bounded. Let $u = (x_1,g_1,y_1)$, $v = (x_2,g_2,y_2)$, $w = (x,g,y)$ be elements in S. Consider

$$||_u f - {}_v f|| = \sup_{w \in S} |f(uw) - f(vw)| =$$

$$= \sup_{(x,g,y) \in S} |f(x_1,g',y) - f(x_2,g'',y)| = |g(x_1) - g(x_2)|$$

and therefore $f \in LUC(S)$. Let $u = (\bar{x}_1,g_1,y_1)$, $v = (\bar{x}_2,g_2,y_2)$, then $f(u.w) = 1$ and $f(v.w) = 0$. This means: $_u f = 1$, $_v f = 0$. Let (μ_α) be an un.as.equ. net then we get

$$0 \leftarrow \delta_u * \mu_\alpha(f) - \delta_v * \mu_\alpha(f) = \mu_\alpha(_u f) - \mu_\alpha(_v f) = 1$$

This is a contradiction and the Proposition 4 is proved.

Proposition 5. Let S be a semigroup which satisfies condition (A) or (B). Then (μ_α) is weakly as.equ. if and only if the net $(\sigma * \mu_\alpha)$ is weakly as.equ. for every $\sigma \in P_a(S)$.

Proof. Let $(\sigma * \mu_\alpha)$ be weakly as.equ. for every $\sigma * P_a(S)$. If S satisfies

(A) then every $\rho \in P_a(S)$ can be written in the form $\rho = \sigma * \tau$, $\sigma, \tau \in P_a(S)$. Hence the net (μ_α) is weakly as.equ.

Now let S satisfy (B), i.e. $M_a(S)$ has a left approximate unit (τ_β). Let $\sigma \in P_a(S)$, let $\varepsilon > 0$ be arbitrary. Then there is an index β_o such that for any $\beta \geq \beta_o$ we have

$$||\sigma - \tau_\beta * \sigma|| < \frac{\varepsilon}{3}$$

Let $\nu \in P(S)$. Since $(\tau_{\beta_o} * \sigma * \mu_\alpha)$ is weakly as.equ. there is also an index α_o such that for any $\alpha \geq \alpha_o$ we have

$$||\nu * \tau_{\beta_o} * \sigma * \mu_\alpha - \tau_{\beta_o} * \sigma * \mu_\alpha|| < \frac{\varepsilon}{3}$$

Therefore we get for every $\alpha \geq \alpha_o$:

$$||\nu * \sigma * \mu_\alpha - \sigma * \mu_\alpha|| \leq ||\nu * \sigma * \mu_\alpha - \nu * \tau_{\beta_o} * \sigma * \mu_\alpha|| +$$

$$+ ||\nu * \tau_{\beta_o} * \sigma * \mu_\alpha - \tau_{\beta_o} * \sigma * \mu_\alpha|| + ||\tau_{\beta_o} * \sigma * \mu_\alpha - \sigma * \mu_\alpha||$$

$$< 2||\sigma - \tau_{\beta_o} * \sigma|| + \frac{\varepsilon}{3} < \varepsilon.$$

<u>Theorem 6.</u>

(a) Let (μ_α) be weakly as.equ. or un.as.equ. Then $(\sigma * \mu_\alpha)$ is un.as.equ. for every $\sigma \in P_a(S)$.

(b) Let S satisfy condition (C). Then (μ_α) is un.as.equ. if and only if the net $(\sigma * \mu_\alpha)$ is un.as.equ. for every $\sigma \in P_a(S)$.

(c) Let S satisfy condition (C). Then every weakly as.equ. net is un.as.equ.

Proof.

<u>(a)</u> If (μ_α) is weakly as.equ. then $(\sigma * \mu_\alpha)$ is as.equ. for every $\sigma \in P_a(S)$ and therefore $(\sigma * \mu_\alpha)$ is un.as.equ. If (μ_α) is un.as.equ. then we get for every $\sigma \in P(S)$:

$$|\nu * (\sigma * \mu_\alpha)(f) - (\sigma * \mu_\alpha)(f)| \leq |(\nu * \sigma) * \mu_\alpha(f) - \mu_\alpha(f)| + |\mu_\alpha(f) - \sigma * \mu_\alpha(f)|$$

<u>(b):</u> Let $(\sigma * \mu_\alpha)$ un.as.equ. for every $\sigma \in P_a(S)$, let $z \in S$, $f \in LUC(S)$, $\varepsilon > 0$. Since LUC(S) is translation invariant it follows that ${}_zf \in LUC(S)$. Let e denote a fixed left unit element of S. Then there exists a compact neighbourhood U of e in S such that:

$$||{}_sf - {}_ef|| < \varepsilon \quad \text{and} \quad ||{}_{zs}f - {}_zf|| < \varepsilon \quad \text{for every } s \in U$$

Because of condition (C) there is a measure $\sigma \in P_a(S)$ such that $\sigma(U) = 1$. Now we have for every $\mu \in P(S)$:

$$|\delta_z * \sigma * \mu(f) - \delta_z * \mu(f)| = |\sigma * \mu(zf) - \mu(zf)| =$$

$$= |\int\int[{}_zf(sy) - {}_zf(y)]\, d\sigma(s)\, d\mu(y)|$$

$$\leq \int_S \mu(dy) \int_U ||{}_{zs}f - {}_zf||\, d\sigma(s)$$

$$< \varepsilon$$

and by the same argument we have also $|\sigma * \mu(f) - \mu(f)| < \varepsilon$. Therefore we get

$$|\delta_z * \mu_\alpha(f) - \mu_\alpha(f)| \leq |\delta_z * \mu_\alpha(f) - \delta_z * \sigma * \mu_\alpha(f)| +$$

$$+ |\delta_z * \sigma * \mu_\alpha(f) - \sigma*\mu_\alpha(f)| + |\sigma * \mu_\alpha(f) - \mu_\alpha(f)|$$

$$< 2\varepsilon + |\delta_z * (\sigma * \mu_\alpha)(f) + (\sigma * \mu_\alpha)(f)|$$

Since $(\sigma * \mu_\alpha)$ is un.as.equ. the statement (b) is proved.

(c): follows from (a) and (b).

Now we shall study some properties of as.equ. nets of probability measures.

Theorem 7. Let (μ_α) be a net in P(S). Consider the following statements:

(a) For every compact $K \subset S$: $\sup \{||\mu_\alpha - \delta_x * \mu_\alpha|| : x \in K\} \to 0$

(b) For every compact $K \subset S$: $\sup \{||\mu_\alpha - \nu * \mu_\alpha|| : x \in P(S),$ $\mathrm{Supp}(\nu) \subset K\} \to 0$

(c) (μ_α) is as.equ.

Then (a) <=> (b) => (c). If $P_a(S) \neq 0$ then (a) <=> (b) <=> (c).

Proof.

(a) => (b): Let K be a compact subset of S and let $\nu \in P(S)$ such that $\mathrm{Supp}(\nu) \subset K$. Then

$$||\mu_\alpha - \nu * \mu_\alpha|| \leq \int ||\mu_\alpha - \delta_x * \mu_\alpha||\, d\nu(x) \leq$$

$$\leq \int \sup_{x \in K} ||\mu_\alpha - \delta_x * \mu_\alpha||\, d\nu(x) \to 0$$

(b) => (a): trivial.

(b) => (c): For every $\nu \in P(S)$ there is a sequence (ν_n) in $P(S)$ such that $K_n = \mathrm{Supp}(\nu_n)$ is compact and $||\nu - \nu_n|| \to 0$. Let $\varepsilon > 0$ be arbitrary. Then there is an integer n and an index α_o such that

$$||\nu - \nu_n|| < \frac{\varepsilon}{2}$$

and

$$\sup \{||\mu_\alpha - \xi * \mu_\alpha|| : \xi \in P(S),\ \mathrm{Supp}(\xi) \subset K_n\} < \frac{\varepsilon}{2}$$

for every $\alpha \leq \alpha_o$. Therefore we have for $\alpha \leq \alpha_o$:

$$||\mu_\alpha - \nu * \mu_\alpha|| \leq || \mu_\alpha - \nu_n * \mu_\alpha|| + ||\nu_n * \mu_\alpha - \nu * \mu_\alpha||$$

$$< \varepsilon .$$

(c) => (a): Let $\sigma \in P_a(S)$, let K be a compact subset of S and let $\varepsilon > 0$ be arbitrary. For every $x \in S$ there is an open neighbourhood U(x) of x such that for every $y \in U(x)$:

$$||\delta_x * \sigma - \delta_y * \sigma|| < \frac{\varepsilon}{2}$$

because the mapping $x \to \delta_x * \sigma$ is norm-continuous. The family $\{U(x) : x \in K\}$ is an open cover of K and by compactness of K there is a finite subcover $\{U(x_i) : x_i \in K,\ 1 \leq i \leq n\}$ of K. Since (μ_α) is as.equ. we can find an index α_o such that for every $\alpha \geq \alpha_o$:

$$||\sigma * \mu_\alpha - \mu_\alpha|| < \frac{\varepsilon}{3}$$

$$||\delta_{x_i} * \sigma * \mu_\alpha - \mu_\alpha|| < \frac{\varepsilon}{3} \qquad 1 \leq i \leq n$$

Let $y \in K$ be arbitrary. Then there is a set $U(x_i)$ such $y \in U(x_i)$. We get for $\alpha \leq \alpha_o$:

$$||\delta_y * \mu_\alpha - \mu_\alpha|| \leq ||\delta_y * \mu_\alpha - \delta_y * \sigma * \mu_\alpha|| + || \delta_y * \sigma * \mu_\alpha - \delta_{x_i} * \sigma * \mu_\alpha||$$

$$+ ||\delta_{x_i} * \sigma * \mu_\alpha - \mu_\alpha|| <$$

$$< ||\mu_\alpha - \sigma * \mu_\alpha|| + ||\delta_y * \sigma - \delta_{x_i} * \sigma|| + \frac{\varepsilon}{3} <$$

$$< \varepsilon$$

This proves (a).

Theorem 8.

Consider the following assertions:

(a) (μ_α) is as.equ.

(b) $||\tau * \mu_\alpha|| \to |\tau(S)|$ for every $\tau \in M(S)^r$

(c) $||\tau * \mu_\alpha|| \to 0$ for every $\tau \in M(S)^r$, $\tau(S) = 0$.

Then (a)<=>(b)=> (c). If S has a left unit element, then (a)<=>(b)<=>(c).

Proof.

(a) =>(c): Let ξ and η be the positiv and negativ part of a measure $\tau = \xi - \eta$ and $\tau(S) = 0$. Since $0 = \tau(S) = \xi(S) - \eta(S) = ||\xi|| = ||\eta||$, we have

$$||\tau * \mu_\alpha|| = ||\xi * \mu_\alpha - \eta * \mu_\alpha|| = ||\xi|| \cdot ||\xi_o * \mu_\alpha - \eta_o * \mu_\alpha||$$

where $\xi_o = \frac{1}{||\xi||}\,\xi$, $\eta_o = \frac{1}{||\eta||}\,\eta \in P(S)$. Therefore $||\tau * \mu_\alpha|| \to 0$ because (μ_α) is as.equ.

(b) =>(c): trivial.

(c) =>(b): Let $\tau \in M(S)^r$ and $\tau = \xi - \eta$ be the Hahn decomposition of τ into the positiv part ξ and negativ part η. Assume that $||\xi|| > ||\eta||$. Let $\xi_1 = \frac{||\eta||}{||\xi||}\,\xi$, $\xi_2 = \xi - \xi_1 = (1 - \frac{||\eta||}{||\xi||})\,\xi$. Therefore $\xi_1 \geq 0$. $\xi_2 \geq 0$. Furthermore

$$|\,||\xi_2 * \mu_\alpha|| - ||(\xi_1 - \eta) * \mu_2||\,| \leq ||(\xi_2 + \xi_1 - \eta) * \mu_\alpha|| \leq$$

$$\leq ||\xi_2 * \mu_\alpha|| + ||(\xi_1 - \eta) * \mu_\alpha||.$$

Since $(\xi_1 - \eta)(S) = 0$ and $||\xi_2 * \mu_\alpha|| = ||\xi_2||$ it follows

$$||(\xi - \eta) * \mu_\alpha|| \to ||\xi_2|| = ||\xi|| - ||\eta|| = \tau(S).$$

(c) =>(a): If S has a left unit element e, then one may consider the measure $\tau = \delta_e - \nu$ ($\nu \in P(S)$) and therefore the equivalenz of (a) and (c) follows.

Remark. To prove the euqivalence of the statement of (a), (b), (c) in Theorem 8 the additional assumption of the existence of a left unit element in S cannot be dropped in generally, see for instance [6] (p. 110) or consider a locally compact left-zero-semigroup.

The following theorem gives an interesting connection between the existence of an as.equ. net in P(S) and the structure of S.

Theorem 9.

(a) If there is an as.equ. net in P(S), then S has at most one minimal right ideal R and $\overline{R}$ (the closure of R) is left amenable (that is there exists a left invariant mean on $M(\overline{R})^*$).

(b) If S has exactly one minimal right ideal R such that $\overline{R}$ is left amenable then there exists an as.equ. net in P(S).

Proof.

(a): Nothing has to be proved when S has no minimal right ideal. Assume S has at least two minimal right ideals R_1 and R_2. Let $x \in R_1$, $y \in R_2$. If (μ_α) is as.equ. then

$$||\delta_x * \mu_\alpha - \delta_y * \mu_\alpha|| \to 0.$$

Now we shall show that there are Borel sets $E_1 \subset R_1$ and $E_2 \subset R_2$ with the property $\delta_x * \mu_\alpha(E_1) = 1$ and $\delta_y * \mu_\alpha(E_2) = 1$. Let $\mu = \mu_\alpha$ there is a σ - compact subset E of S so that $\mu(E) = 1$ (because of the regularity of μ for every $n \in \mathbb{N}$ there is a compact set $E_n \subset S$ such that $\mu(E_n) > 1 - \frac{1}{n}$. Then $E = \cup E_n$ is σ-compact and $\mu(E) = 1$). The sets $E_1 = x . E$ and $E_2 = y . E$ are also σ-compact. Furthermore $1 \geq \delta_x * \mu(E_1) = \mu(x^{-1}xE) \geq$ $\geq \mu(E) = 1$ and so we have $\delta_x * \mu(E_1) = 1$ and likewise $\delta_y * \mu(E_2) = 1$. Since $x \in R_1$ and R_1 is a minimal right ideal we get $E_1 = x . E \subset xS = R_1$ The same argument shows $E_2 \subset R_2$. This facts show that the pair (P_1, P_2), $P_1 = E_1$, $P_2 = S - E_1$, is a Hahn-decomposition for the measure $\delta_x * \mu - \delta_y * \mu$, because $R_1 \cap R_2 = \emptyset$. Therefore

$$2 > ||\delta_x * \mu - \delta_y * \mu|| = \delta_x * \mu(P_1) + \delta_y * \mu(P_2) = 2,$$

which is a contradiction. Thus we have proved that the semigroup has only one minimal right ideal R. Since the closure $\overline{R}$ of R is also a right ideal and so a subsemigroup of S, it is clear by Theorem 1. that $\overline{R}$ is left amenable.

(b): If $\overline{R}$ is left amenable by Theorem 1 there is an as.equ. net (μ_α) in $P(\overline{R})$, that is: for every $\tau \in P(\overline{R})$ we have

$$||\tau * \mu_\alpha - \mu_\alpha|| \to 0.$$

Now every measure $\xi \in M(\overline{R})$ can be extended to a measure ξ' on the whole space S, because $\overline{R}$ is a closed set (see [3] (11.45)).

We shall show that the net (μ_α') is as.equ. net on S.

Since S has exactly one minimal right ideal R then R is also the minimal two-side ideal, the kernel of S, and we get $R.S = S.R = R$. Now let $z \in R$, $\lambda \in P(S)$. Then $\mathrm{Supp}(\lambda*\delta_z) = \overline{\mathrm{Supp}(\lambda).z} \subset \overline{R}$. Therefore $\lambda*\delta_z \in P(\overline{R})$ and we get

$$||\lambda*\mu_\alpha' - \mu_\alpha'|| \leq ||\lambda*\mu_\alpha' - \lambda*\delta_z*\mu_\alpha'|| + ||\delta_z*\lambda*\mu_\alpha' - \mu_\alpha'||$$

$$\leq ||\mu_\alpha - \delta_z*\mu_\alpha|| + ||(\delta_z*\lambda)*\mu_\alpha - \mu_\alpha||$$

This proves Theorem 9.

Using structure theory of semigroups we shall give necessary conditions for an as.equ.net of probability measures. First we need the following

Proposition 10. Let $\eta \in P(S)$. Then the following statements are equivalent:

(a) η is idempotent (i.e. $\eta * \eta = \eta$) and $\mathrm{Supp}(\eta)$ contains a subgroup G which is a left ideal of S .

(b) η is left ivarinat, i.e. $\delta_x * \eta = \eta$ for every $x \in S$.

(c) The semigroup S has a minimal twoside ideal K which is of the form $K = X \times G$ where G is a compact group. Further η is of the form $\eta = \gamma \times \nu$ where γ is the normed Haar measure on G and $\nu \in P(Y)$.

Proof.

(a) => (b) : This is similar to the first part of the proof of Theorem 4. of [8] .

(b) => (c) : Let $L = \mathrm{Supp}(\eta)$. Then L is a left ideal of S and L is a right group: There is an idempotent element $e \in L$ such that L can be written in the form $L = G \times Y$, $G = Le$ being a compact group and Y is a closed right-zero-semigroup consisting of all idempotents of L. The probability measure η decomposes as $\gamma \times \nu$ where γ is the normed Haar measure on G and $\nu \in P(Y)$ (see[15])

Now let R be a right ideal of S, that is $R.S \subset R$. Thus $RL \subset R$ and $(RL).L \subset RL$, this means RL is a right ideal of L. Since L is a right group we have $L = RL \subset R$. Therefore S has a minimal right ideal K and K is the unique minimal right ideal and therefore the minimal two-side ideal because every minimal right ideal R has the form $R = xK$ for some $x \in S$ and all right ideals contain L. Therefore there is no proper right ideal of K and K contains an idempotent element, namely $e \in L$, and this implies that K is a right group $K = G' \times Y'$ where $G' = Ke$ is a locally kompact group and Y' is a closed right-zero-semigroup consisting of all idempotents of K. Now $L = KL \subset K$ implies $G = Le \subset Ke = G'$ and $G' = G'.G = Ke.Le \subset Le = G$ because L is a left ideal of S. This means that the group factor of K is a compact group and equal to the group factor of L.

(c) => (a): Since the group factor G of the kernel K is of the form $G = Ke$ (e being an idempotent of K) we get $S.G = S.Ke \subset \subset Ke = G$, i.e. G is a left ideal of S.

Now consider functions $\varphi \in C_{oo}(G)$, $\phi \in C_{oo}(Y)$, then $f(g,y) = \varphi(g) . \phi(y) \in C_{oo}(K)$. Therefore

$$\begin{aligned}\eta * \eta\,(f) &= \int_K \int_K f((g_1,y_1).(g_2,y_2))\; d\eta(g_1,y_1)\; d\eta(g_2,y_2) \\ &= \int_K d\eta(g_1,y_1) \int_K f(g_1g_2,y_2)\; d\eta(g_2,y_2) \\ &= \int_K d\eta(g_1,y_1) \int_G \varphi(g_1g_2)\; d\gamma(g_2) \int_Y \phi(y_2)\; d\nu(y_2) \\ &= \eta(K) . \gamma(\varphi) . \nu(\phi) \quad = \quad \eta(f)\end{aligned}$$

By the Stone-Weierstraß Theorem ist follows $\eta * \eta(f) = \eta(f)$ for every $f \in C_{oo}(S)$. Therefore the proposition is proved.

Theorem 11.

(a) Let (μ_α) be a net in P(S) and suppose that there exists an $\eta \in P(S)$ such that $\|\mu_\alpha - \eta\| \to 0$. If η satisfies one of the condition in the last proposition then (μ_α) is as.equ.

(b) Let (μ_α) be an as.equ. net and suppose that S has a completely simple minimal ideal K and the group factor G of K is compact. Then there is a $\eta \in P(S)$ and an idempotent element $e \in S$ such that $\|\mu_\alpha - \eta\| \to 0$.

Proof.

(a): Proposition 10 implies that η is a left invariant probability measure and so $\nu * \eta = \eta$ for every $\nu \in P(S)$. Therefore

$$\|\mu_\alpha - \nu*\mu_\alpha\| \leq \|\mu_\alpha - \eta\| + \|\nu*\eta - \nu*\mu_\alpha\| \leq$$

$$\leq \|\mu_\alpha - \eta\| \to 0 .$$

(b): Since (μ_α) is as.equ. by Theorem 9. it follows that $K = G \times Y$, where $G = Ke$ is a compact group and $Y = E(K)$ and $e.e = e$. Let γ' be the normed Haar measure on G and γ be the extension of γ' from G to the whole space S. Now let $\nu_\alpha = \delta_e * \mu_\alpha * \delta_e$ It follows that $\mathrm{Supp}(\nu_\alpha) = \overline{e.\mathrm{Supp}(\nu_\alpha).e} \subset Ke = G$. The net (ν_α) is as.equ. because

$$\|\sigma*\nu_\alpha - \nu_\alpha\| \leq \|(\sigma*\delta_e)*\mu_\alpha - \delta_e*\mu_\alpha\| \to 0$$

for every $\sigma \in P(S)$. The net (ν_α) can be considered as an as.equ. net on the compact group G and therefore $\|\nu_\alpha - \gamma\| \to 0$. This implies

$$\|\mu_\alpha*\delta_e - \gamma\| \leq \|\mu_\alpha*\delta_e - \nu_\alpha\| + \|\nu_\alpha - \gamma\| \leq$$

$$\leq \|\mu_\alpha - \delta_e*\mu_\alpha\| + \|\nu_\alpha - \gamma\| \to 0.$$

§ 4. Asymptotic equidistribution of sequences.

Now in this last section we shall consider arbitrary sequences of probability measures on semigroups and asymptotic properties of such sequences. (For the special sequence of convolution powers see [10]

Proposition 12. A sequence (μ_n) in $P(S)$ is as.equ. if and only if $\| \mu_n - \delta_x * \mu_n \| \to 0$ for every $x \in S$.

Proof: Using Lebesque's Theorem of dominated convergence the proof of the statement is the same as in the group case.

Definition. A sequence (μ_n) is called asymptotic ω-distributed (= as.ω-equ.) when the following conditions are satisfied:

(1) (μ_n) is as.equ.

(2) The sequence (ν_n) of the absolutely continuous parts of (μ_n) has the property that $\|\nu_n\| \to 1$.

Theorem 13. Every as.equ. sequence is as.ω-equ.

Proof.

Let $\lambda \in P_a(S)$ and let B a Borel set in S such that $\lambda(B) > 0$. Let $\varepsilon > 0$ be arbitrary. Then for every $m \in \mathbb{N}$ the set

$$B_m = \{x \in B : \|\mu_n - \delta_x * \mu_n\| < \varepsilon \quad \text{for every } n \geq m\}$$

is a Borel set (because the function $g(x) = \|\mu - \delta_x * \mu\|$ is lower semicontinuous and therefore $L_k = \{x \in S: g(x) \leq \varepsilon - \frac{1}{k}\}$ is closed and $L = \bigcup\{x \in S : g(x) < \varepsilon\}$ is a F_σ-set). Fürther $B = \lim \uparrow B_m$ when (μ_n) is an as.equ. sequence. Now there is an integer $m_o \in \mathbb{N}$ such $\lambda(B_n) > 0$ for every $n \geq m_o$. Define

$$\lambda_n := \lambda(B_n)^{-1} . \lambda(. \cap B_n)$$

$$\nu_n := \lambda_n * \lambda_n * \mu_n$$

Then (ν_n) is a sequence of absolutely continuous probability measures on S $(n \geq m_o)$. We get

$$||\nu_n - \mu_n|| = ||\lambda_n * \lambda_n * \mu_n - \mu_n|| \leq$$
$$\leq \int\int ||\delta_{xy} * \mu_n - \mu_n|| d\lambda_n(x)\ d\lambda_n(y) \leq$$
$$\leq \int\int ||\delta_{xy} * \mu_n - \delta_x * \mu_n||\ d\lambda_n(x) d\lambda_n(y) +$$
$$+ \int\int ||\delta_x * \mu_n - \delta_y * \mu_n|| d\lambda_n(x)\ d\lambda_n(y) \leq$$
$$\leq 2 \int ||\delta_x * \mu_n - \mu_n||\ d\lambda_n(x) =$$
$$= 2 \int_{B_n} ||\delta_x * \mu_n - \mu_n||\ d\lambda_n(x) <$$
$$< 2 \cdot \varepsilon \cdot \lambda_n(B_n) \leq 2\ \varepsilon .$$

Thus $||\nu_n|| \geq |\ ||\mu_n|| - ||\nu_n - \mu_n||\ | \to 1$.

Proposition 14. Let (μ_n) be an un.as.equ. sequence. Then $(\tau * \mu_n)$ is also un.as.equ. sequence for every $\tau \in M(S)$.

Proof. Let $\nu \in P(S)$, $f \in LUC(S)$. Then

$$|\nu * \mu_n * \tau(f) - \mu_n * \tau(f)| =$$
$$= |\int [\nu * \mu_n(f_z) - \mu_n(f_z)]\ d\tau(z)| \leq$$
$$\leq \int |\nu * \mu_n(f_z) - \mu_n(f_z)|\ d|\tau|(z)$$

Now $f \in LUC(S)$ implies $f_z \in LUC(S)$. Furthermore the function which shall be integrated is bounded. Thus the statement follows when we use the Lebesgue's dominated convergence theorem.

R e f e r e n c e s

[1] A.H. CLIFFORD and G.B. PRESTON: The algebraic theory of semigroups. American Mathematical Society: Providence Island 1961.

[2] P. GERL: Gleichverteilung auf lokalkompakten Gruppen. Math. Nachr. 71, 249 - 260 (1976).

[3] E. HEWITT and K.A. ROSS: Abstract harmonic analysis I. Berlin-Göttingen-Heidelberg-New York: Springer 1963.

[4] H. HEYER: Probability measures on locally compact groups. Berlin-Heidelberg-New York: Springer 1977.

[5] J. KERSTAN and K. MATTHES: Gleichverteilungseigenschaften von Verteilungsgesetzen auf lokalkompakten abelschen Gruppen I. Math. Nachr. 37, 267 - 312 (1968).

[6] F. KINZL: Gleichverteilung auf diskreten Halbgruppen. Semigroup Forum 18, 105 - 118 (1979)

[7] F. KINZL: Absolut stetige Maße auf lokalkompakten Halbgruppen Mh. Math. 87, 109 - 121 (1979)

[8] F. KINZL: Convolutions of probability measures on semigroups. Semigroup Forum 20, 369 - 383 (1980).

[9] F. KINZL: Einige Bemerkungen über absolut stetige Maße auf lokalkompakten Halbgruppen. Sitzungsberichte der Österr. Akademie d. Wiss. Mathem.-naturw. Klasse, Abt. II, 189, 361 - 370 (1980).

[10] F. KINZL: Convolution Powers of Probability measures on locally compact semigroups. Lecture Notes in Mathematics 928, 247 - 257, Berlin-Heidelberg-New York: Springer 1982

[11] L. KUIPERS and H. NIEDERREITER: Uniform distribution of sequences. New York-London-Sydney-Toronto: John Wiley & Sons 1974.

[12] W. MAXONES und H. RINDLER: Asymptotisch gleichverteilte Maßfolgen in Gruppen vom Heisenberg-Typ. Sitzungsberichte der Österr. Akademie d. Wiss. Mathem.-naturw. Klasse, Abt. II, 185, 485 - 504 (1976).

[13] W. MAXONES und H. RINDLER: Bemerkungen zu einer Arbeit von P. GERL "Gleichverteilung auf lokalkompakten Gruppen", Math. Nachr. 79, 193 - 199 (1977).

[14] W. MAXONES und H. RINDLER: Asymptotisch gleichverteilte Netze von Wahrscheinlichkeitsmaßen auf lokalkompakten Gruppen. Coll. Math. 40, 131 - 145 (1978).

[15] A. MUKHERJEA and N.A. TSERPES: Measures on topological semigroups. Lecture Notes in Mathematics 547. Berlin-Heidelberg-New York: Springer 1976.

[16] A.B. PAALMAN - DE MIRANDA: Topological semigroups. Mathematisch Centrum Amsterdam: 1970.

[17] G.L.G. SLEIJPEN: Convolution Measure Algebras on Semigroups. Thesis. Nijmegen 1976.

[18] J.C.S. WONG: An ergodic property of locally compact semigroups. Pacific J. Math. 48, 615 - 619 (1973).

Franz Kinzl
Institut für Mathematik
der Universität Salzburg
Petersbrunnstraße 19
A - 5020 Salzburg
Austria

On a formula of N. Ikeda and S. Watanabe concerning the Lévy kernel

Jan Kisyński

1. Introduction

1.1. Semigroups and transition functions. Let $\hat{E}$ be a compact metric space with a distinguished point $\delta \in \hat{E}$ and let $E = \hat{E} \setminus \{\delta\}$. Suppose that $\{S_t : t \geq 0\}$ is a one-parameter strongly continuous semigroup of non-negative linear contractions in the subspace $C_o(E) = \{f \in C(\hat{E}) : f(\delta) = 0\}$ of the space $C(\hat{E})$ of all real continuous functions on $\hat{E}$. For any $f \in C(\hat{E})$ and $t \geq 0$ define

$$\hat{S}_t f = S_t(f - f(\delta)1) + f(\delta)1 .$$

Then $\{\hat{S}_t : t \geq 0\}$ is a one-parameter strongly continuous semigroup of non-negative linear operators in $C(\hat{E})$ such that

$$\hat{S}_t 1 = 1 \qquad \text{and} \qquad \hat{S}_t C_o(E) \subset C_o(E).$$

Denote by $\mathcal{B}(\hat{E})$ the σ-field of all Borel subsets of $\hat{E}$ and by $M(\hat{E})$ the set of all probability measures on $\mathcal{B}(\hat{E})$. For any $t \geq 0$ and $x \in \hat{E}$ there is unique $p_{t,x} \in M(\hat{E})$ such that

$$(S_t f)(x) = p_{t,x}(f) = \int_{\hat{E}} f(y)\, p_{t,x}(dy) .$$

The mapping $[0,\infty) \times \hat{E} \ni (t,x) \longrightarrow p_{t,x} \in M(\hat{E})$ is a Feller transition function on $\hat{E}$, i.e. it is weakly continuous and satisfies the conditions $p_{o,x} = \varepsilon_x$ and $p_{t+s,x} = \int_{\hat{E}} p_{t,x}(dy)\, p_{s,y}$. Moreover, the inclusion $\hat{S}_t C_o(E) \subset C_o(E)$ implies that $p_{t,\delta}(\{\delta\}) \equiv 1$.

1.2. The Lévy kernel. Denote by G and $\hat{G}$ the infinitesimal generators of the semigroups S_t and $\hat{S}_t$. Then $\mathcal{D}(\hat{G}) = \mathcal{D}(G) + R\mathbb{1}$ and $\hat{G}(f + \lambda\mathbb{1}) = Gf$ for every $f \in \mathcal{D}(G)$ and $\lambda \in R$. For any subset F of $\hat{E}$ define $C_c(F) = \{f \in C(\hat{E}) : \operatorname{supp} f \subset F\}$. For compact $K \subset \hat{E}$ we treat $C_c(K)$ as a Banach space with the norm $\|f\|_{C_c(K)} = \sup\{|f(x)| : x \in K\}$, and for open $U \subset \hat{E}$ we treat $C_c(U)$ as the inductive limit of the spaces $C_c(K)$ for K running over compact subsets of U. We denote $C_c^+(U) = \{f \in C_c(U) : f \geq 0\}$.

By the Lévy kernel of $\hat{G}$ we mean a mapping $\eta : x \longrightarrow \eta_x$ which to each $x \in \hat{E}$ assigns the non-negative Randon measure η_x on $\hat{E} \setminus \{x\}$ such that $\eta_x = \lim_{t \downarrow 0} \frac{1}{t} p_{t,x}$ vaguely on $\hat{E} \setminus \{x\}$, i.e. $\eta_x(f) = \lim_{t \downarrow 0} \frac{1}{t} p_{t,x}(f)$ for each $f \in C_c(\hat{E} \setminus \{x\})$.

Consider the four statements:

(A) E is a C^∞-manifold and $C_c^\infty(E) \subset \mathcal{D}(G)$.

(B) $C_c^+(U) \cap \mathcal{D}(\hat{G})$ is sequentially dense in $C_c^+(U)$ for each open $U \subset \hat{E}$.

(C) The Lévy kernel η of $\hat{G}$ exists and $\eta_x(f) = \lim_{t \downarrow 0} \frac{1}{t} p_{t,x}(f)$ uniformly in $x \in \hat{E} \setminus U$ for each open $U \not\subseteq \hat{E}$ and each $f \in C_c(U)$.

(D) The Lévy kernel η of $\hat{G}$ exists and $\eta f = \lim_{t \downarrow 0} \frac{1}{t} p_t f$ uniformly on $\hat{E}$ for each $f \in C(\hat{E}^2)$ having support outside the diagonal of $\hat{E}^2$, where $\eta f \in C(\hat{E})$ and $p_t f \in C(\hat{E})$ are defined by

$$(\eta f)(x) = \int_{\hat{E} \setminus \{x\}} f(x,y)\, \eta_x(dy) \quad , \quad (p_t f)(x) = \int_{\hat{E}} f(x,y)\, p_{t,x}(dy).$$

Lemma 1. Under assumptions of Section 1.1, (A) $\Rightarrow$ (B) $\Rightarrow$ (C) $\Leftrightarrow$ (D) .

The proof of Lemma 1 is postponed to Section 2. Note that the condition A is always true if E is a Lie group and the operators S_t commute with left translations. See [11], [9 ; Sec.4.2], [10;Sec.23.14]. Let us note also that if (A) is satisfied then $\mathcal{D}(G) \supset C_c^2(E)$ and $(Gf)(x)$ has for every $f \in C_c^2(E)$ and $x \in E$ an integro-differential expression of the type of Lévy-Hinčin. See the three preceding references, [13; p.435], [2], [16]. A natural example in which (A) is not satisfied may be found in [15; p.178-179].

1.3. The formula of N.Ikeda and S.Watanabe. The condition A.2 (ii) of N.Ikeda and S.Watanabe, [12; p.80] and [18; p.68], is somewhat weaker then our condition (C). The difference between A.2 (ii) and (C) is that in A.2 (ii) the convergence of $\frac{1}{t} p_{t,x}(f)$ to $\eta_x(f)$ need not be uniform on $\hat{E} \setminus U$ but is assumed to be bounded on $\hat{E} \setminus U$. Under the assumptions as in our Section 1.1 plus the condition A.2 (ii), N.Ikeda and S.Watanabe in [12] and [18] have proved the following probabilistic interpretation of the Lévy kernel. Let X_t be a Hunt process on E /see [1; p.45] or [8; p.92] for the detailed definition/ governed by the transition function $p_{t,x}$. Let τ be a stopping time of X_t and $\mu \in M(\hat{E})$ an initial distribution. Then

$$E_\mu \sum_{t \leqslant \tau} e^{-\lambda t} f(X_{t-0}, X_t) = E_\mu \int_0^\tau e^{-\lambda t} (\eta f) \circ X_t \, dt$$

for each $\lambda \geqslant 0$ and each non-negative Borel function f on $\hat{E}^2$ vanishing on the diagonal of $\hat{E}^2$. The proof of this formula due to N.Ikeda and S.Watanabe is based on advanced potential theory and, at the least in [12], it makes use of some additional hypotheses.

1.4. The subject of the present paper. We shall present a more elementary proof of the formula of N.Ikeda and S.Watanabe, which works under the assumptions of Section 1.1 plus the condition (C) and is based on an idea sketched in the book of Feller [6; Chapter IX, § 5].

The whole probabilistic machinery used in this proof consists only of the material collected in our Section 3 and of the Kinney estimation of distribution of the number of displacements greater then an $\varepsilon > 0$.

2. Proof of the Lemma 1.

(A) $\Rightarrow$ (B). Suppose that (A) is true. Let $U \subset \hat{E}$ be open and $f \in C_c^+(U)$. Define $g = \min(f, f(\delta)\mathbb{1})$, $f_0 = f - g$. Then $g, f_0 \in C_c^+(U)$, $f = f_0 + g$, $\max g = g(\delta) = f(\delta)$ and $f_0(\delta) = 0$. A uniform approximation of f_0 by a non-decreasing sequence of non-negative elements of $C_c^\infty(E)$ proves that f_0 is equal to the limit in the topology of $C_c(U)$ of a sequence of elements of $C_c^+(U) \cap \mathcal{D}(\hat{G})$. We shall prove that it is the same with g. If $f(\delta) = 0$ then $g = 0$ and there is nothing to do. So, suppose that $f(\delta) > 0$. Then $h = g(\delta)\mathbb{1} - g$ has the properties that $0 \leq h \leq g(\delta)$, $h(\delta) = 0$ and $h(x) = g(\delta) = f(\delta) > 0$ on the compact $K = \{x \in E : g(x) = 0\}$. Let V_n and W_n, $n=1,2,\dots$, be open subsets of $\hat{E}$ such that $\overline{V_1} \cap \overline{W_1} = \emptyset$, $V_n \supset \overline{V_{n+1}}$, $W_n \supset \overline{W_{n+1}}$, $\bigcap V_n = K$ and $\bigcap W_n = \{\delta\}$. Let $i_n, j_n \in C(\hat{E})$ be such that $0 \leq i_n \leq 1$, $0 \leq j_n \leq g(\delta)$, $i_n = 1$ outside W_n, $i_n = 0$ on $\overline{W_{n+1}}$, $\operatorname{supp} j_n \subset \overline{V_n}$ and $j_n = g(\delta)$ on $\overline{V_{n+1}}$. Define $k_n = i_n \max(h, j_n)$. Then $k_n \in C(\hat{E})$, $0 \leq k_n \leq g(\delta)$, $k_n = g(\delta)$ on $\overline{V_{n+1}}$, $k_n = 0$ on $\overline{W_{n+1}}$ and $k_n \longrightarrow h$ uniformly on $\hat{E}$. By a standard regularization of the functions k_n we may obtain a sequence of functions $h_1, h_2, \dots$ belonging to $C(\hat{E})$, which are C^∞ on E and are such that $0 \leq h_n \leq g(\delta)$, $h_n = g(\delta)$ on $\overline{V_{n+2}}$, $h_n = 0$ on $\overline{W_{n+2}}$, and $h_n \longrightarrow h$ uniformly on $\hat{E}$. Now define $g_n = g(\delta)\mathbb{1} - h_n$. Then $g_n \geq 0$, $g_n \in C_c^\infty(E) + R\mathbb{1} \subset \mathcal{D}(G) + R\mathbb{1} = \mathcal{D}(\hat{G})$, $\operatorname{supp} g_n \subset \hat{E} \setminus V_{n+2} \subset \hat{E} \setminus K \subset \operatorname{supp} g = \operatorname{supp} f \subset U$ and $g_n \longrightarrow g$ uniformly on $\hat{E}$. Consequently, $g_n \in C_c^+(U) \cap \mathcal{D}(\hat{G})$ and g_n converges to g in the topology of $C_c(U)$, which completes the proof of (A) $\Rightarrow$ (B).

The used above passage from $k_1, k_2, \ldots$ to $h_1, h_2, \ldots$ may be realized as follows. Let d be the distance in $\hat{E}$. For each $r > 0$ choose non-negative $\varphi_r \in C^\infty(\hat{E}^2)$ which is positive on the diagonal of E^2 and is such that $\varphi_r(x,y) = 0$ when $d(x,y) \geq r$. Let ψ be a positive continuous density on E. Take a positive sequence $r_n \longrightarrow 0$ such that $\{x \in E : d(x, V_{n+2}) \leq r_n\} \subset V_{n+1}$ and $\{x \in E : d(x, W_{n+2}) \leq r_n\} \subset W_{n+1}$ and define $h_n(x) = \langle k_n \varphi_{r_n,x}, \psi \rangle / \langle \varphi_{r_n,x}, \psi \rangle$, where $\varphi_{r_n,x}(y) = \varphi_{r_n}(x,y)$.

(B) $\Rightarrow$ (C). If (B) is satisfied then

(B_1) $C_c(U) \cap \mathcal{D}(\hat{G})$ is sequentially dense in $C_c(U)$ for each open $U \subset \hat{E}$ and

(B_2) for each open $U \subset \hat{E}$ and each compact $K \subset U$ there is $f_{K,U} \in C_c^+(U) \cap \mathcal{D}(\hat{G})$ such that $f_{K,U} \geq 1$ on K.

We shall prove that $(B_1) \wedge (B_2) \Rightarrow (C)$. Let $U \subsetneq \hat{E}$ be open and let $f_0 \in C_c(U)$. By (B_1), there is a compact $K \subset U$ and a sequence $f_1, f_2, \ldots$ of functions belonging to $C_c(U) \cap \mathcal{D}(\hat{G})$ such that $\operatorname{supp} f_n \subset K$ for each $n = 0, 1, \ldots$ and $f_n \longrightarrow f_0$ uniformly on $\hat{E}$. Since $f_n \in \mathcal{D}(\hat{G})$ and $f_n = 0$ on $\hat{E} \setminus U$, we have

$$\text{(a)} \qquad \lim_{t \downarrow 0} \max_{x \in \hat{E} \setminus U} \left| \frac{1}{t} p_{t,x}(f_n) - (\hat{G} f_n)(x) \right| = 0$$

for each $n = 1, 2, \ldots$. Moreover, by (B_2),

$$\text{(b)} \qquad \max_{x \in \hat{E} \setminus U} \left| \frac{1}{t} p_{t,x}(f_0) - \frac{1}{t} p_{t,x}(f_n) \right| \leq M \, \| f_n - f_0 \|_{C_c(K)}$$

for each $n = 1, 2, \ldots$ and each $t \in (0, 1]$, where

$$M = \sup \left\{ \frac{1}{t} p_{t,x}(f_{K,U}) : 0 < t \leq 1, \; x \in \hat{E} \setminus U \right\}.$$

The relations (a) and (b) imply that (C) is satisfied

(C) $\Rightarrow$ (D). If $f \in C(\hat{E}^2)$ vanishes in a neighbourhood of the diagonal of $\hat{E}^2$ and f_x is defined by $f_x(y) = f(x,y)$ then $f_x \in C_c(\hat{E} \setminus \{x\})$. So, the pointwise equality $\eta f = \lim_{t \downarrow 0} \frac{1}{t} p_t f$ is nothing but an immediate consequence of the existence of the Lévy kernel. Suppose that the convergence in this equality is not uniform and that (C) is true. Then there must exist a sequence $t_n \downarrow 0$ and a sequence $x_n \longrightarrow x_o$ such that $\frac{1}{t_n} p_{t_n, x_n}(f_{x_n})$ does not convergence to $\eta_{x_o}(f_{x_o})$. But x_o has open neighbourhoods V and W such that $f_x \in C_c(\hat{E} \setminus V)$ for each $x \in W$, and we may assume that $x_n \in W$ for all n's. Then $f_{x_n} \longrightarrow f_{x_o}$ in the sense of the norm in $C_c(\hat{E} \setminus V)$ and, by the uniformity in (C), $\frac{1}{t_n} p_{t_n, x_n}$ is a sequence of continuous linear functionals on $C_c(\hat{E} \setminus V)$ convergent pointwisely to the functional η_{x_o}. Consequently, by an application of the Banach-Steinhaus uniform boundedness theorem, $\frac{1}{t_n} p_{t_n, x_n}(f_{x_n})$ converges to $\eta_{x_o}(f_{x_o})$. This contradiction completes the proof.

(D) $\Rightarrow$ (C). Given an open $U \subsetneq E$ and an $f \in C_c(U)$ take $g \in C(\hat{E})$ such that $g = 1$ on $\hat{E} \setminus U$ and $g = 0$ in a neighbourhood of suppf. Then $g \otimes f \in C(\hat{E}^2)$ has support outside the diagonal of $\hat{E}^2$ and for each $x \in \hat{E} \setminus U$ we have $\eta(g \otimes f)(x) = \eta_x(f)$, $p_t(g \otimes f)(x) = p_{t,x}(f)$.

3. The Markov process governed by a Feller transition function

3.1. The sample space Ω and the probability measures P_μ. As in the Section 1, let $[0,\infty) \times \hat{E} \ni (t,x) \longrightarrow p_{t,x} \in M(\hat{E})$ be a Feller transition function on the compact metric space $\hat{E}$ with a distinguished point $\delta \in \hat{E}$ such that $p_{t,\delta}(\{\delta\}) \equiv 1$. Denote by Ω the set of all the $\hat{E}$-valued functions ω defined on $[0,\infty)$ which are right-continuous on $[0,\infty)$ and have left-side limits everywhere on $(0,\infty)$, and are such that either $\omega^{-1}(\delta) = \emptyset$ or $\omega^{-1}(\delta) = [\zeta, \infty)$ for some $\zeta \in [0,\infty)$. For each $t \in [0,\infty)$ denote by X_t the evaluation mapping $\Omega \ni \omega \longrightarrow \omega(t) \in \hat{E}$. This means that $X_t(\omega) = \omega(t)$. Denote by F^o the σ-field of subsets of Ω generated by the family of sets $\{X_s^{-1}(B): s \geq 0 , B \in \mathcal{B}(\hat{E})\}$. Then for each $\mu \in M(\hat{E})$ there is a unique probability measure P_μ on F^o such that

$$E_\mu f(X_{h_1},\dots,X_{h_1+\dots+h_n}) = \int_{\hat{E}} \mu(dx_o) \int_{\hat{E}} p_{h_1,x_o}(dx_1) \cdots \int_{\hat{E}} p_{h_n,x_{n-1}}(dx_n) f(x_1,\dots,x_n)$$

for each $n=1,2,\dots$, every $h_1,\dots,h_n \geq 0$ and each $f \in C(\hat{E}^n)$, where E_μ is the mathematical expectation with respect to P_μ . See [1; p.46], [8; p.92] or [15; p.246] for a proof.

3.2. The filtration $\{F_t\}$ and first exit times. Put $F = \bigcap_{\mu \in M(\hat{E})} F^\mu$, where F^μ is the completion of F^o with respect to P_μ . Then F is a σ-field of subsets of Ω , $F \supset F^o$, and each P_μ has a unique extension to a probability measure on F . We shall preserve the notations P_μ and E_μ for the probability measures P_μ extended from F^o onto F , and for the corresponding mathematical expectations. We shall use the shortages $P_{\varepsilon_x} = P_x$, $E_{\varepsilon_x} = E_x$.

For each $t \geq 0$ denote by F_t^o the sub-σ-field of F^o generated by the family of sets $\{X_s^{-1}(B): 0 \leq s \leq t, B \in \mathcal{B}(\hat{E})\}$. Let $F_{t+}^o = \bigcap_{h>0} F_{t+h}^o$. Following Dynkin [5; Chapter 4, § 1] define $F_t = \bigcap_{\mu \in M(\hat{E})} F_{t+}^\mu$, where F_{t+}^μ is the completion of F_{t+}^o with respect to $P_\mu|F_{t+}^o$. Then $F_t \supset F_t^o$ and $\{F_t : t \geq 0\}$ is a right-continuous filtration of the measurable space (Ω, F) . Moreover, for each $B \in \mathcal{B}(\hat{E})$ the first exit time of X_t from B is a stopping time of this filtration. See [8; p.17 and 227-232] and [17; p.49].

Let us note that the reasonings as in [1; p. 42-43, proof of the Proposition 8.12] or in [8; p.90-91] show that

$$F_{t+}^\mu = \{A \Delta N : A \in F_t^o , N \in F_{t+}^\mu , P_\mu(N) = 0\}.$$

In this connection see also [3; p. 115].

3.3. The Markov property. Let $\{\theta_t : t \geq 0\}$ be the semigroup of left translations in the function space Ω, i.e. let $(\theta_s \omega)(t) = \omega(t+s)$ or equivalently $X_t \circ \theta_s = X_{t+s}$. The collection $X = (\Omega, F, F_t, X_t, \theta_t, P_x)$ is a Markov process on $\hat{E}$ governed by the transition function $p_{t,x}$. Since the former is a Feller transition function, the collection X has a collection of important properties which qualify X to the Hunt subclass of Markov processes. See [1; p.45] or [8; p.92]. We shall need only the following consequence of the Markov property of X : if $\mu \in M(\hat{E})$, $f \in C(\hat{E}^2)$, $t \geq 0$ and $h > 0$ then

$$E_\mu \{ f(X_t, X_{t+h}) \mid F_t \} = (p_h f) \circ X_t$$

P_μ -almost surely, $p_h f \in C(\hat{E})$ being defined as in the statement (D) of Section 1.2.

4. The processes $P_t(f,\lambda)$ and $Q_t(f,\lambda)$.

4.1. Assumptions and notations. We admitt the assumptions and notations of the Section 1.1. The Markov process $X = (\Omega, F, F_t, X_t, \theta_t, P_x)$ governed by the transition function $p_{t,x}$ is constructed as in the Section 3. Let d be the distance in $\hat{E}$. For each $\varepsilon \geq 0$ define

$$D_\varepsilon = \{ (x,y) \in \hat{E}^2 \ : \ d(x,y) \leq \varepsilon \},$$

so that in particular D_o is the diagonal in $\hat{E}^2$. Denote by $C_c^+(\hat{E}^2 \setminus D_\varepsilon)$ the set of all non-negative finite continuous functions on $\hat{E}^2$ with supports contained in $\hat{E}^2 \setminus D_\varepsilon$, and by $\mathcal{B}^+(E^2 \setminus D_o)$ the set of all non-negative Borel measurable functions on $\hat{E}^2$ vanishing on D_o.

4.2. Definition of $P_t(f,\lambda)$. Since for each $\omega \in \Omega$ the set $\{ s \in [0,\infty) \ : \ X_{s-o}(\omega) \neq X_s(\omega) \}$ is at the most countable, the sum

$$P_t(f,\lambda)(\omega) = \sum_{0<s\leq t} e^{-\lambda s} f(X_{s-o}(\omega), X_s(\omega))$$

makes a sense for every $f \in \mathcal{B}^+(\hat{E}^2 \setminus D_o)$, $\omega \in \Omega$, $t \in [0,\infty]$ and $\lambda \in [0,\infty)$. If $f \in C_c^+(\hat{E}^2 \setminus D_o)$ and $t \in [0,\infty)$ then $P_t(f,\lambda)(\omega)$ is finite because the corresponding sum contains at the most finitely many positive terms.

4.3. An approximation of $P_t(f,\lambda)$. Under the Assumptions 4.1 for every $f \in C_o^+(\hat{E}^2 \setminus D_o)$, $\omega \in \Omega$, $t \in [0,\infty)$ and $\lambda \in [0,\infty)$ we have

$$P_t(f,\lambda)(\omega) = \lim_{h \downarrow 0} P_t(f,\lambda,h)(\omega),$$

where

$$P_t(f,\lambda,h)(\omega) = \sum_{n=0}^{n(t,h)-1} e^{-\lambda nh} f(X_{nh}(\omega), X_{nh+h}(\omega))$$

and

$$n(t,h) = \inf\{n \in N : nh \geqslant t\}.$$

Proof. Let f, ω, t and λ be fixed. Choose an $\varepsilon > 0$ such that $f \in C_c^+(\hat{E}^2 \setminus D_\varepsilon)$ and define

$$T = \{s \in (0,t] : d(\omega(s-o), \omega(s)) \geqslant \varepsilon\}.$$

The set T is finite and

$$P_t(f,\lambda)(\omega) = \sum_{s \in T} e^{-\lambda s} f(\omega(s-o), \omega(s)).$$

Thanks to continuity of f, right-continuity of ω and to existence of left-side limits of ω, we have

$$\sum_{s \in T} e^{-\lambda s} f(\omega(s-o), \omega(s)) = \lim_{h \downarrow 0} \sum_{n \in N_h} e^{-\lambda nh} f(\omega(nh), \omega(nh+h)),$$

where

$$N_h = \{n \in N : T \cap (nh, nh+h] \neq \emptyset\}.$$

So, the proof will be complete if we show that

(*) there is $h_o > 0$ such that $M_h \subset N_h$ for each $h \in (0,h_o]$,

where

$$M_h = \{n \in \{0,\dots,n(t,h)-1\} : f(\omega(nh), \omega(nh+h)) \neq 0\}.$$

If (*) would be not true then there would exist a sequence $h_1, h_2, \dots$ of positive numbers and a sequence $n_1, n_2, \dots$ of integers such that $n_k \in M_{h_k} \setminus N_{h_k}$, $\lim h_k = 0$ and $\lim n_k h_k = t_o \in [0,t]$. But if

$n_k \in M_{h_k}$ then $d(\omega(n_k h_k), \omega(n_k h_k + h_k)) > \varepsilon$, and this inequality implies that $n_k h_k < t_o \leqslant n_k h_k + h_k$ for all sufficiently large k, so that finally $t_o \in T \cap \{n_k h_k, n_k h_k + h_k\}$ for all sufficiently large k. However the former is impossible because $n_k \notin N_{h_k}$.

4.4. Lemma. Let $\mathcal{M}$ be a set of functions on $\hat{E}^2$ with values in $[0,\infty]$ such that

(a) $C_c^+(\hat{E}^2 \setminus D_o) \subset \mathcal{M} \subset \mathcal{B}^+(\hat{E}^2 \setminus D_o)$

(b) if $\lambda, \mu \in [0,\infty)$ and $f, g \in \mathcal{M}$ then $\lambda f + \mu g \in \mathcal{M}$,

(c) if $f, g \in \mathcal{M}$ and $f \geqslant g$ then $f - g \in \mathcal{M}$,

(d) $\lim f_n \in \mathcal{M}$ for each non-increasing sequence $f_1 \geqslant f_2 \geqslant \dots$ of elements of $C_c^+(\hat{E}^2 \setminus D_o)$,

(e) $\lim f_n \in \mathcal{M}$ for each non-decreasing sequence $f_1 \leqslant f_2 \leqslant \dots$ of elements of $\mathcal{M}$.

Then $\mathcal{M} = \mathcal{B}^+(\hat{E}^2 \setminus D_o)$.

Proof. The argumentation is the same as that in the proof of [1; p. 6, Theorem 2.3]. By (a) and (d), $\mathbb{1}_K \in \mathcal{M}$ for each compact subset K of $\hat{E}^2 \setminus D_o$. By (a) and (e), $\mathbb{1}_{\hat{E}^2 \setminus D_o} \in \mathcal{M}$.
So, by an application of (c) and (e) to indicator functions, $\mathcal{D} = \{B \in \mathcal{B}(\hat{E}^2) : \mathbb{1}_B \in \mathcal{M}\}$ is a Dynkin system of subsets of $\hat{E}^2 \setminus D_o$ and $\mathcal{D}$ contains the π-system of all compact subsets of $\hat{E}^2 \setminus D_o$. Consequently, by the monotone class theorem [1; p.5, Theorem 2.2], $\mathcal{D}$ consists exactly of all Borel subsets of $\hat{E}^2 \setminus D_o$. So, by (b), $\mathcal{D}$ contains all the simple functions belonging to $\mathcal{B}^+(\hat{E}^2 \setminus D_o)$. Finally, for each $f \in \mathcal{B}^+(\hat{E}^2 \setminus D_o)$ there is a sequence $0 \leqslant f_1 \leqslant f_2 \dots$ of simple functions such that $f = \lim f_n$ pointwisely on $\hat{E}^2$ and so, by (e), $f \in \mathcal{M}$.

4.5. Optionality of $P_t(f,\lambda)$. Under the Assumptions 4.1 for each

$f \in \mathcal{B}^+(\hat{E}^2 \setminus D_o)$ and each $\lambda \in [0,\infty)$ the process $P_t(f,\lambda)$ is $\{F_t^o\}$-optional. Indeed, denote by $\mathcal{M}$ the set of all the functions f which belong to $\mathcal{B}^+(\hat{E}^2 \setminus D_o)$ and are such that the above optionality holds for all $\lambda \in [0,\infty)$. Then the conditions 4.4(b) and (c), are obvious and, according to the Lemma 4.4, it remains to verify whether 4.4, (a), (d) and (e), are also satisfied.

Ad 4.4(a). If $f \in C_c^+(\hat{E}^2 \setminus D_o)$ then for every $\lambda \in [0,\infty)$ and $n=1,2,\ldots$ the process $P_t(f,\lambda,\frac{t}{n})$ is $\{F_t^o\}$-adapted, right-continuous and has finite left hand limits. So, it follows from 4.3 that $f \in \mathcal{M}$.

Ad 4.4(d). If $f_1 \geq f_2 \geq \ldots$ is a sequence of elements of $C_c^+(\hat{E}^2 \setminus D_o)$ and $f_o = \lim f_n$ then $f_1 \in C_c^+(\hat{E}^2 \setminus D_\varepsilon)$ for some $\varepsilon > 0$ and hence

$$P_t(f_n,\lambda)(\omega) = \sum_{s \in S(\omega,t,\varepsilon)} e^{-\lambda s} f_n(\omega(s-o), \omega(s))$$

for every $\omega \in \Omega$, $t \in (0,\infty)$, $\lambda \in [0,\infty)$ and $n=1,2,\ldots$, where $S(\omega,t,\varepsilon) = \{s \in (0,t] : d(\omega(s-o), \omega(s)) > \varepsilon\}$ is a finite set. From this it is evident that $P_t(f_o,\lambda)(\omega) = \lim_{n\to\infty} P_t(f_n,\lambda)(\omega)$.
But $f_n \in \mathcal{M}$ for each $n=1,2,\ldots$, by the already verified condition 4.4(a). Consequently also $f_o \in \mathcal{M}$.

Ad 4.4(e). Suppose that $f_1 \leq f_2 \leq \ldots$ is a sequence of elements of $\mathcal{M}$ and $f_o = \lim f_n$. For every $\omega \in \Omega$, $t \in (0,\infty)$, $\lambda \in [0,\infty)$ and $n=0,1,\ldots$,

$$P_t(f_n,\lambda)(\omega) = \sum_{s \in S(\omega,t)} e^{-\lambda s} f_n(\omega(s-o), \omega(s)),$$

where the set $S(\omega,t) = \{s \in (0,t] : \omega(s-o) \neq \omega(s)\}$ is at the most countable. Since all the terms of this sum are non-negative and they non-decrease when n increases, it follows that $P_t(f_o,\lambda)(\omega) = \lim_{n\to\infty} P_t(f_n,\lambda)(\omega)$ and so $f_o \in \mathcal{M}$.

4.6. Measurability of ηf for $f \in \mathcal{B}^+(\hat{E}^2 \setminus D_o)$. Under the Assumptions 4.1 suppose moreover that the condition (C) of the Section

1.2 is satisfied. Then ηf is Borel measurable for each $f \in \mathcal{B}^+(\hat{E}^2 \setminus D_o)$. Indeed, according to the Lemma 1, (C) implies the condition (D) of the Section 1.2. So, if $f \in C_c^+(\hat{E}^2 \setminus D_o)$ then ηf is continuous. With this starting point, the measurability of ηf for each $f \in \mathcal{B}^+(\hat{E}^2 \setminus D_o)$ may be easily proved my means of the Lemma 4.4 and of the bounded and the monotone convergence theorems from the theory of the integral.

4.7. Definition of $Q_t(f,\lambda)$. Under the Assumptions 4.1 suppose moreover that the condition (C) of the Section 1.2 is satisfied. Let $f \in \mathcal{B}^+(\hat{E}^2 \setminus D_o)$. Then, according to 4.6, ηf is measurable and hence also $(\eta f) \circ \omega$ is measurable for each $\omega \in \Omega$. Consequently the integral

$$Q_t(f,\lambda)(\omega) = \int_0^t e^{-\lambda s}[(\eta f)\circ\omega](s)\,ds = \int_0^t e^{-\lambda s}[(\eta f)\circ X_s](\omega)\,ds$$

makes a sense for every $f \in \mathcal{B}^+(\hat{E}^2 \setminus D_o)$, $\omega \in \Omega$, $t \in [0,\infty]$ and $\lambda \in [0,\infty)$.

4.8. Predictability of $Q_t(f,\lambda)$. Under the same assumptions as in 4.7, the process $Q_t(f,\lambda)$ is $\{F_t^o\}$-predictable for each $f \in \mathcal{B}^+(\hat{E}^2 \setminus D_o)$ and each $\lambda \in [0,\infty)$. Indeed denote by $\mathcal{M}$ the set of all the functions f which belong to $\mathcal{B}^+(\hat{E}^2 \setminus D_o)$ and are such that the above predictability condition is satisfied for all $\lambda \in [0,\infty)$. By (C)$\Rightarrow$(D) of the Lemma 1 of the Section 1.2, if $f \in C_c^+(\hat{E}^2 \setminus D_o)$ then $\eta f \in C(\hat{E})$ and so, for each $\omega \in \Omega$, the function $(\eta f)\circ\omega$ has only jump discontinuities. Consequently, if $t \in (0,\infty)$ and $f \in C_c^+(\hat{E}^2 \setminus D_o)$ then the integral in the definition of $Q_t(f,\lambda)(\omega)$ may be taken in the sense of Riemann, so that

$$Q_t(f,\lambda)(\omega) = \lim_{h \downarrow 0} Q_t(f,\lambda,h)(\omega) ,$$

where

$$Q_t(f,\lambda,h)(\omega) = h \sum_{n=0}^{n(t,h)-1} e^{-\lambda nh}[(\eta f)\circ\omega](nh) ,$$

$n(t,h)$ being defined as in 4.3, i.e. by

$$n(t,h) = \inf\{n \in N : nh \geqslant t\}.$$

For each $h>0$ the process $Q_t(f,\lambda,h)$ is left-continuous and $\{F_t^o\}$-adapted. Consequently, $Q_t(f,\lambda)$ is $\{F_t^o\}$-predictable. This means that $C_o^+(\hat{E}^2 \setminus D_o) \subset \mathcal{M}$, i.e. that the condition 4.4(a) is satisfied. Obviously the conditions 4.4(b) and (c) are also satisfied. The conditions 4.4(d) and (e) may be easily verified by applications of the bounded and the monotone convergence theorems to the iterated integral

$$Q_t(f,\lambda)(\omega) = \int_0^t e^{-\lambda s}\Big[\int_{\hat{E}\setminus\{\omega(s)\}} f(\omega(s),y)\ \eta_{\omega(s)}(dy)\Big]\,ds .$$

Consequently $\mathcal{M} = \mathcal{B}^+(\hat{E}^2 \setminus D_o)$, by the Lemma 4.4.

5. A proof of the formula of N.Ikeda and S.Watanabe.

5.1. Assumptions. As in the Section 1.1, we assume that $\hat{E}$ is a compact metric space with a distinguished point $\delta \in \hat{E}$, and that $\{S_t : t \geqslant 0\}$ is a one-parameter strongly continuous semigroup of non-negative contractions in the space $C_o(E) = \{f \in C(\hat{E}) : f(\delta) = 0\}$. By definition, the semigroup $\{\hat{S}_t : t \geqslant 0\}$ in $C(\hat{E})$ and the transition function $p_{t,x}$ satisfy the conditions

$$\hat{S}_t f = S_t(f - f(\delta)1) + f(\delta)1 , \quad p_{t,x}(f) = (\hat{S}_t f)(x)$$

for each $f \in C(\hat{E})$ and each $x \in \hat{E}$. The Markov process $X = (\Omega, F, F_t, X_t, \theta_t, P_x)$ governed by the transition function $p_{t,x}$ is constructed as in the Section 3.

5.2. Summary of the measurability results of Section 4. Denote by $\mathcal{B}^+(\hat{E}^2 \setminus D_o)$ the set of all non-negative Borel measurable functions on $\hat{E}^2$ vanishing on the diagonal D_o of $\hat{E}^2$. Under the Assumptions 5.1, for each $f \in \mathcal{B}^+(\hat{E}^2 \setminus D_o)$ and each $\lambda \in [0,\infty)$, the process

$$P_t(f,\lambda) = \sum_{0<s\leqslant t} e^{-\lambda s} f(X_{s-o}, X_s)$$

with values in $[0,\infty]$ is $\{F_t^o\}$ -optional. If moreover the condition (C) of the Section 1.2 is satisfied, then ηf is Borel measurable and the process

$$Q_t(f,\lambda) = \int_0^t e^{-\lambda s} (\eta f)\circ X_s \, ds$$

with values in $[0,\infty]$ is $\{F_t^o\}$ -predictable. Recall that $\eta : x \longrightarrow \eta_x$ is the Lévy kernel of the infinitesimal generator $\hat{G}$ of the semigroup $\{\hat{S}_t\}$ and ηf is the non-negative function of $\hat{E}$ defined by

$$(\eta f)(x) = \int_{\hat{E}\setminus\{x\}} f(x,y)\, \eta_x(dy) \; .$$

5.3. Theorem. Under the Assumptions 5.1 suppose moreover that the Lévy kernel η of the infinitesimal generator $\hat{G}$ of the semigroup $\{\hat{S}_t\}$ exists and satisfies the condition (C) of the Section 1.2. Let τ be a stopping time of the filtration $\{F_t\}$, let $f\in \mathcal{B}^+(\hat{E}^2\setminus D_o)$ and $\lambda\in[0,\infty)$. Then

$$E_\mu \sum_{0<s\leq\tau} e^{-\lambda s} f(X_{s-0},X_s) = E_\mu \int_0^\tau e^{-\lambda s} (\eta f)\circ X_s \, ds$$

for each initial distribution $\mu\in M(\hat{E})$. Moreover, if f is bounded and vanishes in a neighbourhood of the diagonal of $\hat{E}^2$ and either τ is bounded or $\lambda > 0$, then both the sides of this equality are finite.

5.4. Remarks. It follows from the Theorem 5.3 and from the Markov property of the process X that, if $f\in \mathcal{B}^+(\hat{E}^2\setminus D_o)$ is bounded and vanishes in a neighbourhood of D_o , diagonal of $\hat{E}^2$, then for each $\mu\in M(\hat{E})$ the difference $R_t(f,\lambda) = P_t(f,\lambda) - Q_t(f,\lambda)$ is a martingale with respect to the filtration $\{F_t\}$ and to the probability measure P_μ on (Ω, F) , so that

$$P_t(f,\lambda) = Q_t(f,\lambda) + R_t(f,\lambda)$$

is the Doob-Meyer decomposition of $P_t(f,\lambda)$. The existence of such decomposition independent of a particular choice of $\mu \in M(\hat{E})$ was proved by S.Watanabe in [18] without assuming that the Lévy kernel exists and that the process X is governed by a Feller transition function. S.Watanabe proved that in such a general situation $Q_t(f,0)$ may be /not uniquely/ represented in the form $Q_t(f,0) = \int_0^t (\tilde{\eta} f)\circ X_s \, dA_s$, where A_t is a non-decreasing process and $(\tilde{\eta} f)(x) = \int_{\hat{E}\setminus\{x\}} f(x,y)\tilde{\eta}_x(dy)$ for each $x \in \hat{E}$, $\tilde{\eta}_x$ being a non-negative measure on $\hat{E}\setminus\{x\}$. Let us notice that in the terminology of Dynkin [5; Chapter 6, § 3] , $Q_t(f,0)$ is a W-functional of the process X determined by the W-function $F_t(x) = E_x P_t(f,0)$.

5.5. Reduction of the proof to the case of $f \in C_c^+(\hat{E}^2\setminus D_o)$. Denote by $\mathcal{M}$ the set of all the functions f belonging to $\mathcal{B}^+(\hat{E}^2\setminus D_o)$ such that $E_\mu P_\tau(f,\lambda) = E_\mu Q_\tau(f,\lambda)$ for each $\lambda \in [0,\infty)$ each $\mu \in M(\hat{E})$ and each stopping time τ of the filtration $\{F_t\}$. Then the conditions 4.4 (b) and (c), are trivially true. A reasoning based on the monotone convergence theorem of the theory of the integral /similar to that use in 4.5 and in 4.7/ proves that also the condition 4.4 (e) is satisfied. Suppose now that the statement of the Theorem 5.3 is true for each $f \in C_o^+(\hat{E}^2\setminus D_o)$. This means that the condition 4.4 a is satisfied and that $E_\mu P_{\tau\wedge t}(f,\lambda) < \infty$ for each $f \in C_c^+(\hat{E}^2\setminus D_o)$, each $\lambda \in [0,\infty)$ and each $t \in [0,\infty)$. By the bounded convergence theorem of the theory of the integral, it follows that if $f_1 \geqslant f_2 \geqslant \dots$ is a sequence of elements of $C_c^+(\hat{E}^2\setminus D_o)$ and $f_o = \lim f_n$ then $E_\mu P_{\tau\wedge t}(f_o,\lambda) = E_\mu Q_{\tau\wedge t}(f_o,\lambda)$ for each $t \in [0,\infty)$. Passing to the limit as $t \longrightarrow \infty$, by the monotone convergence theorem, we conclude that $E_\mu P_\tau(f_o,\lambda) = E_\mu Q_\tau(f_o,\lambda)$. This means that the condition 4.4 (d) is satisfied. Consequently, it follows from the Lemma 4.4 that if the Theorem 5.3 is true for all $f \in C_o^+(\hat{E}^2\setminus D_o)$ then the equality $E_\mu P_\tau(f,\lambda) = E_\mu Q_\tau(f,\lambda)$ is true for all

$f \in \mathcal{B}^+(\hat{E}^2 \setminus D_o)$. The statement of the Theorem concerning the finiteness of both the sides of this equality follows at once from the similar statement limited to $f \in C_c^+(\hat{E}^2 \setminus D_o)$. Thus we see that if the Theorem 5.3 is true for all $f \in C_c^+(\hat{E}^2 \setminus D_o)$, then it is also true for all $f \in \mathcal{B}^+(\hat{E}^2 \setminus D_o)$.

5.6. Proof for $f \in C_c^+(E^2 \setminus D_o)$. For each $t \in [0,\infty)$ and each $h > 0$ define

$$n(t,h) = \inf\{n \in \mathbb{N} : nh \geqslant t\}, \quad t_h = hn(t,h).$$

Then

$$t \leqslant t_h < t + h .$$

For any fixed $f \in C(\hat{E}^2)$, $\lambda \in [0,\infty)$ and $h \in (0,\infty)$ consider the step process

$$P_t(f, \lambda, h) = \sum_{n=0}^{n(t,h)-1} e^{-\lambda nh} f(X_{nh}, X_{nh+h}),$$

Let τ be a stopping time of the filtration $\{F_t\}$. Then

$$P_{\tau\wedge t}(f, \lambda, h) = \sum_{n=0}^{n(t,h)-1} \mathbb{1}_{\{\tau\wedge t > nh\}} e^{-\lambda nh} f(X_{nh}, X_{nh+h})$$

and

$$\sum_{n=0}^{n(t,h)-1} E_\mu\{\mathbb{1}_{\{\tau\wedge t > nh\}} e^{-\lambda nh} f(X_{nh}, X_{nh+h}) \mid F_{nh}\} = \sum_{n=0}^{n(t,h)-1} \mathbb{1}_{\{\tau\wedge t > nh\}} e^{-\lambda nh} (p_h f) \circ X_{nh}$$

$$= \int_{0+}^{(\tau\wedge t)_h} e^{-\lambda(s_h - h)} \frac{1}{h}(p_h f) \circ X_{s_h - h}\, ds$$

for each $\mu \in M(\hat{E})$ and each $t \in [0,\infty)$, by the mentioned at the end of the Section 3.3 consequence of the Markov property of the process X. Consequently,

(i) $$E_\mu P_{\tau\wedge t}(f, \lambda, h) = E_\mu \int_{0+}^{(\tau\wedge t)_h} e^{-\lambda(s_h - h)} \frac{1}{h}(p_h f) \circ X_{s_h - h}\, ds .$$

Suppose now that $f \in C_c^+(\hat{E}^2 \setminus D_o)$. Then

$$\lim_{h \downarrow 0} \left[\frac{1}{h}(p_h f) \circ X_{s_h - h}\right](\omega) = \left[(\eta f) \circ X_{s-o}\right](\omega)$$

for each $\omega \in \Omega$ and $s \in (0,\infty)$, and

$$\sup\left\{ \left\| \frac{1}{h} p_h f \right\|_{C(\hat{E})} : h \in (0,1] \right\} < \infty ,$$

by the implication (C)$\Rightarrow$(D) of the Lemma 1 of Section 1.2. Consequently, by the Lebesgue bounded convergence theorem, if $f \in C_c^+(\hat{E}^2 \setminus D_o)$ then

$$\lim_{h \downarrow 0} E_\mu \int_{0+}^{(\tau\wedge t)_h} e^{-\lambda(S_h - h)} \frac{1}{h}(p_h f) \circ X_{S_h - h}\, ds = E_\mu \int_0^{\tau\wedge t} e^{-\lambda s} (\eta f) \circ X_{s-0}\, ds$$

and so, by (i),

(ii) $$\lim_{h \downarrow 0} E_\mu P_{\tau\wedge t}(f, \lambda, h) = E_\mu Q_{\tau\wedge t}(f, \lambda)$$

because for each $\omega \in \Omega$ the set $\{s \in (0,\infty) : X_{s-0}(\omega) \neq X_s(\omega)\}$ is at the most countable.

We are going to prove that

(iii) $$\lim_{h \downarrow 0} E_\mu P_{\tau\wedge t}(f, \lambda, h) = E_\mu P_{\tau\wedge t}(f, \lambda)$$

for each $t \in [0,\infty)$. This will complete the proof because (ii) and (iii) imply that $E_\mu P_{\tau\wedge t}(f,\lambda) = E_\mu Q_{\tau\wedge t}(f,\lambda)$ for each $t \in [0,\infty)$ and this implies that $E_\mu P_\tau(f,\lambda) = E_\mu Q_\tau(f,\lambda)$, by passing to the limit as $t \longrightarrow \infty$ and by an application of the monotone convergence theorem of the theory of the integral. To prove (iii) observe that if $f \in C_c^+(\hat{E}^2 \setminus D_o)$ then, according to 4.3,

(α) $$\lim_{h \downarrow 0} P_{\tau\wedge t}(f, \lambda, h) = P_{\tau\wedge t}(f, \lambda)$$

pointwisely on Ω. Moreover, if for a function $f \in C_c^+(\hat{E} \setminus D_o)$ an $\varepsilon > 0$ is choosen so small that $f \in C_c^+(\hat{E}^2 \setminus D_\varepsilon)$, then

(β) $$P_{\tau\wedge t}(f, \lambda, h) \leq \|f\| \mathcal{N}_\varepsilon[0,t]$$

everywhere on Ω, for every $\lambda \in [0,\infty)$ and $h \in (0,\infty)$, where

$$\mathcal{N}_\varepsilon[a,b] = \sup\left\{ \sum_{\nu=1}^{n} \mathbb{1}_{\hat{E}^2 \setminus D_\varepsilon}(X_{t_{\nu-1}}, X_{t_\nu}) : a \leq t_o < \ldots < t_n \leq b ,\ n \in \mathbb{N} \right\}$$

for any interval $[a,b] \subset [0,\infty)$. The quantity $\mathcal{N}_\varepsilon[a,b]$ may be called the number of displacements greater then ε of the process X_t on the interval $[a,b]$. Thanks to right continuity of X_t nothing changes if only rational t_ν's are used in the definition of $\mathcal{N}_\varepsilon[a,b]$ and from this it follows that $\mathcal{N}_\varepsilon[a,b]$ is an F_∞^o-measurable function on Ω with values in $\mathbb{N}\cup\{\infty\}$. Moreover, the distribution of $\mathcal{N}_\varepsilon[a,b]$ satisfies the Kinney estimations

$$P_\mu\left\{\mathcal{N}_\varepsilon[a,b]\geqslant k\right\}\leqslant\left[2\alpha\left(\tfrac{1}{4}\varepsilon,\ b-a\right)\right]^k$$

for each $\mu\in M(\hat{E})$ and $k=1,2,\ldots,$ where

$$\alpha(\varepsilon,h)=\sup\left\{P_{t,x}(\{y\ :\ d(x,y)>\varepsilon\})\ :\ t\in[0,h],\ x\in\hat{E}\right\}.$$

See [14], [4; Lemma 6.4], [7; Chapter III, § 4], [8; p.77]. By the strong continuity of the semigroup $\{\hat{S}_t\ :\ t\geqslant 0\}$ and by the compactness of $\hat{E}$, we have

$$\lim_{h\downarrow 0}\ \alpha(\varepsilon,h)=0 \quad \text{for each} \quad \varepsilon>0,$$

/Let us remark that the condition (C) of the Section 1.2 implies that even $\alpha(\varepsilon,h)=0(h)$ as $h\downarrow 0$, for each $\varepsilon>0$/.
So, for any finite positive ε and t there is a positive integer m such that, for each $\mu\in M(\hat{E})$,

$$P_\mu\left\{\mathcal{N}_\varepsilon\left[\tfrac{i-1}{m}\ t,\ \tfrac{i}{m}\ t\right]\geqslant k\right\}=p_{ik}\leqslant 2^{-k}$$

and consequently

$$E_\mu\mathcal{N}_\varepsilon[0,t]\leqslant m-1+\sum_{i=1}^{m}E_\mu\mathcal{N}_\varepsilon\left[\tfrac{i-1}{m}\ t,\tfrac{i}{m}t\right]$$

$$(\gamma)\qquad \leqslant m-1+\sum_{i=1}^{m}\sum_{k=1}^{\infty}(k+1)(p_{i,k}-p_{i,k+1})$$

$$=m-1+\sum_{i=1}^{m}\left(2p_{i1}+p_{i2}+p_{i3}+\ldots\right)\leqslant\tfrac{5}{2}m-1.$$

Now, by the Lebesgue bounded convergence theorem, the relations (α), (β) and (γ) imply (iii). The proof is complete.

References

[1] R.M.Blumenthal and R.K.Getoor, Markov Processes and Potential Theory, Academic Press, 1968.

[2] Ph.Courrége, Sur la forme intégro-differentielle des operateurs de $\mathcal{C}_K^\infty$ dans $\mathcal{C}$ satisfaisant au principe du maximum, Séminaire Brelot-Choquet-Deny, Théorie du potentiel, 10e année /1965-66/.

[3] C.Dellacherie and P.A.Meyer, Probabilities and Potential, North-Holland Mathematics Studies 29, 1978.

[4] E.B.Dynkin, Fundations of the theory of Markov processes /in russian/, "Fizmatgiz", Moscov, 1059.

[5] E.B.Dynkin, Markov processes /in russian/, "Fizmatgiz", Moscow, 1963.

[6] W.Feller, An introduction to probability theory and its applications, Vol.II /in russian/, "Mir", Moscow, 1967.

[7] I.I.Gihman and A.V.Skorohod, Theory of stochastic processes, Vol.I /in russian/, "Nauka", Moscow, 1971.

[8] H.Heyer, Einführung in die Theorie Markoffscher Prozesse, Bibliographisches Institut, 1979.

[9] H.Heyer, Probability measures on locally compact groups /in russian/, "Mir", Moscow, 1981.

[10] E.Hille and R.S.Phillips, Functional analysis and semigroups, A.M.S. Colloquium Publications, Vol.XXXI, 1957.

[11] G.A.Hunt, Semi-groups of measures on Lie groups, TAMS 81 /1956/, p.264-293.

[12] N.Ikeda and S.Watanabe, On some relations between the harmonic measure and the Lévy measure for a certain class of Markov processes, J.Math.Kyoto Univ. 2 /1962/, p. 79-95.

[13] J.Jacod, Calcul Stochastique et Problèmes de Martingales, Springer Lecture Notes in Mathematics, Vol. 714, 1979.

[14] J.R.Kinney, Continuity properties of sample-functions of Markov processes, TAMS 74 /1953/, p. 280-302.

[15] J.Lamperti, Stochastic Processes, a Survey of the Mathematical Theory, Applied Mathematical Sciences, Vol. 23, Springer-Verlag, 1977.

[16] J.P.Roth, Opérateurs dissipatifs et semi-groupes dans les espaces de fonctions continues, Ann.Inst.Fourier, 26 /1976/, p. 1-97.

[17] S.Saks, Theory of the Integral, Warszawa-Lwów, 1937.

[18] S.Watanabe, On discontinuous additive functionals and Lévy measures of a Markov process, Japanese J.Math. 34 /1964/, p. 53-70.

Institute of Mathematics
Warsaw University
P.K. iN. IX. floor
00-901 WARSAWA
Poland

THE RANDOM SCHRÖDINGER OPERATOR
IN A STRIP

Jean LACROIX

SUMMARY

Let H be the Schrödinger difference operator in a strip. When the sequence of potentials is a family of independant random variables with a common law, H has almost surely a pure point spectrum with exponentialy decaying eigenvectors.

× ×
×

$\mathcal{H}$ is the Hilbert space $\ell^2([1,..d] \times \mathbb{N})$ which we can view as the space of sequences $V : (V_n)_{n\in \mathbb{N}}$ such that $V_n \in \mathbb{C}^d$ and $\sum ||V_n||^2 < +\infty$.

H is the operator acting on $\mathcal{H}$ by :

$(HV)_n = - V_{n-1} - V_{n+1} + A_n V_n \qquad (V_{-1} = 0)$.

(A_n) is a given sequence of $d \times d$ real symmetric matrix.

(When the (A_n) are Jacobi matrix, we say that H is a Schrödinger operator ; the diagonal terms of H_n are called the potentials and the other terms are zero except those of index (i,j) with $|i-j| = 1$ which have the value -1.

E_λ is the resolution of the identity of the self-adjoint operator H on $\mathcal{H}$ and for $u,v \in \mathcal{H}$ we note $\sigma_{u,v}$ the measure $\langle E_\lambda u,v\rangle$. It is easy to see that all the $\sigma_{u,v}$ are absolutely continuous with respect to $\sigma = \sum_{i=1}^{d} \sigma_{(i,o)(i,o)}$ where $\sigma_{(i,n)(j,m)}$ is the spectral measure associated with $u = \varepsilon_{(i,n)}$ $v = \varepsilon_{(j,n)}$ and $\varepsilon_{(i,n)}$, $i \in [1,..,d]$, $n \in \mathbb{N}$, is the usual base of $\mathcal{H}$).

As a consequence of this fact the spectrum of H is the topological support of σ and λ is an eigenvalue of H if and only if $\sigma(\lambda) > 0$.

The proof of the main result is divided in three parts :

. In the first part we establish a criterion implying that σ is a pure point measure (Lemma 3). This is done by constructing a "good" sequence of approximation of σ .

. In the second part, assuming that (A_n) is a sequence of independant identically distributed matrix, we give to the above criterion a new form, using the operators associated with the action of the symplectic group on some compact boundary (Lemma 4).

. In the third part we show how some restrictions on the law of the (A_n) give the expected result. (Theorem 10).

- I -

For $\lambda \in \mathbb{R}$, $P_n(\lambda)$ is the sequence of $d \times d$ matrix solution of :

$$- P_{n-1} - P_{n+1} + A_n P_n = \lambda P_n \qquad P_{-1} = 0 \qquad P_0 = I.$$

The coefficients of $P_n(\lambda)$ are polynoms of degree n in λ.

We note $\mathbb{T}_{m,n}$ the $d\times d$ matrix of measures whose term or order (i,j) is $\sigma_{(i,m)(j,n)}$.

<u>LEMMA 1.</u> $$\mathbb{T}_{(n,o)} = P_n \mathbb{T}_{o,o}$$

<u>Proof</u> : It is easy to prove (by recurrence) that

$$\sum_{k=1}^{d} P_{n\,(i,k)}(H)\ \varepsilon_{(k,o)} = \varepsilon_{(i,n)} \quad \text{so}$$

$$\langle E_\lambda\, \varepsilon_{(i,n)}, \varepsilon_{(j,o)}\rangle = \sum_{k=1}^{d} \langle P_{n(i,k)}(H)\, E_\lambda\, \varepsilon_{k,o}, \varepsilon_{j,o}\rangle \ .$$

$$= \sum_{k=1}^{d} \int_{-\infty}^{\lambda} P_{n(i,k)}(\lambda)\ d\,\sigma_{(k,o)(j,o)}$$

<u>LEMMA 2.</u> *Let for each* n *a sequence of* $d\times d$ *matrix of bounded measures* $({}^{N}\mathbb{T}_n)_{N\in\mathbb{N}}$ *converging to* $\mathbb{T}_{(n,o)}$ *in the tight topology.*

If for an open set I :
$$\sum_{i,j}\ \sum_{n\geq 0}\ \underline{\lim}_{N}\ |{}^{N}\mathbb{T}_{n(i,j)}|\ (I) < +\infty$$

σ *is a pure point measure on* I.

<u>Proof</u> : If f is a positive lower semi-continuous function, so is $\mu \to |\mu|(f)$ therefore :
$$\sum_{n\geq 0}\ \sum_{i,j}\ |\sigma_{(i,n)(j,o)}|\ (I) < +\infty\ ,$$ and the formula

$$\int_I\ \sum_{n\geq 0}\ \sum_{i,j}\ |(P_n\ S)_{ij}|\ d\,\sigma < +\infty$$ where S is a matrix of densities of $\mathbb{T}_{(o,o)}$ with respect to σ , shows that for each column Vector S_j

of S, the sequences $(P_n S_j)_{n\geq 0}$ are for σ almost all λ of I eigenvectors of H.

As they cannot be equal to zero for all $j \in [1,..,d]$ (because tr S = 1 σ a.e.) we deduce that σ almost all λ of I is an eigenvalue of H, so σ is pure point on I.

LEMMA 3. *If for an open set* I :

$$\sum_{n\geq 0} \frac{\lim}{N} \int_I (\mathrm{tr}\ \Delta_N^{-1}\ \mathrm{tr}\ \Delta_N^{-1}\ \Delta_n)^{1/2}\ d\lambda < +\infty$$

σ *is a pure point measure on* I.

(where $\Delta_k = P'_{n+1} P_{n+1} + P'_n P_n$*)*

Proof. Let ${}^N H^X$ be the symmetric operator on $(\mathbb{C}^d)^{N+1}$ defined for a symmetric real $d \times d$ matrix X by :

$$\begin{cases} ({}^N H^X V)_n = (HV)_n & 0 \leq n \leq N-1 \\ ({}^N H^X V)_N = -V_{N-1} + (A_N - X)\, V_N \end{cases}$$

If we note ${}^N\sigma^X_{(r,s)}$ $(r,s \leq N)$ the spectral matrix of ${}^N H^X$ we know that for each X the sequence $({}^N\sigma^X_{r,s})$ converges in the tight topology to $\sigma_{r,s}$. If m is a probability measure on $\mathbb{R}^{\frac{d(d+1)}{2}}$ the sequence ${}^N\sigma_{(r,s)} = \int {}^N\sigma^X_{(r,s)}\, dm(X)$ has the same property.

By a suitable choice of m we obtain ([1], Annexe B)

$$\sum_{i,j} |{}^N\sigma_{(i,r)(j,s)}|\,(I) \leq \frac{d}{\Pi} \int_I \left[\mathrm{tr}\,(\Delta_N^{-1}\ \Delta_r)\ \mathrm{tr}\,(\Delta_N^{-1}\ \Delta_s)\right]^{1/2} d\lambda$$

It may be surprising to approximate σ by absolutely continuous measures in order to prove that σ is a pure point measure, but for X fixed the measures ${}^{N}\sigma^{X}_{(r,s)}$ have no "good" properties of cocycles, and (at least for d>1) it is the only way to handle the problem.

- II -

Let G a localy compact metrisable group, X a compact metrisable homogeneous space of G. We say that a strictly positive continuous function ρ on (G,X) is a cocycle if :

$$\rho(g_1\, g_2,\, x) \;=\; \rho(g_1,\, g_2\, x)\, \rho(g_2, x).$$

If μ is a probability on G and ρ a cocycle such that $\overline{\rho}(g) = \sup_x \rho(g,x)$ is μ-integrable, the operator T_ρ defined by

$T_\rho\, f(x) = \int_G \rho(g,x)\, f(gx)\, d\,\mu(g)$ is a positive bouded operator on the Banach space C(X) of complex continuous functions on X.

Moreover $T^n_\rho\, f(x) \;=\; \int_G \rho(g,x)\; f(gx)\; d\,\mu^{*n}\,(g).$

In our case we take $G = S_p\,(d, \mathbb{R})$ the real symplectic group of matrix of order 2d. A subspace V of $\mathbb{R}^d$ is isotropic if <u, JV> = 0 $\forall (u,v) \in V$ where J is the element $\begin{pmatrix} 0 & I \\ -I & 0 \end{pmatrix}$ of G.

If a is a $[2d,k]$ matrix of rank k we note $\bar{a}$ the k-dimensional subspace of $\mathbb{R}^{2d}$ generated by the columns of a. (For more details see [1]).

Let $X = \{(\bar{b}, \bar{a})\}$ where $\bar{a}$ (resp $\bar{b}$) is a d(resp $d-1$) dimensional isotropic subspace of $\mathbb{R}^{2d}$ such that $\bar{b} \subset \bar{a}$.

X is a compact homogeneous space of G under the action $g(\bar{b},\bar{a}) = (\overline{gb},\overline{ga})$

In the sequel we shall use the cocycle θ on (G,X) :

$\theta(g, (\bar{b},\bar{a})) = \dfrac{\det(a'a)\ \det(b'g'gb)}{\det(a'g'ga)\ \det(b'b)}$ and we note for $t \in [0,1]$,

$\lambda \in \mathbb{R}$, $f \in C(X)$:

$$T_{t,\lambda} f(\bar{b},\bar{a}) = \int_G \theta^t(g,(\bar{b},\bar{a}))\, f(\overline{gb},\overline{ga})\, d\,\mu_\lambda(g)$$

where μ_λ is the image measure of the common law of the independant random variables (A_n) under the transformation

$$A \longrightarrow \begin{pmatrix} A-\lambda I & -I \\ I & 0 \end{pmatrix} \in S_p(d,\mathbb{R}).$$

<u>LEMMA 4.</u> *Let* (A_n) *be a family of independant random matrix with a common law such that* $||A||^{d-2}$ *is integrable, and* H *the random difference operator associated. If for each compact* K *of* $\mathbb{R}$ *there exists* $\rho_K < 1$, C_K, M_K, *such that* $||T^n_{1/2,\lambda}|| \leq C_K\, \rho_K^n$ *and* $||T^n_{1,\lambda}|| \leq M_K$ *for* $\lambda \in K$,

H *has almost surely a pure point spectrum and each eigen-space (of dimension less than* d*) contains an exponentialy decaying eigenvector.*

Proof. If we set $g_n = \begin{pmatrix} A_n - \lambda I & -I \\ I & 0 \end{pmatrix}$ and $S_n = g_n \ldots g_o$,

we have $\binom{P_{n+1}}{P_n} = S_n a_o$ where a_o is the $(2d,d)$ matrix $\binom{I}{0}$ and

$\det \Delta_n = \det(a'_o S'_n S_n a_o)$.

Furthermore if we note b_i $(i = 1,\ldots,d)$ the $(2d, d-1)$ matrix whose column vectors are the usual base vectors of $\mathbb{R}^{2d}$ of index $(1,..,i-1,i+1,..,d)$, we have :

$$\operatorname{tr} \Delta_N^{-1} = \sum_{i=1}^{d} \theta(S_N, (\bar{b}_i, \bar{a}_o)) = \sum_{i=1}^{d} \alpha_i .$$

In the same way, let T_n be a (d,d) matrix such that $\Delta_n^{-1} = T_n T'_n$ and z_i the $(2d, d-1)$ matrix $\binom{T_n}{0}$ without the column of rank i ; we have :

$$\operatorname{tr} \Delta_N^{-1} \Delta_n = \sum_{j=1}^{d} \theta(g_N .. g_{n+1}, (S_n \bar{z}_j, S_n \bar{a}_o)) = \sum_{j=1}^{d} \beta_j .$$

If, in the evaluation of $\mathbb{E}[\alpha_i^{1/2} \beta_j^{1/2}]$, we first integrate with respect to $g_N \ldots g_{n+1}$ using the Schwarz inequality, we obtain :

$$\mathbb{E}[\alpha_i^{1/2} \beta_j^{1/2}] \leq ||T^{n+1}_{1/2,\lambda}|| \; ||T^{N-n}_{1,\lambda}||$$

and if I is an open set contained in a compact set K :

$$\mathbb{E} \int_I (\operatorname{tr} \Delta_N^{-1} \operatorname{tr} \Delta_N^{-1} \Delta_n)^{1/2} d\lambda \leq d^2 |K| C_K M_K \rho_K^{n+1} = D_K \rho_K^n$$

The lemma 3 proves that σ is almost surely pure point. Noting that the last inequality implies that :

$$\mathbb{E}(|\sigma_{(i,n)(j,o)}|(I)) = \mathbb{E}(\int_I |(P^n S)_{ij}| d\sigma) \leq D_K \rho_K^n$$

we obtain for an eigenvalue λ the exponential decaying of $((P^n S)_{i,j}(\lambda))_{n\geq 1}$.

As we do not know the rank of S, we can only say (because the rank is not zero) that there exists an exponentialy decaying eigenvector.

(The integrability of $||A||^{4d-2}$ is required to obtain that the operators $T_{1/2,\lambda}$ and $T_{1,\lambda}$ are well defined on $C(X)$ because : $\theta(g, (\bar{b},\bar{a})) \leq ||g||^{2(d-1)} \; ||g^{-1}||^{2d}$).

- III -

Let T_ρ the operator defined at the beginning of the part II. (We assume that all the cocycles ρ written are such that $\bar{\rho} \in L_1(\mu)$).

We note $\hat{T}_\rho$ the operator associated with the probability $\hat{\mu}$ (image of μ by $g \longrightarrow g^{-1}$) and T^*_ρ the adjoint operator of T_ρ. If m is a probability on X with X as topological support, such that $g^{-1}m$ has the same nul sets as m, and $r(g,x) = \frac{dg^{-1}m}{dm}(x)$ is continuous, it is easy to see that :

$$\forall f \in C(X) \qquad T^*_{r\rho^{-1}}(fm) = (\hat{T}_\rho f)m$$

We can conclude that for $\lambda \in \mathbb{C}$ and $k \in \mathbb{N}$:

$$(\text{Ker } (\hat{T}_\rho - \lambda I)^k)m \subset \text{Ker } (T^*_{r\rho^{-1}} - \lambda I)^k$$

LEMMA 5. *If μ is absolutely continuous with respect to an Haar measure on* G *the operators* $\hat{T}_\rho$ *and* $T_{r\rho^{-1}}$ *are compacts with the same (no nuls) eigenvalues, and the two subspaces*

$$\mathrm{Ker}\,(\hat{T}_\rho - \lambda I)^k \quad \textit{and} \quad \mathrm{Ker}\,(T_{r\rho^{-1}} - \lambda I)^k$$

have the same dimension for all $k \in \mathbb{N}$, $\lambda \in \mathbb{C} - \{0\}$

Proof. It is a well known fact (and easy to prove) that if $\bar{\rho} \in L_1(\mu)$ and μ has a density, the operator T_ρ is compact. Using the fact that for a compact operator T, $\mathrm{Ker}\,(T - \lambda I)^k$ and $\mathrm{Ker}\,(T^* - \lambda I)^k$ have the same (finite) dimension for $\lambda \neq 0$, and that the application $f \longrightarrow fm$ is one to one, we have the desired result (observing that $(\rho^{-1} r)^{-1} r = \rho$).

LEMMA 6. *Let* T *a compact operator on a Banach space whose spectral radius is one.*

$||T^n||$ *is bounded if and only if, for each eigenvalue* λ *of modulus one :* $\mathrm{Ker}\,(T - \lambda I) = \mathrm{Ker}\,(T - \lambda I)^2$.

Proof. Let $P_1, .., P_k$ the spectral projectors associated to the eigenvalues of modulus one. We can write :

$T = \sum_{i=1}^{k} A_i + B$ with $A_i = T P_i$. So we have $T^n = \sum_{i=1}^{k} A_i^n + B^n$

and the spectral radius of B is strictly less than one. The problem is reduced to the Jordan form of T on the finite dimensional subspaces range (P_i)

COROLLARY 7. *If μ is absolutely continuous with respect to an Haar measure on G, the sequence $\|\hat{T}_\rho^{\,n}\|$ is bounded if and only if the sequence $\|T^n_{r\rho-1}\|$ is bounded. (In particular the sequence $\|T^n_r\|$ is always bounded).*

Now if Π is a continuous map of X on to a compact homogeneous metric space $\mathbf{Y}$ such that $g\Pi(x) = \Pi(g\,x)$ and if ρ is a cocycle on (G,Y) we can define the cocycle $\tilde{\rho}$ on (G,X) by $\tilde{\rho}(g,x) = \rho(g,\Pi(x))$.

LEMMA 8. *With the above notations, $\|T^n_{\tilde{\rho}}\| = \|T^n_{\rho}\|$ and each eigenvalue of T_ρ is an eigenvalue of $T_{\tilde{\rho}}$. (In particular T_ρ and $T_{\tilde{\rho}}$ have the same spectral radius).*

Proof. It suffices to remark that :

$$T_{\tilde{\rho}}(f \circ \Pi) = T_\rho(f) \circ \Pi \qquad f \in C(Y).$$

(In general it is all we can say about the spectrum of $T_{\tilde{\rho}}$, take for example Y as a single point).

In the sequel we note T_0 the operator associated to the cocycle $\rho(g,x) \equiv 1$.

Let us return now to the special case $G = S_p(d,\mathbb{R})$, $X = (\bar{b},\bar{a})$ and and $Y = (\bar{b})$.

It is proved in the Annex that the cocycle θ has the form $\theta = r_1 \, \tilde{r}_2^{-1}$ where r_1, (r_2), is the Poisson kernel of X, (Y), and therefore of the form $\frac{dg^{-1}m}{dm}$ where m is the probability on X, (Y), invariant under the action of the compact subgroup of orthogonal elements of G.

<u>LEMMA 9</u>. *We suppose that* $||A||^{d(d+1)}$ *is integrable with respect to the common law of the* (A_n) *and that there exists an integer* k *such that* μ_λ^k *has a density on* $S_p(d,\mathbb{R})$

(*i*) *The spectral radius of* $T_{1/2,\lambda}$ *is strictly less than one ;*

(*ii*) *The sequence* $||T^n_{1,\lambda}||$ *is bounded ;*

(*iii*) *The above majorizations are uniform in* λ *for* λ *belonging to a compact set.*

<u>Proof</u>. The integrability of $||A||^{d(d+1)}$ enables us to use the bounded operators T_{r_2} , T_θ , T_{r_1} and they have a compact k power. It suffices to prove (i) and (ii) for T^k in place of T and we omit the index λ in the proof of (i), (ii).

(ii) The operator T_o^k (on Y) is Markovian, by corollary 7, the sequence $||\hat{T}^{nk}_{r_2}||$ is bounded, by lemma 8, the sequence $||\hat{T}^{nk}_{r_2}||$ has the same property and by corollary 7 $||T_\theta^{nk}||$ is bounded.

(i) Let $r(t)$ be the spectral radius of T_t^k . We know that $r(o) = r(1) = 1$, and it is easy to see that $\text{Log}\, r(t)$ is convex for $t \in [0,1]$.
The properties of $r(t)$ for t near zero are given by the theory of analytical perturbations [3]. The spectral properties of T_o^k (on X) are well known. (See Tutubalin [4]) :
1 is the only eigenvalue of modulus one and is simple ;

So we can conclude that for $t \in [0,\varepsilon]$ there exists for T_t^k an unique eigenfuntion Φ_t and eigen probability ν_t such that $\nu_t(\Phi_t) = 1$ and $T_t^k \Phi_t = r_t \Phi_t$. The continuity of $t \to \Phi_t$ and $t \to \nu_t$ gives :

$$\frac{d}{dt} r_t \Big|_{t=0} = \nu_o \otimes \mu^k (\text{Log } \theta).$$

But this last quantity is equal to $-\gamma_d$ (γ_d is the smallest positive Lyapounoff index of μ^k and is strictly positive, see [2]).
We can conclude that $r(1/2) < 1$.

(iii) For each $t \in [0,1]$ the function $\lambda \to T_{t,\lambda}$ is continuous (in the norm topology of bounded operators on $C(X)$).
This implies that $\lambda \to r_\lambda(1/2)$ is upper semi-continuous and hence attains its maximum on a compact set. For the uniform bound of $T_{1,\lambda}^{k\,n}$ we use the fact that the projection operator associated to the part of modulus one of the spectrum of $T_{1,\lambda}^k$ is continuous with respect to λ [3].

We can now give the main result.

THEOREM 10.

Let H *be the Schrödinger operator associated with a family of independant identically distributed potentials such that their common law has a density on* $\mathbb{R}$ *and* $|x|^{d(d+1)}$ *is integrable.*
H *has almost surely a pure point spectrum and each eigenspace (of dimension less than* d) *contains an exponentialy decaying eigenvector.*

<u>Proof</u>. By Lemmas 4 and 9, it suffices to show that there exists an integer k such that μ_λ^k has a density on $S_p(d, \mathbb{R})$ but this is done in [2] annex A.

<u>Remark</u> : It seems true (but not proved) that under the conditions of theorem 10, the eigenspaces of H are almost surely one dimensional, so H has almost surely a basis of eigenvectors exponentially decaying.

The proof is the same for the schrödinger operator in the whole strip and for more general operators of the form :

$$(HV)_n = -B_n V_{n-1} + A_n V_n - B_n V_{n+1}$$

where A_n and B_n are symmetric real matrix with $0 < \alpha I \leq B_n \leq \beta I$.

ANNEX

BOUNDARIES AND POISSON KERNEL FOR THE SYMPLECTIC GROUP

- I -

$G = S_p(d,\mathbb{R})$ is the group of square real matrix of order 2d such that $g' Jg = J$ (or $g Jg' = J$). The Lie algebra $\mathcal{G}$ of G is the subspace of matrix $\begin{pmatrix} X_1 & X_2 \\ X_3 & -X'_1 \end{pmatrix}$ where X_2 and X_3 are symmetric.

$\mathcal{G}$ admits a Cartan decomposition of the form $\mathcal{G} = \mathcal{K} + \mathcal{P}$ where :

$$\mathcal{K} = \left\{ \begin{pmatrix} X_1 & X_2 \\ -X_2 & X_1 \end{pmatrix} \;/\; X_2 \text{ symmetric and } X_1 \text{ squew symmetric} \right\}$$

$$\mathcal{P} = \left\{ \begin{pmatrix} X_1 & X_2 \\ X_2 & -X_1 \end{pmatrix} \;/\; X_2 \text{ and } X_1 \text{ symmetric} \right\}$$

$$\mathfrak{a} = \left\{ \begin{pmatrix} X & 0 \\ 0 & -X \end{pmatrix} \;/\; X = \operatorname{diag}(x_1,\dots,x_d) \right\}$$ is a maximal abelian sub-algebra of $\mathcal{P}$.

If $\mathfrak{a}^+$ is the Weyl chamber with $x_1 > x_2 \dots > x_d > 0$, the set Δ^+ of positive roots is the set of linear functionals

$$\Phi_{i,j}\begin{pmatrix} X & 0 \\ 0 & -X \end{pmatrix} = x_i - x_j \qquad \text{for} \quad j > i$$

$$\Psi_{i,j}\begin{pmatrix} X & 0 \\ 0 & -X \end{pmatrix} = x_i + x_j \qquad \text{for} \quad j \geq i$$

The associated nilpotent subalgebra $\mathcal{N}$ is given by :

$$\mathcal{N} = \left\{ \begin{pmatrix} X_1 & X_2 \\ 0 & -X'_1 \end{pmatrix} \;/\; X_2 \text{ symmetric, } X_1 \text{ upper-triangular with a nul diagonal.} \right\}$$

So, we have the Iwasawa decomposition $G = KAN$ where $K = G \cap SO(2d) \simeq U(d)$, and the maximal boundary in the Furstenberg theory is the coset space G/H where H is the subgroup of G :

$$H = \left\{ \begin{pmatrix} X & Y \\ 0 & Z \end{pmatrix} \;/\; X \text{ upper triangular} \right\}$$

We note L_k the Lagrangian manifold of the isotropic k dimensional subspaces of $\mathbb{R}^{2d}$, and L the flag manifold $L = (\bar{a}_1,\dots,\bar{a}_d)$ where $\bar{a}_i \in L_i$ and $\bar{a}_i \subset \bar{a}_{i+1}$. L is an homogeneous space of G and H is exactly the subgroup of G leaving the point $(\bar{e}_1,\dots,\bar{e}_d)$ fixed where e_k is the $[2d,k]$ matrix whose column vectors are the first k elements of the usual basis of $\mathbb{R}^{2d}$, so we have $L \simeq G/H$.

In the sequel we note $L_{i_1,\dots,i_r}$ the partial boundary $(\bar{a}_{i_1},\dots,\bar{a}_{i_r})$ with $[i_1,\dots,i_r] \subset [1,\dots,d]$.

It is easy to see that the only cocycles ρ on $L_{i_1,\dots,i_r}$ such that $\rho(g,x) = 1 \;\; \forall g \in K$ are of the form :

$$\rho(g,(\bar{a}_{i_1},\dots,\bar{a}_{i_r})) = \prod_{k=1}^{r} \rho_{i_k}(g,\bar{a}_{i_k})^{\lambda_k} \quad , \qquad \lambda_k \in \mathbb{R}$$

and

$$\rho_k(g,\bar{a}_k) = \frac{\det(a'_k\, g' g\, a_k)}{\det(a'_k\, a_k)} .$$

If $m_{i_1,\dots,i_r}$ is the K invariant probability on $L_{i_1,\dots,i_k}$ the Poisson kernel $P_{i_1,\dots,i_r}(g,x) = \dfrac{dg^{-1} m_{i_1,\dots,i_r}}{m_{i_1\dots i_r}}(x)$ is a cocycle of this kind, and the only thing we have to do, in order to obtain the Poisson kernel, is to compute the exponants λ_k.

This is done by using the trick given by Furstenberg and Tzkoni in [5] which reduces the probelm to find a parametric representation of $L_{i_1,\dots,i_r}$ and to compute $\dfrac{dg^{-1}\ell}{d\ell}(x)$ where ℓ is the image of Lebesgue measure, and this later expression is given by a Jacobian computation.

- II -

THE POISSON KERNEL OF L_k $(1 \leq k \leq d)$

Let a_k be a $[2d,k]$ matrix, we note :

$a_k = \begin{bmatrix} t \\ u \\ v \\ w \end{bmatrix}$ where t and v are $[k,k]$ matrix and u and w are $[d-k,k]$ matrix.

The application Π

$\Pi(\overline{a}_k) = (X,M,Y)$ with $X = u\,t^{-1}$, $Y = w\,t^{-1}$, $M = v\,t^{-1} + X'Y$, is a one to one application of the open submanifold of full dimension of L_k with t regular onto the product of the spaces of $[d-k,k]$ matrix, $[k,k]$ symmetric matrix and $[d-k,k]$ matrix.

Now let h_α $(\alpha > 0)$ the element $\begin{pmatrix} Z & 0 \\ 0 & Z^{-1} \end{pmatrix}$ of H with $Z = \mathrm{diag}(\alpha,1,..,1)$.
$\Pi(h_\alpha^{-1}\,\overline{a}_k) = (X Z_k, Z_k M Z_k, Y Z_k)$ with $Z_k = \mathrm{diag}(\alpha,..,1)$ square matrix of order k.
The Jacobian of the linear transformation induced on (X,M,Y) is equal to α^{2d-k+1}, so we have :

$$P_k(h_\alpha,\overline{e}_k) = \alpha^{-2d+k-1} = [\rho_k\,(h_\alpha,\overline{e}_k)]^{-\frac{2d-k+1}{2}}$$

and therefore :

$$\boxed{P_k(g,\overline{a}_k) = [\rho_k\,(g,\overline{a}_k)\,]^{-\frac{2d-k+1}{2}}}$$

- III -

THE POISSON KERNEL OF $L_{(k,d)}$ $(1 \leq k \leq d-1)$

If a_k is a $[2d,k]$ matrix with $k \leq d-1$, we note

$a_k = \begin{bmatrix} u \\ v \\ w \end{bmatrix}$ where u is a $[k,k]$ matrix, v is a $[d-k,k]$ matrix and w a $[d,k]$ matrix.

If a_d is a $[2d,d]$ matrix, we note :

$a_d = \begin{bmatrix} x \\ y \end{bmatrix}$ where x and y are $[d,d]$ matrix.

The application Π

$\Pi(\bar{a}_k, \bar{a}_d) = (X, M)$ with $X = v u^{-1}$, $M = y x^{-1}$ is a one to one application of the open submanifold of full dimension of $L_{(k,d)}$ with u and x regular onto the product of the spaces of $[d-k, k]$ matrix and $[d,d]$ symmetric matrix.

Now let $h_{\alpha,\beta}$ $(\alpha > 0, \beta > 0)$ the element $\begin{pmatrix} Z & 0 \\ 0 & Z^{-1} \end{pmatrix}$ of H with $Z = \mathrm{diag}(\alpha, 1, \ldots, 1, \beta)$.

$\Pi(h_{\alpha,\beta}^{-1}(\bar{a}_k, \bar{a}_d)) = (Z_k X \tilde{Z}_k, \ Z M Z)$

with $Z_k = \mathrm{diag}(1, .., 1, \frac{1}{\beta})$ of order $d-k$

$\tilde{Z}_k = \mathrm{diag}(\frac{1}{\alpha}, 1, .., 1)$ of order k.

The Jacobian of the linear transformation induced on (X,M) is equal to $\alpha^{2d-k+1} \beta^{d-k+1}$. Using the fact that $\rho_k(h_{\alpha,\beta}, \bar{e}_k) = \alpha^2$ and $\rho_d(h_{\alpha,\beta}, \bar{e}_d) = \alpha^2 \beta^2$, we obtain :

$$P_{k,d}(g, (\bar{a}_k, \bar{a}_d)) = [\rho_k(g, \bar{a}_k)]^{-\frac{d}{2}} [\rho_d(g, \bar{a}_d)]^{-\frac{d+1-k}{2}}$$

In particular, we remark that on $L_{(d-1,d)}$ the cocycle $\theta(g, (\bar{a}_{d-1}, \bar{a}_d)) = P_{d-1}^{-1}(g, \bar{a}_{d-1}) P_{d-1,d}(g, (\bar{a}_{d-1}, \bar{a}_d))$ is equal to $\rho_{d-1}(g, \bar{a}_{d-1}) \rho_d^{-1}(g, \bar{a}_d)$

(For more details see [1] Annex A).

Jean LACROIX
U.E.R. Math-Informatique
UNIVERSITE DE RENNES I
35042 RENNES CEDEX - France

REFERENCES :

[1] LACROIX J. "Localisation pour l'opérateur de Schrödinger aléatoire dans un ruban". Annales de l'I.H.P. Section A 83 - à paraître.

[2] LACROIX J. "Singularité du spectre de l'opérateur de Schrödinger aléatoire dans un ruban ou un demi-ruban". Annales de l'I.H.P. Section A Vol.38 n° 4 83.

[3] KATO T. "Perturbation theory for linear operators". Springer Verlag 1966.

[4] TUTUBALIN "On limit theorems for the product of random matrices". Theory of Proba. Applic. 10 (1965) p. 15-27.

[5] H. FURSTENBERG and I. TZKONI "Spherical functions and integral geometry". Israël J. Math - 10 (1971) 327-338.

IRMAR
Université de Rennes 1
Campus de Beaulieu
35042 R E N N E S Cédex
France

On the Lévy-Hinčin formula for commutative hypergroups

R. Lasser
Mathematisches Institut
der Technischen Universität München

Arcisstraße 21
8 München 2

1. Introduction.

The Lévy-Hinčin formula is studied in the set up of Gelfand pairs in [5]. Homogeneous stochastic processes associated with ultraspherical polynomials are characterized by means of a Lévy-Hinčin formula in [1]. A dual concept is considered in [10]. For many related approaches to the subject we refer to the survey article of Heyer [8]. Now these contributions may be viewed as a Lévy-Hinčin formula for negative definite functions defined on certain commutative hypergroups. Thus it seems worthwile to study negative definite functions and their Levy-Hinčin representation in the general framework of commutative hypergroups.

Let K be a locally compact Hausdorff space. M(K) denotes the space of all bounded Radon measures and $M^1(K)$ the subset of all probability measures. The support of a measure μ is denoted by supp μ. K is called a hypergroup if the following conditions are satisfied:

(H1) There exists a map $*$: $K \times K \to M^1(K)$, $(x,y) \mapsto p_x * p_y$, called convolution, which is continuous, where $M^1(K)$ bears the vague topology. The linear extension to M(K), see [9, Lemma 2.4B], satisfies $p_x*(p_y*p_z) = (p_x*p_y)*p_z$.
(H2) supp p_x*p_y is compact.
(H3) There exists a homeomorphism $^-$: $K \to K$, $x \mapsto \bar{x}$, called involution, such that $x = \bar{\bar{x}}$ and $(p_x*p_y)^- = p_{\bar{y}}*p_{\bar{x}}$.
(H4) There exists an element $e \in K$, called unit element, such that $p_e*p_x = p_x*p_e = p_x$, where p_x denotes the point measure of $x \in K$.
(H5) $e \in$ supp $p_x*p_{\bar{y}}$ if and only if $x = y$.
(H6) The map $(x,y) \mapsto$ supp p_x*p_y of $K \times K$ into the space of nonvoid compact subsets of K is continuous, the latter space with the topology as given in [9, 2.5].

Here we only deal with commutative hypergroups, i.e. $p_x * p_y = p_y * p_x$. The theory of hypergroups is developed separately by Dunkl [4], Jewett [9] and Spector [15]. We have chosen Jewett's axioms. Now we present some examples:

(i) Let G be a locally compact group, H a compact subgroup. Consider the double coset space G//H equipped with the quotient topology. With HeH as unit element, with $HxH^- = Hx^{-1}H$ as involution and with

$$p_{HxH} * p_{HyH} = \int_H p_{HxtyH} dt$$

(dt the normalized Haar measure on H) as convolution, the space K = G//H is a hypergroup, see [9, Theorem 8.2B]. An important subclass consists of the $\bar{B}$-orbit spaces G_B. Here G is a locally compact group and B is a subgroup of the automorphism group Aut(G) having compact closure $\bar{B}$, where Aut(G) bears the Birkhoff topology. (G is called a $[FIA]_B^-$ group.) The $\bar{B}$-orbit space G_B is a hypergroup with $\{e\}$ as unit element, $(\bar{B}x)^- = \bar{B}x^{-1}$ as involution and

$$p_{\bar{B}x} * p_{\bar{B}y} = \int_{\bar{B}} p_{\bar{B}\beta x y} d\beta$$

($d\beta$ the normalized Haar measure on $\bar{B}$) as convolution. We refer to [9, Theorem 8.3B] and [14, §1] and we note that $K = G_B$ is commutative if B contains the group I(G) of all inner automorphisms.

(ii) Let $(P_n(x))_{n=0}^{\infty}$ be a sequence of random walk polynomials. That is: $P_n(x)$ are orthogonal polynomials determined by

$$P_0(x) = 1, \qquad a_0 P_1(x) = x - b_0,$$
$$a_n P_{n+1}(x) = P_1(x)P_n(x) - b_n P_n(x) - c_n P_{n-1}(x), \tag{R}$$

where $a_0 > 0$, $b_0 \in \mathbb{R}$ with $a_0 + b_0 = 1$ and $a_n > 0$, $b_n \geq 0$, $c_n > 0$ with $a_n + b_n + c_n = 1$, $n \in \mathbb{N}$. Further write the linearization of the product of two orthogonal polynomials $P_m(x)$ and $P_n(x)$:

$$P_m(x)P_n(x) = \sum_{k=0}^{2\min(n,m)} g(m,n,m+n-k) P_{m+n-k}(x) \tag{L}$$

We shall say that $(P_n(x))_{n=0}^{\infty}$ satisfies property (P) if each linearization coefficient $g(m,n,m+n-k)$ in (L) is nonnegative. The coefficients $g(m,n,m+n-k)$ are completely determined by the a_n, b_n, c_n, $n \in \mathbb{N}$, see [11,(1)]. In [11] we have established that there is an intimate relation between hypergroup structures on $\mathbb{N}_0 = \{0,1,2,\ldots\}$ and orthogonal polynomial sequences $(P_n(x))_{n=0}^{\infty}$ having the positivity property (P). In fact $\mathbb{N}_0$ is a hypergroup with 0 as unit element, the identity map as involution and

$$p_n * p_m = \sum_{k=0}^{2\min(m,n)} g(m,n,m+n-k)p_{m+n-k}$$

as convolution.

We have to set up some basic facts of harmonic analysis for commutativ hypergroups K. The dual space $\hat{K}$ is the set of all continuous bounded complex-valued functions α such that $\alpha(e) = 1$, $\alpha(\bar{x}) = \overline{\alpha(x)}$ and $p_x * p_y(\alpha) = \alpha(x)\alpha(y)$. Equipped with the topology of uniform convergence on compacta $\hat{K}$ is a locally compact space. $\hat{K}$ is rich enough to separate points of K. One may ask whether there exists a dual hypergroup structure on $\hat{K}$, i.e.: For $\alpha, \beta \in \hat{K}$ there exists a probability measure $p_\alpha * p_\beta \in M^1(\hat{K})$ such that

$$\alpha(x)\beta(x) = \int_{\hat{K}} \gamma(x)dp_\alpha * p_\beta(\gamma) \quad \text{for } x \in K$$

and $\hat{K}$ is a hypergroup with the constant function 1 as unit element, with $\alpha \mapsto \bar{\alpha}$ as involution and $(\alpha,\beta) \mapsto p_\alpha * p_\beta$ as convolution. If these conditions are satisfied we shall say that $\hat{K}$ is the dual hypergroup of K. In general $\hat{K}$ is not a dual hypergroup. If $\hat{K}$ is a dual hypergroup only a weak Pontryagin duality $K \subseteq \hat{\hat{K}}$ is valid. Positively one knows tha there exists a Haar measure m on K, see [16]. The Fourier transformati is given by

$$\hat{\mu}(\alpha) = \int_K \overline{\alpha(x)}d\mu(x), \quad \mu \in M(K)$$

and $L^1(K) = L^1(K,m)$ is viewed as a subspace of M(K). A Plancherel formula is valid: There exists a unique positive measure π on $\hat{K}$ such that

$$\int_K |f(x)|^2 dm(x) = \int_{\hat{K}} |\hat{f}(\alpha)|^2 d\pi(\alpha) \quad \text{for } f \in L^1(K) \cap L^2(K).$$

We note that $\operatorname{supp}\pi \subseteq \hat{K}$, where proper inclusion is possible. The invers Fourier transform is given by

$$\check{\nu}(x) = \int_{\hat{K}} \alpha(x)d\nu(\alpha), \quad \nu \in M(\hat{K}).$$

One has to distinguish exactly between Fourier transformation and inverse Fourier transformation, since $\hat{K}$ is in general not a hypergroup. For these results and further facts concerning Fourier analysis on hypergroups we refer to [9].

2. Positive definite and negative definite functions.

In this section and the following one there are mainly summarized recent results concerning a Levy-Hinčin formula for commutative hypergroups, which we have established in [13]. A continuous function $\psi: K \longrightarrow \mathbb{C}$ is called positive definite resp. negative definite, if for $c_1,\ldots,c_n \in \mathbb{C}$, $x_1,\ldots,x_n \in K$ the following holds:

$$\sum_{i,j=1}^{n} c_i \overline{c_j}\, p_{x_i} * p_{\overline{x_j}}(\psi) \geqslant 0$$

resp.

$$\sum_{i,j=1}^{n} c_i \overline{c_j} (\psi(x_i) + \overline{\psi(x_j)} - p_{x_i} * p_{\overline{x_j}}(\psi)) \geqslant 0.$$

Bochner's theorem is proved in [9, Theorem 12.3B]:

Satz 1. Let f be a function on K. Then $f = \check{\nu}$ for a positive measure $\nu \in M(\hat{K})$ if and only if f is a bounded positive definite function.

If $\hat{K}$ is a dual hypergroup one may ask for a characterization of the Fourier transforms $\hat{\mu}$ of positive measures $\mu \in M(K)$ by terms of positive definiteness on $\hat{K}$. Note that a Pontryagin duality may not hold. A partial answer gives the following, see [12, Theorem 2]:

Satz 2. Let K be a compact hypergroup such that $\hat{K}$ is a dual hypergroup. Let f be a function on $\hat{K}$. Then $f = \hat{\mu}$ for a positive measure $\mu \in M(K)$ if and only if $\widehat{p_x} f$ is a bounded positive definite function for each $x \in K$.

Negative definite functions ψ on K satisfy the following:
$\psi(e) \geqslant 0$, $p_x * p_{\bar{x}}(\psi) \in \mathbb{R}$, $\overline{\psi(x)} = \psi(\bar{x})$, $\psi(x) + \psi(\bar{x}) \geqslant p_x * p_{\bar{x}}(\psi)$.
But in general $\operatorname{Re} \psi(x) \geqslant 0$ does not hold, see [13, Remark]. Bounded negative definite functions ψ satisfy $\operatorname{Re} \psi \geqslant 0$, see [13, Proposition 1.3]. Concerning a generalization of Schoenberg's theorem we only know one direction, see [13, Theorem 1.5]:

<u>Proposition 1</u>. Let ψ be a continuous function on K with $\psi(e) \geq 0$. Assu that the functions φ_t: $x \longmapsto \exp(-t\psi(x))$ are positive definite for each $t > 0$. Then ψ is negative definite.

A continuous function l: $K \longrightarrow \mathbb{R}$ is called homomorphism if $l(\bar{x}) = -l(x)$ and $p_x * p_y(l) = l(x) + l(y)$. A quadratic form is a continuous function q: $K \longrightarrow \mathbb{R}$ such that $p_x * p_y(q) + p_x * p_{\bar{y}}(q) = 2(q(x) + q(y))$ holds. It is easily shown that $\psi = il$ is negative definite, if l is a homomorphism. For quadratic forms a characterizing identity is valid:

<u>Lemma 1</u>. Let q be a quadratic form on K and let $\mu \in M(K)$. Then for eac $n \in \mathbb{N}$

$$\mu^n(q) = n^2 \mu(K)^{n-1} \mu(q) - \frac{n(n-1)}{2} \mu(K)^{n-2} \mu * \bar{\mu}(q)$$

holds.

This statement is proved by means of induction, see [13, Lemma 1.7]. Using this identity for quadratic forms one obtains:

<u>Theorem 3</u>. A nonnegative quadratic form on K is a negative definite function.

In contrast to the group case, where $q(x^n) = n^2 q(x)$ holds, for hypergroups a quadratic form q satisfies

$$\lim_{n \to \infty} \frac{p_x{}^n(q)}{n^2} = q(x) - \tfrac{1}{2} p_x * p_{\bar{x}}(q).$$

If we consider the case where $K = G/\!/H$ is a double coset space an H-biinvariant function on G may be viewed as a function on K and conversel Now it may happen that there exists an H-biinvariant function on G, which is not negative definite on the group G but which is negative definite on the hypergroup $G/\!/H$, see [13, Remark].

3. <u>The Lévy-Hinčin formula</u>.

A family $(\mu_t)_{t>0}$ of positive measures μ_t is called a convolution semi group on K, if

(a) $\mu_t(K) \leq 1$ for each $t > 0$,

(b) $\mu_t * \mu_s = \mu_{t+s}$ for $t, s > 0$,

(c) $\lim_{t \to 0} \mu_t = p_e$ with respect to the vague topology on M(K).

If $\hat{K}$ is a dual hypergroup a description of convolution semigroups on K may be given by means of negative definite functions on $\hat{K}$, see [13, Theorem 2.2].

Theorem 4. Assume that $\hat{K}$ is a dual hypergroup. If $(\mu_t)_{t>0}$ is a convolution semigroup on K, then there exists exactly one negative definite function $\psi: \hat{K} \to \mathbb{C}$ with $\operatorname{Re}\psi \geq 0$ such that

$$\widehat{\mu_t}(\alpha) = \exp(-t\psi(\alpha))$$

for each $\alpha \in \hat{K}$, $t > 0$.

This negative definite function $\psi: \hat{K} \to \mathbb{C}$ with $\operatorname{Re}\psi \geq 0$ is called associated to (μ_t). Given a convolution semigroup (μ_t) the net $(t^{-1}\mu_t|K\setminus\{e\})$ converges vaguely. The measure $\mu = \lim_{t\to 0} t^{-1}\mu_t|K\setminus\{e\}$ is called the Lévy-measure of (μ_t).

We shall say that K satisfies property (F) if the following holds: If $C \subseteq \hat{K}$ is compact then there exist a constant $M_C \geq 0$, a neighbourhood U_C of e in K and a finite subset N_C of C such that for each $x \in U_C$

$$\sup\{1 - \operatorname{Re}\alpha(x) : \alpha \in C\} \leq M_C \cdot \sup\{1 - \operatorname{Re}\alpha(x) : \alpha \in N_C\}$$

holds. Obviously each compact or discrete hypergroup satisfies the property (F).

Theorem 5. Let $\hat{K}$ be a dual hypergroup and suppose that K satisfies property (F). Let $(\mu_t)_{t>0}$ be a convolution semigroup on K with symmetric Levy-measure μ, i.e. $\bar{\mu} = \mu$, and associated negative definite function $\psi: \hat{K} \to \mathbb{C}$.

(a) Then ψ can be written

$$\psi(\alpha) = c + il(\alpha) + q(\alpha) + \int_{K\setminus\{e\}} (1 - \operatorname{Re}\alpha(x))d\mu(x), \qquad (*)$$

where c is a nonnegative constant, l is a homomorphism and q is a nonnegative quadratic form.

(b) Moreover c, l, q in $(*)$ are determined uniquely by $(\mu_t)_{t>0}$:

$$c = \psi(1),\ l = \operatorname{Im}\psi \text{ and } q(\alpha) = \lim_{n\to\infty}\left[\frac{p_\alpha{}^n(\psi)}{n^2} + \frac{p_\alpha{}^n * p_{\bar{\alpha}}{}^n(\psi)}{2n}\right].$$

For the proof we refer to [13, Theorem 3.9].

Theorem 5 applies for the following examples:

(a) Consider the hypergroup $K = \mathbb{N}_0$ with the structure which correspond to the Jacobi polynomials $P_n^{(\alpha,\beta)}(x)$, $\alpha \geq \beta > -1$, $\alpha+\beta+1 \geq 0$, see (ii) of ch.1 and [11]. If we assume that in addition $\beta \geq -1/2$ or $\alpha+\beta \geq 0$ then $\hat{K}$ is a hypergroup and may be identified with $[-1,1]$, see [11] and ch.4 below. Since $\hat{K}$ is compact, the Lévy-Hinčin formula for a given convolution semigroup (μ_t) on $\mathbb{N}_0$ writes

$$\hat{\mu}_t(x) = \exp(-t\psi(x)) \quad \text{for } x \in [-1,1], \text{ where}$$

$$\psi(x) = c + \sum_{n=1}^{\infty} (1 - P_n^{(\alpha,\beta)}(x))\mu(n).$$

Note that the Levy measure μ is here always symmetric and bounded. Compare [10] for the ultraspherical case $\alpha = \beta$.

(b) Consider the dual hypergroup $K = [-1,1]$ corresponding to the Jacob polynomials $P_n^{(\alpha,\beta)}(x)$, where (α,β) belongs to the same region as in (a $\hat{K}$ is a dual hypergroup and may be identified with $\mathbb{N}_0$, see [11]. Given a convolution semigroup (μ_t) on $[-1,1]$ the Lévy measure μ is symmetric The Lévy-Hinčin formula is given by

$$\hat{\mu}_t(n) = \exp(-t\psi(n)) \quad \text{for } n \in \mathbb{N}_0, \text{ where}$$

$$\psi(n) = c + a\frac{n(n+\alpha+\beta+1)}{\alpha+\beta+2} + \int_{-1}^{1-0} (1 - P_n^{(\alpha,\beta)}(x))d\mu(x).$$

Concerning the quadratic form see ch.4 below. Compare [1] and [6], too

(c) Consider $K = G_B$, where G is a $[FIA]_B^-$ group with $\bar{B} \supseteq I(G)$, see (i) above. $\widehat{G_B}$ is a dual hypergroup [7]. Further one can show that G_B satisfies (F), see [13]. We mention that the class of $[FIA]_B^-$ groups covers for instance orbit spaces of compact groups or for $B = I(G)$ of locally compact groups having relatively compact conjugacy classes and having small invariant neighbourhoods of the identity. A prominent example we meet with $G = \mathbb{R}^n$, $B = SO(n)$ the special orthogonal group, where we get involved with Bessel functions. In fact G_B and $\widehat{G_B}$ may be identified with $[0,\infty[$. Each non-constant character $\alpha \in \widehat{G_B}$ is given by $\alpha = \alpha_y$, $y \in]0,\infty[$, where

$$\alpha_y(x) = \Gamma(\nu+1)2^{\nu} \frac{J_\nu(yx)}{(yx)^\nu} \quad \text{for } x \in]0,\infty[\text{ and } \alpha_y(0) = 1,$$

where J_ν is the Bessel function of the first kind of order ν and $\nu = (n-2)/2$.

4. Quadratic forms for hypergroup structures on $\mathbb{N}_0$.

In the final section we consider the quadratic forms on $\mathbb{N}_0$ with respect to a hypergroup structure presented above in (ii) of ch.1. Given an orthogonal polynomial sequence $(P_n(x))_{n=0}^\infty$ satisfying property (P) denote

$$D_S = \{x \in \mathbb{R}: (P_n(x))_{n=0}^\infty \text{ is bounded}\}.$$

For $x \in \mathbb{R}$ denote $\alpha_x: \mathbb{N}_0 \longrightarrow \mathbb{R}$, $\alpha_x(n) = P_n(x)$. D_S is a compact set on the real line and $x \longmapsto \alpha_x$, $D_S \longrightarrow \widehat{\mathbb{N}}_0$ is a homeomorphism. Further

$$\operatorname{supp}\pi \subseteq D_S \subseteq [1-2a_0, 1],$$

where the Plancherel measure π is the orthogonalization measure of $(P_n(x))_{n=0}^\infty$ (up to normalization). Let $(a_n)_{n=0}^\infty$, $(b_n)_{n=0}^\infty$, $(c_n)_{n=1}^\infty$ determine the $P_n(x)$ by means of (R). Define

$$s_0 = 0, \; s_1 = 1, \; s_{n+1} = (1 + (1-b_n)s_n - c_n s_{n-1})/a_n \qquad \text{(S)}$$

In [13, Proposition 1.11] we determined the nonnegative quadratic forms on $\mathbb{N}_0$:

Proposition 2. The nonnegative quadratic forms on $K = \mathbb{N}_0$ are exactly given by the functions $q(n) = as_n$, where $a \geq 0$.

Differentiating the recursion formula (R) for the $P_n(x)$ and setting $x = 1$ one notes that the $a_0 P_n'(1)$ satisfy the same formula as the s_n. Hence we obtain:

Theorem 6. The nonnegative quadratic forms on $K = \mathbb{N}_0$ are exactly given by the functions $q(n) = aP_n'(1)$, where $a \geq 0$.

If the product of two characters α_x and α_y, $x, y \in D_S$ is a positive definite function on $\mathbb{N}_0$, and if $\alpha_1 \cong 1$ is not isolated in D_S, then we can show that there exists a family $(\mu_t)_{t>0}$ of positive measures on D_S

such that $\check{\mu}_t(n) = \exp(-tq(n))$.

Theorem 7. Consider $K = \mathbb{N}_0$. Assume that the products $\alpha_x\alpha_y$, $x, y \in D_S$, are positive definite functions on $\mathbb{N}_0$ and assume that 1 is not isolate in D_S. Given a nonnegative quadratic form q on $\mathbb{N}_0$ there exists a convolution semigroup $(\mu_t)_{t>0}$ on D_S such that $\check{\mu}_t(n) = \exp(-tq(n))$.

Proof. By the assumption one easily obtains that for $x \in D_S$ and $m \in \mathbb{N}$ the functions $(\alpha_x)^m$ are positive definite. Using the exponential power series we see that for $t > 0$ the funtion $\exp(t\alpha_x)$ is positive definite and then $\exp(-t(1-\alpha_x))$, too. If $x \neq 1$ replace t by $t(1-x)^{-1}$ obtaining that $n \mapsto \exp(-t(1-P_n(x))(1-x)^{-1})$ is positive definite. Now 1 is not isolated in D_S. Hence by $\lim_{x\to 1} (1-P_n(x))(1-x)^{-1} = P_n'(1$ and Theorem 6 we have $n \mapsto \exp(-tq(n))$ is positive definite for $t > 0$. By Theorem 1 there exist positive measures $\check{\mu}_t \in M(D_S)$ such that $\check{\mu}_t(n) = \exp(-tq(n))$.

Remark. The condition that $\alpha_x\alpha_y$ $(x, y \in D_S)$ is positive definite impli that D_S is a weak hypergroup, see [11, Proposition 1]. The family (μ_t) of Theorem 7 is a convolution semigroup on the weak hypergroup D_S. To that compare [13, Theorem 2.3].

Examples.

(a) Consider $K = \mathbb{N}_0$ with the hypergroup structure corresponding to the Jacobi polynomials $P_n^{(\alpha,\beta)}(x)$, where $\alpha \geq \beta > -1$, $\alpha+\beta+1 \geq 0$. For the defining sequences (a_n), (b_n), (c_n) see [11, 3(a)]. The nonnegative quadratic forms are $q(n) = as_n$, $a \geq 0$, where

$$s_n = \frac{n(n+1+\alpha+\beta)}{\alpha+\beta+2}.$$

We have $D_S = [-1,1]$. If $\beta \geq -1/2$ or $\alpha+\beta \geq 0$ then D_S is a dual hypergrou Hence by Theorem 7 there exists a Gaussian convolution semigroup (μ_t) such that $\check{\mu}_t(n) = \exp(-tq(n))$.

(b) Consider on $\mathbb{N}_0$ the structure which corresponds to q-ultraspherical polynomials $P_n(x;\beta|q)$, where $-1 < \beta < 1$, $0 < q < 1$, see [11, 3(c)]. We hav $D_S = [-1,1]$. In order to calculate s_n use a famous identity of Rogers, see e.g. [2, Proposition 1.1]. One obtains:

$$s_n = \sum_{k=0}^{n} \frac{(\beta;q)_k(\beta;q)_{n-k}}{(q;q)_k(q;q)_{n-k}} (n - 2k)^2 \Big/ \sum_{k=0}^{n} \frac{(\beta;q)_k(\beta;q)_{n-k}}{(q;q)_k(q;q)_{n-k}},$$

where $(a;q)_n = (1 - a)(1 - aq)\dots(1 - aq^{n-1})$, $(a;q)_0 = 1$.

(c) If $\mathbb{N}_0$ bears the hypergroup structure of the associated Legendre polynomials $L_n^{\nu}(x)$, $\nu \geq 0$, see [11, 3(b)], we have $D_S = [-1,1]$. Using formula (12.6) of [3, p.202] we obtain

$$s_n = \left[\frac{n(n+1)}{2} + \sum_{k=1}^{n} \frac{\nu}{k+\nu}\left(\frac{k(k+1)}{2} + \frac{(n-k)(n-k+1)}{2}\right)\right] \Big/ \left(1 + \sum_{k=1}^{n} \frac{\nu}{k+\nu}\right).$$

Here as well as in (b) we do not know whether there exists a corresponding family $(\mu_t)_{t>0}$.

(d) Let $\beta > -1$, $\alpha \geq \beta + 1$. The generalized Tchebichef polynomials $T_n^{(\alpha,\beta)}(x)$ bear a hypergroup structure [11, 3(f)]. The dual space D_S equals $[-1,1]$. Using formula (2.40) of [3. p.156] one calculates that

$$s_n = \begin{cases} \dfrac{n(n+2\alpha+2\beta+2)}{2\alpha+2} & \text{if } n \text{ even} \\[2ex] \dfrac{n(n+2\alpha+2\beta+2) - (2\beta+1)}{2\alpha+2} & \text{if } n \text{ odd.} \end{cases}$$

If $\beta \geq -1/2$ we may apply Theorem 7 obtaining a Gaussian convolution semigroup $(\mu_t)_{t>0}$. In fact for $\beta = -1/2$ we have ultraspherical polynomials, see (a). If $\beta > -1/2$ then D_S is a weak hypergroup, see [11, 4, ad(f)].

(e) Let $a \geq 1$. Define $a_n = \frac{a}{a+1}$, $b_n = 0$, $c_n = \frac{1}{a+1}$, $a_0 = 1$, $b_0 = 0$. The polynomials $P_n^a(x)$ determined by (R) are connected with homogeneous trees, see [11, 3(d)]. One easily obtains by (S) that

$$s_n = n + 2 \sum_{k=1}^{n-1} k a^{k-n}.$$

References

1. Bochner, S.: Sturm-Liouville and heat equations whose eigenfunctions are ultraspherical polynomials or associated Bessel functions. In: Proceedings of the Conference on Differential Equations,

pp. 23-48. University of Maryland 1955.

2. Bressoud, D.M.: On partitions, orthogonal polynomials and the expansion of certain infinite products. Proc. London Math. Soc. 42, 478-500 (1981)

3. Chihara, T.S.: An Introduction to Orthogonal Polynomials. New Yor Gordon and Breach 1978.

4. Dunkl, C.F.: The measure algebra of a locally compact hypergroup. Trans. Amer. Math. Soc. 179, 331-348 (1973)

5. Faraut, J., Harzallah, K.: Distances hilbertiennes invariantes sur un espace homogène. Ann. Inst. Fourier 24, 171-217 (1974)

6. Gasper, G.: Banach algebras for Jacobi series and positivity of a kernel. Ann. of Math. 95, 261-280 (1972)

7. Hartmann, K., Henrichs, R.W. and Lasser, R.: Duals of orbit space in groups with relatively compact inner automorphism groups are hypergroups. Mh. Math. 88, 229-238 (1979)

8. Heyer, H.: Convolution semigroups of probability measures on Gelfand pairs. Expo. Math. 1, 3-45 (1983)

9. Jewett, R.I.: Spaces with an abstract convolution of measures. Advances in Math. 18, 1-101 (1975)

10. Kennedy, M.: A stochastic process associated with ultraspherical polynomials. Proc. Royal Irish Acad. 61, Sec.A, 89-100 (1961)

11. Lasser, R.: Orthogonal polynomials and hypergroups. to appear in Rend. Mat.

12. Lasser, R.: Bochner theorems for hypergroups and their applicatio to orthogonal polynomial expansions. J. Approx. Theory 37, 311-325 (1983)

13. Lasser, R.: Convolution semigroups on hypergroups. Preprint

14. Ross, K.A.: Centers of hypergroups. Trans. Amer. Math. Soc. 243, 251-269 (1978)

15. Spector, R.: Aperçu de la théorie des hypergroupes. In: Analyse Harmonique sur les Groupes de Lie. Lecture Notes in Math. Vol.497 pp. 643-673. Springer 1975.

16. Spector, R.: Mesures invariantes sur les hypergroupes. Trans. Ame Math. Soc. 239, 147-165 (1978)

REPARTITION D'ETAT

D'UN OPERATEUR DE SCHRÖDINGER ALEATOIRE

DISTRIBUTION EMPIRIQUE DES VALEURS PROPRES

D'UNE MATRICE DE JACOBI

Emile LE PAGE

I - INTRODUCTION

Soit $(X_n)_{n \in Z}$ une suite de variables aléatoires indépendantes et de même loi μ à support compact définie sur un espace probabilisé $(\Omega, \mathcal{F}, P)$.

Considérons l'opérateur aux différences aléatoires défini sur

$$\ell^2(Z) = \{u = (u_n) \in C^Z / \sum_{n \in Z} |u_n^2| < +\infty\}$$

$$(H(\omega)\, u)_n = -u_{n+1} - u_{n-1} + X_n(\omega)\, u_n \qquad n \in Z.$$

L'opérateur $H(\omega)$ est auto-adjoint. Notons E_t^ω la résolution de l'identité de $H(\omega)$.

D'autre part, pour tout entier $L \geq 0$, soit ${}^L H(\omega)$ l'opérateur défini sur $\ell^2(Z)$ par la restruction de la matrice de $H(\omega)$ à $[-L, L]$ c'est-à-dire pa la matrice de Jacobi :

$$J_L(\omega) = \begin{pmatrix} X_{-L} & -1 & & & (0) \\ -1 & X_{-L+1} & \ddots & & \\ & \ddots & \ddots & \ddots & \\ & & \ddots & & -1 \\ (0) & & & -1 & X_L \end{pmatrix}$$

Notons alors $N_L^{(\omega)}$ la fonction de répartition de la distribution empirique des valeurs propres $(\lambda_i^L(\omega))_{-L\leq i\leq L}$ de la matrice $J_L(\omega)$:

$$N_L^{(\omega)}(t) = \frac{1}{2L+1} \sum_{i=-L}^{L} 1_{[\lambda_i^L(\omega)\leq t]}$$

On peut alors énoncer en notant $(e_k)_{k\in\mathbb{Z}}$ la base canonique de $\ell^2(\mathbb{Z})$, et $<u,v> = \sum_{n\in Z} u_n \overline{v}_n$ où $u = (u_n)_{n\in Z} \in \ell^2(Z)$ et $v = (v_n)_{n\in\mathbb{Z}} \in \ell^2(\mathbb{Z})$ le

théorème (1-1)

1) *Il existe un ensemble* $\Omega_o \in \mathcal{F}$ *tel que* $P(\Omega_o) = 1$ *et tel que pour tout* $\Omega \in \Omega_o$ *on ait :*

$$\lim_L \sup_{t\in\mathbb{R}} |N_L^{(\omega)}(t) - N(t)| = 0$$

où $N(t) = E < E_t e_o, e_o >$, *est continue.*

2) *Le spectre de presque tout opérateur* $H(\omega)$ *est égal au support de la probabilité de fonction de répartition* $N(t) = E < E_t e_o, e_o >$ *et ce support est* $[-2,2]$ + *support de* μ.

La fonction de répartition $N(t)$ figurant dans l'énoncé précédent est appelée répartition d'état de l'opérateur H.
Les résultats figurant dans le théorème (1-1) sont contenus dans [15] et [7].

Soit J un intervalle compact de $\mathbb{R}$ tel que $[-2,2]$ + support $\mu \subset \overset{\circ}{J}$.
Considérons la suite de processus $Y_L(t) = \sqrt{2L+1}\,[N_L(t)-N(t)]$ $L\geq 0$, $t\in\mathbb{R}$. Pour tout $L\geq 1$, $t\notin J$ on a $Y_L(t) = 0$.
L'objet du présent travail est de préciser la convergence figurant au 1) du théorème (1-1), en étudiant la suite de processus $Y_L(t)$ $L\geq 1$ $t\in J$ à valeurs dans l'espace $\mathfrak{D}(J)$ des fonctions de J dans $\mathbb{R}$ continues à

droite et possédant une limite à gauche, muni de la topologie de Skorokhod [2] 7

Cette étude est menée au paragraphe 1, sous l'hypothèse supplémentaire que μ admette une densité continue : on établit la convergence en loi de la suite de processus $(Y_L)_{L\geq 1}$ vers un processus gaussien Y, presque sûrement à trajectoire continue. Notons que le résultat obtenu est analogue au théorème classique concernant la distribution empirique de variables aléatoires réelles indépendantes, qui correspond à l'étude de la distribution empirique d'une matrice diagonale.

La méthode utilisée pour établir le résultat précédent nécessite l'étude de certains produits de matrices aléatoires indépendantes 2×2 de déterminant un dépendant d'un paramètre, et de chaînes de Markov définies à partir de ces produits. Les techniques mises en place à cette occasion nous permettront de prouver la formule de Thouless [6] au paragraphe 2. L'étude de la suite de processus $(Y_L)_{L\geq 0}$ a été abordée dans [15].

Paragraphe 2 - DISTRIBUTION EMPIRIQUE DES VALEURS PROPRES D'UNE MATRICE de Jacobi :

2-1 Avec les notations définies au paragraphe 1, et en appelant de plus S_μ le support de μ nous pouvons énoncer le :

Théorème (2-1) : *Si* μ *admet une densité continue* p *à support compact :*

La suite de processus $(Y_L)_{L\geq 0}$ *converge en loi dans* $\mathfrak{D}(J)$ *vers un processus gaussien centré* Y.

De plus si $t_1<t_2<\ldots<t_k\in\,]-2,2[\,+S_\mu$

la matrice de covariance de $(Y(t_1),Y(t_2),\ldots,Y(t_k))$ *est positive non dégénérée.*

Enfin pour tout $\gamma<1/2$ *on a :* $$\text{p s}\ \lim_{h\downarrow 0} h^{-\gamma} \sup_{\substack{|t-s|\leq h\\ t,s\in J}} |Y(t)-Y(s)| = 0$$

et en particulier Y *est presque sûrement à trajectoires continues.*

2-2) <u>Démonstration du théorème (2-1)</u> :

Il est clair que l'étude des distributions empiriques des valeurs propres des matrices $J_L(\omega)$ $L \geq 1$ est équivalente à l'étude des distributions empiriques des matrices

$$\tilde{J}_{L+1}(\omega) = \begin{pmatrix} X_o(\omega) & -1 & & (0) \\ -1 & X_1(\omega) & \ddots & -1 \\ (0) & & -1 & X_L(\omega) \end{pmatrix} \qquad L \geq 0$$

Nous nous placerons désormais dans ce cadre, les notations adoptées précédemment étant modifiées de façon convenable par l'adjonction d'un tilde.

Soit $p_{L+1}(t)$ le polynome caractéristique de $\tilde{J}_{L+1}$

on a la relation :

$$p_{L+1}(t) = (X_L - t)\, p_L(t) - p_{L-1}(t) \qquad L \geq 0$$

avec $p_{-1}(t) = 0$ $\quad p_o(t) = 1$

c'est-à-dire que pour tout $L \geq 0$:

$$\begin{pmatrix} p_{L+1}(t) \\ p_L\ (t) \end{pmatrix} = g_L^t \ g_{L-1}^t \ --- \ g_o^t \begin{pmatrix} 1 \\ 0 \end{pmatrix}$$

où $g_k^t = \begin{pmatrix} X_k - t & -1 \\ 1 & 0 \end{pmatrix}$ $\qquad 0 \leq k \leq L.$

Les valeurs propres de $\tilde{J}_{L+1}$ sont les zéros de $p_{L+1}(t)$. De plus nous pouvons énoncer la proposition suivante dont la justification figure dans [1].

Proposition (2-1) :

a) $p_{L+1}(t)$ *possède* (L+1) *zéros réels deux à deux distincts* $t_o<t_1<..\leq t_L$

b) Le nombre de changements de signes de la suite $\{p_o(t),p_1(t),..,p_{L+1}(t)\}$ *est égal à* 0 *si* $t<t_o$, $r+1$ *si* $t_r<t<t_{r+1}$ $\quad 0\leq t\leq L$, *et à* (L+1) *si* $t>t_L$.

Si l'on note $P(\mathbb{R}^2)$ l'espace projectif de $\mathbb{R}^2$,

$I = \{\overline{x}\in P(\mathbb{R}^2) \quad x = (\cos\theta,\sin\theta) \; -\frac{\Pi}{2}<\theta\leq 0\}$

et $\overline{x_o}$ l'image dans $P(\mathbb{R}^2)$ du vecteur $x_o = (1,o)$. On en déduit le :

corollaire (2-1) :

1) $\forall t\in\mathbb{R}$

$$\left|\tilde{N}_L(t) - \frac{1}{L+1}\sum_{k=0}^{L} 1_I(g_k^t\, g_{k-1}^t \quad g_o^t.\overline{x_o})\right| \leq \frac{1}{L+1}$$

2) Dans le cas où la loi de μ *est diffuse, on a de plus :*

$$\forall t\in\mathbb{R} \quad ps \quad \tilde{N}_L(t) = \frac{1}{L+1}\sum_{k=0}^{L} 1_I(g_k^t\dots g_o^t.\overline{x}_o)$$

où $g_\circ\overline{x}$ désigne l'action de la matrice g de $SL(2,\mathbb{R})$ sur l'élément $\overline{x}$ de $P(\mathbb{R}^2)$.

L'assertion 1 du corollaire (2-1) est une conséquence immédiate de la proposition (2-1), et l'assertion 2 résulte du fait que si la loi de μ est diffuse, on a pour tout $k\in\mathbb{N}$ et pour tout $t\in\mathbb{R}$ $P(p_k(t)=0) = 0$

Remarquons que les énoncés précédents sont analogues au théorème d'oscillation de Sturm dans la théorie des équations différentielles du second ordre.

Le corollaire 1 permet de ramener l'étude du comportement asymptotique de $\tilde{N}_L(t), L\geq 1$ à celui d'une fonctionnelle additive des états d'une chaîne de Markov.

Nous allons scinder la preuve du théorème (2-1) en deux parties :

I) On commence par étudier pour un nombre fini de réels $t_1<t_2<\ldots<t_k$ le comportement en loi de la suite de vecteurs aléatoires $(\tilde{Y}_n(t_1), \tilde{Y}_n(t_2),\ldots,\tilde{Y}_n(t_k))$, où $\tilde{Y}_n(t) = \sqrt{n+1}\,[\tilde{N}_n(t) - N(t)]$ $n\geq 1$, $t\in\mathbb{R}$.

II) On établit la relative compacité faible des lois de la suite de processus $\tilde{Y}_n(t)$ $t\in J$, $n\geq 1$, à valeurs dans $\mathfrak{D}(J)$ muni de la métrique de Skorokhod [2].

2-2-1) PREMIERE PARTIE :

On étudie pour $t_1<t_2<\ldots<t_k$ le comportement en loi de la suite de vecteurs aléatoires $(\tilde{Y}_n(t_1), \tilde{Y}_n(t_2), \ldots, \tilde{Y}_n(t_k))$ qui en raison du corollaire (2-1) est le même que celui du vecteur aléatoire.

$$\left(\sum_{j=0}^{n} \frac{1_I\,(g_j^{t_i}\, g_{j-1}^{t_i} \cdots g_1^{t_i}\cdot x_0) - (n+1)\,N(t_i)}{\sqrt{n+1}}\right)_{1\leq i\leq k}$$

Pour cela considérons la chaîne de Markov $(\tilde{x}_j^t)$ $j\geq o$, à valeurs dans $X^k = [P(\mathbb{R}^2)]^k$ de probabilité de transition définie de la façon suivante :

si $\underline{t} = (t_1,t_2,\ldots,t_k)\in\mathbb{R}^k$ et $\overline{x} = (x_1,x_2,\ldots,x_k)\in X^k$

$$\begin{aligned} P_{\underline{t}}\, f(\overline{x}) &= \int f(g_1^{t_1}(\omega).x_1,\, g_1^{t_2}(\omega).x_2,\ldots,g_1^{t_k}(\omega).x_k)\, dP(\omega) \\ &= \int f(g.\overline{x})\, d\mu_{\underline{t}}(g) \end{aligned}$$

où $\mu_{\underline{t}}$ désigne la loi du vecteur aléatoire $(g_{t_1}(\omega), g_{t_2}(\omega),\ldots,g_{t_k}(\omega))\in[SL(2,\mathbb{R})]^k$ et f une fonction borélienne bornée sur X^k.

Nous noterons $P_{x,\underline{t}}$ $x \in X$, (resp. $P_{\nu,\underline{t}}$ où ν est une probabilité sur X) la loi sur $X^{\mathbb{N}}$ des trajectoires de $(\tilde{X}_j^t \ j \geq 0)$ issues de x (resp. de loi de départ ν).

Soit $\mathcal{C}(X^k)$ l'espace de Banach des fonctions continues sur X^k muni de la norme de la convergence uniforme sur X^k $|f| = \sup_{x \in X^k} |f(x)|$

$P_{\underline{t}}$ opère sur $\mathcal{C}(X^k)$ et possède les propriétés suivantes :

<u>Proposition (2-2)</u> *Si* μ *admet une densité, l'opérateur* $P_{\underline{t}}$ *est quasi compact sur* $\mathcal{C}(X^k)$ *et l'on a :*

$\forall\, n \geq 1 \quad \forall\, f \in \mathcal{C}(X^k)$

$$P_{\underline{t}}^n f = \nu_{\underline{t}}(f) + Q_t^n(f)$$

où $\nu_{\underline{t}}$ *est l'unique probabilité* $P_{\underline{t}}$ *invariante portée par* X^k *et* $Q_{\underline{t}}$ *est un opérateur de rayon spectral strictement inférieur à 1 et tel que*

$$\nu_{\underline{t}}\, Q_{\underline{t}} = 0$$

<u>Démonstration de la proposition (2-2)</u> :

Commençons par énoncer des résultats utiles pour cette preuve :

a)
<u>Proposition (2-3)</u> *Si* μ *admet une densité, il existe un entier* $n_o \geq 1$ *tel que la probabilité* $\mu_{\underline{t}}^{n_o}$ *obtenue en convolant* n_o *fois* $\mu_{\underline{t}}$ *dans le groupe* $[SL(2,\mathbb{R})]^k$ *possède une densité dans* $[SL(2,\mathbb{R})]^k$ qui sera justifiée à la fin du paragraphe (2-2-1).

b)
<u>Proposition (2-4)</u> *Si* μ *charge au moins 2 points ;* $\mu_{\underline{t}}$ *admet une unique probabilité invariante* $\nu_{\underline{t}}$ *sur* X^k *et l'on a*

$$\forall\, f \in \mathcal{C}(X^k) \ \sup_{\bar{x} \in X^k} |P_{\underline{t}}^n f(\bar{x}) - \nu_{\underline{t}}(f)| = 0$$

Démonstration de la proposition (2-4) :

Pour $1 \leq j \leq k$ notons μ_{t_j} la loi de la matrice aléatoire $g^{t_j}(\omega) \in SL(2, \mathbb{R})$. Si μ charge au moins deux points, le sous-groupe fermé H_j de $SL(2, \mathbb{R})$ engendré par le support de μ_{t_j} est non compact et ne contient pas de sous-groupe d'indice fini ayant une action irréductible sur $\mathbb{R}^2$ [13] . Munissons $\mathbb{R}^2$ de la structure euclidienne définie par la norme $\|x\| = (x_1^2 + x_2^2)^{1/2}$ où $x = \begin{pmatrix} x_1 \\ x_2 \end{pmatrix}$, et définissons la distance d sur $P(\mathbb{R}^2)$ par

$$d(x,y) = |\sin \theta(\tilde{x},\tilde{y})|$$

où $\tilde{x},\tilde{y}$ sont deux vecteurs de $\mathbb{R}^2$ de norme 1, d'image x,y dans $P(\mathbb{R}^2)$, et $\theta(\tilde{x},\tilde{y})$ est l'angle de ces deux vecteurs.

On a alors : [9]

$$\forall \ 1 \leq j \leq k \quad \lim_n \sup_{x,y \in P(\mathbb{R}^2)} \int d(g.x,g.y) \ \mu_{t_j}^n(dg) = 0$$

Par conséquent si maintenant d_k est la distance sur X^k définie par

$$d_k(\bar{x},\bar{y}) = \sum_{i=1}^{k} d(x_i,y_i) \qquad \text{où } \bar{x} = (x_1,x_2,\ldots,x_k) \in X^k$$
$$\bar{y} = (y_1,y_2,\ldots,x_k) \in X^k$$

il en résulte que

$$\lim_n \delta_n = \lim_n \sup_{\bar{x},\bar{y} \in X^k} \int d_k(g.\bar{x},g.\bar{y}) \ \mu_{\underline{t}}^n(dg) = 0$$

Pour toute fonction f Lipschitzienne sur X^k et toute suite $(\bar{x}_n)_{n\geq 1}$ de X^k on a : $\forall \ n,m \geq 1$

$$|P_{\underline{t}}^{n+m} f(\bar{x}_{n+m}) - P_{\underline{t}}^n f(\bar{x}_n)| \leq 2 \ \delta_n$$

Il en résulte que pour toute fonction f Lipschitzienne, la suite de fonctions $(P_{\underline{t}}^n f)_{n\geq 1}$, converge uniformément vers une constante $L(f)$ sur X^k. Par densité, le résultat reste vrai pour toute fonction f de $\mathcal{C}(X^k)$. Clairement la fonctionnelle $f \to L(f)$ de (X^k) dans $\mathbb{R}$ définit l'unique probabilité $P_{\underline{t}}$ invariante portée par X^k ; d'où la proposition (2-4).

c)

X^k est un espace homogène compact de $[SL(2,\mathbb{R})]^k$; comme $\mu_{\underline{t}}^{no}$ admet une densité dans $[SL(2,\mathbb{R})]^k$, il en résulte que l'opérateur P_t^{no} est un opérateur compact sur $\mathcal{C}(X^k)$ [17] ; par conséquent, $P_{\underline{t}}$ est une contraction, quasi-compacte sur $\mathcal{C}(X^k)$. Les seules valeurs spectrales de $P_{\underline{t}}$ de module 1 sont alors des pôles simples. Le résultat de la proposition (2-4) montre que 1 est la seule valeur spectrale de module 1. Les autres valeurs spectrales de $P_{\underline{t}}$ ayant toutes un module - majoré par un nombre strictement inférieur à 1, on en déduit la proposition (2-2).
La chaîne de Markov définie par $P_{\underline{t}}$ satisfait donc à la condition de Doeblin.

Adoptons maintenant les notations : pour $\lambda = (\lambda_i)_{1\leq i\leq k}$

$\lambda' = (\lambda'_i)_{1\leq i\leq k} \in \mathbb{R}^k$ $\qquad \langle\lambda,\lambda'\rangle = \sum_{i=1}^{k} \lambda_i \lambda'_i$

et pour $\bar{x} \in X^k$, $\bar{x} = (x_1,\ldots,x_k)$ $\qquad F_I(\bar{x}) = (1_I(x_1), 1_I(x_2),\ldots,1_I(x_k))$

et considérons alors l'opérateur $P_t^{no}(\lambda)$ défini par

$$P_{\underline{t}}^{no}(\lambda)\, f(\bar{x}) = \int e^{i\langle\lambda, F_I(g\bar{x})\rangle} f(g\bar{x})\, p_{\underline{t}}^{no}(dg) \qquad \text{où} \quad f \in \mathcal{C}(X^k),\ \lambda \in \mathbb{R}^k$$

$P_{\underline{t}}^{no}$ opère sur $\mathcal{C}(X^k)$ car $p_{\underline{t}}^{no}$ a une densité dans $[SL(2,\mathbb{R})]^k$.

Par ailleurs $P_{\underline{t}}^{no}(\lambda)$ est un opérateur qui est obtenu par perturbation analytique de l'opérateur $P_{\underline{t}}^{no}(o) = P_{\underline{t}}^{no}$. Si l'on désigne par r(L) le rayon spectral d'un opérateur L sur $\mathcal{C}(X^k)$, on peut alors en tenant compte de la proposition (2-2) et des résultats de la théorie des perturbations analytiques d'opérateurs [3], [10] conclure que : il existe un voisinage V de o dans $\mathbb{R}^k$ tel que :

$$\forall f \in \mathcal{C}(X^k) \quad \forall \lambda \in V \quad \forall p \geq 1$$

(1) $$[P_{\underline{t}}^{no}(\lambda)]^p f = \tilde{k}_{\underline{t}}^p(\lambda)\, \tilde{N}_{\underline{t}}(\lambda) f + \tilde{Q}_{\underline{t}}^p(\lambda) f$$

où $\tilde{k}_t(\lambda)$ est l'unique valeur propre de plus grand module de $P_{\underline{t}}^{no}(\lambda)$ et vérifie $|\tilde{k}_t(\lambda)| > \frac{2+r(Q^{no})}{3}$

$\tilde{N}_{\underline{t}}(\lambda)$ est la projection sur le sous-espace propre de dimension 1 correspondant à $\tilde{k}_t(\lambda)$

$\tilde{Q}_{\underline{t}}(\lambda)$ est un opérateur de $\mathcal{C}(X^k)$ de rayon spectral $r(\tilde{Q}_t(\lambda)) \leq \frac{1+r(Q^{no})}{3}$ vérifiant $\tilde{Q}_{\underline{t}}(\lambda)\, \tilde{N}_{\underline{t}}(\lambda) = 0$.

De plus les applications $\lambda \to \tilde{k}_{\underline{t}}(\lambda)$, $\lambda \to \tilde{N}_{\underline{t}}(\lambda)$, $\lambda \to \tilde{Q}_{\underline{t}}(\lambda)$ sont analytiques.

La décomposition précédente va nous permettre de justifier la

Proposition (2-5) :

Sous les hypothèses du théorème (2-1) *pour tout* $x = (x_1, x_2, \ldots, x_k) \in X^k$ *on a* :

1) $$\text{ps}\lim_n \Big(\frac{1}{n} \sum_{p=1}^{n} 1_I(g_p^{t_j} g_{p-1}^{t_j} \cdots g_1^{t_j} x_j)\Big)_{1\leq j\leq k} = (\nu_{t_j}(I))_{1\leq j\leq k} = (N(t_j))_{1\leq j\leq k}$$

$$= \Big(\frac{1}{n_o}\Big[\frac{1}{i}\frac{\partial}{\partial\lambda_j}[k_{\underline{t}}](o)\Big]\Big)_{1\leq j\leq k}$$

2) *Le vecteur aléatoire* $\left(\frac{1}{\sqrt{n}} \sum_{p=1}^{n} 1_I(g_p^{t_j} g_{p-1}^{t_j} \cdots g_1^{t_j} x_j) - n\, N(t_j)\right)_{1\leq j\leq k}$

converge en loi vers une loi de Gauss $Y_{\underline{t}}$ *centrée portée par* $\mathbb{R}^k$ *et de matrice de covariance* $\Sigma = (\sigma_{i,j})_{1\leq i,j\leq k}$

définie par : $1\leq i,j\leq k$

$$\sigma_{i,j} = \frac{1}{n_o 2} \left\{- \frac{\partial^2}{\partial \lambda_i \partial \lambda_j} [k_{\underline{t}}](o) + \frac{\partial}{\partial \lambda_i} [k_{\underline{t}}](o) \frac{\partial [k_{\underline{t}}]}{\partial \lambda_j}(o)\right\}$$

$$= \int_{X^2} [f_i\, f_j - P_{t_i} f_i P_{t_j} f_j]\, d\nu_{t_i,t_j}$$

où $\quad f_i = \sum_{n=o}^{\infty} P_{t_i}^n (1_I - \nu_{t_i}(I)) = (I - P_{t_i})^{-1}(1_I - \nu_{t_i}(I))$

<u>Démonstration de la proposition (2-5).</u>

L'existence des limites précédentes se déduisent de résultats connus [10] concernant le comportement asymptotique des fonctionnelles additives d'états de chaînes de Markov satisfaisant à la condition de Doeblin. Par ailleurs l'expression de ces limites à l'aide de dérivées de $k_{\underline{t}}(\lambda)$ s'obtient en passant à la limite selon la sous-suite $(n\ no)_{n\geq 1}$ et en utilisant la décomposition (1) : (voir [10] et [18] pour une étude détaillée).

Nous achèverons la première partie en prouvant la

<u>proposition (2-6)</u> :

Sous les hypothèses du théorème (2-1) *pour tous* $t_1 < t_2 < \ldots < t_k \in\,]-2,2[\, + S_{\bar{\mu}}$ $Y_{\underline{t}}$ *est non dégénérée.*

<u>Démonstration de la proposition (2-6) :</u>

Elle repose sur plusieurs lemmes.

1) lemme (2-1) :

Sous les hypothèses du théorème (2-1), *si* $t_1 < t_2 < .. < t_k \in]-2,2[+ S_\mu$ *le support de* $\nu_{\underline{t}}$ *est égal à* X^k.

dont nous donnerons la preuve à la fin du paragraphe 2-2-1.

2) lemme (2-2) :

sous les hypothèses du théorème (2-1), *si* $t_1 < t_2 < .. < t_k \in]-2,2[+ S_\mu$ *il existe une boule* B *centrée en* o *dans* $\mathbb{R}^k$ *telle que pour* $\lambda \in B - \{o\}$ *on ait* $|k_{\underline{t}}(\lambda)| < 1$.

Démonstration du lemme (2-2) :

Pour tout $\lambda \in V$, il existe une fonction $\Phi_\lambda \in \mathcal{C}(X^k)$ telle que pour tout $n \geq 1$:

(2) $\quad [P_{\underline{t}}^{no}(\lambda)]^n \Phi_\lambda = [k_{\underline{t}}(\lambda)]^n \Phi_\lambda \qquad \Phi_\lambda \neq 0.$

Supposons que $\lambda \neq 0$ et $|k_t(\lambda)| = 1$; il résulte alors de (2) que pour tout $n \geq 1$

$$P_{\underline{t}}^{n\,no} |\Phi_\lambda| \geq |\Phi_\lambda|$$

d'où l'on obtient en faisant tendre n vers $+\infty$, que :

(3) $\quad \nu_{\underline{t}} |\Phi_\lambda| \geq |\Phi_\lambda|$

De (3) il résulte puisque $|\Phi_\lambda| \in \mathcal{C}(X^k)$ et que $\text{supp}\,\nu_{\underline{t}} = X^k$ d'après le lemme (2-1) que :

(4) $\quad \forall\ \overline{x} \in X^k \qquad |\Phi_\lambda(\overline{x})| = \sup_{\overline{y} \in X^k} |\Phi_\lambda(\overline{y})| > 0$

en tenant compte de (2) et (4) on a :

$\forall\ \overline{x} \in X^k, \quad \forall n \geq 1 \qquad \forall\ g \in \sup \mu_{\underline{t}}^{n\,no}$

$$e^{i\langle \lambda, F_I(g.\overline{x})\rangle} \Phi_\lambda(g\overline{x}) = [k_{\underline{t}}(\lambda)]^n \Phi_\lambda(\overline{x})$$

soit

$$(5)\qquad \frac{\Phi_\lambda(g\overline{x})}{\Phi_\lambda(x)} = [k_{\underline{t}}(\lambda)]^n \quad e^{-i<\lambda F_I(g\overline{x})>}$$

En utilisant le fait que le premier membre de (5) est un cocycle multiplicatif sur $[SL(2,\mathbb{R})]^k \times X^k$, on voit que

$$\forall\, n\geq 1,\quad \forall\, \overline{x}\in X^k,\quad \forall\, g \in \text{supp } P_{\underline{t}}^{n\, no}$$

$$e^{i<\lambda\ F_I(g\,\overline{x})>} = 1$$

ce qui dès que $\|\lambda\| < \frac{\Pi}{2\sqrt{k}}$ entraîne que $<\lambda, F_I(g\,\overline{x})> = 0$

c'est-à-dire que $\forall\, \overline{x} \in X^k \qquad <\lambda, F_I(\overline{x})> = 0$,

ce qui est impossible si $\lambda \neq 0$, d'où le lemme (2-2).

3) Lemme (2-3) :

Si $\lambda \in \mathbb{R}^k-\{o\}$ *et si* ${}^t\lambda\ \Sigma\,\lambda = 0$ *il existe une mesure de probabilité* $\theta_\lambda^{\underline{t}}$ *sur* $\mathbb{R}$ *telle que pour toute fonction* Φ *continue bornée sur* $\mathbb{R}$ *on ait :*

$$\lim_n \quad E_{\nu_{\underline{t}}}[\Phi(S_n^{\underline{t}}(\lambda))] = \theta_\lambda^{\underline{t}} * \overset{\vee}{\theta}{}_\lambda^{\underline{t}}\ (\Phi)$$

où $S_n^{\underline{t}}(\lambda) = \sum_{j=1}^{n} \{<\lambda, F_I(\tilde{X}_j^{\underline{t}})> - \int <\lambda, F_I(\overline{x})>\nu_{\underline{t}}(d\overline{x})\}$

et $\overset{\vee}{\theta}{}_\lambda^{\underline{t}}$ *est la probabilité symétrique de* $\theta_\lambda^{\underline{t}}$

Démonstration du lemme (2-3) :

Posons $F_\lambda = <\lambda, F_I> - \int <\lambda, F_I(\overline{x})>\nu_{\underline{t}}(d\overline{x})$;

comme $\nu_{\underline{t}}(F_\lambda) = 0$ et en raison du fait que $P_{\underline{t}}$ satisfait à la condition de Doeblin, nous pouvons également définir :

$$h_\lambda = \sum_{n\geq o} P_{\underline{t}}^n(F_\lambda) = (I - P_{\underline{t}})^{-1}(F_\lambda)$$

On a alors :

$$\sigma_\lambda^2 = {}^t\lambda\Sigma\lambda = \int\{P_{\underline{t}}(h_\lambda)^2(\bar{x}) - [P_{\underline{t}}h_\lambda(\bar{x})]^2\} \quad \nu_{\underline{t}}(d\bar{x})$$

D'autre part on a : pour tout $\bar{x} \in X^k$ et tout $j \geq 0$:

$$E_{\bar{x}}\ [h_\lambda(\tilde{x}_{j+1}) - P_{\underline{t}}\ h_\lambda(\tilde{x}_j)]^2 = P_{\underline{t}}^j\ [P_{\underline{t}}\ h_\lambda^2 - (P_t h_\lambda)^2]\ (\bar{x})$$

Par conséquent si $\sigma^2_\lambda = 0$, il existe une constante $K>0$ telle que pour tout $j \geq 0$:

$$\sup_{\bar{x}\in X^k} E_{\bar{x}}\ [h_\lambda(\tilde{x}_{j+1}) - P_{\underline{t}}h_\lambda(\tilde{x}_j)]^2 \leq K\ \rho^j$$

c'est-à-dire que pour tout $j \geq 0$

$$\sup_{\bar{x}\in X^k} E_{\bar{x}}\ [h_\lambda(\tilde{x}_{j+1}) - h_\lambda(\tilde{x}_j) + F_\lambda(\tilde{x}_j)]^2 \leq K\ \rho^j.$$

d'où l'on déduit à l'aide de l'inégalité de Schwarz que pour tout $j \geq 0$

$$(6) \quad \sup_{\bar{x}\in X^k} E_{\bar{x}}\ |h_\lambda(\tilde{X}_{j+1}) - h_\lambda(\tilde{X}_j) + F_\lambda(\tilde{X}_j)| \leq \sqrt{K}\ \rho^{j/2}$$

Posons alors $T_{n,n} = \sum_{j=n}^{2n-1} F_\lambda(\tilde{X}_j)$

De (6) il résulte que :

$$\sup_{\bar{x}\in X^k} E_{\bar{x}}\ |T_{n,n} + [h_\lambda(\tilde{X}_{2n+1}) - h_\lambda(\tilde{X}_n)]| \leq \sqrt{K} \sum_{j=n}^{2n} \rho.^{j/2}$$

On en déduit que la suite de variables aléatoires $(T_{n,n} - [h_\lambda(\tilde{X}_n) - h_\lambda(\tilde{X}_{2n})])$ $n \geq 0$ converge vers 0 en $P_{\nu_{\underline{t}}}$ probabilité quand n tend vers $+\infty$.

Par ailleurs, $\forall\ k,m \in \mathbb{N}$ on a

$$\int h_\lambda^k\ (\tilde{X}_{2n}(\omega))\ h_\lambda^m(\tilde{X}_n(\omega)) \quad P_{\nu_t}(d\omega) = \int P_{\underline{t}}^n\ (h_\lambda^k)(\bar{x})h_\lambda^m(\bar{x})\cdot \quad \nu_{\underline{t}}(d\bar{x})$$

d'où $\forall\ k,m \in \mathbb{N}$.

$$\lim_n \int h_\lambda^k(\tilde{X}_{2n+1}(\omega))\, h_\lambda^m(\tilde{X}_n(\omega))\, dP_{\nu_{\underline{t}}}(\omega) = \nu_{\underline{t}}(h_\lambda^k)\, \nu_{\underline{t}}(h^m).$$

La fonction h_λ étant bornée, il existe sur $\mathbb{R}$ une unique probabilité $\theta_\lambda^{\underline{t}}$ telle que pour tout $k\geq 0$ $\int x^k\, \theta_\lambda^{\underline{t}}(dx) = \nu_{\underline{t}}(h_\lambda^k)$ et la suite des lois par rapport à $P_{\nu_{\underline{t}}}$ de la suite des variables aléatoires $(h_\lambda(\tilde{X}_n) - h_\lambda(\tilde{X}_{2n}))$ $n\geq 1$ converge vers $\theta_\lambda^{\underline{t}} * \check{\theta}_\lambda^{\underline{t}}$.

L'assertion du lemme (2-3) résulte immédiatement de ce qui précède car les variables aléatoires $T_{n,n}$ et $S_n^{\underline{t}}(\lambda)$ ont même loi par rapport à $P_{\nu_{\underline{t}}}$.

4) Terminons maintenant la démonstration de la proposition (2-6).

Soit ψ la fonction continue bornée sur $\mathbb{R}$ définie par $\psi(x) = \frac{1}{\Pi}\frac{1-\cos x}{x^2}$;

elle a pour transformée de Fourier

$$\hat{\psi}(u) = \frac{1}{\Pi}\int_{-\infty}^{+\infty} e^{iux}\,\psi(x)\,dx = \begin{cases} 1-|u| & \text{si } |u|\leq 1 \\ 0 & \text{sinon.}\end{cases}$$

Pour tout $n\geq 1$, on a :

$$E_{\nu_{\underline{t}}}(\psi(S^{\underline{t}}_{nno}(\lambda))) = \frac{1}{2\pi}\int_{-1}^{1}\int_{X^k}(1-|u|)\, e^{-iunno\nu_{\underline{t}}(<\lambda,F_I>)}(P_{\underline{t}}^{no}(\lambda u))^n e(\overline{x})\,du\, d\nu_{\underline{t}}(\overline{x})$$

où e est la fonction identique à 1 sur X^k.

Si $\lambda \in B-\{o\}$ et si $u \in [-1,1]-\{0\}$, on a en raison du lemme (2-2) et de (1) :

$$\forall\, \overline{x} \in X^k \qquad \lim_n [P_{\underline{t}}^{n_o}(\lambda u)]^n e(\overline{x}) = 0.$$

En utilisant le théorème de Lebesgue, on peut alors conclure que si $\lambda \in B-\{o\}$

$$\lim_n E_{\nu_{\underline{t}}}(\psi(S^{\underline{t}}_{n\,n_o}(\lambda))) = 0.$$

Comme pour toute probabilité θ sur $\mathbb{R}$, on a $\theta * \check{\theta}(\psi)>0$, le lemme (2-3) montre alors que si $\lambda \in B-\{o\}$

$${}^t\lambda\,\Sigma\,\lambda \neq 0$$

ce qui établit la proposition (2-6).

Les conclusions des propositions (2-5) et (2-6) justifient toutes les assertions contenues dans (2-1) concernant le comportement en loi du vecteur aléatoire $(Y_n(t_1),Y_n(t_2),\dots,Y_n(t_k))_{n\geq 1}$ lorsque $n \to +\infty$ où $t_1<t_2<\dots< t_k$.

Il nous reste cependant à donner les démonstrations de la proposition (2-3), et du lemme (2-1).

Démonstration de la proposition (2-3)

Elle repose sur les lemmes suivants :

1) lemme (2-4) :

Si μ admet une densité, le sous groupe fermé $G_{\underline{t}}$ de $[SL(2,\mathbb{R})]^k$ engendré par le support de $\mu_{\underline{t}}$ est égal à $[SL(2,\mathbb{R})]^k$

Démonstration du lemme (2-4) :

On raisonne par récurrence sur k.

a) Supposons que $k=1$;

pour $t \in \mathbb{R}$, $q \in \operatorname{supp}\mu$ on a

$$g_t(q) = \begin{pmatrix} q-t & -1 \\ 1 & 0 \end{pmatrix} = \exp(q-t)X \quad g_0$$

où $X = \begin{pmatrix} 0 & 1 \\ 0 & 0 \end{pmatrix}$ et $g_o = \begin{pmatrix} 0 & -1 \\ 1 & 0 \end{pmatrix}$

On en déduit que si $q,q' \in \operatorname{supp} \mu$

$$g_t^{-1}(q)\, g_t(q') = \exp(-q'+q)Y \in G_t$$

où $Y = \begin{pmatrix} 0 & 0 \\ 1 & 0 \end{pmatrix}$

et $g_t(q)\, g_t^{-1}(q') = \exp(q-q')\, X \in G_t$

Comme le support de $\mu * \check{\mu}$ ($\check{\mu}$ étant le probabilité symétrique de μ) contient un voisinage ouvert de 0, l'algèbre de Lie de G_t contient X et Y. Or X et Y engendrent l'algèbre de Lie de $SL(2,\mathbb{R})$ qui est connexe, par conséquent $G_t = SL(2,\mathbb{R})$.

) Supposons le résultat du lemme (2-4) établi pour tout entier $1 \leq k \leq k_o$ Soient alors $t_1 < t_2, .. < t_{k_o+1} \in \mathbb{R}^{k_o+1}$;

notons $\underline{t}^{(k_o+1)} = (t_1, t_2, .., t_{k_o+1})$ et pour $1 \leq p \leq k_o+1$

$H_p = I_{p-1} \times SL(2,\mathbb{R}) \times I_{k_o+1-p} \cap G_{\underline{t}}(k_o+1)$ où $I_k = (I, I, .., I)$ I étant la matrice identité de $SL(2,\mathbb{R})$. H_p est un sous-groupe distingué de $G_{\underline{t}}(k_o+1)$ et en raison du a) également du groupe $I_{p-1} \times SL(2,\mathbb{R}) \times I_{k_o+1-p}$.

Or ce dernier groupe est semi simple, donc on a :

soit : $H_p = I_{k_o+1}$

soit : $H_p = I_{p-1} \times SL(2,\mathbb{R}) \times I_{k_o+1-p}$

Distinguons deux cas :

α) S'il existe un entier p_o tel que $H_{p_o} = I_{p_o-1} \times SL(2,\mathbb{R}) \times I_{k_o+1-p_o}$, on peut affirmer compte-tenu du fait que grâce à l'hypothèse de récurrence $G_{\underline{t}}(k_o+1)$ se projette surjectivement sur le groupe

$[SL(2,\mathbb{R})]^{p_o-1} \times I \times [SL(2,\mathbb{R})]^{k_o+1-p_o}$

que $G_{\underline{t}(k_o+1)} = [SL(2,\mathbb{R})]^{k_o+1}$

β) Si $\forall p,\ 1\le p\le k_0+1$ H_p est l'identité de $[SL(2,\mathbb{R})]^k$, alors compte-tenu de a), il existe des isomprphismes (χ_{t_i}) $2\le i\le k_0+1$ de $SL(2,\mathbb{R})$ sur lui-même tels que tout élément g de $G_{\underline{t}}(k_0+1)$ s'écrive sous la forme :

$$(g_1, \chi_{t_2}(g_1), \chi_{t_3}(g_1), .., \chi_{t_{k_0+1}}(g_1)) \qquad g_1 \in SL(2,\mathbb{R})$$

Montrons que ceci n'est pas possible, ce qui assurera que seule la situation α) se présente et achèvera la preuve du lemme (2-4).

Pour tout $q \in S_\mu$ on a nécessairement

(7) $\chi_{t_2}(g_{t_1}(q)) = \exp(t_1-t_2)X\, g_{t_1}(q)$

il en résulte que pour tous $q,q' \in S_\mu$ on a :

$$\chi_{t_2}(g_{t_1}(q)\, g_{t_1}^{-1}(q') = \chi_{t_2}(\exp\,(q-q')X) = \exp(q-q')X$$

$$\chi_{t_2}(g_{t_1}^{-1}(q)\, q_{t_1}(q')) = \chi_{t_2}(\exp\,(q-q')Y) = \exp(q-q')Y$$

Le support de $\mu * \check{\mu}$ contient un voisinage ouvert de o, par conséquent pour tout $h\in\mathbb{R}$,

$$\chi_{t_2}(\exp hX) = \exp hX\ , \qquad \chi_{t_2}(\exp hY) = \exp hY$$

d'où il résulte que pour tout $g \in SL(2,\mathbb{R})$ $\chi_{t_2}(g) = g$, ce qui en contradiction avec (7) puisque $t_1 \neq t_2$.

2) Soit $\mathcal{G}_k$ l'algèbre de Lie de $[SL(2,\mathbb{R})]^k$ et notons Ad l'action adjointe de $[SL(2,\mathbb{R})]^k$ sur $\mathcal{G}_k$. L'algèbre de Lie $\mathcal{G}_k$ est égale à la puissance kième de l'algèbre de Lie de $SL(2,\mathbb{R})$.

Appelons $\tilde{X}$ l'élément $(X,X,..,X)$ de $\mathcal{G}_k$.

Nous pouvons alors énoncer le

Lemme (2-5) :

Soit U *une partie de* $[SL(2,\mathbb{R})]^k$ *engendrant topologiquement* $[SL(2,\mathbb{R})]^k$, *il existe alors un entier* n_o *et des éléments* $g_1, g_2 \dots g_{n_o}$ *de* U *tels que le sous-espace vectoriel de* $\mathfrak{g}_k$ *engendré par les éléments* $\{Ad(g_1 \cdot g_2 \cdot g_p)(\tilde{X}) \quad 1 \leq p \leq n_o\}$ *soit égal à* $\mathfrak{g}_k$.

Démonstration du lemme (2-5):

Pour $n \geq 1$ et $\omega = (g_1, g_2, \dots, g_n) \in U^n$ notons $H_n(\omega)$ le sous-espace vectoriel de $\mathfrak{g}_k$ engendré par les éléments $(Ad(g_1 \cdot g_2 \cdots g_p)(\tilde{X}) \quad 1 \leq p \leq n)$.

On a pour tous $m, n \geq 1$ la relation :

(8) $H_{m+n}(\omega\omega') = H_n(\omega) + Ad(g_1 \cdot g_2, \dots, g_n)(H_m(\omega'))$ où $\omega' \in U^m$.

Si $\omega_o = (g_1, g_2, \dots, g_{n_o}) \in U^{n_o}$ est tel que la dimension de $H_{n_o}(\omega_o)$ soit maximum il résulte alors de la relation (8) que $Ad(g_1 \cdot g_2, \dots g_{n_o})(H_{n_o}(\omega_o)) = H_{n_o}(\omega_o)$ en faisant $\omega = \omega' = \omega_o$; de plus en faisant $\omega = \omega_o$ on voit que $Ad(g_1 \cdot g_2, \dots, g_{n_o})(H_n(\omega') \subset H_{n_o}(\omega_o)$

et donc d'après la remarque précédente on a $H_n(\omega') \subset H_{n_o}(\omega_o)$.

Par conséquent pour tout $g \in U$, on a $Ad(g)(H_{n_o}(\omega_o)) \subset H_{n_o}(\omega_o)$.

Comme U engendre $[SL(2,\mathbb{R})]^k$, $H_{n_o}(\omega_o)$ est un idéal de $\mathfrak{g}_k$; de plus $H_{n_o}(\omega_o)$ a une projection non réduite à 0 sur chacun des facteurs de $\mathfrak{g}_k$ et donc $H_{n_o}(\omega_o) = \mathfrak{g}_k$

3) Terminons maintenant la démonstration de la proposition (2-3) :

Soit $q \to g_{\underline{t}}(q)$ l'application de $\mathbb{R}$ dans $[SL(2,\mathbb{R})]^k$ définie par

$g_{\underline{t}}(q) = (g_{t_1}(q), g_{t_2}(q), \dots, g_{t_k}(q))$

Pour tout $n \geq 1$ l'application M_n analytique de $\mathbb{R}^n$ dans $[SL(2,\mathbb{R})]^k$ définie par

$$M_n(q_1,q_2,..,q_n) = g_{\underline{t}}(q_1)\, g_{\underline{t}}(q_2)..g_{\underline{t}}(q_n)$$

a en tout point $(q_1,q_2,..,q_n) \in \mathbb{R}^n$ un rang égal à la dimension du sous-espace vectoriel engendré par les éléments

$\{Ad\, [g_{\underline{t}}(q_1)\, g_{\underline{t}}(q_2)..g_{\underline{t}}(g_p)]^{-1}\, (\tilde{X}) \quad 1 \leq p \leq n\}$ de $\mathcal{G}_k$.

En appliquant le lemme (2-5) à $U = [g_{\underline{t}}(S_\mu)]^{-1}$, ce qui est possible d'après le lemme (2-4), il existe un entier n_o et des réels $q_1,q_2,..,q_{n_o} \in S_\mu$ tels que le rang de M_{n_o} au point $(q_1,q_2,...,q_{no}) \in S_\mu$ soit égal à la dimension de $\mathcal{G}_k$. L'ensemble des points où le rang de M_{n_o} est strictement inférieur est, puisque M_{n_o} est analytique, une réunion dénombrable de sous-variétés de $\mathbb{R}^{no}$ de dimension inférieure ou égale à $n_o - 1$, et donc négligeable. On en déduit que si μ a une densité, il en est de même pour $\mu_{\underline{t}}^{n_o}$

Donnons maintenant la :

<u>Démonstration du lemme (2-1)</u> :

Si x_o est un point de support de $\nu_{\underline{t}}$, en raison du fait que $\nu_{\underline{t}}$ est l'unique probabilité invariante par $\mu_{\underline{t}}$, le support de $\nu_{\underline{t}}$ est égal à $\overline{T_{\mu_{\underline{t}}}.x_o}$ où $T_{\mu_{\underline{t}}}$ est le semi-groupe fermé engendré par le support de $\mu_{\underline{t}}$. La démonstration du lemme (2-1) sera donc achevée si l'on prouve que pour tout élément x de X^k on a $\overline{T_{\mu_{\underline{t}}}.x} = X^k$.

Justifions ce dernier résultat en raisonnant par récurrence sur k.

1) La propriété est vraie si k=1 ; en effet sous les hypothèses du lemme (2-1), il existe un élément q de S_μ tel que la matrice $g_{t_1}(q)$ soit elliptique et possède pour valeurs propres $e^{i2\pi\alpha}$ et $e^{-i2\pi\alpha}$ avec $\alpha \notin Q$, et l'on a alors $\overline{T_{\mu_{t_1}}.x} \supset \overline{\{g_{t_1}^n(q).x \quad n \geq o\}} = X$ d'où le résultat.

2) Supposons le résultat établi jusqu'à l'ordre k_o, et montrons qu'il reste alors vrai à l'ordre $k = k_o+1$.

Soit $q_o \in \overset{\circ}{S}_\mu - F$ où $F = \{q / \exists 1 \leq i \leq k_o+1 \ ; \ |q-t_i| = 2\}$

et soient $T(q_o) = \{i \ ; \ 1 \leq i \leq k_o+1 \quad |q_o-t_i| < 2\}$ et

$T'(q_o) = \{i \ ; \ 1 \leq i \leq k_o+1 \quad |q_o-t_i| > 2\}$.

On peut supposer que $T(q_o) \neq \phi$ d'après les hypothèses du lemme ; on peut également admettre sans restriction que les matrices $g_{t_i}(q_o)$ $t_i \in T(q_o)$ possèdent les valeurs propres $e^{i2\pi\alpha_i}$, et $e^{-i2\pi\alpha_i}$ telles que $(1,\alpha_i)$ $i \in T(q_o)$ soient linéairement indépendants sur les rationnels (en effet sinon il suffit de substituer à q_o un élément $q \in S_\mu$ voisin de q_o, choisi de sorte que $T(q) = T(q_o)$, $T'(q) = T'(q_o)$.)

Notons $\mathrm{card}(T(q_o))$ (resp card $T'(q_o)$) le cardinal de $T(q_o)$ (resp de $T'(q_o)$).
L'orbite de tout élément de $X^{\mathrm{card}(T(q_o))}$ sous les puissances positives de $(g_{t_i}(q_o))_{i \in T(q_o)} \in [SL(2,\mathbb{R})]^{\mathrm{card}(T(q_o))}$ est dense dans $X^{\mathrm{card}(T(q_o))}$.
D'autre part si card $T'(q_o) \neq o$ pour tout $y = (y_i)_{i \in T'(q_o)} \in X^{\mathrm{card}(T'(q_o))}$ la suite $(g^n_{t_i}(q_o).y_i)_{i \in T'(q_o)}$ $n \geq 0$ converge vers un point z_o.

Si card $T'(q_o) = o$ le résultat cherché est obtenu immédiatement en raisonnant comme dans le 1) ; si card $T'(q_o) \neq o$ les considérations précédentes montrent que pour tout $x \in X^{k_o+1}$ $\overline{T_{\mu_{\underline{t}}}.x}$ contient la fibre $z_o \times X^{\mathrm{card}T(q_o)}$ l'hypothèse de récurrence montre de plus que $\overline{T_{\mu(\underline{t_i})_{i \in T'(q_o)}}.z_o} = X^{\mathrm{card}(T'(q_o))}$, d'où le résultat cherché.

La démonstration du lemme (2-1) est ainsi achevée.

2-2-2) DEUXIEME PARTIE :

A) Pour terminer la démonstration du théorème (2-1), il reste à établir la relative compacité faible des lois de la suite de processus $\tilde{Y}_n(t)$ $t \in J$ $n \geq 1$ à valeurs dans $\mathcal{D}(J)$.

Précisons une condition suffisante permettant d'assurer cette compacité [2].

Donnons tout d'abord quelques notations :
Pour tout intervalle K de $\mathbb{R}$ et toute fonction α de J dans $\mathbb{R}$ on pose :

$$W(\alpha,K) = \sup_{s,t \in K} |\alpha(s) - \alpha(t)|$$

et l'on définit alors pour toute fonction α de J dans $\mathbb{R}$ et $\delta > o$

$$W_J(\alpha,\delta) = \text{Sup } \{W(\alpha,[t,t+\delta]) \; ; \; [t,t+\delta] \subset J\}$$

si les conditions (i) et (ii) suivantes sont vérifiées :

(i) il existe un réel $t_o \in J$, tel que quel que soit $\varepsilon > o$, on puisse trouver un réel $a > o$ tel que : pour tout $n \geq 1$

$$P(|\tilde{Y}_n(t_o)| > a) \leq \varepsilon$$

(ii) Pour tous $\eta > o$, et $\varepsilon > o$ il existe un δ, $o < \delta < 1$ et un entier n_o tels que pour $n \geq n_o$

$$P(W_J(\tilde{Y}_n,\delta) \geq \eta) \leq \varepsilon$$

alors les lois de la suite $(\tilde{Y}_n)_{n \geq 1}$ sont relativement compactes pour la topologie de la convergence faible sur $\mathcal{D}(J)$; de plus toute valeur d'adhérence de cette suite est une loi de processus continu.

La condition (i) résulte immédiatement de la convergence en loi de la suite de variables aléatoires $\tilde{Y}_n(t)$ $n \geq 1$, $t \in \mathbb{R}$, établie dans la première partie.

D'après le corollaire (2-1), on a pour tout $t \in \mathbb{R}$ $\quad n \geq 1$

ps $\quad \tilde{Y}_n(t) = \tilde{Z}_n(t)$

où $\quad \tilde{Z}_n(t) = \frac{1}{\sqrt{n+1}} \left(\sum_{j=1}^{n+1} 1_I (\tilde{X}_j^t) - (n+1) N(t) \right).$

Pour démontrer (ii), nous utiliserons en outre les deux propositions suivantes qui seront justifiées ultérieurement.

Proposition (2-7) :

Pour tous $0<\alpha<1$ *et* $p=1,2$ *il existe une constante* $A_k(\alpha)$ *telle que pour tous* $n \geq 1$ *et* $t,s \in J$

$$E(\tilde{Z}_{2n-1}(t) - \tilde{Z}_{2n-1}(s))^{2p} \leq A_p(\alpha)\left(|t-s|^{p\alpha} + \frac{|t-s|^{\alpha}}{2n-1}\right)$$

Proposition (2-8) :

Sous les hypothèses du théorème (2-1) la répartition d'état N *admet une densité continue.*

ainsi que lemme suivant [2].

lemme (2-6) :

Soient $(\xi_i)_{1 \leq i \leq m}$ *des variables aléatoires,* $\quad S_k = \xi_1 + \xi_2 + \ldots + \xi_k \quad 1 \leq k \leq m$

et $\quad M_m = \sup_{1 \leq k \leq m} |S_k|$

S'il existe des réels positifs $(u_i)_{1 \leq i \leq m}$, $\gamma \geq 0$ *et* $\beta > 1$ *tels que*

$$E|S_i - S_j|^{\gamma} \leq \left(\sum_{i < \ell \leq j} u_\ell\right)^{\beta} \qquad 1 \leq i \leq j \leq m$$

alors pour tout $\lambda > 0$ *on a*

$$P(M_m \geq \lambda) \leq \frac{K(\gamma,\beta)}{\lambda^{\gamma}} (u_1 + u_2 + \ldots + u_m)^{\beta}$$

où $K(\gamma,\beta)$ *est une constante.*

Pour tout $n \geq 1$ l'application $t \to \tilde{Y}_{2n-1}(t) + \sqrt{2n}\, N(t)$ est croissante ; donc pour $s \leq t \leq s+h \in \mathbb{R}$ on a

$$-\sqrt{2n}\,[N(s+h) - N(s)] \leq \tilde{Y}_{2n-1}(t) - \tilde{Y}_{2n-1}(s) \leq \tilde{Y}_{2n-1}(s+h) - \tilde{Y}_{2n-1}(s) + \sqrt{2n}\,[N(s+h) - N(s)].$$

On en déduit, en tenant compte du fait (proposition (2-8)) que N a une densité continue à support compact, qu'il existe une constante $c>o$ telle que pour $s \leq t \leq s+h \in \mathbb{R}$, on ait :

$$|\tilde{Y}_{2n-1}(t) - \tilde{Y}_{2n-1}(s)| \leq |\tilde{Y}_{2n-1}(s+h) - \tilde{Y}_{2n-1}(s)| + c\sqrt{2n}\, h$$

et par conséquent pour tous $m \geq 1$ $h>o$ $s \in \mathbb{R}$

$$(9) \quad \sup_{t \in [s,s+mh]} |\tilde{Y}_{2n-1}(t) - \tilde{Y}_{2n-1}(s)| \leq 3 \sup_{i \leq m} |\tilde{Y}_{2n-1}(s+ih) - \tilde{Y}_{2n-1}(s)| + c\, h \sqrt{2n}$$

soit $\frac{1}{2} < \alpha < 1$; posons $\varepsilon_1 = \left(\frac{n}{3+c}\right)^{\alpha}$

De la proposition (2-7) on déduit que

si $\left|\frac{t-s}{2n-1}\right|^{\alpha} \leq \frac{1}{\varepsilon_1} |t-s|^{2\alpha}$ et si $t,s \in J$

$$(10) \quad E[\tilde{Y}_{2n-1}(t) - \tilde{Y}_{2n-1}(s)]^4 = E[\tilde{Z}_{2n-1}(t) - \tilde{Z}_{2n-1}(t)]^4 \leq A_2(\alpha)\left(1+\frac{1}{\varepsilon_1}\right)|t-s|^{2\alpha}$$

Par conséquent si h vérifie les inégalités $\left(\frac{\varepsilon_1}{2n-1}\right)^{1/\alpha} \leq h \leq \frac{(\varepsilon_1)^{1/\alpha}}{\sqrt{2n}}$,

si $[s,s+mh] \subset J$ où m est un entier ≥ 1, il vient en tenant compte du (9) et de (10) et en appliquant le lemme (2-6) aux variables aléatoires $\xi_k = \tilde{Y}_{2n-1}(s+kh) - \tilde{Y}_{2n-1}(s+(k-1)h)$ $1 \leq k \leq m$:

(11) $P(\sup_{t\in[s,s+mh]} |\tilde{Y}_{2n-1}(t)-\tilde{Y}_{2n-1}(s)| \geq (3+c)(\varepsilon_1)^{1/\alpha})$

$$\leq P(\sup_{i\leq m} |\tilde{Y}_{2n-1}(s+ih) - \tilde{Y}_{2n-1}(s)| > (\varepsilon_1)^{1/\alpha})$$

$$\leq K(4,2\alpha)\ A_2\ (\alpha)(1+\frac{1}{\varepsilon_1})\ \frac{(mh)^{2\alpha}}{\varepsilon_1^{4/\alpha}}$$

Choisissons maintenant $0<\delta<1$ de sorte que

(12) $$K(4,2\alpha)A_2(\alpha)\ (1+\frac{1}{\varepsilon_1})\ \frac{\delta^{2\alpha}}{\varepsilon_1^{4/\alpha}} < \varepsilon$$

Comme $\alpha>1/2$ il existe un entier n_1 tel que pour $n\geq n_1$ il existe un réel h_n et un entier m_n tels que $m_n\ h_n = \delta$ et $\frac{(\varepsilon_1)^{1/\alpha}}{(2n-1)^{1/\alpha}} \leq h_n \leq \frac{(\varepsilon_1)^{1/\alpha}}{\sqrt{2n}}$

Donc d'après (11) et (12) pour $n\geq n_1$ on a

$$P(W_J\ (\tilde{Y}_{2n-1}\ \delta) > \eta) < \varepsilon$$

Comme pour tous $n\geq 1$ $t\in\mathbb{R}$ on a de plus :

$$\tilde{Y}_{2n}\ (t) = \frac{\sqrt{2n}}{\sqrt{2n+1}}\ \tilde{Y}_{2n-1}(t) + \frac{\Delta_n(t)}{\sqrt{2n+1}} \quad \text{où} \quad |\Delta_n(t)| \leq 2$$

on en déduit facilement (ii).

Les résultats de lapremière partie et la relative compacité pour la topologie de la convergence faible sur $\mathcal{D}(J)$, des lois $\mathcal{L}(\tilde{Y}_n)$ de la suite de processus $(\tilde{Y}_n(t)\ t \in J)\ n\geq 1$ permettent de conclure que la suite $(\tilde{Y}_n)\ n\geq 1$ converge faiblement vers la loi d'un processus gaussien Y, centré presque sûrement à trajectoire continue et ayant les propriétés de non-dégénérescence précisés dans l'énoncé du théorème (2-1).

De plus, de la proposition (2-8), il résulte que pour tous $s,t \in J$ $0<\alpha<1$, on a :

$$E\,[Y(s) - Y(t)]^2 \leq A_2(\alpha)\ |t-s|^\alpha$$

d'où découle [12] la dernière affirmation du théorème (2-1).
La suite de ce paragraphe va être consacrée à la preuve des propositions (2-7) et (2-8). Pour cela il nous sera nécessaire d'introduire et d'étudier les propriétés de certains opérateurs :

B) Pour toute fonction f mesurable bornée sur X^2 tout $u \in \mathbb{R}$, $s,t \in J$, $x = (x_1,x_2) \in X^2$ on définit :

$$P_{s,t}(u)\ f(x_1,x_2) = \int e^{i\,u[1_I(g^s(\omega).x_1)-1_I(g^t(\omega).x_2)]}\ f(g^s(\omega).x_1,\ g^t(\omega).x_2)\ P(d\omega)$$

Notons que $P_{t,t}(o)$ n'est plus un opérateur quasi-compact sur $\mathcal{C}(X^2)$ ainsi que l'est $P_{s,t}(o)$ $s \neq t$. Nous ferons donc agir $P_{s,t}(u)$ ou les puissances de $P_{s,t}(u)$ sur un autre espace de Banach.

Pour $0<\alpha\leq 1$ et $f \in \mathcal{C}(X^2)$

soit $$m_\alpha(f) = \sup_{\substack{x,y\in X^2\\ x\neq y}} \frac{|f(x) - f(y)|}{d_2^\alpha\ (x,y)}$$

et $$\mathcal{L}_\alpha = \{f \in \mathcal{C}(X^2)\ ;\ \|f\|_\alpha = |f| + m_\alpha(f) < +\infty\}$$

$\mathcal{L}_\alpha$ est une algèbre de Banach unitaire munie de la norme $\|\ \|_\alpha$.

Désignons par $\mathcal{L}(\mathcal{L}_\alpha,\mathcal{L}_\alpha)$ l'espace des applications linéaires continues de $\mathcal{L}_\alpha$ dans $\mathcal{L}_\alpha$. Si $T \in \mathcal{L}(\mathcal{L}_\alpha,\mathcal{L}_\alpha)$ on note $\|T\|_\alpha = \sup_{\|f\|_\alpha=1} \|Tf\|_\alpha$. Nous pouvons alors énoncer les deux propositions suivantes :

Proposition (2-9)

1) *Pour tout* $0<\alpha<1$ $P^2_{s,t}(u)$ *est un opérateur continu de* $\mathcal{L}_\alpha$ *dans* $\mathcal{L}_\alpha$ *l'application* $u \to P^2_{s,t}(u)$ *de* $\mathbb{R}$ *dans* $\mathcal{L}(\mathcal{L}_\alpha,\mathcal{L}_\alpha)$ *est analytique ; de plus on a pour tout* $k\geq 0$

$$\sup_{\substack{s,t\in J\\ u\in\mathbb{R}}} \left\|\frac{d^k}{du^k}(P^2_{s,t}(u))\right\|_\alpha < +\infty$$

et il existe une constante $C(\alpha)$ *telle que*

$$\sup_{s,t\in J} \|P^2_{s,t}(u) - P^2_{s,t}(u_1)\|_\alpha \leq C(\alpha)\ |u-u_1| \quad u,u_1 \in \mathbb{R}.$$

2) *Pour tous* $0<\alpha<1$ $0<r<1$ *il existe une constante* $C'(\alpha)$ *telle que pour toute fonction* $\Phi \in \mathcal{L}_\alpha$ *et tous* $s,t \in J$ *on ait pour* $u \in \mathbb{R}$, $k\geq 0$

$$\left\|\frac{d^k}{du^k}[P^2_{s,t}(u)]\Phi - \frac{d^k}{du^k}[P^2_{s,s}(u)]\Phi\right\|_{(1-r)\alpha} \leq C'(\alpha)\ |t-s|^{r\alpha}\|\Phi\|_\alpha$$

Proposition (2-10)

Pour tout $0<\alpha<1$, $P^2_{s,t}(u)$ *est un opérateur quasi-compact de* $\mathcal{L}_\alpha$ *dans* $\mathcal{L}_\alpha$. *Il existe de plus un voisinage compact* W *de* o *dans* $\mathbb{R}$ *tel que : pour tous* $u \in W, f \in \mathcal{L}_\alpha$, $n\geq 1$ *on ait :*

$$P^{2n}_{s,t}(u)f = [k_{s,t}(u)]^n \quad N_{s,t}(u)f + Q^n_{s,t}(u)f$$

où $k_{s,t}(u)$ *est l'unique valeur propre de plus grand module de* $P^2_{s,t}(u)$
$N_{s,t}(u)$ *est la projection sur le sous-espace propre de dimension* 1 *correspondant à* $k_{s,t}(u)$
$Q_{s,t}(u)$ *est un opérateur de* $\mathcal{L}_\alpha$ *dans* $\mathcal{L}_\alpha$ *vérifiant* $Q_{s,t}(u)\ N_{s,t}(u) = 0$.

Il existe par ailleurs une constante $o<\rho_2'(\alpha)<1$ *telle que* $Q_{s,t}(u)$ *ait un rayon spectral* $r_\alpha(Q_{s,t}(u))$ *satisfaisant à*

$$\sup_{\substack{(s,t)\in J\times J\\ u\in W}} r_\alpha(Q_{s,t}(u)) \leq \frac{1+2\rho_2'(\alpha)}{3} < 1$$

et telle que

$$\inf_{\substack{(s,t)\in J\times J \\ u\in W}} |k_{s,t}(u)| \geq \frac{2+\rho_2'(\alpha)}{3}$$

Par ailleurs les applications $u \to k_{s,t}(u)$, $u \to N_{s,t}(u)$ $u \to Q_{s,t}(u)$ *sont analytiques sur* $\overset{\circ}{W}$

Donnons tout d'abord la

démonstration de la proposition (2-9) :

Soient R_k $k = 1,2,3,4$ les sous-ensembles de X^2 définis par $R_1 = I \times CI$, $R_2 = CI \times I$, $R_3 = CI \times CI$, $R_4 = I \times I$

Pour toute fonction f mesurable bornée sur X^2, $s,t \in J$, $u \in \mathbb{R}$ on a l'égalité :

$$P_{s,t}(u)f(s) = e^{iu}\, T_{1,s,t}f(x) + e^{-iu}T_{2,s,t}f(x) + T_{3,s,t}f(x) + T_{4,s,t}f(x)$$

$x \in X^2$

où $T_{k,s,t}f(x) = \int 1_{R_k}(gx)f(gx)\ \mu^2_{s,t}(dg)$ $\qquad k = 1,2,3,4.$

Pour établir la proposition (2-9) il suffit de montrer que pour tout $k = 1,2,3,4$ $\quad T_{k,s,t}$ vérifie les propriétés suivantes :

(i) Pour tout $0<\alpha<1$, $T^2_{k,s,t}$ est un opérateur borné sur $\mathcal{L}_\alpha$, et

$$\sup_{s,t\in J\times J} \|T^2_{k,s,t}\|_\alpha < +\infty$$

(ii) Pour tout $0<\alpha<1$, il existe une constante $C_k(\alpha)$ telle que pour toute fonction $f \in \mathcal{L}_\alpha$, tout $0<r<1$ $s,t \in J$, on ait :

$$\|T^2_{k,s,t} - T^2_{k,s,s} f\|_{(1-r)\alpha} \leq C_k(\alpha)\ |t-s|^\alpha\ \|f\|_\alpha$$

Nous nous contenterons d'établir (i) et (ii) dans le cas où k=4, la démonstration étant analogue pour les autres valeurs de k. Nous nous appuierons pour cela sur le lemme (2-6).

Lemme (2-6) :

1) *Il existe une constante* C_4 *telle que pour* $x,y \in X^2$

$$\sup_{s,t\in J} \int |1_{R_4}(gx) - 1_{R_4}(gy)| \ \mu^2_{s,t}(dg) \leq C_4 \ d_2(x,y)$$

2) *Il existe une constante* $C'_4(\alpha)$ *telle que pour* $f \in \mathcal{L}_\alpha$, $s,t \in J$ *on ait :*

$$\sup_{x\in X^2} |T^2_{4,s,t} f(x) - T^2_{4,s,s} f(x)| \leq C'_4(\alpha) \ \|f\|_\alpha \ |s-t|^\alpha$$

que nous prouverons par la suite.

Soit $f \in \mathcal{L}_\alpha$ on a facilement : pour $x,y \in X^2$, $s,t \in J$

$$|T^2_{4,s,t} f(x) - T^2_{4,s,t} f(y)| \leq m_\alpha(f) \int d_2^\alpha(gx,gy) \ \mu^2_{s,t}(dy)$$

$$+ |f| \int |1_{R_4}(gx) - 1_{R_4}(gy)| \ \mu^2_{s,t}(dg)$$

d'où il résulte puisque $K(\alpha) = \sup_{\substack{x,y\in X^2 \\ s,t\in J}} \int \frac{d_2^\alpha(gx,gy)}{d_2^\alpha(x,y)} \ \mu^2_{s,t}(dg) < +\infty$

et en raison du lemme (2-6) que pour $s,t \in J$

(13) $m_\alpha \ (T^2_{4,s,t}(f)) \leq \|f\|_\alpha \ (K(\alpha) + C_4)$

Comme d'autre part on a:

$|T^2_{4,s,t} f| \leq |f|$

la propriété (i) est démontrée pour $T_{4,s,t}$.

Prouvons maintenant (ii). Pour $x,y \in X^2$ $s,t \in J$ on a, en tenant compte de (13) et du lemme (2-6) 2)

$$|[T^2_{4,s,t}f(x) - T^2_{4,s,s}f(x)] - [T^2_{4,s,t}f(y) - T^2_{4,s,s}f(y)]|$$

$$\leq 2 \|f\|_\alpha \ \mathrm{Inf}(C'_4(\alpha)\ |s-t|^\alpha\ ,\ (K(\alpha) + C_4)\ d_2^\alpha(x,y))$$

$$\leq 2 \|f\|_\alpha \ \sup(C'_4(\alpha),\ K(\alpha) + C_4)\ \mathrm{Inf}(|s-t|^\alpha,\ d_2^\alpha(x,y))$$

d'où il résulte que pour $0<r<1$

$$(14) \quad m_{(1-r)\alpha}(T^2_{4,s,t}f - T^2_{4,s,s}f) \leq 2\|f\|_\alpha \sup(C'_4(\alpha),\ K(\alpha) + C_4)\ |s-t|^{r\alpha}$$

Cette inégalité, jointe au 2) du lemme (2-6) permet de conclure.

Pour terminer, donnons la

<u>démonstration du lemme (2-6)</u>

Pour $x \in P(\mathbb{R}^2)$ notons $\theta(x)$ l'angle de l'intervalle $]-\pi/2, \pi/2]$ tel que le vecteur $\binom{\cos\theta(x)}{\sin\theta(x)} \in \mathbb{R}^2$ ait pour image x dans $P(\mathbb{R}^2)$. Soit de plus $F(x) = \int_{-\infty}^{x} p(q)\,dq$ la fonction de répartition de μ.

On a alors : si $x = (x_1,x_2)$, $y = (y_1,y_2) \in X^2$

$$\int |1_{R_4}(gx) - 1_{R_4}(gy)|\ \mu^2_{s,t}(dg) = \iint \Big| 1_{\{q_1 \leq \mathrm{Inf}(s + \frac{1}{q_2-s-tg\theta(x_1)},\ t + \frac{1}{q_2-t-tg\theta(x_2)})\}}$$

$$- 1_{\{q_1 \leq \mathrm{Inf}(s + \frac{1}{q_2-s-tg\theta(y_1)},\ t + \frac{1}{q_2-t-tg\theta(y_2)})\}} \Big|\ p(q_1)p(q_2)dq_1dq_2$$

d'où il résulte que

$$\int |1_{R_4}(gx) - 1_{R_4}(gy)|\, \mu^2_{s,t}(dg)$$

$$= \int p(q_2)\, |\, \text{Inf}\{F(s + \frac{1}{q_2-s-tg\theta(x_1)}),\ F(t + \frac{1}{q_2-t-tg\theta(x_2)})\}$$

$$- \text{Inf}\ \{F(s + \frac{1}{q_2-s-tg\theta(y_1)}),\ F(t + \frac{1}{q_2-t-tg\theta(y_2)})\}|\, dq_2$$

$$\leq \int |F(s + \frac{1}{q_2-s-tg\theta(x_1)}) - F(s + \frac{1}{q_2-s-tg\theta(y_1)})|\, p(q_2)\, dq_2$$

$$+ \int |F(t + \frac{1}{q_2-t-tg\theta(x_2)}) - F(t + \frac{1}{q_2-t-tg\theta(y_2)})|\, p(q_2)\, dq_2$$

Or on a :

$$\int |F(s + \frac{1}{q_2-s-tg\theta_1}) - F(s + \frac{1}{q_2-s-tg\theta_1'})|\quad p(q_2)\, dq_2$$

$$= \int |\int_{tg\theta_1}^{tg\theta_2} p(s + \frac{1}{u-y})\, \frac{du}{(u-y)^2}|\, p(y+s)\, ds$$

d'où l'on déduit, puisque p est à support compact, qu'il existe une constante K_1 telle que :

$$\sup_{s\in S} \int |F(s + \frac{1}{q_2-s-tg\theta_1}) - F(s + \frac{1}{q_2-s-tg\theta_1'})|\, p(q_2)\, dq_2 \leq K_1\ |\sin(\theta_1-\theta_1')|$$

On a donc :

$$(15)\quad \sup_{s,t\in J} \int |1_{R_4}(gx) - 1_{R_4}(gy)|\, d\mu^2_{s,t}(g) \leq K_1\quad d(x,y)$$

d'où le 1) du lemme (2-6).

Par ailleurs, on a pour $x = (x_1,x_2)\in X^2$ $s,t\in J$, $f \in \mathcal{L}_\alpha$

$$|T^2_{4,s,t}f(x) - T^2_{4,s,s}f(x)| \leq m_\alpha(f) \int d_1^\alpha(g_2^t(\omega)g_1^t(\omega).x_2, g_2^s(\omega)g_1^s(\omega)x_2)P(d\omega)$$

$$+ \quad |f| \int |1_I(g_2^t(\omega)g_1^t(\omega)x_2) - 1_I(g_2^s(\omega)g_1^s(\omega)x_2)|\ P(d\omega).$$

Or il existe une constante K_2 telle que

$$\sup_{x_2 \in X} \int d_1^{\alpha}(g_2^t(\omega)\, g_1^t(\omega)\, x_2, g_2^s(\omega) g_1^s(\omega) x_2) P(d\omega) \leq K_2 |t-s|^{\alpha}$$

et une constante K_3 telle que

$$\sup_{x_2 \in X} \int | 1_I(g_2^t(\omega) g_1^t(\omega) x_2) - 1_I(g_2^s(\omega) g_1^s(\omega) x_2) | \, P(d\omega)$$

$$= \sup_{x_2 \in X} \int p(q) \left| F(t + \frac{1}{q-t-tg\theta(x_2)}) - F(s + \frac{1}{q-s-tg\theta(x_2)}) \right| dq$$

$$\leq K_3 \, |t-s|$$

F admettant une dérivée p à support compact.

Par conséquent on a :

$$(16) \quad \sup_{x \in X^2} |T^2_{4,s,t} f(x) - T^2_{4,s,s} f(x)| \leq \|f\|_{\alpha} \, (K_2 |t-s|^{\alpha} + K_3 \, |t-s|)$$

ce qui achève la preuve du lemme (2-6) et de la proposition (2-9).

Démonstration de la proposition (2-10):

1) On commence par établir que les conclusions de la proposition (2-10) sont vérifiées pour la valeur u=0.
   Pour cela, on utilise le

   lemme (2-7) : *Pour tout* $0<\alpha<1$ *il existe une constante* $0<\rho(\alpha)<1$ *telle que*

$$\lim_n \left[\sup_{\substack{s,t \in J \\ x \neq y \in X^2}} \int \frac{d_2^{\alpha}(gx, gy)}{d_2^{\alpha}(x,y)} \, \mu^n_{s,t}(dg) \right]^{1/n} = \rho(\alpha).$$

<u>démonstration du lemme (2-7)</u> :

on a l'inégalité :

$$(17)\quad \sup_{\substack{s,t\in J\\ x\neq y\in X2}} \int \frac{d_2^{\alpha}(gx,gy)}{d_2^{\alpha}(x,y)}\ \mu^n_{s,t}(dy) \leq 2 \sup_{\substack{s\in J\\ x_1\neq y_1\in X}} \int \frac{d_1^{\alpha}(gx_1,gy_1)}{d_1^{\alpha}(x_1,y_1)}\ \mu^n_s(dy)$$

et de plus

$$(18)\quad \sup_{\substack{s\in J\\ x_1\neq y_1\in X}} \int \frac{d_1^{\alpha}(gx_1,gy_1)}{d_1^{\alpha}(x_1,y_1)}\ \mu^n_s(dg) \leq \sup_{\substack{s\in J\\ x_1\in X}} \int \frac{1}{\|gx_1\|^{2\alpha}}\ \mu^n_s(dg)$$

Considérons alors l'opérateur $T_s(2\alpha)$ sur $\mathcal{C}(X)$ défini par

$$(19)\quad T_s(2\alpha)f(\bar{x}) = \int \sigma^{2\alpha}(g,\bar{x})\ f(g.\bar{x})\ \mu_s(dg) \qquad \bar{x}\in X,\ f\in\mathcal{C}(X),\ s\in J$$

où $\sigma(g,\bar{x}) = \frac{1}{\|gx\|}$, $\|\ \|$ étant la norme euclidienne associée au produit scalaire canonique sur $\mathbb{R}^2$, et x un vecteur de norme 1 d'image $\bar{x}$ dans $P(\mathbb{R}^2)$.

Pour $0<\alpha<1$, l'opérateur $T_s(2\alpha)$ est de rayon spectral $r_s(2\alpha)$ strictement inférieur à 1 (voir l'appendice). L'application $s \to T_s(2\alpha)$ est de plus continue ; il en résulte alors que l'application $s \to r_s(2\alpha)$ est semi-continue supérieurement et donc que pour $0<\alpha<1$

$$\sup_{s\in J} r_s(2\alpha) < 1$$

d'où l'on déduit que (20) $\lim_n \left[\sup_{\substack{s\in J\\ x\in X}} T^n_s(2\alpha)e(x)\right]^{1/n} < 1.$

On obtient le lemme (2-7) en tenant compte de (17), (18), (19), (20).

Nous en déduirons alors immédiatement le

lemme (2-8) : *Pour tout* $0<\alpha<1$ *il existe un entier* $n_o(\alpha)\geq 1$ *et une constante* $0<\rho_1(\alpha)<1$ *tels que :*

$\forall f\in\mathcal{L}_\alpha$, $\forall$ $s,t\in J$

$$\|P^{n_o(\alpha)}_{s,t}(0)\, f\|_\alpha \leq \rho_1(\alpha)\, \|f\|_\alpha + |f|$$

Par ailleurs si L est une partie bornée de $(\mathcal{L}_\alpha, \|\ \|_\alpha)$ $P^n_{s,t}(0)(L)$ est une partie bornée et équicontinue de $\mathcal{C}(X^2)$ et donc d'après le théorème d'Ascoli une partie compacte de $(\mathcal{C}(X^2), |\ |)$.

En tenant compte du fait que $P_{s,t}(0)$ est une contraction de $(\mathcal{C}(X^2), |\ |$, on déduit alors du lemme (2-8), et de la remarque précédente, à l'aide du théorème de Ionescu-Tulcea et Marinescu [13] que $P_{s,t}(0)$ est un opérateur quasi-compact sur $\mathcal{L}_\alpha$. Comme d'après la proposition (2-4), 1 est la seule valeur propre de module 1 de $P_{s,t}(0)$, nous pouvons conclure que

(21) $\forall n>0$ $P^n_{s,t}(0) = \nu_{s,t} + (Q'_{s,t})^n$

où $Q'_{s,t}$ est un opérateur de $(\mathcal{L}_\alpha, \|\ \|_\alpha)$ de rayon spectral strictement inférieur à 1, et tel que $Q'_{s,t}1 = 0$.

Nous achèverons la première partie de la démonstration de la proposition (2-10) en montrant que $\sup\limits_{s,t\in J} r_\alpha(Q'_{s,t})<1$.

Pour λ tel que $|\lambda| = 1$ et $t,s\in J$ notons $R(\lambda,Q'_{s,t})$ la résolvante $(I - \lambda Q'_{s,t})^{-1}$ sur $\mathcal{L}_\alpha$. On a alors le

lemme (2-9) :

$$\sup_{\substack{t,s\in J\times J\\ |\lambda|=1}} \|R(\lambda,Q'_{s,t})\|_\alpha = C_5(\alpha) < +\infty$$

<u>démonstration du lemme (2-9)</u> :

Il suffit d'établir que pour toute suite $(\lambda_n, s_n t_n)_{n\geq 1} \in U\times J\times J$ où $U = \{z\in C \; ; \; |z|=1\}$ on a $\sup_{n\geq 1} \|R(\lambda_n, Q_{s_n,t_n}\|_\alpha < +\infty$

Soit $g \in \mathcal{L}_\alpha - \{0\}$; comme aucun nombre complexe de module 1 n'est dans le spectre de Q'_{s_n,t_n}, pour tout $n\geq 1$ il existe une fonction $f_n \in \mathcal{L}_\alpha$ telle que :

$\lambda_n f_n - Q_{s_n,t_n} f_n = g.$

Posons $f'_n = \frac{f_n}{|f_n|}$; on a :

$$(22) \quad f'_n = \frac{1}{\lambda_n} Q_{s_n,t_n} f'_n + \frac{g}{|f_n|}$$

d'où il résulte d'après le lemme (2-8) et (21) :

$$\|f'_n\|_\alpha \leq \rho_1(\alpha) \|f'_n\|_\alpha + 2 + \frac{\|g\|_\alpha}{|f_n|}$$

c'est-à-dire :

$$(23) \quad \|f'_n\|_\alpha \leq \frac{2 + \frac{\|g\|_\alpha}{|f_n|}}{1 - \rho_1(\alpha)} \qquad n \geq 1.$$

Montrons que $\sup_n |f_n| < +\infty$. Sinon d'après (23), la suite $\|f'_n\|_\alpha \; n\geq 1$ est bornée, et il existe une sous-suite $(n_k)_{k\geq 1}$ de $\mathbb{N}$ telle que la suite $(f_{n_k})_{k\geq 1}$ de $\mathcal{L}_\alpha$ converge vers une fonction f de $\mathcal{L}_\alpha$. On peut de plus supposer sans inconvénient que $\lim_k \lambda_{n_k} = \lambda_o \in U$ et $\lim_k (s_{n_k}, t_{n_k}) = (s_o, t_o) \in J\times J$.

On a alors :

$$(2\text{-}4) \quad \lim_k \; |Q'_{s_{n_k}, t_{n_k}} f_{n_k} - Q'_{so,to} f| = 0.$$

En effet, on peut écrire :

$$|Q'_{s_{n_k}, t_{n_k}} f_{n_k} - Q'_{s_o,t_o} f| \leq |Q'_{s_{n_k} t_{n_k}} f - Q'_{s_o,t_o} f| + |Q'_{s_{n_k},t_{n_k}} (f_{n_k} - f)|$$

$$\leq |P_{s_{n_k},t_{n_k}}(o) f - P_{s_o,t_o} f| + |\nu_{s_{n_k},t_{n_k}}(f) - \nu_{s_o,t_o}(f)| + 2|f_{n_k} - f|$$

Comme $\lim_k |P_{s_{n_k},t_{n_k}}(0) f - P_{s_o,t_o} f| = 0$ et que puisque pour tous $s,t \in J$ $\mu_{s,t}$ admet une unique probabilité invariante $\nu_{s,t}$ sur X^2 on a également $\lim_k |\nu_{s_{n_k},t_{n_k}}(f) - \nu_{s_o,t_o}(f)| = 0$.

(24) est donc établie.

En passant à la limite suivant la sous-suite $(n_k)_{k\geq 1}$ dans l'égalité (22) on obtient alors :

$f = \frac{1}{\lambda_o} Q'_{t_o,s_o} f$ avec $|f| = 1$, ce qui est impossible car $r(Q'_{t_o,s_o})<1$

Par conséquent on a $d = \sup_{n\geq 1} |f_n| < +\infty$

L'inégalité (23) peut alors s'écrire :

$$||f_n||_\alpha \leq \frac{2d + ||g||_\alpha}{1 - \rho_1(\alpha)} \quad n\geq 1$$

ou encore

$$||R(\lambda_n, Q'_{s_n,t_n})g||_\alpha \leq \frac{2d + ||g||_\alpha}{1 - \rho_1(\alpha)} \quad n\geq 1$$

Le théorème de Banach-Steinhaus permet alors de conclure que :

$$(25) \quad \sup_{n\geq 1} ||R(\lambda_n, Q'_{s_n,t_n})||_\alpha < +\infty$$

ce qui établit le lemme (2-9).

Du lemme (2-9), il résulte que pour tout $\lambda \in \Pi$ et tout $\mu \in \mathbb{C}$ tel que $|\mu| < \frac{1}{C_5(\alpha)}$ $\lambda + \mu$ appartient à l'ensemble résolvant de $Q'_{s,t}$ pour tous $s,t \in J$. Il existe donc un réel $\rho'_2(\alpha)<1$ tel que pour tous $s,t \in J$, le spectre de $Q'_{s,t}$ soit contenu dans $\{z \in \mathbb{C} / |z| \leq \rho'_2(\alpha)\}$. Les conclusions de la proposition (2-10) sont donc prouvées pour la valeur $u=0$ (on pose $Q_{s,t} = Q'^2_{s,t}$).

2) La théorie des perturbations analytiques d'opérateurs [10],[18] va nous permettre de terminer la preuve de la proposition (2-10).

Soit $N_{1,s,t}$ l'opérateur défini par $N_{1,s,t}(f) = \nu_{s,t}(f)\, e$. $f \in \mathcal{L}_\alpha$, où e est la fonction identique à 1 sur X^2.

Pour $|z| > \rho'_2(\alpha)$ et $z \neq 1$, la résolvante de $P^2_{s,t}(0)$ est :

(26) $R_{s,t}(z) = \frac{1}{z-1} N_{1,s,t} + \sum_{n \geq 0} \frac{Q^{n+1}_{s,t}}{z^{n+1}}$

(27) Si $\|P^2_{s,t}(0) - P^2_{s,t}(u)\|_\alpha < \frac{1}{\|R_{s,t}(z)\|_\alpha}$

La série $\sum_{n=0}^{+\infty} R_{s,t}(z) \{[P^2_{s,t}(u) - P^2_{s,t}(0)]\, R_{s,t}(z)\}^k$ converge dans $\mathcal{L}(\mathcal{L}_\alpha, \mathcal{L}_\alpha)$ et détermine la résolvante $R_{s,t}(z,u)$ de $P^2_{s,t}(u)$.

Considérons alors les cercles I_1 et I_2 de centre 1, et 0 respectivement et de rayons :

$$r_1(\alpha) = \frac{1 - \rho'_2(\alpha)}{3},$$

$$r_2(\alpha) = \frac{1 + 2\,\rho'_2(\alpha)}{3}$$

De plus soit $\delta > 0$ tel que $\delta < r_1(\alpha)$ et $\rho'_2(\alpha) + \delta < r_2(\alpha)$;

et soit $M_\delta = \sup_{s,t \in J} \ \sup_{z \in \{z;|z|>\rho'_2(\alpha)+\delta;\ |z-1|>\delta\}} \|R_{s,t}(z)\|_\alpha < +\infty$ d'après (26).

D'après la proposition (2-9) 1) si $|u| < \frac{1}{C(\alpha) M_\delta}$

on a $\|P^2_{s,t}(u) - P^2_{s,t}(0)\|_\alpha < \frac{1}{M_\delta}$ pour $s,t \in J$ et donc les cercles I_1 et I_2 appartiennent à l'ensemble résolvant de $P_{s,t}(u)$ $s,t \in J$.

Considérons alors les projections :

(28) $N_{1,s,t}(u) = \frac{1}{2i\pi} \int_{I_1} R_{s,t}(z,u)\, dz$

(29) $N_{2,s,t}(u) = \frac{1}{2i\pi} \int_{I_2} R_{s,t}(z,u)\, dz$

Dès que (30) $\|N_{1,s,t}(u) - N_{1,s,t}(0)\|_\alpha < 1$ l'image $E_{s,t,u}$ de $N_{1,s,t}(u)$ est comme celle $N_{1,s,t}(0)$ de dimension 1 [10] et on a :

(31) $P_{s,t}(u)\ N_{1,s,t}(u)\ e_{s,t,u} = N_{1,s,t}(u)\ P_{s,t}(u)\ e_{s,t,u} = k_{s,t}(u)\ e_{s,t,u}$
où $e_{s,t,u} \in \mathcal{L}_\alpha$ engendre $E_{s,t,u}$

D'après (26), (28) et la proposition (29) 1) on a pour $s,t \in J$ et $|u| < \frac{1}{C(\alpha)M_\delta}$

$$\|N_{1,s,t}(u) - N_{1,s,t}(0)\|_\alpha \leq r_1(\alpha) \sum_{k\geq 1} M_\delta (C_\alpha |u| M_\delta)^k = r_1(\alpha) M_\delta \frac{C_\alpha |u| M_\delta}{1 - C_\alpha |u| M_\delta}$$

et donc (30) est réalisée dès que $u \in W$ où :

$$W = \{u \in \mathbb{R} \ ; \ |u| < \frac{1}{C(\alpha)M_\delta [2r_1(\alpha)M_\delta + 1]}\}$$

Par ailleurs on a pour tout $n \geq 1$ $s,t \in J$ $u \in W$

$$(P^2_{s,t}(u))^n = [P^2_{s,t}(u)]^n\ N_{1,s,t}(u) + [P^2_{s,t}(u)]^n\ N_{2,s,t}(u)$$

$$= [k_{s,t}(u)]^n\ N_{1,s,t}(u) + Q^n_{s,t}(u)$$

où $Q^n_{s,t} = \frac{1}{2i\pi} \int_{I_2} z^n R_{s,t}(z,u)\, dz$

et où donc $r_\alpha(Q_{s,t}(u)) \leq r_2(\alpha) < 1$

En outre $k_{s,t}(u) = \dfrac{P^2_{s,t}(u)\ N_{1,s,t}(u)\ e(x_o,x_o)}{N_{1,s,t}(u)\ e\ (x_o,x_o)}$

et $|k_{s,t}(u)| > 1 - r_1(\alpha) = \dfrac{2 + \rho_2(\alpha)}{3}$

L'analyticité des applications $u \to N_{1,s,t}(u)$, $u \to Q_{s,t}(u)$, $u \to k_{s,t}(u)$ résulte de l'analyticité de l'application $u \to P_{s,t}(u)$ de $\mathring{W}$ dans $\mathcal{L}(\mathcal{L}_\alpha, \mathcal{L}_\alpha)$, qui entraîne celle de $u \to R_{s,t}(z,u)$, et des formules précédentes.
La preuve de laproposition (2-10) est ainsi terminée.

Donnons maintenant la

C) DEMONSTRATION DE LA PROPOSITION (2-7).

Considérons la famille d'opérateurs

$$\tilde{P}_{s,t}(u) = e^{-iu[N(s)-N(t)]} P_{s,t}(u) \qquad s,t \in J \qquad u \in W$$

En raison de la proposition (2-8) (admise pour le moment mais indépendante de la proposition (2-7)) l'application $s \to N(s)$ est Lipschitzienne sur $\mathbb{R}$. Il en résulte que les opérateurs $\tilde{P}_{s,t}(u)$ possèdent les mêmes propriétés que celles décrites dans les propositions (2-9) et (2-10) pour les opérateurs $P_{s,t}(u)$. On modifie les notations adoptées pour décrire les propriétés des opérateurs $P_{s,t}(u)$ en leur adjoignant un tilde. On a alors :

(32) $\quad \tilde{k}'_{s,t}(0) = 0$

De plus la transformée de Fourier de $Z_{2n-1}(t) - Z_{2n-1}(s)$

$n \geq 1 \quad t,s \in J$ est :

(33) $\quad E(e^{iu(\tilde{Z}_{2n-1}(t) - \tilde{Z}_{2n-1}(s))}) = \tilde{P}^{2n}_{s,t}(\frac{u}{\sqrt{2n}})\ e(x_o,x_o) \qquad n \geq 1$

d'où il résulte que :

(34) $\quad E(\tilde{Z}_{2n-1}(t) - \tilde{Z}_{2n-1}(s))^{2p} = \dfrac{1}{(2n)^p} \dfrac{d^{2p}}{du^{2p}} [P^{2n}_{s,t}(0)]\, e\ (x_o,x_o)$

Pour terminer la preuve de la proposition (2-7), nous utiliserons le

lemme (2-10) : *Soient* $0<\alpha<1$, $0<r<1$, $p \geq 0$ *et* K *un compact dans l'ensemble résolvant de* $\tilde{P}^2_{s,t}(u)$ *opérant sur* $\mathcal{L}_\alpha$ *et* $\mathcal{L}_{(1-r)\alpha}$.
Il existe alors une constante $C_p(\alpha,r,K)$ *telle que pour toute fonction* $\Phi \in \mathcal{L}_\alpha$ *on ait l'inégalité :*

$$\sup_{\substack{s,t \in J \\ z \in K}} \left\| \frac{d^p}{du^p} (\tilde{R}_{s,t}(z,0))\Phi - \frac{d^p}{du^p} (\tilde{R}_{s,s}(z,o))\Phi \right\|_{(1-r)\alpha}$$

$$\leq C_p(\alpha,r,K)\ |t-s|^{r\alpha}\ \|\Phi\|_\alpha$$

<u>démonstration du lemme (2-10)</u> :

a) On traite tout d'abord le cas p=0.

Dans $\mathcal{L}(\mathcal{L}_\alpha,\mathcal{L}_\alpha)$ et $\mathcal{L}(\mathcal{L}_{(1-r)\alpha},\mathcal{L}_{(1-r)\alpha})$ on a la relation :

$$(35)\quad \tilde{R}_{s,t}(z,o) - \tilde{R}_{s,s}(z,o) = \tilde{R}_{s,t}(z,o)\,[\tilde{P}^2_{s,t}(o) - \tilde{P}^2_{s,s}(o)]\,\tilde{R}_{s,s}(z,o)$$

$s,t\in J,\quad z\in K$

D'autre part en raisonnant comme dans la preuve du lemme (2-9) on a :

$$(36)\quad \sup_{\substack{s,t\in J\\ z\in K}} \|\tilde{R}_{s,t}(z,o)\|_{(1-r)\alpha} < \infty \quad \text{et}$$

$$\sup_{\substack{s,t\in J\\ z\in K}} \|\tilde{R}_{s,t}(z,o)\|_{\alpha} < +\infty$$

De (35) et de l'analogue de la proposition (2-9) pour $\tilde{P}_{s,t}(u)$ on déduit alors que si $\Phi\in\mathcal{L}_\alpha$, $s,t\in J$ et $z\in K$, on a :

$$(37)\quad \|\tilde{R}_{s,t}(z,o)\Phi - \tilde{R}_{s,s}(z,o)\Phi\|_{(1-r)\alpha}$$

$$\leq \sup_{\substack{s,t\in J\\ z\in K}} \|\tilde{R}_{s,t}(z,o)\|_{(1-r)\alpha}\ \tilde{C}'(\alpha)\ |t-s|^{r\alpha} \times \sup_{\substack{s,t\in J\\ z\in K}} \|\tilde{R}_{s,t}(z,o)\|_\alpha$$

c'est-à-dire le résultat souhaité.

b) Pour $\|\tilde{P}^2_{s,t}(u) - \tilde{P}^2_{s,t}(o)\|_\alpha < 1$ et $\|\tilde{P}^2_{s,t}(u) - \tilde{P}^2_{s,t}(0)\|_{(1-r)\alpha} < 1$ on a l'égalité suivante valable dans $\mathcal{L}(\mathcal{L}_\alpha,\mathcal{L}_\alpha)$ et $\mathcal{L}(\mathcal{L}_{(1-r)\alpha},\mathcal{L}_{(1-r)\alpha})$

$$(38)\quad \tilde{R}_{s,t}(z,u) = \sum_{n\geq 0} \tilde{R}_{s,t}(z)\ \{[\tilde{P}^2_{s,t}(u) - \tilde{P}^2_{s,t}(0)]\ \tilde{R}_{s,t}(z)\}^n \qquad z\in K$$

On en déduit que $\frac{d^k}{du^k}(\tilde{R}_{s,t}(z,o))$ est une somme finie d'opérateurs de la forme :

$$\tilde{R}_{s,t}(z)\ \left(\frac{d^{i_1}}{du^{i_1}}(\tilde{P}^2_{s,t}(o))\ \tilde{R}_{s,t}(z)\right)\left(\frac{d^{i_2}}{du^{i_2}}(\tilde{P}^2_{s,t}(o)\ \tilde{R}_{s,t}(z)\right)\cdots$$

$$\left(\frac{d^{i_j}}{du^{i_j}}(\tilde{P}_{s,t}(o))\ \tilde{R}_{s,t}(z)\right)$$

où $i_1 \geq 1$, $i_2 \geq 1$, .. $i_j \geq 1$ $\qquad i_1+i_1+..+i_j = p$, $\quad 1 \leq j \leq k$.

Les considérations de la partie a) et l'analogue de la proposition (2-9) pour $\tilde{P}_{s,t}(u)$ permettent alors de conclure immédiatement à la validité du lemme (2-10) pour $p \geq 1$.

Soient $0<\alpha<1$, $0<r<1$. L'analogue de la proposition (2-10) pour $\tilde{P}_{s,t}(u)$ permet d'obtenir la décomposition suivante de $\tilde{P}^{2n}_{s,t}(u)$ $s,t \in J$, $u \in W$, valable dans $\mathcal{L}(\mathcal{L}_\alpha,\mathcal{L}_\alpha)$ et $\mathcal{L}(\mathcal{L}_{(1-r)\alpha},\mathcal{L}_{(1-r)\alpha})$.

(39) $\tilde{P}^{2n}_{s,t}(u) = [\tilde{k}_{s,t}(u)]^n \, \tilde{N}_{s,t}(u) + \tilde{Q}^n_{s,t}(u) \qquad n \geq 1$

où (40) $\tilde{N}_{s,t}(u) = \dfrac{1}{2i\pi} \displaystyle\int_{\tilde{\Gamma}_1} \tilde{R}_{s,t}(z,u)\, dz$

(41) $\tilde{Q}^n_{s,t}(u) = \dfrac{1}{2i\pi} \displaystyle\int_{\tilde{\Gamma}_2} z^n \, \tilde{R}_{s,t}(z,u)\, dz$

(42) $\tilde{k}_{s,t}(u) = \dfrac{\tilde{P}^2_{s,t}(u)\, N_{s,t}(u)\, e(x_o,x_o)}{N_{s,t}(u)\, e(x_o,x_o)} = \dfrac{\displaystyle\int_{\tilde{\Gamma}_1} z \; \tilde{R}_{s,t}(z,u) e(x_o,x_o) dz}{\displaystyle\int_{\tilde{\Gamma}_1} \tilde{R}_{s,t}(z,u) e(x_o,x_o)\, dz}$

Du lemme (2-9) et des formules précédentes résulte alors le

<u>lemme (2-11)</u> : *Pour tout* $p \geq 0$ *il existe une constante* $C'_p(r,\alpha)$ *telle que pour toute fonction* $\Phi \in \mathcal{L}_\alpha$

1) $\sup_{s,t \in J} \left\| \dfrac{d^p}{du^p}(\tilde{N}_{s,t}(o))\Phi - \dfrac{d^p}{du^p}(\tilde{N}_{t,t}(o))\Phi \right\|_{(1-r)\alpha} \leq C'(r,\alpha)\, |t-s|^{r\alpha} \, \|\Phi\|_\alpha$

2) $\sup_{s,t \in J} \left\| \dfrac{d^p}{du^p}(\tilde{Q}^n_{s,t}(o))\Phi - \dfrac{d^p}{du^p}(\tilde{Q}^n_{t,t}(o))\Phi \right\|_{(1-r)\alpha}$

$\leq C'_p(r,\alpha)\, \tilde{\rho}^n_2(r,\alpha)\, |t-s|^\alpha \, \|\Phi\|_\alpha$

où $\tilde{\rho}_2(r,\alpha) < 1$ *est le rayon du cercle* $\tilde{\Gamma}_2$.

3) $|\tilde{k}^{(p)}_{s,t}(o) - \tilde{k}^{(p)}_{t,t}(o)| \leq C'_p(r,\alpha)\, |t-s|^\alpha$

En raison de (39) et (32), on a :

$$(43)\quad \frac{d^2}{du^2}[\tilde{P}^{2n}_{s,t}](o)e(x_o,x_o) = n\,\tilde{k}''_{s,t}(o) + \frac{d^2}{du^2}[\tilde{Q}^n_{s,t}](o)e(x_o,x_o)$$

et

$$(44)\quad \frac{d^4}{du^4}[\tilde{P}^{2n}_{s,t}](o)e(x_o,x_o) = 3n(n-1)\,[\tilde{k}''_{s,t}(o)]^2 + n\{k^{(4)}_{s,t}(o)$$

$$+\ 3k^{(3)}_{s,t}(o)\frac{d}{du}[\tilde{N}_{s,t}](o)e(x_o,x_o)+4k^{(2)}_{s,t}(o)\frac{d^2}{du^2}[\tilde{N}_{s,t}](o)e(x_o,x_o)$$

$$+\ \frac{d^4}{du^4}[\tilde{N}_{s,t}](o)e(x_o,x_o) + \frac{d^4}{du^4}[\tilde{Q}^n_{s,t}](o)(x_o,x_o).$$

De plus on a (45) $k''_{t,t}(o) = \lim_n \frac{1}{n}\frac{d^2}{du^2}[\tilde{P}^{2n}_{t,t}](o)e(x_o,x_o)$

$$= 2\lim_n E(\tilde{Z}_{2n-1}(t) - \tilde{Z}_{2n-1}(t))^2 = 0.$$

Des égalités précédentes, du lemme (2-10) et du fait que d'après (34) on a :

$$E(\tilde{Z}_{2n-1}(t) - \tilde{Z}_{2n-1}(s))^{2p} = \frac{1}{2^p n^p}\Big[\frac{d^{2p}}{du^{2p}}(\tilde{P}^{2n}_{s,t})(o)e(x_o,x_o) - \frac{d^{2p}}{du^{2p}}(\tilde{P}^{2n}_{t,t})(o)\,e(x_o,xo)\Big]$$

on déduit que pour tous $0<\alpha_o<1$ $p = 1,2$
Il existe des constantes $C'_p(\alpha,r)$ telles que pour tous $s,t\in I$

$$(46)\quad E(\tilde{Z}_{2n-1}(t) - \tilde{Z}_{2n-1}(s))^{2p} \le C'_p(\alpha,r)\ \{|t-s|^{pr\alpha} + \frac{|t-s|^{r\alpha}}{n}\}$$

Tout $\alpha_o, o<\alpha_o<1$ s'écrivant sous la forme $\alpha_o = r\alpha$

$0<r<1$ $0<\alpha<1$, la proposition (2-7) est ainsi démontrée.

D) <u>DEMONSTRATION DE LA PROPOSITION (2-8)</u>

On commence par construire une approximation de la probabilité $\overline{\sigma}_{o,o}$ de fonction de répartition $N(t) = E(\langle E_t\, e_o,e_o\rangle)$.

Pour cela on considère les opérateurs symétriques sur $[-L,L]$ définis par

$$L_{H_x} = L_H - x\pi_L \qquad x \in \mathbb{R}.$$

π_L étant la projection définie par $\pi_L(u) = u_L$ où $u = (u_n)_{n\in Z} \in \ell^2(Z)$

Soit $(p_n(L,\lambda))_{n\geq 1}$ la solution de l'équation :

$$- u_{n+1} - u_{n-1} + X_n u_n = \lambda u_n \qquad n \in \mathbb{Z}$$

égale à 0 en $-L-1$ et à 1 en $-L$.

Les valeurs propres de la matrice symétrique définissant L_{H_x} sont les zéros du polynome $P_{L+1}(L,\lambda) - x\, P_L(L,\lambda)$.

La relation [1] :

$$(47) \quad 0 < \sum_{k=-L}^{L} p_k^2(L,\lambda) = p_L(L,\lambda) \frac{d}{d\lambda} (p_{L+1}(L,\lambda)) - p_{L+1}(L,\lambda)\frac{d}{d}(p_L(L,\lambda))$$

établit que les racines de $p_{L+1}(L,\lambda) - x\, p_L(L,\lambda)$ sont simples.
Les sous-espaces propres sont donc de dimension 1 et si λ est une valeur propre, le vecteur

$$\left(\frac{p_n(L,\lambda)}{(\sum_{k=-L}^{L} p_k^2(L,\lambda))^{1/2}}\right)_{-L\leq n\leq L}$$

est propre de norme 1.

Par conséquent si ${}^{L}E_x(t)$ est la résolution de l'identité de ${}^{L}H_x$ la probabilité ${}^{L}\sigma_{o,o}^{x}$ de fonction de répartition $\langle {}^{L}E_x(t)e_o,e_o\rangle$ est égale à :

$${}^{L}\sigma_{o,o}^{x} = \sum_{\{\lambda\,;\, p_{L+1}(L,\lambda) = x\, p_L(L,\lambda)\}} \frac{p_o^2(L,\lambda)}{(\sum_{k=-L}^{L} p_k^2(L,\lambda))} \varepsilon_\lambda$$

Montrons que pour tout $x \in \mathbb{R}$ la suite de probabilités $({}^L\sigma^x_{o,o})_{L\geq 1}$ converge étroitement vers la probabilité $\sigma_{o,o}$ de fonction de répartition $\langle F_t\, e_o, e_o\rangle$.

Pour cela on utilise la méthode des moments ; pour tout $p\geq 1$ on a :

$$\int \lambda^p \; {}^L\sigma^x_{o,o}\,(d\lambda) = \langle ({}^L H_x)^p e_o, e_o\rangle$$

Or si $L \geq p+1$ on a : $({}^L H_x)^p e_o = ({}^L H)^p e_o = H^p e_o$

et par conséquent : si $L \geq p+1$

$$\int \lambda^p \; {}^L\sigma^x_{o,o}(d\lambda) = \langle H^p e_o, e_o\rangle = \int \lambda^p \; \sigma_{o,o}(d\lambda)$$

ce qui établit le résultat cherché.

De même si τ est une probabilité sur $\mathbb{R}$ la suite de probabilités

${}^L\sigma^{\tau}_{o,o} = \int {}^L\sigma^x_{o,o}\, d\tau(x) \quad L\geq 1$ converge étroitement vers $\sigma_{o,o}$,

et on a également $\lim\limits_{L} E\,({}^L\sigma^{\tau}_{o,o}) = \bar{\sigma}_{o,o}$.

En choisissant τ égale à la loi de Cauchy sur $\mathbb{R}$, on peut obtenir une expression assez simple de ${}^L\sigma^{\tau}_{o,o}$. En effet pour toute fonction Φ continue à support compact on a, en tenant compte de (47) :

$${}^L\sigma^{\tau}_{o,o}(\Phi) = \frac{1}{\pi}\int \frac{dx}{1+x^2} \sum_{\lambda\in\{p_{L+1}(\lambda)=xp_L(\lambda)\}} \frac{p_o^2(L,\lambda)}{p_L^2(L,\lambda)} \left\{\frac{d}{d\lambda}\left(\frac{p_{L+1}(L,\lambda)}{p_L\;(L,\lambda)}\right)\right\}^{-1} \Phi(\lambda)$$

$$= \frac{1}{\pi}\int_{-\infty}^{+\infty} \frac{p_o^2\;(L,\lambda)}{p_L^2(L,\lambda) + p_{L+1}^2(L,\lambda)}\;\Phi(\lambda)\; d\lambda$$

La dernière égalité est obtenue par changement de variable en remarquant que dans chaque intervalle ouvert défini par la partition de $\mathbb{R}$ par les zéros de $p_L(L,\lambda)$ l'application $\lambda \to \dfrac{p_{L+1}(L,\lambda)}{p_L(L,\lambda)}$ croit strictement de $-\infty$ à $+\infty$ d'après (47), et que ${}^L\sigma_{o,o}$ ne change les zéros de $p_L(L,\lambda)$ car $p_{L+1}(L,\lambda)$ et $P_L(L,\lambda)$ n'ont pas de zéro commun.

Le résultat précédent peut encore s'écrire sous la forme

(48) $$ {}^{L}\sigma^{T}_{o,o}(\Phi) = \frac{1}{\pi}\int_{-\infty}^{\infty}\frac{(<g_o^{\lambda}\, g_1^{\lambda}\, \cdots\, g_{-L}^{\lambda}\, x_o, x_o>)^2}{\|g_L^{\lambda}\, g_{L-1}^{\lambda}\, g_{-L}^{\lambda}\, x_o\|^2}\,\Phi(\lambda)\,d\lambda $$

$< >$ désigne le produit scalaire canonique sur $\mathbb{R}^2$.

Il en résulte que :

(49) $$ E\,({}^{L}\sigma^{T}_{o,o}(\Phi)) = \frac{1}{\pi}\int_{-\infty}^{+\infty}\Phi(\lambda)\int\frac{|<gx_o, x_o>|^2}{\|gx_o\|^2}\,T_{\lambda}^{L}(2)e(g\bar{x}_o)\mu_{\lambda}^{L+1}(dg)\,d\lambda $$

or on peut écrire (voir l'appendice)

$T_{\lambda}^{2L}(2)e = e_{\lambda,2} + R_{\lambda,2}^{L}e \qquad L\geq 1$

où $e_{\lambda,2}\in \mathcal{C}(X)$ est telle que $T_{\lambda}^{2}(2)e_{\lambda,2} = e_{\lambda,2}$ et $R_{\lambda,2}^{L}$ est un opérateur de norme spectrale, strictement inférieure à 1.
Pour $\bar{x}\in X$ on définit $\cos^2\bar{x}$ par $\cos^2\bar{x} = \cos^2 x$ où x est un vecteur de norme 1, d'image $\bar{x}$ dans X.

On a alors :

(50) $$ E({}^{2L}\sigma^{T}_{o,o}(\Phi)) = \frac{1}{\pi}\int_{-\infty}^{+\infty}\Phi(\lambda)\Big[\int\cos^2(g\bar{x}_o)\; e_{\lambda,2}(g\bar{x}_o)\mu_{\lambda}^{2L+1}(dg)\Big]\,d\lambda $$
$$ +\frac{1}{\pi}\int_{-\infty}^{+\infty}\Phi(\lambda)\Big[\int\cos^2(g\bar{x}_o)\; R_{\lambda,2}^{2L}e(g\bar{x}_o)\mu_{\lambda}^{(2L+1)}(dg)\Big]\,d\lambda $$

Comme $\lim\limits_{L}\int\cos^2(g\bar{x}_o)\; e_{\lambda,2}(gx_o)\;\mu_{\lambda}^{2L+1}(dg) = \int\cos^2\bar{x}\; e_{\lambda,2}(\bar{x})\nu_{\lambda}(d\bar{x})$

et $\overline{\lim\limits_{L}}\Big[\int\cos^2(g\bar{x}_o)\; R_{\lambda,2}^{2}e(g.\bar{x}_o)\;\mu_{\lambda}^{2L+1}(dg)\Big] \leq \overline{\lim\limits_{L}}\,\|R_{\lambda,2}^{2L}\| = 0.$

on déduit de (50) que :

$$ \lim_{L} E({}^{2L}\sigma^{T}{}_{o,o}(\Phi)) = \bar{\sigma}_{o,o}(\Phi) = \frac{1}{\pi}\int\Phi(\lambda)\Big[\int\cos^2\bar{x}\; e_{\lambda,2}(\bar{x})\nu_{\lambda}(d\bar{x})\Big]\,d\lambda $$

ce qui établit que $\overline{\sigma_{o,o}}$ admet pour densité la fonction

$\lambda \to \frac{1}{\pi}\int\cos^2\bar{x}\;\; e_{\lambda,2}(\bar{x})\nu_{\lambda}(d\bar{x}).$

La continuité de l'application $\lambda \to T_\lambda^2(2)$ de $\mathbb{R}$ dans l'espace $\mathcal{L}(\mathcal{C}(X),\mathcal{C}(X))$ des applications linéaires continues de $\mathcal{C}(X)$ dans $\mathcal{C}(X)$ entraîne la continuité de $\lambda \to e_{\lambda,2}$ et de $\lambda \to \nu_\lambda$, et par conséquent la densité de $\overline{\sigma_{o,o}}$ est continue.

Paragraphe 3 - LA FORMULE DE THOULESS

En reprenant les notations déjà définies, nous pouvons énoncer le

Théorème (3-1) :

Si μ *est à support compact et charge au moins deux points :*

1) *Pour tout* $t \in \mathbb{R}$, *et tout* $x \in \mathbb{R}^2 - \{o\}$, *on a*

$$P\ p\ s \quad \lim_L \frac{1}{L+1} \operatorname{Log}|P_{L+1}(t)| = \lim_L \frac{1}{L+1} \operatorname{Log} \|g_L^t \cdots g_o^t x\| = \gamma(t)$$

où $$o<\gamma(t) = -\int \log \sigma(g,x)\ \mu_t(dg)\ \nu_t(dx).$$

2) *Pour tout* $t \in \mathbb{R}$, *on a l'égalité suivante "Formule de Thouless"*

$$\gamma(t) = \int \operatorname{Log}|t-x|\ N(dx)$$

Démonstration du théorème (3-1) :

1) Justifions tout d'abord l'affirmation 1)
Si μ charge au moins deux points, le sous-groupe fermé engendré par μ_t est non compact et ne contient pas de sous-groupe d'indice fini ayant une action réductible sur $\mathbb{R}^2$ [14] .

Donc d'après le théorème du Fürstenberg [4] pour tout $x \in R^2-\{0\}$ on a

$$(51) \quad P\ p\ s \lim_{L \to +\infty} \frac{1}{L+1} \log \| g_L^t\, g_{L-1}^t \cdots g_o^t x \| = \gamma(t)$$

$$\text{où } (52) \quad \gamma(t) = -\int \log\ \sigma(g,x)\ \mu_t(dg)\ \nu_t(dx) > 0$$

Par ailleurs, $p_{L+1}(t)$ est un coefficient de la matrice $g_L^t\, g_{L-1}^t \ldots g_o^t$, il résulte alors de [5] que l'on a également

$$(53) \quad P\ p\ s \lim_{L} \frac{1}{L+1} \log |p_{L+1}(t)| = \gamma(t)$$

2) La justification du 2) sera basée sur plusieurs résultats préliminaires :

<u>lemme (3-1)</u> : *Pour tout* $t \in \mathbb{R}$, *l'intégrale* $\int \log |t-x|\ N(dx)$ *est convergente.*

<u>Démonstration du lemme (3-1)</u> :

Pour tous $L \geq 1$, $t \in \mathbb{R}$ $M \in \mathbb{N}$ on a :

$$(54) \quad \frac{1}{L+1} \log |p_{L+1}(t)| = \int \log |t-x|\ \tilde{N}_L(dx) \leq \int \sup(\log |t-x|, -M)\ \tilde{N}_L(dx).$$

P-presque sûrement les probabilités de fonction de répartition $(\tilde{N}_L)_{L \geq 1}$, convergent étroitement vers la probabilité de fonction de répartition N (théorème 1-1) ; elles ont de plus leur support dans un compact fixe J. Par conséquent on déduit de l'inégalité précédente et en tenant compte du 1) que pour tout $M \in \mathbb{N}$:

$$(55) \quad 0 < \gamma(t) \leq \int \sup(\log |t-x|, -M)\ N(dx)$$

ce qui assure puisque N est à support compact
l'intégrabilité par rapport à la probabilité dN
de la fonction $x \to |\log||t-x|$

Soit λ la mesure de Lebesgue sur $\mathbb{R}$; on a alors le

lemme (3-2) :

Pour λ *presque tout* t *on a :*

$$\gamma(t) = \int \log |t-x| \, N(dx)$$

Démonstration du lemme (3-2) :

Comme la suite de fonctions de répartition $(N_L)_{L \geq 1}$, converge étroitement vers la probabilité de fonction de répartition N, on en déduit que pour tout $b > 0$, on a :

$$(56) \quad P\ p\ s \quad \lim_{L} \int_{-b}^{b} dt \left| \int \log |t-x| \, \tilde{N}_L(dx) - \int \log |t-x| N(dx) \right| = 0$$

Le lemme (3-2) se déduit alors de (53) et de (56).

Avant d'énoncer une proposition qui nous sera également utile, précisons une notation :

Pour tout intervalle ouvert $]a,b[$ de $\mathbb{R}$, nous noterons $\mathcal{L}_{loc}(]a,b[)$ l'espace des fonctions f de $]a,b[$ dans $\mathbb{R}$ telle que pour tout intervalle compact $T \subset]a,b[$, il existe une constante $0<\alpha(T)<1$, de sorte que l'application $t \in T \to f(t)$ soit Lipschitzienne d'ordre $\alpha(T)$.
On a alors la

proposition (3-1) : *l'application* $t \to \gamma(t)$ *appartient à* $\mathcal{L}_{loc}(\mathbb{R})$

dont nous donnerons la preuve à la fin du paragraphe 3.

Achevons alors la preuve du théorème (3-1).
Soit $A>0$ tel que $]-A,A[\supset J$.
La fonction $x \to \psi_A(x) = 1_{]-A,A[}(x)\, N(x)$ appartient à l'espace $\mathcal{L}^2(\mathbb{R})$ des fonctions de carré intégrable sur $\mathbb{R}$. Elle admet donc [11] une transformée de Hilbert $\tilde{\psi}_A \in \mathcal{L}^2(\mathbb{R})$ définie par $\tilde{\psi}_A = \lim_{\varepsilon \to 0} \tilde{\psi}_{A,\varepsilon}(t)$
λ presque sûrement, où

$$\tilde{\psi}_{A,\varepsilon}(t) = \frac{1}{\pi} \int_{|t-x|>\varepsilon} \frac{\psi_A(x)\, dx}{t-x}$$

Il résulte du lemme (3-2) par intégration par parties que :

$$\lambda p s \quad \gamma(t) = \log |A-t| - \pi\tilde{\psi}_A(t),$$

c'est-à-dire

$$(57) \quad \lambda p s \quad \tilde{\psi}_A(t) = \Phi_A(t)$$

$$\text{où} \quad (58) \quad \Phi_A(t) = \frac{1}{\pi} (\log |A-t| - \gamma(t)) \in \mathcal{L}^2(\mathbb{R}).$$

D'après la proposition (3-2) et l'égalité précédente, la fonction Φ_A appartient à $\mathcal{L}_{loc}(]-A,A[)$. Si l'on pose

$$\tilde{\Phi}_{A,\varepsilon}(x) = \frac{1}{\pi} \int_{|x-t|>\varepsilon} \frac{\Phi_A(t)}{x-t}\, dt$$

on déduit de la théorie de la transformée de Hilbert [11] dans $\mathcal{L}^2(\mathbb{R})$ que $\lambda p s$ la limite $\tilde{\Phi}_A(x) = \lim_{\varepsilon \to 0} \tilde{\Phi}_{A,\varepsilon}(x)$ existe ; de plus comme Φ_A appartient à $\mathcal{L}_{loc}(]-A,A[)$ la limite précédente existe pour tout $x \in]-A,A[$ et $\check{\Phi}_A \in \mathcal{L}_{loc}(]-A,A[)$.

La formule d'inversion de la transformée de Hilbert permet de conclure que :

λ p s $\quad 1_{]-A,A[}N(x) = \check{\Phi}_A(x)$.

Mais comme $x \to N(x)$ est continue et de plus $\check{\Phi}_A \in \mathcal{L}_{loc}(]-A,A[)$, on a :

$\forall x \in]-A,A[\quad \psi_A(x) = 1_{]-A,A[}\, N(x) = \check{\Phi}_A(x)$.

Ceci établit en particulier puisque A peut être choisi arbitrairement que $N \in \mathcal{L}_{loc}(\mathbb{R})$.

Comme $\psi_A \in \mathcal{L}_{loc}(]-A,A[)$, on en déduit que $\check{\psi}_A(t)$ est définie pour tout $t \in]-A,A[$ et que $\check{\psi}_A \in \mathcal{L}_{loc}(]-A,A[)$.

Les deux membres de (57) sont alors des fonctions continues de t sur $]-A,A[$ et l'on a donc :

(59) $\forall\, t \in]-A,A[\qquad \gamma(t) = \log|A-t| - \pi\, \check{\psi}_A(t)$

Comme on a $\log |A-t| - \pi\check{\psi}_A(t) = \int \log|t-x|\, dN(x)$, par intégration par parties, on obtient donc :

(60) $\forall\, t \in]-A,A[\qquad \gamma(t) = \int \log\, |t-x|\, dN(x)$

et le théorème (3-1) est ainsi démontré.

Avant de prouver la proposition (3-1), énonçons sous forme de proposition un résultat acquis au cours de la démonstration précédente.

Proposition (3-2) : *L'application* $t \to N(t)$ *appartient à* $\mathcal{L}_{loc}(\mathbb{R})$.

Démonstration de la proposition (3-1) : elle est calquée sur la preuve de la proposition (2-10).

1) On commence par établir le

lemme (3-1) : *sous les hypothèses du théorème* (3-1) *pour tout compact*

$T \subset \mathbb{R}$ *il existe un réel* $0<\alpha(T)<1$ *tel que pour* $0<\alpha\leq\alpha(T)$

$$\lim_n \Big[\sup_{\substack{t\in T\\ x\neq y\in X}} \int \frac{d^\alpha(gx,gy)}{d^\alpha(x,y)} \mu_t^n(dg) \Big]^{1/n} = \rho(T,\alpha)$$

avec $\rho(T,\alpha)<1$

<u>démonstration du lemme (3-1)</u> :

Pour $0<\alpha<1$ on a l'inégalité

$$\sup_{\substack{x\neq y\in X\\ t\in T}} \int \frac{d^\alpha(gx,gy)}{d^\alpha(x,y)} \mu_t^n(dg) \leq \sup_{\substack{x\in X\\ t\in T}} \int \frac{1}{\sigma^{2\alpha}(g,x)} \mu_t^n(dg), \qquad n\geq 1.$$

Comme la suite de fonctions $\frac{1}{n}\int \log\sigma(g,x)\,\mu_t^n(dg)$ $n\geq 1$, converge uniformément sur $X\times T$ vers $\gamma(t)>0$ (2°) du théorème (3-1))
il existe un entier N_o tel que :

$$\beta = \operatorname*{Inf}_{\substack{t\in T\\ x\in X}} \int \log\sigma(g,x)\,\mu_t^{N_o}(dg) > 0$$

Si $\delta(g) = \sup_{x\in X} \sigma(g,x)$ on a l'inégalité :

$$\sup_{\substack{t\in T\\ x\in X}} \int \frac{1}{\sigma^{2\alpha}(g,x)} \mu_t^{N_o}(dg) \leq 1 - 2\alpha\beta + 2\alpha^2 \int \delta^2(g)\, e^{2\alpha\delta(g)}\, \mu_t^{N_o}(dg)$$

d'où puisque $\beta>0$, il existe un réel $0<\alpha(T)<1$ tel que pour $0<\alpha\leq\alpha(T)$:

$$(61) \qquad \sup_{\substack{t\in T\\ x\in X}} \int \frac{1}{\sigma^{\alpha}(g,x)} \mu_t^{N_o}(dg) < 1$$

Par ailleurs la suite $\Big(\sup_{\substack{t\in T\\ x\in X}} \int \frac{1}{\sigma^{\alpha}(g,x)} \mu_t^n(dg)\Big)$ $n\geq 1$

est sous multiplicative, donc on a :

$$\lim_n \Big(\sup_{\substack{t\in T\\ x\in X}} \int \frac{1}{\sigma^\alpha(g,x)} \mu_t^n(dg)\Big)^{1/n} = \operatorname*{Inf}_n \Big(\sup_{t\in T} \int \frac{1}{\sigma^\alpha(g,x)} \mu_t^n(dg)\Big)^{1/n}$$

ce qui - compte-tenu de (61) - établit le lemme (3-1).

Le rôle de ce lemme est analogue à celui du lemme (2-7).

2) On considère de même que dans le paragraphe 2 B) pour $0<\alpha<1$ l'algèbre de Banach $\mathcal{L}'_\alpha$ des fonctions Lipschitziennes d'ordre α sur X ; les notations sont identiques à celles du paragraphe 2 B) et uniquement modifiées par l'adjonction d'un prime.

On étudie alors la famille des opérateurs P_t : $t \in \mathbb{R}$

(62) $$P_t f(x) = \int f(gx)\, \mu_t(dg) \quad x \in X,$$

où f est mesurable bornée sur X.

Ces opérateurs satisfont alors aux propriétés énoncées dans les propositions suivantes et dont la preuve est la même que celle des propositions (2-9) et (2-10) (cas où $u=0$), en substituant le lemme (3-1) au lemme (2-7).

Proposition (3-2) :

1) *Pour tout* $0<\alpha<1$ P_t *est un opérateur borné sur* $\mathcal{L}'_\alpha$, *et si* T *est un compact de* $\mathbb{R}$ *on a :*

$$\sup_{t\in T} \|P_t\|'_\alpha < +\infty$$

2) *Si* $0<\alpha<1$, $0<r<1$ *et si* T *est un compact de* $\mathbb{R}$, *il existe une constante* $C(T,\alpha)$ *telle que pour toute fonction* $\Phi \in \mathcal{L}'_\alpha$ *et tous* $s,t \in T$

$$\|P_t\Phi - P_s\Phi\|_{(1-r)\alpha} \leq C(T,\alpha)\ |t-s|^{t\alpha}\ \|\Phi\|_\alpha$$

Proposition 3-3 : *Soit* T *un intervalle compact de* $\mathbb{R}$. *Pour tous* $t \in T$ *et* $0<\alpha\leq (T)$, P_t *est un opérateur quasi-compact de* $\mathcal{L}'_\alpha$ *dans* $\mathcal{L}'_\alpha$ *et tel que pour tout* $n\geq 1$ *on ait :*

$$P_t^n = \nu_t + Q_t^n$$

où Q_t *est un opérateur de* $\mathcal{L}'_\alpha$ *dans* $\mathcal{L}'_\alpha$ *vérifiant*

$$Q_t e = 0,\ \nu_t Q_t = 0$$

De plus si $r'_\alpha(Q_t)$ *est le rayon spectral de* Q_t *dans* $\mathcal{L}'_\alpha$ *on a :*

$$\sup_{t\in T} r_\alpha(Q_t) < 1$$

On en déduit le

<u>corollaire (3-1)</u> : *Si* T *est un intervalle compact de* $\mathbb{R}$, *si* $0<\alpha\leq\alpha(T)$, *si* $0<r<1$, *il existe une constante* $C(r,\alpha,T)$ *telle que pour toute fonction* $\Phi\in\mathcal{L}'_\alpha$ *on ait pour tous* s,t *de* T

$$|\nu_t(\Phi) - \nu_s(\Phi)| \leq C(r,\alpha,T) \quad |t-s|^{r\alpha} \; ||\Phi||'_\alpha$$

dont la preuve est la même que celle du lemme (2-11) 1).

Ce corollaire nous permet de terminer la preuve de la proposition (3-1). En effet, on a :

$$(63) \quad \gamma(t) - \gamma(t_o) = \int \log\sigma(g,x)\ \mu_t(dg)\nu_t(dx) - \int \log\sigma(g,x)\mu_{t_o}(dg)\nu_{t_o}(dg)$$

$$= \nu_t(\Phi_t) - \nu_{t_o}(\Phi_t) + \nu_{t_o}(\Phi_t - \Phi_{t_o})$$

$$\text{où} \quad \Phi_t(x) = \int \log\ \sigma(g,x)\ d\mu_t(g).$$

Comme l'application $(t,x)\to\Phi_t(x)$ de $\mathbb{R}\times X$ dans $\mathbb{R}$ est continument différentiable, on en déduit que pour tout intervalle compact T de $\mathbb{R}$, on a :

$$\sup_{t\in T} ||\Phi_t||'_\alpha < +\infty , \qquad \sup_{x\in X} |\Phi_t(x) - \Phi_{t_o}(x)| \leq k(T)|t-t_o|$$

où $k(T)$ est une constante.

Par conséquent en tenant compte du corollaire (3-1), on obtient que si $0<\alpha\leq\alpha(T)$, $0<r<1$, $t,t_o\in T$

$$(64) \quad |\gamma(t) - \gamma(t_o)| \leq C(r,\alpha,T) \sup_{t\in T} ||\Phi'_t||_\alpha \quad |t-t_o|^{r\alpha} + k(T)\ |t-t_o|$$

ce qui prouve bien que la fonction $t\to\gamma(t)$ appartient à $\mathcal{L}_{loc}(\mathbb{R})$.

APPENDICE

Dans cet appendice, nous précisons des résultats concernant certains opérateurs, utilisés au cours des démonstrations du lemme (2-7), et de la proposition (2-8). Les notations utilisées sont les mêmes que précédemment.

Soit p une probabilité sur $SL(2,\mathbb{R})$ ayant une densité à support compact. On considère l'opérateur $T(\alpha)$ défini sur $\mathcal{C}(X)$ par

$$T(2\alpha)\, f(\bar{x}) = \int \sigma^{2\alpha}(g,\bar{x})\, f(g\bar{x})\, p(dg)$$

$f \in \mathcal{C}(X)$, $\bar{x} \in X$, $\alpha \in \mathbb{R}$.

On a alors la

Proposition A_1.

1) *Pour tout* $\alpha \in \mathbb{R}$, $T(\alpha)$ *est un opérateur compact positif sur* $\mathcal{C}(X)$; *Son rayon spectral* $r(\alpha)$ *est tel que la fonction* $\alpha \to \log r(\alpha)$ *soit convexe, et que*

$r(0) = r(2) = 1, \quad r(\alpha) < 1$ *pour* $0<\alpha<2$

2) *L'opérateur* $T(2)$ *peut s'écrire sous la forme :*

$T(2)\,[f\,] = m(f)\, e_2 + R_2(f)$

où m *est la probabilité sur* X *invariante par les rotations.*

e_2 *l'unique fonction positive de* $\mathcal{C}(X)$ *telle que* $T(2)e_2 = e_2$ *et* $m(e_2) = 1$.

$\mathbb{R}_2$ *est un opérateur de* $\mathcal{C}(X)$ *de norme spectrale strictement inférieure à* 1 *et tel que :*

$m\, R_2 = 0 \qquad R_2(e_2) = 0.$

Démonstration de la proposition A_1.

1) Pour établir la compacité de l'opérateur $T(\alpha)$, on commence par supposer que p a une densité continue : dans ce cas $T(\alpha)$ est défini par un noyau continu [18], d'où le résultat.
Le cas général s'obtient en approximant la probabilité p en variation à l'aide d'une suite de probabilités p_n $n\geq 1$, admettant des densité continues dans un même compact fixe.

L'inégalité de Hölder permet facilement d'obtenir que pour

$\alpha \in \mathbb{R}$, $\beta \in \mathbb{R}$, $0<t<1$ $n\geq 1$ on a

$$\text{Log } ||T^n(t\alpha+ (1-t)\beta)|| \leq t \log||T^n(\alpha)|| + (1-t)\log ||T^n((1-t)\beta)||$$

d'où la convexité de la fonction $\alpha \to \log r(2\alpha) = \lim_n \log||T^n(\alpha)||^{1/n}$

2) Pour terminer la démonstration, nous utiliserons la propriété suivante des opérateurs $T(\alpha)$:

<u>lemme A</u> : *Il existe un réel* $\alpha_0>0$ *tel que pour* $|\alpha|<\alpha_0$ *l'opérateur* $T^n(\alpha)$ $n\geq 1$ *peut s'écrire sous la forme*

$$T^n(\alpha) = \lambda^n(\alpha)\, \nu_\alpha(\alpha)\, e_\alpha + R^n(\alpha)$$

où : $\lambda(\alpha) > 0$ *est l'unique valeur propre de plus grand module de* $T(\alpha)$

ν_α *l'unique probabilité sur* X *telle que* $\nu_\alpha T(\alpha) = \lambda(\alpha)\nu_\alpha$

$e_\alpha \geq 0$ *l'unique élément de* $\mathcal{C}(X)$ *tel que*
$T(\alpha)\, e_\alpha = \lambda(\alpha)\, e_\alpha$, $\nu_\alpha(e_\alpha) = 1$
$R(\alpha)$ *est un opérateur de* $\mathcal{C}(X)$ *de rayon spectral strictement inférieur à* $\lambda(\alpha)$ *tel que*

$\nu_\alpha\, Q(\alpha) = 0$ $\qquad Q(\alpha)\, e_\alpha = 0$

De plus les applications $\alpha \to \nu_\alpha$, $\alpha \to e_\alpha$, $\alpha \to R(\alpha)$ *sont analytiques.*

<u>Démonstration du lemme A</u> :

L'énoncé du lemme A est satisfait pour l'opérateur $T(o)$; on a :

$$T^n(o) = \nu e + R^n \qquad n \geq 1$$

où ν est l'unique probabilité p invariante portée par X.

D'autre part l'application $\alpha \to T(\alpha.)$ est analytique. La théorie des perturbations analytiques d'opérateurs assure qu'il existe un réel α_o tel que pour $|\alpha| < \alpha_o$ $T(\alpha.)$ admette une unique valeur simple isolée de module égal au rayon spectral de $T(\alpha.)$, et de sous-espace propre de dimension 1.

Comme d'autre part, $T(\alpha.)$ est un opérateur positif, cette valeur propre est égale au rayon spectral de $T(\alpha.)$ et admet une fonction propre positive [17].

L'énoncé du lemme A se déduit alors immédiatement de ce qui précède et de la théorie des perturbations analytiques d'opérateurs.

3) Etablissons le 2) de la proposition A_1. Pour cela on raisonne par dualité. Nous notons $\check{T}(\alpha.)$ l'opérateur défini de façon analogue à $T(\alpha.)$, mais en remplaçant la probabilité p par sa symétrisée $\check{p}$. Il est clair que $\check{T}(2\alpha)$ a les mêmes propriétés que $T(\alpha)$.

Si θ est une mesure bornée sur X, et si $\Psi \in \mathcal{C}(X)$, on note :

$$(\theta,\Psi) = \int \Psi(x)\, d\theta(x).$$

On a alors la relation :

$$(1') \quad (\Phi.m,\ T^k(2)\,[\Psi]) = (\Psi.m,\ \check{T}^k(o)\,[\Phi]) = ((\Psi.m)^{\vee k}T(o),\Phi)$$

où $\Phi,\ \Psi \in \mathcal{C}(X)$, $k \geq 0$.

Cette relation résulte du fait que $\sigma^2(g,\bar{x}) = \frac{dg^{-1}m}{dm}(\bar{x})$

$\bar{x} \in X$, $g \in SL(2,\mathbb{R})$.

Si λ est une valeur spectrale non nulle de l'opérateur compact $T(2)$, il existe un entier $k \geq 1$ et une fonction $\Psi_\lambda \in \mathcal{C}(X)$ telle que

$$[\lambda I - T(2)]^k \psi_\lambda = 0$$

$$m(\psi_\lambda) = 1.$$

On en déduit d'après (1') que :

$$(\psi_\lambda . m)[\lambda I - \check{T}(o)]^k = 0.$$

L'adjoint de $\check{T}(o)$, admet 1 pour unique valeur propre de module supérieur ou égal à 1 ; cette valeur propre est simple, et le sous-espace propre correspondant est engendré par l'unique probabilité $\check{p}$ invariante $\check{\nu}$ portée par X. Il en résulte que si $|\lambda| \geq 1$ on a $\lambda=1$, $k=1$ et $\Psi_1 m = \check{\nu}$. Posons $e_2 = \Psi_1 \geq 0$.

Comme la dimension de l'espace des mesures bornées θ sur X satisfaisant à $\theta[I - T(2)] = 0$, est la même que celle de l'espace vectoriel des fonctions $f \in \mathcal{C}(X)$ satisfaisant à $[I - T(2)] f = 0$, c'est-à-dire 1, et que d'après (1') on a $m\, T(2) = m$, on en déduit le 2) de la proposition A_1.

4) L'application $\alpha \to \lambda(\alpha)$ est dérivable et l'on a [4], [18]

$$\lambda'(o) = - \int \log \sigma(g,x)\, p(dg)\, \nu(dx) < 0.$$

Comme d'autre part on a d'après ce qui précède $r(o) = r(2) = 1$, et que l'application $\alpha \to \log r(\alpha)$ est convexe, il est clair que pour $0<\alpha<2$ on a nécessairement : $0<r(\alpha)<1$

REFERENCES

[1] Atkinson — *Discrete and continuous boundary - Problems* Academic Press New-York (1964).

[2] Billingsley — *Convergence of probability measures* - John Wiley and sons - New-york (1968).

[3] Dunford - Schwartz — *Linear operators - Part 1.* Interscience Publishers - New-York (1958).

[4] Furstenberg — *Non-commuting random products* - Trans. American Math Soc 108 (1963) p. 337-428.

[5] Guivarc'h - Raugi — *Frontière de Furstenberg - Propriétés de contractio et théorèmes de convergence.* Séminaire de probabilités - Rennes 1981 et 1982.

[6] Ishii — *Localisation of eigenstates and transport phenomena in the one dimensional disorded system.* Suppl. Progr. Theor. Phys. 53, 77 (1963).

[7] Kunz et Souillard — *Sur le spectre des opérateurs aux différences finie aléatoires.* Commun. Math Phys. 78 (1980) p. 201-246.

[8] Lacroix — *Problèmes probabilistes liés à l'étude des opérateur aux différences aléatoires.* Annales del'Institut Elie Cartan(Nancy) n°7

[9] Le Page — *Théorèmes limites pour les produits de matrices aléatoires.* Probability measures on groups - Springer Lect.Note 528.

[10] Nagaev — *Some limit theorems for stationary Markov chains* - Theory of Proba and applications 2 (1957) p.378-40

[11] Neri — *Singular Integrals* Lecture Notes in Mathematics 200.

[12] Neveu — *Bases mathematiques du calcul des probabilités* - Masson et Cie.

[13] Norman — *Markov processes and learning models.* Academic Press New-York vol. 84

[14] O. Connor — *Disordered Harmonic Chain.* Commun. Math. Physics 45 (1975) p. 63-77.

.../...

...

5 Pastur *Spectral properties of disordered systems in the one body approximation.* Commun. Math Physics 75(1980) p. 179-196.

6 Reznikova *The central limit theorem for the spectrum of random Jacobi matrices.* Theory of Proba and its applications - volume XXV number 3 (1980).

7 Schaefer *Topological vector spaces.* The Mac Millan Company New-York (1967).

8 Tutubalin *On limit theorems for the product of random matrices.* Theory of Proba. Applic. 10 (1965) p. 15-27.

RMAR
niversité de Rennes 1
ampus de Beaulieu
5042 Rennes Cédex
rance

ASYMPTOTICALLY CENTRAL FUNCTIONS AND INVARIANT EXTENSIONS OF DIRAC MEASURE

by V.Losert and H.Rindler

Let G be a locally compact group with unit element e and left Haar measure dx. $L^1(G)$, $L^\infty(G)$ have the usual meaning, for $f \in L^1(G)$, we put $\|f\|_1 = \int |f(x)|dx$. M(G) denotes the space of (complex, bounded) Radon measures on G, δ_x stands for the Dirac measure concentrated in $x \in G$. For $f, g \in L^1(G)$, convolution is defined by $f*g(x) = \int f(y)g(y^{-1}x)dy$, similarly in the case of measures. $\mathcal{K}(G)$ denotes the space of continuous functions with compact support, $C_o(G)$ its uniform closure.
For $u \in L^1(G)$, $f \in L^\infty(G)$, we write $\langle u,f\rangle = \int_G u(x)f(x)dx$ (see [10] for general references).

A linear functional M on $L^\infty(G)$ is called a <u>mean</u>, if M is non-negative and satisfies M(1)=1 (where on the left-hand side 1 denotes the function with constant value 1), see [2]. For $x \in G$, the corresponding inner automorphism (<u>conjugation</u>) induces a mapping τ_x' on $L^\infty(G)$ by $\tau_x' f(y) = f(xyx^{-1})$. The adjoint map τ_x on $L^1(G)$ is given by $\tau_x u(y) = u(x^{-1}yx)\Delta(x)$ (Δ denotes the Haar modulus of G). This can also be written as $\tau_x u = \delta_x * u * \delta_{x^{-1}}$. A net (u_σ) in $L^1(G)$ is called a left <u>approximate unit</u> (a.u.), if $\lim \|u_\alpha * f - f\|_1 = 0$ for all $f \in L^1(G)$.

Similary for right and two-sided a.u. Unless otherwise stated, all a.u. will be two-sided.

<u>Definition</u>: A net (u_α) in $L^1(G)$ is called <u>asymptotically central</u> (a.c.), if $\lim \|u_\alpha\|^{-1}(\tau_x u_\alpha - u_\alpha) = 0$ weakly (i.e. for $\sigma(L^1, L^\infty)$), for all $x \in G$.
The center of $L^1(G)$ consists of all u satisfying $\tau_x u = u$ for all $x \in G$.
By a result of Mosak [9] (see also [11]), $L^1(G)$ admits a central a.u. if G has a neighbourhood basis at the identity consisting of conjugation - invariant sets.
In the course of investigations of the relations between double multipliers, and the bidual of a Banach algebra by Tomiuk [15], [16] and

Grosser [3] they made use of a.u. satisfying the condition $\lim \mu * u_\alpha - u_\alpha * \mu = 0$ weakly for each $\mu \in M(G)$. (This is clearly satisfied for $\mu \in L^1(G)$, but it was falsely statet in [15] that it always holds for general μ see also [3] and [4]).

In [14]p.47 an a.u. (u_α) is called quasi central for M(G), if $\|\mu * u_\alpha - u_\alpha * \mu\|_1 \to 0$ for all $\mu \in M(G)$. Sinclair asks: when does $L^1(G)$ have a quasi central a.u.? ([14]p.137).

RESULTS:

(These have partly been announced in [4]).

THEOREM 1:

(i) If G is abelian or discrete, then any a.u. is a.c.

(ii) If G is neither abelian nor discrete, $\epsilon > 0$, then there exists an a.u. (u_α) such that $\|u_\alpha\| \le 1+\epsilon$, but (u_α) is not a.c.

PROPOSITION 1: If the centralizer $C(x) = \{y \in G: y^{-1}xy = x\}$ is open for all $x \in G$, then any a.u. (u_α) with $\|u_\alpha\|_1 \le 1$ is a.c. and conversely, if all such a.u. are a.c., then C(x) is open for all x.

REMARK: Thm. 1 and Prop. 1 are also true for left or right a.u. The condition a.c. can be replaced by $\lim \|\mu * u_\alpha - u_\alpha * \mu\|_1 = 0$ for all $\mu \in M(G)$. Theorem 1 describes those groups G for which $L^1(G)$ is Arens semi-regular in the sense of [3]. If G is first-countable, one can consider sequences instead of nets.

THEOREM 2: If $L^1(G)$ has a bounded a.c.a.u., then the Dirac measure in e admits an extension to a conjugation-invariant mean M on $L^\infty(G)$, i.e. $M(\tau_x' f) = M(f)$ for all $f \in L^\infty(G)$, $x \in G$ and $M(f) = f(e)$ if f is continuous and bounded.

Conversely, if δ_e has a conjugation invariant extension to $L^\infty(G)$, Then there exists an a.u. (u_α) satisfying $u_\alpha \ge 0$, $\|u_\alpha\|_1 = 1$ and $\lim \|\tau_x u_\alpha - u_\alpha\|_1 = 0$ uniformly on compact subsets of G.

REMARK: If $\lim \|\tau_x u_\alpha - u_\alpha\|_1 = 0$ uniformly on compact subsets (and the u_α

are bounded in $L^1(G)$), then $\lim \|\mu * u_\alpha - u_\alpha * \mu\|_1 = 0$ for all $\mu \in M(G)$. This means that the existence of an a.c.a.u. is equivalent to the existence of a quasi central a.u. If G is second countable, one can consider sequences instead of nets.

THEOREM 3: If G is amenable, then δ_e admits a conjugation invariant extension to $L^\infty(G)$, hence there exists an a.c. (or quasi central) a.u.

EXAMPLE 1: $G = T^n \times_s SL(n,\mathbb{Z})$ (semi-direct product), where $T^n = (\mathbb{R}/\mathbb{Z})^n$ denotes the n-dimensional torus, $SL(n,\mathbb{Z})$ acts by matrix multiplication mod. 1. For $n>1$ $L^1(G)$ admits no bounded a.c.a.u.

EXAMPLE 2: $G = K^H \times_s H$, where K is any non-trivial compact group and the discrete group H acts on K^H by coordinate translation: $s(h_o)(k_h)_{h\in H} = (k_{h_o^{-1}h})$. $L^1(G)$ has a bounded a.c.a.u. iff H is amenable.

EXAMPLE 3: If A is abelian and H is discrete, then any a.u. (u_α) in $L^1(A\times H)$ satisfying $\|u_\alpha\|_1 \leq 1$ is a.c.

THEOREM 4: If $L^1(G)$ admits a bounded a.c.a.u. then G_o (the connected component of e) is amenable.

COROLLARY: If G is connected, then $L^1(G)$ admits a bounded a.c.a.u. iff G is amenable.

EXAMPLE 4: If G is a non-compact semi-simple Lie group (e.g. $G=SL(2,\mathbb{R})$), then $L^1(G)$ has no a.c.a.u.

THEOREM 5: The following conditions are equivalent:

(i) δ_e admits a conjugation invariant extension to a mean on $L^\infty(G)$.

(ii) There exists a functional M_o on $L^\infty(G)$ and $\alpha<2$ such that

$|M_o(\tau'_x f - f)| \leq \alpha \|f\|_\infty$ for all $f \in L^\infty(G)$ and $x \in G$

$M_o(f) = f(e)$ if f is continuous and bounded.

(iii) There exists $\alpha<2$ and an a.u. (u_α) such that $\|u_\alpha\|_1 \leq 1$ and

$\lim |\langle \tau_x u_\alpha - u_\alpha, f\rangle| \leq \alpha \|f\|_\infty$ for all $x \in G$, $f \in L^\infty(G)$.

(compare [5]Thm.1)

PROPOSITION 2: Let Ω be a compact space, G a group of homeomorphisms of Ω. Assume that there exists a Radon probability measure μ on Ω and $\sigma < 2$ such that $\|g(\mu) - \mu\|_1 \leq \sigma$ for all $g \in G$. Then there exists a G-invariant probability measure on Ω. ($g(\mu)$ denotes the image measure of μ under g).

PROOFS:

LEMMA 1: (i) Let (U_α) be a base of neighbourhoods at e. Assume that $u_\alpha \in L^1(G)$, $u_\alpha \geq 0$, $\|u_\alpha\|_1 = 1$, supp $u_\alpha \subseteq U_\alpha$.
Then (u_α) is an a.u. for $L^1(G)$.

(ii) Let G be non-discrete, $\epsilon > 0$, U a non-empty open subset, $V \subseteq U$ such that $\overline{V^o} \cap \overline{(U \setminus V)^o} \neq \emptyset$. Then there exist $v_\alpha \in L^1(G)$ with supp $v_\alpha \subseteq U$, $\|v_\alpha\|_1 = \epsilon$, $\int_V v_\alpha \, dx = \epsilon/2$ and $\lim \|v_\alpha * f\|_1 = \lim \|f * v_\alpha\|_1 = 0$ for all $f \in L^1(G)$. (^{o}denotes the interior, $^-$ the closure).

PROOF: (i) is easy and standard.

(ii): Choose $y \in \overline{V^o} \cap \overline{(U \setminus V)^o}$. Let (U_α) be as in (i). If $y_\alpha \in V^o$ is sufficiently close to y, then $y_\alpha U_\alpha$ intersects $(U \setminus V)^o$. It follows that there exists $x_\alpha \in U_\alpha$ and a non-empty open set $W_\alpha \subseteq U_\alpha$ such that $y_\alpha W_\alpha \subseteq V$ and $y_\alpha x_\alpha W_\alpha \subseteq U \setminus V$. Choose u_α as in (i) with supp $u_\alpha \subseteq W_\alpha$ and put $v_\alpha = (\epsilon/2)\delta_{y_\alpha} * (u_\alpha - \delta_{x_\alpha} * u_\alpha)$.

Then supp $v_\alpha \subseteq y_\alpha W_\alpha \cup y_\alpha x_\alpha W_\alpha \subseteq U$, $\|v_\alpha\| = \epsilon$, $\int_V v_\alpha \, dx = (\epsilon/2) \int_{W_\alpha} u_\alpha \, dx = \epsilon/2$ and

$$\|v_\alpha * f\|_1 = (\epsilon/2)\|(u_\alpha * f - f) - \delta_{x_\alpha} * (u_\alpha * f - f) + (\delta_{x_\alpha} * f - f)\|_1 \to 0$$

$$\|f * v_\alpha\|_1 = (\epsilon/2)\|(f * \delta_{y_\alpha} - f * \delta_{y_\alpha x_\alpha}) * u_\alpha\|_1 \to 0.$$

LEMMA 2: Let (u_α) be an a.u. for $L^1(G)$. Then the following holds:

(i) $\liminf \|u_\alpha\|_1 \geq 1$.

(ii) If H is an open subgroups of G, $u_\alpha = v_\alpha + w_\alpha$ where supp $v_\alpha \subseteq H$, supp $w_\alpha \subseteq G \setminus H$ <u>and</u> (v_α) <u>is bounded in</u> $L^1(G)$, <u>then</u> (v_α) <u>is an a.u. for</u> $L^1(G)$.

<u>PROOF:</u> (i) follows from $\|f\|_1 - \|f-u_\alpha*f\|_1 \leq \|u_\alpha*f\|_1 \leq \|u_\alpha\|_1 \ \|f\|_1$.
(ii) If supp $f \subseteq H$, then $\|u_\alpha*f-f\|_1 = \|v_\alpha*f-f\|_1+\|w_\alpha*f\|_1$, therefore $\|v_\alpha*f-f\|_1 \to 0$. Since the elements $f*\delta_x$ where supp $f \subseteq H$, $x \in G$, generate a dense subspace of $L^1(G)$ and (v_α) is bounded, we obtain that (v_α) is an a.u. for $L^1(G)$.

<u>PROOF of THEOREM 1</u>:

(i) is trivial if G is abelian. If G is discrete, then $\|u_\alpha - \delta_e\|_1 = \|u_\alpha * \delta_e - \delta_e\|_1 \to 0$.

(ii): Assume that G is not abelian and not discrete, $\epsilon > 0$. Then there are $x,y \in G$ such that $xyx^{-1} \neq y$. Choose a non-empty open set U such that $xUx^{-1} \cap U = \emptyset$. If G is not discrete, it has a non-discrete metrizable quotient G/H (H a closed subgroup of G). From this one can easily construct a set V as in (ii) of Lemma 1 (consider any convergent sequence (t_n) in G/H with t_n pairwise different and also different from the limit; then take for V the intersection of U with the preimage of the union of balls around t_{2n} with sufficiently small radii). Choose $(u_\alpha),(v_\alpha)$ as in Lemma 1. Then both (u_α) and $(u_\alpha+v_\alpha)$ are a.u. for $L^1(G)$, $\|u_\alpha\|_1=1$, $\|u_\alpha+v_\alpha\|_1 \leq 1+\epsilon$. Since supp $\tau_x v_\alpha \subseteq xUx^{-1}$ we have $<v_\alpha - \tau_x v_\alpha, c_V> = <v_\alpha, c_V> = \epsilon/2$ (c_V denotes the characteristic function of V). Hence it is impossible that both (u_α) and $(u_\alpha+v_\alpha)$ are a.c.

<u>PROOF of PROPOSITION 1</u>:

Assume that (u_α) is an a.u., $\|u_\alpha\|_1 \leq 1$. If $x \in G$, then by assumption C(x) is open. Put $u_\alpha = v_\alpha + w_\alpha$ as in Lemma 2(ii) (applied to C(x)). Then (v_α) is again an a.u. Since τ'_x is the identity for functions supported on C(x), the same is true for τ_x, consequently $\tau_x v_\alpha = v_\alpha$. Since $\|u_\alpha\|_1 \leq 1$, it follows from Lemma 2(i) that $\|w_\alpha\|_1 = \|u_\alpha\|_1 - \|v_\alpha\|_1 \to 0$. Thus we obtain that

$$\|\tau_x u_\alpha - u_\alpha\|_1 \leq \|\tau_x u_\alpha - \tau_x v_\alpha\|_1 + \|\tau_x v_\alpha - v_\alpha\|_1 + \|v_\alpha - u_\alpha\|_1 \to 0,$$

i.e. (u_α) is a.c.

Converse: if $C(x)$ is not open for some $x \in G$, then for any neighbourhood U_α of e there exists $y_\alpha \in U_\alpha$ such that $xy_\alpha x^{-1} \neq y_\alpha$.
Therefore, we find an open subset W_α of U_α satisfying $xW_\alpha x^{-1} \cap W_\alpha = \emptyset$.
Choose (u_α) as in Lemma 1 with supp $u_\alpha \subseteq W_\alpha$. Then (u_α) is an a.u. but $\|\tau_x u_\alpha - u_\alpha\|_1 = \|\tau_x u_\alpha\|_1 + \|u_\alpha\|_1 = 2$ for all α.

LEMMA 3: Assume that M is a mean on $L^\infty(G)$ that satisfies $M(g) = g(e)$ for all $g \in \mathcal{K}(G)$. The the following holds:

(i) If $f \in L^\infty(G)$ and $f = 0$ in a neighbourhood of e, then $M(f) = 0$.

(ii) If $f \in L^\infty(G)$ is continuous in a neighbourhood of e, then $M(f) = f(e)$.

PROOF: (i) We may assume that $\|f\|_\infty = 1$. Put $U = \{x \in G: f(x) = 0\}$.
There exists $g \in \mathcal{K}(G)$ such that $0 \leq g \leq 1$, $g = 0$ outside $U, g(e) = 1$.
Then $\| |f| + g\|_\infty = 1$ and $g \leq |f| + g$. Consequently,

$$1 = g(e) = M(g) \leq M(|f| + g) \leq \| |f| + g\|_\infty \leq 1.$$

It follows that $M(|f|) = 0$, hence $M(f) = 0$.

(ii) is an easy consequence of (i).

PROOF of THEOREM 2:

Assume that (u_α) is a bounded a.c.a.u. We may assume that the u_α are real valued. We embed $L^1(G)$ into the dual $L^\infty(G)'$ of $L^\infty(G)$. Since (u_α) is a bounded net, it has a $\sigma(L^{\infty\prime}, L^\infty)$-accumulation point $M_0 \in L^\infty(G)'$. We may assume that $\lim u_\alpha = M_0$. For $f \in L^\infty(G)$, $x \in G$, we get:

$$M_0(\tau_x' f - f) = \lim < u_\alpha, \tau_x' f - f > = \lim < \tau_x u_\alpha - u_\alpha, f > = 0.$$

Consequently M_0 is conjugation invariant.
For $v \in L^1(G)$, we have (put $\tilde{v}(x) = v(x^{-1})\Delta(x^{-1})$, compare [10]p.81)

$$M_0(v * f) = \lim < u_\alpha, v * f > = \lim < \tilde{v} * u_\alpha, f > = < \tilde{v}, f > = v * f(e).$$

If $g \in C_0(G)$, then $\lim \|u_\alpha * g - g\|_\infty = 0$, hence $M_0(g) = g(e)$. Similarly $M_0(1) = 1$.
Let Ω be the spectrum of the commutative C*-algebra $L^\infty(G)$, i.e. $L^\infty(G) =$

$= C(\Omega)$ ([13]13. 3,7 p.233). Since $C_0(G)$ is a subalgebra, we have a continuous surjection $\pi: \Omega \to G^*$ (where G^* denotes the Alexandroff compactification of G, [13]7.7.1.p.127). M_0 corresponds to a Radon measure μ_0 on Ω. Since the u_α are real-valued, the same is true for M_0 and μ_0. $M_0(g) = g(e)$ for $g \in C_0(G) \cup \{1\}$ implies that $\pi(u_0) = \delta_e$, hence $\mu_0(\pi^{-1}(e))=1$ Let μ_1 be the restriction of $|\mu_0|$ to $\pi^{-1}(e)$ ($\mu_1 \neq 0$ by the last equation). Finally put

$$\mu = \mu_1/\|\mu_1\|$$

and let M be the functional on $L^\infty(G)$ corresponding to μ.

M is a mean since μ is a probability measure, $M(g) = g(e)$ for $g \in C_0(G)$, since μ is concentrated on $\pi^{-1}(e)$. By Lemma 3, M extends δ_e.

For $x \in G$, τ_x' induces a homeomorphism of Ω ([13]7.7.1) and π commutes with these homeomorphisms. Since M_0 is τ_x'-invariant, the same is true for μ_0, μ_1, μ and M.

Converse: If δ_e has a conjugation-invariant extension to $L^\infty(G)$, then, as above, there is a conjugation-invariant mean M on $L^\infty(G)$ extending δ_e. Choose any $w \in L^1(G)$ with $\int w(x)dx = 1$ and put (the integral converges for $\sigma(L^\infty, L^1)$):

$$M_1(f) = M(\int w(x)\tau'_{x^{-1}} f dx).$$

Then one gets as in [2] p.28:

$$M_1(\int v(x)\tau'_{x^{-1}} f\, dx) = \int v(x)dx\ M_1(f) \qquad \text{for all } v \in L^1(G).$$

It follows from similar arguments as in [2]2.4.2 and 2.4.3 that there exists a net (u_α) in $L^1(G)$ such that $u_\alpha \geq 0$, $\|u_\alpha\|_1 = 1$, $\lim u_\alpha = M_1$ (for $\sigma(L^{\infty\prime}, L^\infty)$) and $\lim_\alpha \|\tau_x u_\alpha - u_\alpha\|_1 = 0$ uniformly on compact subsets of G (see also [1] or [10]).

For $f \in L^\infty(G)$, $v \in L^1(G)$ we have (put $\check{v}(x) = v(x^{-1})$ and observe that $f * \check{v}$ is bounded and continuous):

$$\lim \langle u_\alpha * v, f\rangle = \lim \langle u_\alpha, f * \check{v}\rangle = M_1(f * \check{v}) = f * \check{v}(e) = \langle v, f\rangle.$$

Consequently, $\lim u_\alpha * v = v$ for $\sigma(L^1, L^\infty)$ and similarly $\lim v * u_\alpha = v$.

Replacing (u_α) by appropriate convex combinations, we obtain by standard arguments (for convex sets weak and strong closure coincide, see e.g. [2] p.34) an a.c.a.u.

PROOF of THEOREM 3:

Let (u_α) be any bounded a.u. for $L^1(G)$, $\varepsilon > 0$ and C be a compact subset of G. Since G is amenable, there exists $w \in \mathcal{K}(G)$, with $w \geq 0$, $\|w\|_1 = 1$ and $\|\delta_x * w - w\|_1 < \varepsilon$ for all $x \in C$ ([10] 8.6.1). Put $u'_\alpha = \int w(x)\tau_x u_\alpha dx$ (Bochner integral). Then $\|\tau_x u'_\alpha - u'_\alpha\|_1 < \varepsilon\|u_\alpha\|_1$ for $x \in C$. We have for $v \in L^1(G)$:

$$u'_\alpha * v = \int_G w(x)(\tau_x u_\alpha) * v \, dx \quad \text{and}$$

$$\lim_\alpha(\tau_x u_\alpha) * v = \lim_\alpha \tau_x(u_\alpha * \tau_{x^{-1}} v) = \tau_x \tau_{x^{-1}} v = v \text{ uniformly on}$$

compact subsets of G. Hence $\lim u'_\alpha * v = v$. This shows that given $v_1, \ldots,$ $v_n \in L^1(G)$, there exists α such that $\|u'_\alpha * v_i - v_i\| < \varepsilon$ for all i.
Now the result follows easily.

PROOF for EXAMPLE 1:

Let $G = T^n \times_s SL(n,\mathbb{Z})$. First we give an elementary argument for the case $n = 2$. Assume M is a conjugation-invariant mean on $L^\infty(G)$ extending δ_e. T^2 is an open subgroup which will be identified with $[-1/2,1/2[^2$. For $S \in SL(2,\mathbb{Z})$, the corresponding automorphism of T^2 is given by matrix multiplication. If A is a measurable subset of T^2, we write M(A) instead of $M(c_A)$. If $A' = A$ in a neighbourhood of 0, then by Lemma 3 $M(A) = M(A')$. The same is true, if A and A' differ only by a set of Lebesgue measure zero (since M is defined on $L^\infty(G)$).
For $0 \leq \alpha < \beta \leq 2\pi$ let $A_{\alpha,\beta} = \{x \in T^2 : \alpha < \arg x < \beta\}$.

$$\text{Put} \quad P = \begin{pmatrix} 0 & -1 \\ 1 & 0 \end{pmatrix}, \quad Q = \begin{pmatrix} 1 & -1 \\ 1 & 0 \end{pmatrix}.$$

Then $PA_{0,\pi/2}$ equals $A_{\pi/2,\pi}$ near zero, hence $M(A_{0,\pi/2}) = A_{\pi/2,\pi}$.
Similarly $M(A_{\pi/2,\pi}) = M(A_{\pi,3\pi/2}) = M(A_{3\pi/2,2\pi})$. Since

$$\bigcup_{i=0}^{3} A_{i\pi/2,(i+1)\pi/2} = T^2 \quad \text{apart from a negligible set,}$$

we get $M(A_{o,\pi/2}) = M(A_{\pi/2,\pi}) = 1/4$.

Similary, $QA_{o,\pi/4} = A_{\pi/4,\pi/2}$ and $QA_{\pi/4,\pi/2} = A_{\pi/2,\pi}$ near zero. Hence $M(A_{o,\pi/4}) = M(A_{\pi/4,\pi/2}) = 1/4$. On the other hand $M(A_{o,\pi/4}) + M(A_{\pi/4,\pi/2}) = M(A_{o,\pi/2}) = 1/4$, a contradiction.

In the general case, we argue as follows: Assume M is a conjugation invariant mean on $L^\infty(G)$ extending δ_e. By Lemma 3 (or Lemma 2), it defines an SL $(n,\mathbb{Z})$ - invariant mean on $L^\infty(T^n)$, extending δ_e. But by [6] Ex. a,b, [12] Thm.3.4. or [8] Prop.2 for $n \geq 2$ the only SL $(n,\mathbb{Z})$ - invariant mean on $L^\infty(T^n)$ is the ordinary Lebesgue measure.

PROOF for EXAMPLE 2:

This can be settled as above (compare [6] Ex.d.).

PROOF for EXAMPLE 3: This is clear by Proposition 1.

PROOF of THEOREM 4:

Assume that G_o is non-amenable. If M is a conjugation-invariant mean on $L^\infty(G)$ extending δ_e, then M is "supported" by each open subgroup of G (Lemma 3). Thus we may assume that G/G_o is compact. In this case, G is the projective limit of Lie groups. Thus there exists a compact normal subgroup K such that G/K is a non-amenable Lie group and again by Lemma 3, we can reduce the problem to the connected case. This is done elsewhere. For connected non-amenable groups there are no a.c. sequences at all. Hence there can be no (bounded or unbounded) a.c.a.u.

PROOF of PROPOSITION 2:

It is clearly sufficient to assume that G is finitely generated, hence countable. Write $G = \{g_1, g_2, \dots\}$ and put $\nu = \Sigma\, 2^{-n} g_n(\mu)$. Then ν is a probability measure on Ω, μ is absolutely continuous with respect to ν and ν is quasi-invariant, i.e. $g(\nu)$ is equivalent to ν for all $g \in G$. Now the same arguments as in [5] produce a G-invariant mean on $L^\infty(\Omega,\nu)$. Its restriction to $C(\Omega)$ defines a G-invariant probability measure on Ω.

PROOF of THEOREM 5:

We argue as in Theorem 2. Conditions (ii) and (iii) both yield a functional M_0 on $L^\infty(G)$ with $M_0(g) = g(e)$ for $g \in C_0(G)$, $M_0(1) = 1$, $|M_0(\tau_x' f - f)| \leq \alpha \|f\|_\infty$ for $x \in G$, $f \in L^\infty(G)$. Hence, the corresponding measure μ_0 on Ω satisfies $\|\tau_x \mu_0 - \mu_0\|_1 \leq \alpha$ for $x \in G$, where τ_x denotes the homeomorphism of Ω induced by τ_x'. The measures μ_1 and μ have the same property ($\|\mu_1\|_1 \geq 1$ since $\pi(\mu_1) \geq \delta_e$). This gives us a measure μ as in Proposition 2 and therefore there exists even an invariant measure on $\pi^{-1}(e)$. As in Theorem 2, we get (i). The converse follows again from Theorem 2.

Addendum:

We give some structural description of the groups appearing in Prop. 1. If $C(x)$ is open, then it contains G_0, the connected component of e. Now assume that G is almost connected (i.e. G/G_0 is compact). Then $C(x)$ has finite index in G for all $x \in G$. Equivalently all conjugacy classes are finite. By a result of Schlichting [17], G has an open abelian subgroup. Since any subgroup of finite index contains a normal subgroup of finite index, G has even an open central subgroup. Clearly, the converse holds too, i.e. this characterizes the groups of Prop. 1 in the almost connected case. In the general case, it follows at least that G has an open abelian subgroup. This can be used to prove our remark above that any a.u. bounded by 1 is quasi central for M(G).

REFERENCES

[1] P.EYMARD, Moyennes invariantes et représentation unitaires, Lecture Notes in Math. 300, Springer, Berlin 1972.

[2] F.P.GREENLEAF, Invariant means on topological groups, van Nostrand Math. Studies, New York 1969.

[3] M.GROSSER, Arens semi-regular Banach algebras, preprint.

[4] M.GROSSER, V.LOSERT, H.RINDLER, 'Double multipliers' und asymptotisch invariante approximierende Einheiten, Anz.Österr.Akad.d. Wiss., Math.-naturw. Kl. 1980, 7 - 11.

[5] V.LOSERT, Some properties of groups without the property P_1, Comm.Math.Helv. 54 (1979), 133 - 139.

[6] V.LOSERT, H.RINDLER, Almost invariant sets, Bull.London Math. Soc. 13 (1981), 145 - 148.

[7] V.LOSERT, H.RINDLER, Conjugation-invariant means, preprint.

[8] G.A.MARGULIS, Some remarks on invariant means, Monatsh.Math. 90 (1980), 233 - 235.

[9] R.D.MOSAK, Central functions in group algebras, Proc.AMS 29 (1971), 613 - 616.

[10] H.REITER, Classical harmonic analysis and locally compact groups, Oxford Univ.Press. 1968.

[11] H.RINDLER, Approximate units in ideals of group algebras, Proc.AMS 71 (1978), 62 - 64.

[12] J.ROSENBLATT, Uniqueness of invariant means for measure-preserving transformations, Trans.AMS 265 (1981), 623 - 636.

[13] Z.SEMADENI, Banach spaces of continuous functions, Monografie Matematyczne, Warszawa, PWN 1971.

[14] A.M.SINCLAIR, Continuous semigroups in Banach algebras, London Math.Soc. Lecture Notes 63, Cambridge Univ.Press 1982.

[15] B.J.TOMIUK, Multipliers on Banach algebras, Studia Math. 54 (1976), 267 - 283.

[16] B.J.TOMIUK, Arens regularity and the algebra of double multipliers, Proc.AMS 81 (1981), 293 - 298.

[17] G.SCHLICHTING, Topologische Gruppen mit endlichen Klassen konjugierter Elemente, Math.Z. 142, 15 - 17 (1975).

V.LOSERT, H.RINDLER

INSTITUT FÜR MATHEMATIK, UNIVERSITÄT WIEN

STRUDLHOFGASSE 4, A - 1090 WIEN, AUSTRIA

On the support of absolutely continuous Gauss measures on $SL(2,\mathbb{R})$

M. McCRUDDEN AND R.M. WOOD

Department of Mathematics
University of Manchester
Manchester M13 9PL
ENGLAND.

Introduction For any locally compact topological group G let $M(G)$ denote the topological semigroup of all probability Borel measures on G, furnished with the weak topology and with convolution as the multiplication. By a Gauss semigroup on G we mean a homomorphism $t \to \mu_t$ of $\mathbb{R}_+^*$ (the strictly positive reals under addition) into $M(G)$ such that

(i) no μ_t is a point measure

(ii) for each neighbourhood V of $1 \in G$, we have

$$\frac{1}{t}\mu_t(G\backslash V) \to 0 \quad \text{as} \quad t \downarrow 0.$$

Now let G be a connected Lie group, and let $(\mu_t)_{t>0}$ be a Gauss semigroup on G. It has recently been shown by Siebert [2] that either every μ_t is absolutely continuous w.r.t. Haar measure on G, or every μ_t is singular w.r.t. Haar measure on G. In the first case we say that $(\mu_t)_{t>0}$ is an absolutely continuous Gauss semigroup on G.

For an absolutely continuous Gauss semigroup $(\mu_t)_{t>0}$ on a connected Lie group G, it can happen that the support of μ_t, denoted by $S(\mu_t)$, fails to be the whole of G. This has been demonstrated by Siebert [2] who has given an example on the two-dimensional affine group A of an absolutely continuous Gauss semigroup $(\mu_t)_{t>0}$ such that for every $t > 0$, $S(\mu_t)$ is a proper subsemigroup of A. However what is true in Siebert's example is that every μ_t has the same support i.e. $\forall\, t, s$, $S(\mu_s) = S(\mu_t)$. In the same paper [2] Siebert has posed the question whether every absolutely continuous Gauss semigroup $(\mu_t)_{t>0}$ on an arbitrary connected Lie group has a common support i.e. whether we have $S(\mu_t) = S(\mu_s)$ $\forall\, s, t > 0$.

To investigate Siebert's question in a particular case, we set ourselves the task of calculating all possible supports of absolutely continuous Gauss semigroups on the specific Lie group $SL(2,\mathbb{R})$. It turns out that in this special case the computations can be accomplished without great difficulty and an explicit description of all possible supports is obtained, as follows.

Theorem. Let $G = SL(2,\mathbb{R})$ and for $A = \begin{pmatrix} a & b \\ c & d \end{pmatrix} \in G$, write

$$\tau(A) = ad \qquad \text{and} \qquad \sigma(A) = a^2 + b^2 + c^2 + d^2 .$$

(a) Let $(\mu_t)_{t>0}$ be an absolutely continuous Gauss semigroup on G, and let $\{S(\mu_t) : t > 0\}$ be the family of supports of $(\mu_t)_{t>0}$.

Then either

(i) $\forall t > 0,\ S(\mu_t) = G$, or

(ii) $\exists \alpha \neq 0$ and $B \in G$ such that

$$\forall t > 0,\ BS(\mu_t)B^{-1} = \left\{A = \begin{pmatrix} a & b \\ c & d \end{pmatrix} \in G : a,b,c,d \geq 0,\ \tau(A) \geq \cosh^2(\alpha t)\right\}$$

or

(iii) $\exists \alpha \neq 0$ and $B \in G$ such that

$$\forall t > 0,\ BS(\mu_t)B^{-1} = \left\{A = \begin{pmatrix} a & b \\ c & d \end{pmatrix} \in G : a,d \geq 0,\ b,c \leq 0,\ \tau(A) \geq \cosh^2(\alpha t)\right\}$$

or

(iv) $\exists \alpha \neq 0$ and $B \in G$ such that

$$\forall t > 0,\ BS(\mu_t)B^{-1} = \left\{A \in G : \sigma(A) \leq 2\cosh(\alpha t)\right\}.$$

(b) Conversely, every one-parameter family of subsets of G listed under (a)(i)(ii)(iii)(iv) is the family of supports of some absolutely continuous Gauss semigroup on G.

There are some features of the above classification that are worth noting.

(I) The possibilities given in (a)(ii) and a(iii) are inverse to each other. By this we mean that the set on the right of the equality sign in a(ii) is the inverse set of the corresponding set in a(iii). This corresponds to the fact that if $(\mu_t)_{t>0}$ is an absolutely continuous Gauss semigroup on G, then so is $(\mu_t^*)_{t>0}$, where $\mu_t^*(E) = \mu_t(E^{-1})$ for all Borel sets $E \subseteq G$.

(II) The situation given in (a)(iv) is self-inverse. i.e. in this case $S(\mu_t)^{-1} = S(\mu_t)$.

(III) An absolutely continuous Gauss semigroup $(\mu_t)_{t>0}$ with supports as in a(ii) (or a(iii)) has the properties that

$$\forall t > 0, \qquad 1 \notin S(\mu_t), \text{ and}$$

$$\forall 0 < t < s, \qquad S(\mu_s) \subsetneqq S(\mu_t)$$

(IV) An absolutely continuous Gauss semigroup $(\mu_t)_{t>0}$ with supports as in a(iv) has the property that

$$\forall t > 0 \qquad 1 \in S(\mu_t)$$

$$\forall 0 < t < s, \qquad S(\mu_t) \subsetneqq S(\mu_s).$$

Note also that in this case we have $S(\mu_t)$ compact for all t.

In view of (III) and (IV) above it is now clear that an absolutely continuous Gauss semigroup on a connected Lie group does not necessarily possess a common support, nor indeed are the supports always decreasing or always increasing with increasing t.

§1. Carriers, supports and absolute continuity.

In this section we briefly recall some ideas and results from [2] which enable us to calculate the supports which interest us without having to concern ourselves with the actual measures that are supported. Besides [2] all other directly relevant background material on Gauss semigroups may be found in the book of Heyer [1].

Let G be a connected Lie group whose Lie algebra is $\mathfrak{g}$. For any Gauss semigroup $(\mu_t)_{t>0}$ on G we have its associated infinitesimal generator N and we may choose a basis $X_1, \ldots, X_n$ for $\mathfrak{g}$ such that N has the form

$$N = \sum_{i=1}^{n} a_i \tilde{X}_i + \sum_{i=1}^{r} \tilde{X}_i^2$$

for some $1 \leq r \leq n$, where for each $X \in \mathfrak{g}$ the symbol $\tilde{X}$ denotes the left-invariant Lie derivative determined by X [2]. Hence there is associated with each Gauss semigroup $(\mu_t)_{t>0}$ a pair $(\mathfrak{m}, X_0)$, consisting of $\mathfrak{m}$, the smallest subalgebra of $\mathfrak{g}$ containing $\{X_1, \ldots, X_r\}$, and the element $X_0 = \sum_{i=0}^{n} a_i X_i$. We shall call the pair $(\mathfrak{m}, X_0)$ the carrier of the semigroup $(\mu_t)_{t>0}$.

Conversely if $\mathfrak{m}$ is any subalgebra of $\mathfrak{g}$ and $X_0 \in \mathfrak{g}$, there exists at least one (highly non-unique) Gauss semigroup on G whose carrier is $(\mathfrak{m}, X_0)$ ([1], page 446).

The notion of the carrier of a Gauss semigroup $(\mu_t)_{t>0}$ is important here because it determines the support $S(\mu_t)$. To explain this connection we define, for every subalgebra $\mathfrak{m}$ of $\mathfrak{g}$ and every element $X \in \mathfrak{g}$, and every $t \in \mathbb{R}_+^*$, the set

$$R(\mathfrak{m}, X, t) = \overline{\bigcup_{n=1}^{\infty} \left(M \exp \frac{tX}{n}\right)^n} \subseteq G.$$

In this formula M denotes the analytic subgroup of G corresponding to the subalgebra $\mathfrak{m}$, exp is the usual exponential map from $\mathfrak{g}$ to G, the line denotes topological closure of the set below it, and for $A \subseteq G$ the notation A^n denotes $\{a_1 \ldots a_n : a_i \in A,\ 1 \leq i \leq n\}$ i.e. the set of all products in G of length n of elements from the set A. We shall need the following result of Siebert.

Theorem A [2]. Let $(\mu_t)_{t>0}$ be a Gauss semigroup on a connected Lie group G and let $(\mathfrak{m}, X)$ be the carrier of $(\mu_t)_{t>0}$. Then

$$\forall\, t \in \mathbb{R}_+^*, \quad S(\mu_t) = R(\mathfrak{m}, X, t).$$

The carrier of a Gauss semigroup also determines whether the measures μ_t are absolutely continuous or singular w.r.t. Haar measure on G. To explain this it is convenient to introduce the following notion. By an S-pair in $\mathfrak{g}$ (the Lie algebra of G) we shall mean a pair $(\mathfrak{m}, X)$, such that

(i) $\mathfrak{m}$ is a Lie subalgebra of $\mathfrak{g}$ and $X \in \mathfrak{g}$,

and (ii) the smallest subalgebra of $\mathfrak{g}$ which contains $\mathfrak{m}$ and is adX-invariant is $\mathfrak{g}$ itself.

By a <u>proper S-pair</u> in $\mathfrak{g}$ we shall mean an S-pair $(\mathfrak{m}, X)$ such that $\mathfrak{m} \neq \mathfrak{g}$.

<u>Theorem B</u> [2]. Let $(\mu_t)_{t>0}$ be a Gauss semigroup on the connected Lie group G, whose Lie algebra is $\mathfrak{g}$. Then either

(i) every μ_t is absolutely continuous w.r.t. Haar measure on G

or (ii) every μ_t is singular w.r.t. Haar measure on G.

Case (i) occurs if and only if $(\mathfrak{m}, X)$ is an S-pair in $\mathfrak{g}$, where $(\mathfrak{m}, X)$ is the carrier of $(\mu_t)_{t>0}$.

In view of theorems A and B we see that our original problem of calculating all possible supports for absolutely continuous Gauss semigroups on $G = SL(2, \mathbb{R})$ is equivalent to the problem of finding all possible proper S-pairs $(\mathfrak{m}, X)$ in $\mathfrak{g} = s\ell(2, \mathbb{R})$ and then computing the corresponding sets $R(\mathfrak{m}, X, t)$.

§2. <u>S-pairs in $s\ell(2, \mathbb{R})$</u>.

In this section $\mathfrak{g} = s\ell(2, \mathbb{R})$ and $G = SL(2, \mathbb{R})$. We want to give a description of all possible S-pairs in $\mathfrak{g}$. We begin by giving some examples of S-pairs in $\mathfrak{g}$.

<u>Proportion 1.</u> (i) Let $\mathfrak{h}$ be the one-dimensional subalgebra of $\mathfrak{g}$ generated by $\begin{pmatrix} 1 & 0 \\ 0 & -1 \end{pmatrix}$ and let $X = \begin{pmatrix} 0 & a \\ b & 0 \end{pmatrix}$ with $ab \neq 0$. Then $(\mathfrak{h}, X)$ is an S-pair in $\mathfrak{g}$.

(ii) Let $\mathfrak{k}$ be the one-dimensional subalgebra of $\mathfrak{g}$ generated by $\begin{pmatrix} 0 & 1 \\ 0 & 0 \end{pmatrix}$ and let $X = \begin{pmatrix} a & 0 \\ b & -a \end{pmatrix}$ with $b \neq 0$. Then $(\mathfrak{k}, X)$ is an S-pair in $\mathfrak{g}$.

(iii) Let $\mathfrak{l}$ be the one-dimensional subalgebra of $\mathfrak{g}$ generated by $\begin{pmatrix} 0 & 1 \\ -1 & 0 \end{pmatrix}$ and let $X = \begin{pmatrix} a & b \\ 0 & -a \end{pmatrix}$ with either $b \neq 0$ or $a \neq 0$. Then $(\mathfrak{l}, X)$ is an S-pair in $\mathfrak{g}$.

(iv) Let $\mathfrak{n}$ be the two-dimensional subalgebra $\{\begin{pmatrix} a & b \\ 0 & -a \end{pmatrix} : a, b \in \mathbb{R}\}$ and let $X = \begin{pmatrix} 0 & 0 \\ c & 0 \end{pmatrix}$ with $c \neq 0$. Then $(\mathfrak{n}, X)$ is an S-pair in $\mathfrak{g}$.

<u>Proof.</u> In each of the four cases let $\mathcal{A}$ be the smallest subalgebra of $\mathfrak{g}$ which contains the given subalgebra and is adX-invariant.

(i) $\mathcal{A}$ contains

$$[\begin{pmatrix} 0 & a \\ b & 0 \end{pmatrix}, \begin{pmatrix} \frac{1}{2} & 0 \\ 0 & -\frac{1}{2} \end{pmatrix}] = \begin{pmatrix} 0 & -a \\ b & 0 \end{pmatrix} \text{ and hence contains } [\begin{pmatrix} 0 & -a \\ b & 0 \end{pmatrix}, \begin{pmatrix} \frac{1}{2} & 0 \\ 0 & -\frac{1}{2} \end{pmatrix}] = \begin{pmatrix} 0 & a \\ b & 0 \end{pmatrix}.$$

By addition and subtraction, $\mathcal{A}$ also contains $\begin{pmatrix} 0 & a \\ 0 & 0 \end{pmatrix}$ and $\begin{pmatrix} 0 & 0 \\ b & 0 \end{pmatrix}$. Since $\mathcal{A}$ contains $\begin{pmatrix} 1 & 0 \\ 0 & -1 \end{pmatrix}$ also, it follows that $\mathcal{A} = \mathfrak{g}$.

(ii) For each $d \in \mathbb{R}$, $\mathcal{A}$ contains

$$[\begin{pmatrix} 0 & d \\ 0 & 0 \end{pmatrix}, \begin{pmatrix} a & 0 \\ b & -a \end{pmatrix}] + \begin{pmatrix} 0 & 2ad \\ 0 & 0 \end{pmatrix} = \begin{pmatrix} db & 0 \\ 0 & -db \end{pmatrix}.$$

As $b \neq 0$ it follows that $\mathcal{A}$ contains the subalgebra $\mathfrak{h}$ of (i). Hence $\mathcal{A}$ is invariant under $\mathrm{ad}\begin{pmatrix} a & 0 \\ 0 & -a \end{pmatrix}$, and so is invariant under $\mathrm{ad}\begin{pmatrix} 0 & 0 \\ b & 0 \end{pmatrix}$. But $\mathcal{A}$ is clearly also invariant under $\mathrm{ad}\begin{pmatrix} 0 & 1 \\ 0 & 0 \end{pmatrix}$, hence $\mathcal{A}$ is invariant under $\mathrm{ad}\begin{pmatrix} 0 & 1 \\ b & 0 \end{pmatrix}$. By

(i) it follows that $\mathcal{A} = \mathfrak{g}$.

(iii) $\mathcal{A}$ contains

$$\left[\begin{pmatrix} 0 & 1 \\ -1 & 0 \end{pmatrix}, \begin{pmatrix} a & b \\ 0 & -a \end{pmatrix}\right] + \begin{pmatrix} 0 & -2a \\ 2a & 0 \end{pmatrix} = \begin{pmatrix} b & -4a \\ 0 & -b \end{pmatrix}$$

hence $\mathcal{A}$ also contains

$$\left[\begin{pmatrix} b & -4a \\ 0 & -b \end{pmatrix}, \begin{pmatrix} a & b \\ 0 & -a \end{pmatrix}\right] = \begin{pmatrix} 0 & 2b^2+8a^2 \\ 0 & 0 \end{pmatrix} .$$

It follows that $\mathcal{A}$ contains the algebra $\mathfrak{k}$ of (ii) and as $\begin{pmatrix} 0 & 1 \\ -1 & 0 \end{pmatrix}$ lies in $\mathcal{A}$, $\mathcal{A}$ is invariant under $\mathrm{ad}\begin{pmatrix} 0 & 1 \\ -1 & 0 \end{pmatrix}$ and hence invariant under $\mathrm{ad}\begin{pmatrix} 0 & 0 \\ -1 & 0 \end{pmatrix}$. By (ii) we conclude that $\mathcal{A} = \mathfrak{g}$.

(iv) Follows at once from (ii) as $\mathfrak{k} \subseteq \mathfrak{n}$.

From now on the letters $\mathfrak{h}$, $\mathfrak{k}$, $\mathfrak{l}$, $\mathfrak{n}$ will be used to denote the specific subalgebras of $\mathfrak{g}$ given in (i)(ii)(iii)(iv) of Proposition 1. The S-pairs appearing in (i) of Proposition 1 will be called <u>S-pairs of type I</u>. Similarly we define <u>S-pairs of type II, III, IV</u> corresponding to (ii),(iii)(iv) respectively of Proposition I.

Given subalgebras $\mathfrak{m}$, $\mathfrak{m}'$ of $\mathfrak{g}$ and $X, X' \in \mathfrak{g}$, we say that $(\mathfrak{m}, X)$ and $(\mathfrak{m}', X')$ are <u>conjugate</u> if and only if there exists some $A \in G$ such that $A\mathfrak{m}A^{-1} = \mathfrak{m}'$ and $AXA^{-1} = X'$. We say that $(\mathfrak{m}, X)$ and $(\mathfrak{m}', X')$ are <u>equivalent</u> to mean that there exists $Y \in \mathfrak{g}$ such that $(\mathfrak{m}, X)$ and $(\mathfrak{m}', Y)$ are conjugate, and $(X' - Y) \in \mathfrak{m}'$.

<u>Proposition 2</u>. Let $(\mathfrak{m}, X)$ be a proper S-pair in $\mathfrak{g}$. Then $(\mathfrak{m}, X)$ is equivalent to an S-pair of type I, II, III or IV.

<u>Proof.</u> (a) Assume $\dim \mathfrak{m} = 1$. Pick $B \in \mathfrak{m}$ such that $\mathfrak{m} = \mathbb{R}B$ and consider the eigenvalues of B. The possibilities are

(i) B has real non-zero eigenvalues λ and $-\lambda$.

(ii) Zero is the only eigenvalue of B

(iii) the eigenvalues of B are non-real and conjugate, so are pure imaginary.

In case (i) there exists $A \in SL(2, \mathbb{R})$ such that $A\mathfrak{m}A^{-1} = \mathfrak{h}$, and if $Y = AXA^{-1}$ then $(\mathfrak{h}, Y)$ is an S-pair conjugate to $(\mathfrak{m}, X)$. If $Y = \begin{pmatrix} a & b \\ c & -a \end{pmatrix}$ let $Z = \begin{pmatrix} 0 & b \\ c & 0 \end{pmatrix}$ and note that $(\mathfrak{h}, Z)$ is an S-pair that is equivalent to $(\mathfrak{m}, X)$. If $b = 0$ then the subalgebra of lower-triangular matrices contains $\mathfrak{h}$ and is adZ-invariant, contradicting the fact that $(\mathfrak{h}, Z)$ is an S-pair. Hence $b \neq 0$ and for similar reasons $c \neq 0$. Hence $(\mathfrak{h}, Z)$ is of type I.

In case (ii) there exists $A \in SL(2, \mathbb{R})$ such that $A\mathfrak{m}A^{-1} = \mathfrak{k}$ and if $Y = AXA^{-1}$, then $(\mathfrak{k}, Y)$ is an S-pair conjugate to $(\mathfrak{m}, X)$. If $Y = \begin{pmatrix} a & b \\ c & -a \end{pmatrix}$ we let $Z = \begin{pmatrix} a & 0 \\ c & -a \end{pmatrix}$ and then clearly $(\mathfrak{k}, Z)$ is an S-pair equivalent to $(\mathfrak{m}, X)$ and $c \neq 0$, since otherwise the subalgebra of upper-triangular matrices contains $\mathfrak{k}$ and is adZ-invariant, which is impossible as $(\mathfrak{k}, Z)$ is an S-pair. Hence $(\mathfrak{k}, Z)$ is of type II.

In case (iii) we can find some $A \in SL(2, \mathbb{R})$ such that $A\mathfrak{m}A^{-1} = \mathfrak{l}$ and we have $(\mathfrak{m}, X)$ conjugate to an S-pair $(\mathfrak{l}, Y)$ for some $Y \in \mathfrak{g}$. If $Y = \begin{pmatrix} a & b \\ c & -a \end{pmatrix}$ we take $Z = \begin{pmatrix} a & d \\ 0 & -a \end{pmatrix}$ where $d = b + c$ and then $(\mathfrak{l}, Z)$ is an S-pair equivalent to $(\mathfrak{m}, X)$. Clearly $Z \neq 0$ and so $(\mathfrak{l}, Z)$ is an S-pair of type III.

(b) Assume $\dim \mathfrak{m} = 2$. By replacing $\mathfrak{m}$ by a conjugate if necessary, we may assume that either (i) $\mathfrak{h} \subseteq \mathfrak{m}$ or (ii) $\mathfrak{k} \subseteq \mathfrak{m}$ or (iii) $\mathfrak{l} \subseteq \mathfrak{m}$.

Case (iii) cannot arise because it is clear from Proposition 1 (iii) that the only subalgebra of $\mathfrak{g}$ which properly contains $\mathfrak{l}$ is $\mathfrak{g}$ itself.

In case (i) either $\mathfrak{m} = \mathfrak{n}$ or $\mathfrak{m}$ equals the algebra of lower-triangular matrices (which is conjugate to $\mathfrak{n}$) or $\mathfrak{m}$ contains an element of the form $\begin{pmatrix} 0 & b \\ c & 0 \end{pmatrix}$ with $bc \neq 0$, which is impossible by Proposition 1 (i), since $\mathfrak{m} \neq \mathfrak{g}$. So in case (i), $\mathfrak{m}$ is conjugate to $\mathfrak{n}$.

In case (ii) either $\mathfrak{m} = \mathfrak{n}$ or $\mathfrak{m}$ contains a matrix of the form $\begin{pmatrix} a & 0 \\ b & -a \end{pmatrix}$ with $b \neq 0$, which is impossible by Proposition 1 (ii).

So in all three cases we see that $(\mathfrak{m}, X)$ is conjugate to an S-pair $(\mathfrak{n}, Y)$ for some $Y = \begin{pmatrix} a & b \\ c & -a \end{pmatrix}$ and so is equivalent to the S-pair $(\mathfrak{n}, Z)$ where $Z = \begin{pmatrix} 0 & 0 \\ c & 0 \end{pmatrix}$. Clearly $c \neq 0$ and $(\mathfrak{n}, Z)$ is an S-pair of type IV.

§3. Properties of R-sets.

Now let G be any connected Lie group, let $\mathfrak{g}$ be its Lie algebra, let $\mathfrak{m}$ be a subalgebra of $\mathfrak{g}$ and let $X \in \mathfrak{g}$. Recall from section 1 the sets

$$R(\mathfrak{m}, X, t) = \overline{\bigcup_{n=1}^{\infty} \left(M \exp \frac{tX}{n}\right)^n}$$

where $t \in \mathbb{R}_+^*$ and M is the analytic subgroup of G corresponding to the subalgebra $\mathfrak{m}$.

Proposition 3. $R(\mathfrak{m}, X, t) = \overline{\left\{ \prod_{i=1}^{r} (m_i \exp t_i X) : r \geq 1, m_1, \ldots, m_r \in M, t_i > 0, \sum_{i=1}^{r} t_i = t \right\}}$

Proof It is immediate that L.H.S. $\subseteq$ R.H.S.
Now let $x = \prod_{i=1}^{r} (m_i \exp t_i X)$, where $t_i > 0$, $m_i \in M\ \forall\ 1 \leq i \leq r$, and $\sum_{i=1}^{r} t_i = t$. For each $1 \leq i \leq n$ pick a sequence p_{ij} of positive rationals such that $p_{ij} t \uparrow t_i$. Let $x_j = \prod_{i=1}^{r} (m_i \exp(p_{ij} t X))$ then clearly $x_j \to x$ as $j \to \infty$. If $p_{ij} = m_{ij}/n_{ij}$, and $n_j = n_{1j} n_{2j} \ldots n_{rj}$, then clearly $x_j \exp((1 - \sum_{i=1}^{r} p_{ij}) tX)$ lies in $(M \exp \frac{tX}{n_j})^{n_j}$. Hence $x = \lim_{j\to\infty} x_j$ lies in $R(\mathfrak{m}, X, t)$. Now take limits to obtain R.H.S. $\subseteq$ L.H.S.

Corollary For all $t, s \in \mathbb{R}_+^*$, $R(\mathfrak{m}, X, t) R(\mathfrak{m}, X, s) \subseteq R(\mathfrak{m}, X, t+s)$.

Proof Immediate from Proposition 3.

Remark The above corollary is hardly surprising, since the sets $R(\mathfrak{m}, X, t)$ are supports of some Gauss semigroup $(\mu_t)_{t>0}$, and so the corollary just asserts that $S(\mu_t) S(\mu_s) \subseteq S(\mu_{t+s})$. But it seems worthwhile to give a direct proof that is independent of the connection with measures.

<u>Proposition 4.</u> $R(\mathfrak{m}, X, t)$ is invariant under left and right multiplication by elements of M.

<u>Proof</u> Invariance on the left is clear, so we prove invariance on the right. Let $a \in \bigcup_{n=1}^{\infty} (M\exp\frac{tX}{n})^n$, then $\exists$ $k \geq 1$ and $m_1, \ldots, m_k \in M$ such that $a = m_1(\exp\frac{tX}{k})m_2(\exp\frac{tX}{k}) \ldots m_k(\exp\frac{tX}{k})$.

Let $m \in M$ and set $a_r = m_1(\exp\frac{tX}{k})m_2(\exp\frac{tX}{k}) \ldots m_k(\exp\frac{t(r-1)X}{kr})m(\exp\frac{tX}{kr})$ then clearly $a_r \in (M\exp\frac{tX}{kr})^{kr} \subseteq R(\mathfrak{m}, X, t)$, and $a_r \to am$ as $r \to \infty$. We conclude that $\forall\, m \in M$, $am \in R(\mathfrak{m}, X, t)$. We can now pass to the case when a is an arbitrary point of $R(\mathfrak{m}, X, t)$ by taking limits.

<u>Corollary 1.</u> $R(\mathfrak{m}, X, t) = \overline{\left\{\prod_{i=1}^{r}(m_i \exp t_i X): r\geq 1, m_1,\ldots,m_r \in M,\ t_i \geq 0 \text{ and } \sum_{i=1}^{r} t_i = t\right\}}$

<u>Proof.</u> Follows from Proposition 3 and 4.

<u>Corollary 2.</u> $R(\mathfrak{m}, X, t) = R(\mathfrak{m}, -X, t)^{-1}$.

<u>Proof</u> Immediate from Corollary 1.

<u>Corollary 3.</u> Let $\mathfrak{g} = s\ell(2, \mathbb{R})$, let $\mathfrak{m}$ be a subalgebra of $\mathfrak{g}$, let $X \in \mathfrak{g}$ and let $A \in M$ (the analytic subgroup of $SL(2, \mathbb{R})$ corresponding to $\mathfrak{m}$). Then $\forall\, t \in \mathbb{R}_+^*$, $R(\mathfrak{m}, X, t) = R(\mathfrak{m}, AXA^{-1}, t)$.

<u>Proof</u>
$$\begin{aligned} R(\mathfrak{m}, X, t) &= \overline{\bigcup_{n=1}^{\infty} (M\exp\tfrac{tX}{n})^n} \\ &= \overline{\bigcup_{n=1}^{\infty} (MA(\exp\tfrac{tX}{n})A^{-1})^n}\, A \\ &= \overline{\bigcup_{n=1}^{\infty} (M\exp\tfrac{t}{n}(AXA^{-1}))^n}\, A \\ &= R(\mathfrak{m}, AXA^{-1}, t)A = R(\mathfrak{m}, AXA^{-1}, t) \end{aligned}$$

by Proposition 4.

<u>Proposition 5.</u> Let G be a connected Lie group with Lie algebra $\mathfrak{g}$, let $\mathfrak{m}$ be a subalgebra of $\mathfrak{g}$ and let $X \in \mathfrak{g}$ and $Y \in \mathfrak{m}$. Then $\forall\, t \in \mathbb{R}_+^*$,

$$R(\mathfrak{m}, X, t) = R(\mathfrak{m}, X+Y, t).$$

<u>Proof.</u> For $n \geq 1$ let $a \in (M\exp\frac{t(X+Y)}{n})^n$. Then

$$a = m_1 \exp\frac{t(X+Y)}{n}\, m_2 \exp\frac{t(X+Y)}{n} \,\ldots\ldots\, m_n \exp\frac{t(X+Y)}{n}$$

for some $m_1, \ldots, m_n \in M$. For $k \geq 1$ we write

$$b_k = m_1(\exp\frac{tY}{kn}\exp\frac{tX}{kn})^k\, m_2(\exp\frac{tY}{kn}\exp\frac{tX}{kn})^k \,\ldots\, m_n(\exp\frac{tY}{kn}\exp\frac{tX}{kn})^k$$

and we note that since $\exp\frac{tY}{kn} \in M$, we have

$$\forall\, k \geq 1, \qquad b_k \in (M\exp\frac{tX}{kn})^{kn} \subseteq R(\mathfrak{m}, X, t).$$

But in view of the well-known relation

$$(\exp\frac{tY}{kn}\exp\frac{tX}{kn})^k \to \exp\frac{t(Y+X)}{n} \quad \text{as} \quad k\to\infty,$$

we see that $b_k \to a$ as $k\to\infty$ and so $a \in R(\mathfrak{m}, X, t)$. Hence $\forall\, t > 0$,

$$R(\mathfrak{m}, X+Y, t) = \overline{\bigcup_{n=1}^{\infty}(M\exp\frac{t(X+Y)}{n})^n} \subseteq R(\mathfrak{m}, X, t).$$

But the same argument shows that

$$\forall\, t > 0,\ R(\mathfrak{m}, X, t) = R(\mathfrak{m}, (X+Y)-Y, t)$$
$$\subseteq R(\mathfrak{m}, X+Y, t) \qquad \text{as} \quad -Y \in \mathfrak{m}.$$

The result follows.

Corollary Suppose $(\mathfrak{m}, X)$ and $(\mathfrak{m}', X')$ are equivalent S-pairs in $\mathfrak{g} = s\ell(2,\mathbb{R})$. Then $\exists\, A \in SL(2,\mathbb{R})$ such that

$$\forall\, t \in \mathbb{R}_+^*, \qquad AR(\mathfrak{m}, X, t)A^{-1} = R(\mathfrak{m}', X', t).$$

Proof. Since $(\mathfrak{m}, X)$ and $(\mathfrak{m}', X')$ are equivalent, there exists $Y \in \mathfrak{g}$ such that $(\mathfrak{m}, X)$ is conjugate to $(\mathfrak{m}', Y)$ and $X' - Y \in \mathfrak{m}'$. Hence by proposition 5,

$$\forall\, t \in \mathbb{R}_+^*, \qquad R(\mathfrak{m}', X', t) = R(\mathfrak{m}', Y, t).$$

But there exists $A \in SL(2,\mathbb{R})$ such that $A\mathfrak{m}A^{-1} = \mathfrak{m}'$ and $AXA^{-1} = Y$, so for all $t \in \mathbb{R}_+^*$

$$AR(\mathfrak{m}, X, t)A^{-1} = \overline{\bigcup_{n=1}^{\infty} A(M\exp\frac{tX}{n})^n A^{-1}}$$
$$= \overline{\bigcup_{n=1}^{\infty}(AMA^{-1}\exp\frac{t}{n}(AXA^{-1}))^n}$$
$$= \overline{\bigcup_{n=1}^{\infty}(M'\exp\frac{tY}{n})^n} = R(\mathfrak{m}', Y, t),$$

where M' is the analytic subgroup of $SL(2,\mathbb{R})$ corresponding to $\mathfrak{m}'$.

§4. R-sets for S-pairs of type I.

From this point on, G will always denote $SL(2,\mathbb{R})$ and $\mathfrak{g}$ will always denote $s\ell(2,\mathbb{R})$. In view of Proposition 2 and Proposition 5, Corollary, the problem of finding all possibilities for the families $\{R(\mathfrak{m}, X, t) : t \in \mathbb{R}_+^*\}$ as $(\mathfrak{m}, X)$ runs over all possible proper S-pairs in $\mathfrak{g}$ reduces to the problem of determining these families for S-pairs of type I, II, III, IV. In this section we solve this problem for S-pairs of type I.

We begin with a further reduction of the problem.

Proposition 6. Suppose $(\mathfrak{h}, X)$ is an S-pair in $\mathfrak{g}$ of type I, so that

$X = \begin{pmatrix} 0 & a \\ b & 0 \end{pmatrix}$ with $ab \neq 0$.

(i) If $a > 0$, $b > 0$ then

$$\forall\, t > 0,\ R(\mathfrak{h}, X, t) = R(\mathfrak{h}, E_\alpha, t)$$

where $E_\alpha = \begin{pmatrix} 0 & \alpha \\ \alpha & 0 \end{pmatrix}$ with $\alpha^2 = ab$ and $\alpha > 0$.

(ii) If $a > 0$, $b < 0$ then

$$\forall\, t > 0,\ R(\mathfrak{h}, X, t) = R(\mathfrak{h}, F_\alpha, t)$$

where $F_\alpha = \begin{pmatrix} 0 & \alpha \\ -\alpha & 0 \end{pmatrix}$ with $\alpha^2 = |ab|$ and $\alpha > 0$.

(iii) If $a < 0$, $b < 0$, then

$$\forall\, t > 0,\ R(\mathfrak{h}, X, t) = R(\mathfrak{h}, -E_\alpha, t)$$

with E_α as in (i) and $\alpha^2 = ab$ with $\alpha > 0$.

(iv) If $a < 0$, $b > 0$, then

$$\forall\, t > 0,\ R(\mathfrak{h}, X, t) = R(\mathfrak{h}, -F_\alpha, t)$$

with F_α as in (ii) and $\alpha^2 = |ab|$ with $\alpha > 0$.

Proof. All the above assertions follow at once from Proposition 4, Corollary 3, and the observation that

$$\begin{pmatrix} \lambda & 0 \\ 0 & \lambda^{-1} \end{pmatrix} \begin{pmatrix} 0 & a \\ b & 0 \end{pmatrix} \begin{pmatrix} \lambda & 0 \\ 0 & \lambda^{-1} \end{pmatrix}^{-1} = \begin{pmatrix} 0 & \lambda^2 a \\ \lambda^{-2} b & 0 \end{pmatrix}$$

So we obtain the result in all cases by taking $\lambda = \left|\frac{b}{a}\right|^{\frac{1}{4}}$.

In view of Proposition 4, Corollary 2 and Proposition 6, it is sufficient to compute the families $\{R(\mathfrak{h}, E_\alpha, t) : t \in \mathbb{R}_+^*\}$ and $\{R(\mathfrak{h}, F_\alpha, t) : t \in \mathbb{R}_+^*\}$ for all $\alpha \in \mathbb{R}_+^*$.

For a matrix $A = \begin{pmatrix} a & b \\ c & d \end{pmatrix}$ we write $\tau(A)$ for ad, and we let H be the analytic subgroup of G corresponding to $\mathfrak{h}$, so that H is the subgroup of diagonal matrices with positive entries.

Proposition 7. Let $t \in \mathbb{R}_+^*$ and write $a(t) = \begin{pmatrix} \cosh t & \sinh t \\ \sinh t & \cosh t \end{pmatrix}$. Then $\forall\, n \geq 1$ and $\forall\, A \in (Ha(t))^n$ we have

(i) all entries of A are (strictly) positive,

(ii) $\tau(A) \geq \cosh^2(nt)$.

Proof. (i) follows because the product of positive matrices remains positive and clearly any matrix in $Ha(t)$ has positive entries. To prove (ii) we proceed by induction on n, observing first that the result is clear when $n = 1$. So we assume true for $n = r$ and let $A \in (Ha(t))^{r+1}$. Then $\exists\, \lambda \in \mathbb{R}_+^*$ and $B = \begin{pmatrix} \alpha & \beta \\ \gamma & \delta \end{pmatrix} \in (Ha(t))^r H$ such that

$$A = \begin{pmatrix} \alpha & \beta \\ \gamma & \delta \end{pmatrix} \begin{pmatrix} \cosh t & \sinh t \\ \sinh t & \cosh t \end{pmatrix}$$

$$= \begin{pmatrix} \alpha \cosh t + \beta \sinh t & \alpha \sinh t + \beta \cosh t \\ \gamma \cosh t + \delta \sinh t & \gamma \sinh t + \delta \cosh t \end{pmatrix}$$

But clearly τ is unchanged by multiplication on either side by elements of H, so $\exists\, C \in (Ha(t))^r$ such that $\tau(B) = \tau(C)$, hence by inductive hypothesis, $\alpha\delta \geq \cosh^2(rt)$. Further, by (i), $\alpha, \beta, \gamma, \delta > 0$. Hence

$$\begin{aligned}\tau(A) &= \alpha\delta\cosh^2 t + \beta\gamma\sinh^2 t + (\alpha\gamma + \beta\delta)\sinh t\cosh t\\ &\geq \alpha\delta\cosh^2 t + \beta\gamma\sinh^2 t + 2(\alpha\gamma\beta\delta)^{\frac{1}{2}}\sinh t\cosh t\\ &= ((\alpha\delta)^{\frac{1}{2}}\cosh t + (\beta\gamma)^{\frac{1}{2}}\sinh t)^2\end{aligned}$$

As $\alpha\delta - \beta\gamma = 1$ we have

$$\beta\gamma = \alpha\delta - 1 \geq \cosh^2(rt) - 1 = \sinh^2(rt)$$

hence

$$\begin{aligned}\tau(A) &\geq (\cosh(rt)\cosh t + \sinh(rt)\sinh t)^2\\ &= \cosh^2(r+1)t.\end{aligned}$$

By induction (ii) is shown.

Corollary. Let $t \in \mathbb{R}_+^*$ and $A \in R(\mathfrak{h}, E_\alpha, t)$. Then A has all its entries non-negative and $\tau(A) \geq \cosh^2(\alpha t)$.

Proof. We have $\exp\frac{tE_\alpha}{n} = \begin{pmatrix}\cosh(\frac{t\alpha}{n}) & \sinh(\frac{t\alpha}{n})\\ \sinh(\frac{t\alpha}{n}) & \cosh(\frac{t\alpha}{n})\end{pmatrix}$, so if $A \in (H\exp\frac{tE_\alpha}{n})^n$, Proposition 7 implies that A has positive entries and $\tau(A) \geq \cosh^2(\alpha t)$. The result for general elements of $R(\mathfrak{h}, E_\alpha, t)$ follows by taking limits and using the continuity of τ.

Theorem 1. For all $t, \alpha \in \mathbb{R}_+^*$

$$R(\mathfrak{h}, E_\alpha, t) = \{A \in G : \tau(A) \geq \cosh^2(\alpha t) \text{ and all entries of } A \text{ non-negative}\}.$$

Proof. In view of Proposition 7, Corollary, it suffices to show that if $A = \begin{pmatrix}\varepsilon & \beta\\ \gamma & \delta\end{pmatrix} \in G$, with $\varepsilon, \beta, \gamma, \delta > 0$ and $\varepsilon\delta \geq \cosh^2(\alpha t)$, then $A \in R(\mathfrak{h}, E_\alpha, t)$.

Note that for each $\lambda > 0$, $R(\mathfrak{h}, E_\alpha, t)$ contains

$$B_\lambda = \begin{pmatrix}\cosh\frac{\alpha t}{2} & \sinh\frac{\alpha t}{2}\\ \sinh\frac{\alpha t}{2} & \cosh\frac{\alpha t}{2}\end{pmatrix}\begin{pmatrix}\lambda & 0\\ 0 & \lambda^{-1}\end{pmatrix}\begin{pmatrix}\cosh\frac{\alpha t}{2} & \sinh\frac{\alpha t}{2}\\ \sinh\frac{\alpha t}{2} & \cosh\frac{\alpha t}{2}\end{pmatrix}$$

and we may calculate that

$$\tau(B_\lambda) = (\lambda^2 + \lambda^{-2})\sinh^2(\tfrac{\alpha t}{2})\cosh^2(\tfrac{\alpha t}{2}) + \cosh^4(\tfrac{\alpha t}{2}) + \sinh^4(\tfrac{\alpha t}{2}).$$

Since $\tau(B_1) = \cosh^2(\alpha t)$ and $\tau(B_\lambda) \to \infty$ as $\lambda \to \infty$, and as τ is continuous and $R(\mathfrak{h}, E_\alpha, t)$ is clearly connected, we conclude that

$$\{\tau(B) : B \in R(\mathfrak{h}, E_\alpha, t)\} = [\cosh^2(\alpha t), \infty).$$

Hence there exists $B \in R(\mathfrak{h}, E_\alpha, t)$ such that $\tau(B) = \tau(A)$. But in view of the relations

$$\begin{pmatrix} \lambda & 0 \\ 0 & \lambda^{-1} \end{pmatrix}\begin{pmatrix} \varepsilon & \beta \\ \gamma & \delta \end{pmatrix}\begin{pmatrix} \lambda & 0 \\ 0 & \lambda^{-1} \end{pmatrix} = \begin{pmatrix} \lambda^2\varepsilon & \beta \\ \gamma & \lambda^{-2}\delta \end{pmatrix}$$

$$\begin{pmatrix} \lambda & 0 \\ 0 & \lambda^{-1} \end{pmatrix}\begin{pmatrix} \varepsilon & \beta \\ \gamma & \delta \end{pmatrix}\begin{pmatrix} \lambda^{-1} & 0 \\ 0 & \lambda \end{pmatrix} = \begin{pmatrix} \varepsilon & \lambda^2\beta \\ \lambda^{-2}\gamma & \delta \end{pmatrix}$$

we may find positive diagonal matrices D_1, D_2 such that

$$D_1 AD_2 = \begin{pmatrix} x & y \\ y & x \end{pmatrix}$$

for some $x, y > 0$, and as multiplication by diagonal matrices does not change τ, we have $\tau(A) = x^2$ and $y^2 = x^2 - 1 = \tau(A) - 1$. But the same reasoning applies to B, and there exist positive diagonal matrices D_1', D_2' such that

$$D_1' B D_2' = \begin{pmatrix} x & y \\ y & x \end{pmatrix}.$$

Hence $A = D_1^{-1}D_1' B D_2' D_2^{-1}$ and since $B \in R(\mathfrak{h}, E_\alpha, t)$ and the latter set is H-invariant by Proposition 4, we conclude that $A \in R(\mathfrak{h}, E_\alpha, t)$ as required.

We now turn to the determination of $R(\mathfrak{h}, F_\alpha, t)$ for all $t, \alpha \in \mathbb{R}_+^*$.

Proposition 8. Let $x = \begin{pmatrix} a & b \\ c & d \end{pmatrix} \in G$ with $abcd \neq 0$, $bc < 0$ and $(\frac{a}{b} + \frac{c}{d}) > 0$. Then $\begin{pmatrix} 1 & 1 \\ -1 & 0 \end{pmatrix} \in (Hx)^2H$.

Proof. For every $\lambda, \mu > 0$, $(Hx)^2$ contains

$$\begin{pmatrix} \lambda & 0 \\ 0 & \lambda^{-1} \end{pmatrix}\begin{pmatrix} a & b \\ c & d \end{pmatrix}\begin{pmatrix} \mu & 0 \\ 0 & \mu^{-1} \end{pmatrix}\begin{pmatrix} a & b \\ c & d \end{pmatrix}$$

$$= \begin{pmatrix} \lambda a^2\mu + \lambda bc\mu^{-1} & \lambda ab\mu + \lambda bd\mu^{-1} \\ \lambda^{-1}ca\mu + \lambda^{-1}dc\mu^{-1} & \lambda^{-1}bc\mu + \lambda^{-1}d^2\mu^{-1} \end{pmatrix}.$$

If we take $\mu = (-d^2b^{-1}c^{-1})^{\frac{1}{4}}$, $\lambda = (-cb^{-1})^{\frac{1}{2}}$, this product becomes the matrix $\begin{pmatrix} \frac{a}{b}+\frac{c}{d} & 1 \\ -1 & 0 \end{pmatrix}$. Taking $f = (\frac{a}{b}+\frac{c}{d})^{\frac{1}{2}}$ we may postmultiply and premultiply this last matrix by the matrix $\begin{pmatrix} f^{-1} & 0 \\ 0 & f \end{pmatrix}$ to obtain the matrix $\begin{pmatrix} 1 & 1 \\ -1 & 0 \end{pmatrix}$ as an element of $(Hx)^2H$.

Corollary 1. Let x satisfy the conditions of Proposition 8. Then $1 \in (Hx)^{12}H$.

Proof. Follows from Proposition 8 because $\begin{pmatrix} 1 & 1 \\ -1 & 0 \end{pmatrix}^6 = 1$.

Corollary 2. Suppose $x = \begin{pmatrix} \cos t & \sin t \\ -\sin t & \cos t \end{pmatrix}$ for some $0 < t < \frac{\pi}{4}$. Then $1 \in (Hx)^{12}H$.

Proof. Immediate from Corollary 1.

Corollary 3. For all $t \in \mathbb{R}_+^*$, $I \in R(\mathfrak{h}, F_1, t)$.

<u>Proof.</u> Note that $\exp(\frac{tF_1}{n}) = \begin{pmatrix} \cos\frac{t}{n} & \sin\frac{t}{n} \\ -\sin\frac{t}{n} & \cos\frac{t}{n} \end{pmatrix}$. So if $0 < t < 3\pi$, Corollary 2 implies that

$$1 \in (H\exp\frac{tF_1}{12})^{12} H \subseteq R(\mathfrak{h}, F_1, t)$$

by Proposition 4. But by Proposition 3, Corollary, for all $m \geq 1$,

$$R(\mathfrak{h}, F_1, t)^m \subseteq R(\mathfrak{h}, F_1, mt)$$

so it follows that $1 \in R(\mathfrak{h}, F_1, s)$ for all $s \in \mathbb{R}_+^*$.

<u>Corollary 4.</u> For all $t < s \in \mathbb{R}_+^*$, $R(\mathfrak{h}, F_1, t) \subseteq R(\mathfrak{h}, F_1, s)$.

<u>Proof.</u> Follows from Proposition 3, Corollary, and Corollary 3 above.

<u>Proposition 9.</u> Suppose $0 < t < \frac{\pi}{2}$ and $t \leq u < \frac{\pi}{2}$. Then

$$\begin{pmatrix} \cos u & \sin u \\ -\sin u & \cos u \end{pmatrix} \in (Hx)^2 H, \text{ where } x = \begin{pmatrix} \cos\frac{t}{2} & \sin\frac{t}{2} \\ -\sin\frac{t}{2} & \cos\frac{t}{2} \end{pmatrix}.$$

<u>Proof.</u> We note that for all $b > 0$, xHx contains

$$C_b = \begin{pmatrix} \cos\frac{t}{2} & \sin\frac{t}{2} \\ -\sin\frac{t}{2} & \cos\frac{t}{2} \end{pmatrix} \begin{pmatrix} b & 0 \\ 0 & b^{-1} \end{pmatrix} \begin{pmatrix} \cos\frac{t}{2} & \sin\frac{t}{2} \\ -\sin\frac{t}{2} & \cos\frac{t}{2} \end{pmatrix}$$

$$= \begin{pmatrix} b\cos^2\frac{t}{2} - b^{-1}\sin^2\frac{t}{2} & (b+b^{-1})\sin\frac{t}{2}\cos\frac{t}{2} \\ -(b+b^{-1})\sin\frac{t}{2}\cos\frac{t}{2} & -b\sin^2\frac{t}{2} + b^{-1}\cos^2\frac{t}{2} \end{pmatrix}.$$

Then

$$\tau(C_b) = \sin^4\frac{t}{2} + \cos^4\frac{t}{2} - \cos^2\frac{t}{2}\sin^2\frac{t}{2}(b + b^{-1})$$

and as b goes from 1 to ∞ $\tau(C_b)$ decreases from $(\cos t)^2$ to $-\infty$. But $0 < \cos u \leq \cos t$, so there exists $b_0 \geq 1$ such that $\tau(C_{b_0}) = (\cos u)^2$.

So if $C_{b_0} = \begin{pmatrix} \alpha & \beta \\ -\beta & \gamma \end{pmatrix}$ (the off-diagonal elements in the matrix C_b are always mutual negatives) we have $\alpha\gamma = \cos^2 u$, and note also that as $0 < t < \frac{\pi}{2}$ we have $\cos^2\frac{t}{2} > \sin^2\frac{t}{2}$, hence as $b_0 \geq 1$ we must have $\alpha > 0$.

It follows that $\alpha^{-1}\cos u > 0$, and so $(Hx)^2H$ contains

$$\begin{pmatrix} (\alpha^{-1}\cos u)^{\frac{1}{2}} & 0 \\ 0 & (\alpha^{-1}\cos u)^{-\frac{1}{2}} \end{pmatrix} \begin{pmatrix} \alpha & \beta \\ -\beta & \gamma \end{pmatrix} \begin{pmatrix} (\alpha^{-1}\cos u)^{\frac{1}{2}} & 0 \\ 0 & (\alpha^{-1}\cos u)^{-\frac{1}{2}} \end{pmatrix} = \begin{pmatrix} \cos u & \beta \\ -\beta & \cos u \end{pmatrix}$$

As we are in $SL(2, \mathbb{R})$, we conclude that $\beta^2 = \sin^2 u$, and so $\beta = \sin u$, because $0 < u < \frac{\pi}{2}$ and the top right-hand corner of C_b is clearly always non-negative. The proof is now complete.

Corollary. If $0 < t < 2\pi$ then $\forall\, u \in \mathbb{R}$, $\begin{pmatrix} \cos u & \sin u \\ -\sin u & \cos u \end{pmatrix} \in R(\mathfrak{h}, F_1, t)$.

Proof. If $0 \leqslant u \leqslant t$ the conclusion follows at once from Proposition 8 Corollary 4, since $\begin{pmatrix} \cos u & \sin u \\ -\sin u & \cos u \end{pmatrix} \in R(\mathfrak{h}, F_1, u)$.

If $0 < t < \frac{\pi}{2}$ and $t \leqslant u \leqslant \frac{\pi}{2}$ then by Proposition 4 and Proposition 9,

$$\begin{pmatrix} \cos u & \sin u \\ -\sin u & \cos u \end{pmatrix} \in \overline{(H \exp \frac{tF_1}{2})^2}\, H \subseteq R(\mathfrak{h}, F_1, t).$$

Combining the two we see that if $0 < t < \frac{\pi}{2}$ and $0 \leqslant u \leqslant \frac{\pi}{2}$, then

$$\begin{pmatrix} \cos u & \sin u \\ -\sin u & \cos u \end{pmatrix} \in R(\mathfrak{h}, F_1, t).$$

Hence if $0 < t < 2\pi$ and $0 \leqslant u \leqslant \frac{\pi}{2}$,

$$\begin{pmatrix} \cos u & \sin u \\ -\sin u & \cos u \end{pmatrix} \in R(\mathfrak{h}, F_1, \tfrac{t}{4}).$$

Hence by Proposition 3 corollary it follows that if $0 < t < 2\pi$ and $v \in [0, 2\pi]$

$$\begin{pmatrix} \cos v & \sin v \\ -\sin v & \cos v \end{pmatrix} = \begin{pmatrix} \cos \frac{v}{4} & \sin \frac{v}{4} \\ -\sin \frac{v}{4} & \cos \frac{v}{4} \end{pmatrix}^4 \in R(\mathfrak{h}, F_1, \tfrac{t}{4})^4 \subseteq R(\mathfrak{h}, F_1, t).$$

Theorem 2. For all $\alpha, t \in \mathbb{R}_+^*$, $R(\mathfrak{h}, F_\alpha, t) = G$.

Proof. Let L be the circle subgroup of G, then given $A \in G$, there exist $D \in H$ and $U, V \in L$ such that $A = UDV$. If $0 < t < 4\pi$ then $U, V \in R(\mathfrak{h}, F_1, \frac{t}{2})$ by Proposition 9, corollary, so by Proposition 4 and Proposition 3 corollary,

$$A = UDV \subseteq R(\mathfrak{h}, F_1, \tfrac{t}{2}) H R(\mathfrak{h}, F_1, \tfrac{t}{2}) \subseteq R(\mathfrak{h}, F_1, t).$$

Hence $\forall\, t \in (0, 4\pi]$, $R(\mathfrak{h}, F_1, t) = G$. Hence by Proposition 3, corollary, for $t \in (0, 4\pi]$ and $m \geqslant 1$, $R(\mathfrak{h}, F_1, mt) = G^m = G$, which gives the result for $\alpha = 1$.

Finally, for any $\alpha, t \in \mathbb{R}_+^*$, $R(\mathfrak{h}, F_\alpha, t) = R(\mathfrak{h}, F_1, \alpha t) = G$.

§5. R-sets for S-pairs of type II and IV.

In this section we compute the R-sets in G for S-pairs of type II and IV. The calculations in this case are very easy. We denote by K the analytic subgroup of G corresponding to $\mathfrak{k}$, so that

$$K = \left\{ \begin{pmatrix} 1 & a \\ 0 & 1 \end{pmatrix} : a \in \mathbb{R} \right\}.$$

Proposition 10. Let $x = \begin{pmatrix} \alpha & 0 \\ \beta & \alpha^{-1} \end{pmatrix}$ with $\alpha\beta \neq 0$. Then for each $d \in \mathbb{R}$ $\begin{pmatrix} 0 & -\beta^{-1} \\ \beta & d \end{pmatrix}$ lies in KxK.

Proof. For each $c \in \mathbb{R}$, KxK contains

$$\begin{pmatrix} 1 & -\alpha\beta^{-1} \\ 0 & 1 \end{pmatrix}\begin{pmatrix} \alpha & 0 \\ \beta & \alpha^{-1} \end{pmatrix}\begin{pmatrix} 1 & c \\ 0 & 1 \end{pmatrix} = \begin{pmatrix} 0 & -\beta^{-1} \\ \beta & \beta c+\alpha^{-1} \end{pmatrix}$$

and we can choose c so that $\beta c + \alpha^{-1}$ has any desired value.

Corollary 1. Let x be as in Proposition 10. Then for each $c \in \mathbb{R}\setminus\{0\}$, $\begin{pmatrix} -1 & c \\ -c^{-1} & 0 \end{pmatrix}$ lies in $(Kx)^2K$.

Proof. By Proposition 10, $(Kx)^2K$ contains

$$\begin{pmatrix} 0 & -\beta^{-1} \\ \beta & -\beta^{-1}c^{-1} \end{pmatrix}\begin{pmatrix} 0 & -\beta^{-1} \\ \beta & -\beta c \end{pmatrix} = \begin{pmatrix} -1 & c \\ -c^{-1} & 0 \end{pmatrix}.$$

Corollary 2. Let x be as in Proposition 10. Then for all $d \in \mathbb{R}$ and all $c \in \mathbb{R}\setminus\{0\}$, the matrix $\begin{pmatrix} d & c \\ -c^{-1} & 0 \end{pmatrix}$ lies in $(Kx)^2K$.

Proof. By Proposition 10, corollary 1, $(Kx)^2K$ contains

$$\begin{pmatrix} 1 & -(d+1)c \\ 0 & 1 \end{pmatrix}\begin{pmatrix} -1 & c \\ -c^{-1} & 0 \end{pmatrix} = \begin{pmatrix} d & c \\ -c^{-1} & 0 \end{pmatrix}.$$

Corollary 3. Let x be as in Proposition 10. Then $(Kx)^2K$ contains all elements of G of the form $\begin{pmatrix} \alpha & \beta \\ \gamma & \delta \end{pmatrix}$, with $\gamma \neq 0$.

Proof. By Proposition 10, Corollary 2, $(Kx)^2K$ contains

$$\begin{pmatrix} \alpha & -\gamma^{-1} \\ \gamma & 0 \end{pmatrix}\begin{pmatrix} 1 & \delta\gamma^{-1} \\ 0 & 1 \end{pmatrix} = \begin{pmatrix} \alpha & \beta \\ \gamma & \delta \end{pmatrix}.$$

Corollary 4. Let x be as in Proposition 10. Then $\overline{(Kx)^2K} = G$.

Theorem 3. Let $(\mathfrak{m}, X)$ be an S-pair in $\mathfrak{g}$ of type II or IV. Then for all $t \in \mathbb{R}_+^*$, $R(\mathfrak{m}, X, t) = G$.

Proof. We note that if $X \in \mathfrak{g}$ has the form $\begin{pmatrix} a & 0 \\ b & -a \end{pmatrix}$ with $b \neq 0$, then $\exp\frac{tX}{2}$ has the form $\begin{pmatrix} \alpha & 0 \\ \beta & \alpha^{-1} \end{pmatrix}$ with $\alpha\beta \neq 0$. So by Proposition 10 Corollary 4, and Proposition 4,

$$G \subseteq \overline{(K\exp\tfrac{tX}{2})^2 K} \subseteq R(\mathfrak{k}, X, t).$$

Finally if $(\mathfrak{n}, X)$ is an S-pair of type IV, then $(\mathfrak{k}, X)$ is an S-pair of type II, and as $\mathfrak{k} \subseteq \mathfrak{n}$, we have

$$G = R(\mathfrak{k}, X, t) \subseteq R(\mathfrak{n}, X, t),$$

for every $t \in \mathbb{R}_+^*$.

§6. R-sets for S-pairs of type III.

In this section we calculate the R-sets corresponding to S-pairs of type III. So we let $(\mathfrak{l}, X)$ be such an S-pair, and we note that the analytic subgroup of G corresponding to $\mathfrak{l}$ is

$$L = \left\{ \begin{pmatrix} \cos t & \sin t \\ -\sin t & \cos t \end{pmatrix} : t \in \mathbb{R} \right\}.$$

We treat first the case when X is diagonal.

For a matrix $A \in G$ we use $\sigma(A)$ to denote the sum of the squares of the entries in A and we note that σ is invariant under left and right multiplication by elements of L, and that $\sigma(A) = \sigma(A^{-1})$.

Proposition 11. If $A = \begin{pmatrix} a & b \\ c & d \end{pmatrix} \in G$ with $\sigma(A) \leq 2\cosh\alpha$ for some $\alpha > 0$, then $(a^2 - b^2) + (c^2 - d^2) \leq 2\sinh\alpha$.

Proof. By the Cauchy-Schwartz inequality

$$1 = ad - bc \leq (a^2 + c^2)^{\frac{1}{2}}(b^2 + d^2)^{\frac{1}{2}}$$

Then

$$(a^2 + c^2 - (b^2 + d^2))^2 = (a^2 + b^2 + c^2 + d^2)^2 - 4(a^2 + c^2)(b^2 + d^2)$$

$$\leq 4\cosh^2\alpha - 4 = (2\sinh\alpha)^2$$

and the result follows because $\sinh\alpha \geq 0$.

Proposition 12. Suppose $\alpha > 0$ and let $y = \begin{pmatrix} e^{\alpha} & 0 \\ 0 & e^{-\alpha} \end{pmatrix}$. Then $\forall n \geq 1$ and $\forall z \in (Ly)^n$ we have $\sigma(z) \leq 2\cosh(2n\alpha)$.

Proof. Clearly the result is true when $n = 1$ by invariance of σ under multiplication by elements of L.

Now suppose it is true for some n and let $z \in (Ly)^{n+1}$. Then

$$z = \begin{pmatrix} a & b \\ c & d \end{pmatrix}\begin{pmatrix} e^{\alpha} & 0 \\ 0 & e^{-\alpha} \end{pmatrix} = \begin{pmatrix} ae^{\alpha} & be^{-\alpha} \\ ce^{\alpha} & de^{-\alpha} \end{pmatrix}$$

where $\begin{pmatrix} a & b \\ c & d \end{pmatrix} \in (Ly)^n L$ and so by the inductive hypothesis and invariance of σ under L we have $a^2 + b^2 + c^2 + d^2 \leq 2\cosh(2n\alpha)$.

Then by Proposition 11 we have

$$\begin{aligned} \sigma(z) &= e^{2\alpha}(a^2 + c^2) + e^{-2\alpha}(b^2 + d^2) \\ &= \cosh(2\alpha)(a^2 + b^2 + c^2 + d^2) + \sinh(2\alpha)(a^2 + c^2 - (b^2 + d^2)) \\ &\leq 2\cosh(2\alpha)\cosh(2n\alpha) + 2\sinh(2\alpha)\sinh(2n\alpha) \\ &= 2\cosh((2n+1)\alpha). \end{aligned}$$

The result follows by induction.

Corollary 1. The statement of Proposition 12 holds good for any $\alpha \neq 0$.

Proof. Follows because for every $A \in G$, $\sigma(A) = \sigma(A^{-1})$ and because σ is invariant under multiplication by elements of L.

Corollary 2. If $X = \begin{pmatrix} a & 0 \\ 0 & -a \end{pmatrix}$ with $a \neq 0$, then for all $t \in \mathbb{R}_+^*$ and for all $z \in R(\mathfrak{l}, X, t)$ we have $\sigma(z) \leq 2\cosh(2at)$. The upper bound is attained at the point $\exp tX \in R(\mathfrak{l}, X, t)$.

Proof. We have $\exp \frac{tX}{n} = \begin{pmatrix} e^{at/n} & 0 \\ 0 & e^{-at/n} \end{pmatrix}$ so by Proposition 12, Corollary 1 we have $\sigma(z) \leq 2\cosh(2at)$ for all $z \in (L \exp \frac{tX}{n})^n$. Now take limits to obtain the inequality for an arbitrary point of $R(\mathfrak{l}, X, t)$.

Proposition 13. If x is a diagonal matrix then $1 \in (Lx)^4$.

Proof. The matrix $J = \begin{pmatrix} 0 & 1 \\ -1 & 0 \end{pmatrix}$ lies in L and computation yields $(Jx)^4 = 1$.

Corollary. If $X = \begin{pmatrix} a & 0 \\ 0 & -a \end{pmatrix}$ with $a \neq 0$, then for all $t \in \mathbb{R}_+^*$, $1 \in R(\mathfrak{l}, X, t)$.

Theorem 4. If $X = \begin{pmatrix} a & 0 \\ 0 & -a \end{pmatrix}$ with $a \neq 0$, then for all $t \in \mathbb{R}_+^*$,

$$R(\mathfrak{l}, X, t) = \left\{ A \in G : \sigma(A) \leq 2\cosh(2at) \right\}.$$

Proof. Note first that for $A \in G$ we have $\sigma(A) \geq 2$. This is obvious for diagonal matrices and follows for general A by using the invariance of σ under L, and the decomposition $G = LHL$. Since σ is continuous and $R(\mathfrak{l}, X, t)$ is connected, contains 1 (Proposition 13, Corollary) and contains $\exp tX$, we see that

$$\left\{ \sigma(A) : A \in R(\mathfrak{l}, X, t) \right\} \supseteq [2, 2\cosh(2at)].$$

So if $A \in G$ with $\sigma(A) \leq 2\cosh(2at)$ it follows that there exists $B \in R(\mathfrak{l}, X, t)$ such that $\sigma(B) = \sigma(A)$. We may write

$$A = U_1 D_1 V_1 \quad \text{and} \quad B = U_2 D_2 V_2$$

where $U_i, V_i \in L$ for $i = 1, 2$ and $D_i = \begin{pmatrix} e^{\lambda_i} & 0 \\ 0 & e^{-\lambda_i} \end{pmatrix}$ with $\lambda_i \geq 0$, $i = 1, 2$. Then by invariance of σ under L,

$$2\cosh(2\lambda_1) = \sigma(A) = \sigma(B) = 2\cosh(2\lambda_2)$$

so $D_1 = D_2$ and so $A = U_1 U_2^{-1} B V_2^{-1} V_1$. By Proposition 4 we conclude that $A \in R(\mathfrak{l}, X, t)$. Hence

$$\left\{ A \in G : \sigma(A) \leq 2\cosh(2at) \right\} \subseteq R(\mathfrak{l}, X, t)$$

and since the opposite inequality is Proposition 12 Corollary 2, the proof is complete.

We now go on to determine the sets $R(\mathfrak{l}, X, t)$ in the case when X has the form $\begin{pmatrix} a & b \\ 0 & -a \end{pmatrix}$ with $b \neq 0$. The method is to reduce this case to the diagonal case already dealt with. We note that when $X = \begin{pmatrix} a & b \\ 0 & -a \end{pmatrix}$, then

$$\exp\frac{tX}{n} = \begin{cases} \begin{pmatrix} 1 & \frac{bt}{n} \\ 0 & 1 \end{pmatrix} & \text{if } a = 0. \\ \begin{pmatrix} e^{at/n} & \frac{b}{a}\sinh(\frac{ta}{n}) \\ 0 & e^{-at/n} \end{pmatrix} & \text{if } a \neq 0. \end{cases}$$

Proposition 14. Let $X = \begin{pmatrix} a & b \\ 0 & -a \end{pmatrix}$ with $b \neq 0$. Then $\forall n \geq 1$ and $\forall t \in \mathbb{R}_+^*$, there exist $U_n, V_n \in L$ such that $\exp\frac{tX}{n} = U_n D_n V_n$ where $D_n = \begin{pmatrix} e^{\lambda_n} & 0 \\ 0 & e^{-\lambda_n} \end{pmatrix}$, with

$$\lambda_n = \begin{cases} \frac{1}{2}\cosh^{-1}(1 + \frac{b^2t^2}{2n^2}) & \text{if } a = 0 \\ \frac{1}{2}\cosh^{-1}\left\{(1 + \frac{b^2}{4a^2})\cosh(\frac{2at}{n}) - \frac{b^2}{4a^2}\right\} & \text{if } a \neq 0 \end{cases}$$

Proof. In both cases we may certainly write $\exp\frac{tX}{n} = U_n D_n V_n$ with $U_n, V_n \in L$ and $D_n \in H$ (of course U_n, V_n depend also on t). We may assume that $D_n = \begin{pmatrix} e^{\lambda_n} & 0 \\ 0 & e^{-\lambda_n} \end{pmatrix}$ with $\lambda_n > 0$, and using invariance of σ under L gives

$$e^{2\lambda_n} + e^{-2\lambda_n} = \begin{cases} 2 + \frac{b^2t^2}{n^2} & \text{if } a = 0 \\ e^{\frac{2at}{n}} + e^{-\frac{2at}{n}} + \frac{b^2}{a^2}\sinh^2(\frac{ta}{n}) & \text{if } a \neq 0. \end{cases}$$

Calculation now yields the values of λ_n given above.

Proposition 15. (a) For all $b \in \mathbb{R}$ and $a, t \in \mathbb{R}_+^*$,

$$\sup_{n\geq 1}\left(2\cosh\{n\cosh^{-1}[(1 + \frac{b^2}{4a^2})\cosh(\frac{2at}{n}) - \frac{b^2}{4a^2}]\}\right) = 2\cosh(t(b^2+4a^2)^{\frac{1}{2}})$$

(b) For all $b \in \mathbb{R}$ and $t \in \mathbb{R}_+^*$

$$\sup_{n\geq 1}\left(2\cosh\{n\cosh^{-1}[1 + \frac{b^2t^2}{2n^2}]\}\right) = 2\cosh(tb) .$$

Proof. (a)(i) Let $\alpha = 1 + \frac{b^2}{4a^2} > 1$ and consider the function

$$g(x) = \cosh(\sqrt{\alpha}x) - \alpha\cosh x + (\alpha - 1).$$

We note that $g(0) = 0$ and

$$g'(x) = \sqrt{\alpha}\sinh(\sqrt{\alpha}x) - \alpha\sinh x = \sum_{n=1}^{\infty}\frac{x^{2n+1}}{(2n+1)}(\alpha^{n-1} - \alpha) > 0$$

whenever $x > 0$, because $\alpha > 1$. Hence $\forall x \geq 0$,

$$\cosh(\sqrt{\alpha}x) \geq \alpha \cosh x + (1 - \alpha)$$

hence $\forall x > 0$

$$\frac{1}{x}\cosh^{-1}(\alpha\cosh x + (1-\alpha)) \leq \sqrt{\alpha}.$$

(ii) Using L'Hopital's Rule we may show that as $x \downarrow 0$,

$$\frac{1}{x}\cosh^{-1}(\alpha\cosh x + (1-\alpha)) \to \sqrt{\alpha}.$$

(iii) Let $a_n = n\cosh^{-1}[(1 + \frac{b^2}{4a^2})\cosh(\frac{2at}{n}) - \frac{b^2}{4a^2}]$

$$= 2at(\frac{n}{2at})\cosh^{-1}[\alpha\cosh(\frac{2at}{n}) + (1-\alpha)]$$

Then by (i) and (ii), if $a, t \in \mathbb{R}_+^*$,

$$\sup_{n\geq 1}(a_n) = 2at\sqrt{\alpha} = t(b^2 + 4a^2)^{\frac{1}{2}}.$$

Taking cosh on both sides gives the result. Note that the conclusion also holds when $a < 0$ because cosh is even.

(b)(i) Let $\alpha = b^2t^2$, then clearly

$$\forall x \geq 0, \qquad 1 + \tfrac{1}{2}\alpha x^2 \leq \cosh(\sqrt{\alpha}x).$$

Hence $\forall x \geq 0$,

$$\frac{1}{x}\cosh^{-1}(1 + \tfrac{1}{2}b^2t^2x^2) \leq \sqrt{\alpha}.$$

(ii) We may use L'Hopital's Rule to show that as $t \downarrow 0$,

$$\frac{1}{x}\cosh^{-1}(1 + \tfrac{1}{2}b^2t^2x^2) \to \sqrt{\alpha}.$$

(iii) So if $b_n = 2\cosh(n\cosh^{-1}(1 + \frac{1}{2}b^2t^2/n^2))$, then

$$\sup_{n\geq 1} b_n = 2\cosh\sqrt{\alpha} = 2\cosh(bt).$$

Theorem 5. Let $X = \begin{pmatrix} a & b \\ 0 & -a \end{pmatrix}$ with $b \neq 0$. Then for all $t \in \mathbb{R}_+^*$,

$$R(\mathfrak{l}, X, t) = \left\{A \in G : \sigma(A) \leq 2\cosh(t(b^2 + 4a^2)^{\frac{1}{2}})\right\}.$$

Proof. By the argument in the proof of theorem 4, it is sufficient to show that

$$\{\sigma(A) : A \in R(\mathfrak{l}, X, t)\} = [2, 2\cosh(t(b^2 + 4a^2)^{\frac{1}{2}})].$$

Clearly $\forall t > 0$ and all $n \geq 1$ we have

$$\left(L \exp \frac{tX}{n}\right)^n L = (L D_n)^n L$$

where D_n is as in Proposition 14, so as σ is invariant under L we have

$$\left\{\sigma(A) : A \in \left(L \exp \frac{tX}{n}\right)^n\right\} = \left\{\sigma(A) : A \in (LD_n)^n\right\} \ldots\ldots \quad (*)$$

Then by Proposition 12 and Proposition 15, we have

$$\forall n \geq 1 \quad \text{and} \quad \forall A \in (LD_n)^n, \quad \sigma(A) \leq 2\cosh(2n\lambda_n) \leq 2\cosh(t(b^2 + 4a^2)^{\frac{1}{2}}).$$

Note also that $(LD_n)^n$ contains D_n^n and $\sigma(D_n^n) = 2\cosh(2n\lambda_n)$. Hence $\forall A \in R(\mathfrak{l}, X, t)$, $\sigma(A) \leq 2\cosh(t(b^2 + 4a^2)^{\frac{1}{2}})$ so that $R(\mathfrak{l}, X, t)$ is compact, and further

$$\forall n \geq 1, \quad 2\cosh(2n\lambda_n) \in \left\{\sigma(A) : A \in R(\mathfrak{l}, X, t)\right\}.$$

Hence by Proposition 15,

$$2\cosh(t(b^2 + 4a^2)^{\frac{1}{2}}) \in \left\{\sigma(A) : A \in R(\mathfrak{l}, X, t)\right\}.$$

But $*$ and Proposition 12 show that $1 \in R(\mathfrak{l}, X, t)$, hence

$$\left\{\sigma(A) : A \in R(\mathfrak{l}, X, t)\right\} \subseteq [2, 2\cosh(t(b^2 + 4a^2)^{\frac{1}{2}}]$$

and the set on the left is connected and contains both end-points of the interval on the right. Hence these two sets are equal and the proof is complete.

Acknowledgement It is a pleasure to record our thanks to Professor E. Siebert, both for bringing to the attention of the first-named author the problem of the existence of a common support for an absolutely continuous Gauss semigroup on a connected Lie group, and for providing us with a preprint of his paper [2].

References

[1] Heyer, H. "Probability measures on locally compact groups", Ergebnisse der Mathematik und ihrer Grenzgebiete 94, Berlin-Heidelberg-New York, Springer, 1977.

[2] Siebert, E. "Absolute continuity, singularity, and supports of Gauss semi-groups on a Lie group", Monatsh. für Math. 93, 239-253 (1982).

INFINITE CONVOLUTION VIA REPRESENTATIONS

Imre Z. Ruzsa
Mathematical Institute of the
Hungarian Academy of Sciences,
Budapest, Hungary

ABSTRACT. Kloss' "general principle of convergence" is extended to all locally compact groups.

1. Introduction

This paper continues my work on infinite convolutions presented at the previous conference (Ruzsa (1982), but the reader is not assumed to have read it; every necessary concept and result will be restated. In doing so, I shall refer to the above paper as *(I)*.

Let $\mu_1, \mu_2, \ldots$ be *distributions* (tight probability measures) on a Hausdorff topological group G. We are interested in the convergence of the infinite convolution

$$\mu_1 \mu_2 \mu_3 \cdots \quad \delta(g_n)$$

for some choice of $g_n \in G$ ($\delta(g)$ denotes the point mass at g and we write simply $\mu\nu$ for the convolution of μ and ν). This clearly cannot happen if the partial products

$$(1.1) \qquad \nu_n = \mu_1 \mu_2 \cdots \mu_n$$

are "too spread". To formulate this exactly we introduce the (left) *concentration function*

$$Q_\ell(\mu, X) = \sup_{g\in G} \mu(Xg) = \sup_{g\in G} (\mu\delta(g))(X) .$$

So, (1.1) cannot converge to a tight measure if

$$Q_\ell(\nu_n, K) \to 0$$

for every compact $K \subset G$, in which case we call the product *dispersing*. That an infinite convolution must be either dispersing, or convergent under a suitable centering (a modified form of Kloss' "general principle of convergence") was established after the works of Kloss (1961) and Csiszár (1966)), by Tortrat (1970) for first countable (M_1) groups. I applied a Fourier analytic method to obtain it for compact commutative groups, and then combining it with Csiszár's ideas in (I) I could prove it for a class of groups that included e. g. all commutative locally compact groups. I beleive it holds for all groups. I am still far from achieving this, but now, using noncommutative Fourier analysis, I shall prove it for the widely renowned class of locally compact groups.

This problem is closely connected with the so called *problem of shift convergence*. In a not completely general form, this is the following: suppose (μ_n) is a tight sequence of distributions and all its cluster points are of the form $\mu\delta(g)$. Does it follow that $\mu_n\delta(g_n) \to \mu$ for suitable g_n ? This is easy in M_1 groups and a positive answer would immediately yield the result on products. Namely if our ν_n of (1.1) is not dispersing, then by Csiszár's method one can easy find g_n such that $\nu_n\delta(g_n)$ is tight and all its cluster points are translates of each other.

I think that a negative answer to the problem of shift convergence would mean that we must change our image of the convolution. In this paper a partial positive answer will be given which is still strong enough to be applied for infinite convolutions.

2. The main result

We shall consider a generalized version of infinite products. By a *D-product*, where D is a directed set, we mean a system $(a^n_m)_{m<n,\ m,\ n\in D}$ of elements (from an arbitrary topological semigroup) satisfying

$$a_k^m \, a_m^n = a_k^n \qquad (k < m < n) \, .$$

We call the above D-product *convergent*, if the net $(a_m^n)_{n \in D}$ is convergent for every fixed $m \in D$. (For the concept and basic properties of nets see Kelley (1955).) Observe that for ordinary products this is a bit stronger than usual convergence (the "composition convergence" of Heyer (1977), def. 2.3.3).

Two D-products of measures, (μ_m^n) and (ν_m^n), will be called *associates*, if for suitable $g_n \in G$ the relation

$$\mu_m^n = \delta(g_m^{-1}) \, \nu_m^n \, \delta(g_n)$$

holds for all m, n. A D-product (ν_m^n) of distributions is *dispersing*, if

$$\lim_n Q_\ell(\nu_m^n, K) = 0$$

for every compact $K \subset G$ and fixed $m \in D$.

We call a group G *compactly countable* if it has a compact normal subgroup G_1 such that G/G_1 is first countable.

THEOREM 1. *If G has the property that every σ-compact subset is containted in a compactly countable topologically normal subgroup (e. g. if G is locally compact), then every D-product of distributions on G is either dispersing or associate to a convergent product.*

This seems to be near to the limit of possibilities of my method. Perhaps the requirement that the subgroup G_1 must be normal can be omitted, but more general groups are out of the range.

That locally compact groups have this property can be proved as follows. First we show that a σ-compact locally compact group G is countably compact. Let

$$G = \bigcup_{j=1}^{\infty} K_j, \; K_1 \subset K_2 \quad \ldots$$

with compact K_j. Let U_1 be any compact neighbourhood of the unity e. Given U_n, let U_{n+1} be a compact neighbourhood of e satisfying

$$U_{n+1} \, U_{n+1}^{-1} \subset U_n$$

and $x U_{n+1} x^{-1} \subset U_n$ for all $x \in K_n$. Then clearly $G_1 = \cap U_n$

is a compact normal subgroup, and it is also of type G_δ , which shows that G/G_1 is M_1 . This proof was communicated to me by Prof. H. Rindler with the additional information that it is probably due to Kakutani; we have not been able to find Kakutani's paper.

Now let $X \subset G$ be σ-compact in the locally compact group G; we may assume that it is a neighbourhood of e . Then

$$G_o = \cup (XX^{-1})^n$$

is an open σ-compact subgroup containing X , thus it is also locally compact and an application of the above result completes the proof.

3. Connection with the shift-convergence

To describe this we first quote a number of concepts and results from *(I)*, sections 4-7.

We use $\mathcal{D}(G)$ to denote the convolution semigroup of distributions on the group G . We say that μ is a right divisor of ν and write $\mu |_r \nu$ if $\nu = \rho\mu$ for some $\rho \in \mathcal{D}(G)$. If both $\mu |_r \nu$ and $\nu |_r \mu$, we call μ and ν right associates and write $\mu \sim_r \nu$; is known to be equivalent to the existence of a $g \in G$ such that $\mu = \delta(g)\nu$. (Clearly these concepts have their "left" analogues. Why we now use "right" things though we started with a "left" one is motivated in *(I)* . The right and left structures of $\mathcal{D}(G)$ are, by the way, isomorphic.)
By considering the equivalence classes of $\sim_r$ we obtain the right factor space

$$\mathcal{D}_r^* = \mathcal{D}/\sim_r \quad .$$

Equivalence classes can be shown to be closed, hence $\mathcal{D}_r^*$ can be endowed with the factor topology. We write

$$\psi_r : \mathcal{D} \to \mathcal{D}_r^*$$

for the natural homeomorphism.

$\sim_r$ is not a congruence relation, thus we cannot define multiplication in $\mathcal{D}_r^*$ (unless G is commutative), but we can define divisibility: for $\alpha, \beta \in \mathcal{D}_r^*$ we write $\alpha | \beta$ if there are

$\mu, \nu \in \mathcal{D}$ such that $\mu|_r\nu$ and $\psi_r(\mu) = \alpha$, $\psi_r(\nu) = \beta$.

A net $(\alpha_n)_{n\in D}$, $\alpha_n \in \mathcal{D}_r^*$ is *decreasing* if $\alpha_m|\alpha_n$ whenever $m > n$. If $\alpha_n = \psi_r(\mu_n)$, this is equivalent to

$$\mu_m|_r\mu_n \quad \text{for} \quad m > n ,$$

in which case we call the net (μ_n) *right-decreasing.*

We say that a pair (G, D), where G is a topological group and D a directed set, has the *property of shift-convergence* if the following proposition holds:

(SC) If $(\mu_n)_{n\in D}$ is a net of distributions on G and $\psi_r(\mu_n) \to \psi_r(\mu)$, then $\delta(g_n)\mu_n \to \mu$ for suitable $g_n \to G$.

We shall consider also two weaker versions of (SC).

(SC): (SC) when μ is a Haar measure on a compact subgroup of G.

(SCHD): (SC) when μ is a Haar measure and moreover (μ_n) is right-decreasing.

Obviously (SC)$\Rightarrow$(SCH)$\Rightarrow$(SCHD). In *(I)* I proved (Theorem 2, section 7) the following:

(3.1) LEMMA. *If* (SCHD) *holds for a pair (G,D), then every D-product of distribution on G is either disperising or associate to a convergent product.*

The remainder of the paper is devoted to establishing a shift-convergence property for certain groups.

4. Shift-convergence in compactly countable groups

Here we prove our

THEOREM 2. (SCH) *holds in topologically normal compactly countable groups.*

(4.1) LEMMA. (SC) *holds in M_1 groups.*

This is Proposition (8.1) of *(I)*.

PROOF of Theorem 2. Let G be our group, G_1 its compact normal subgroup such that G/G_1 is M_1, $\omega_1 = \omega(G_1)$ the Haar measure on G_1. Let $(\mu_n)_{n\in D}$ be the net of distributions considered; we know

$$\psi_r(\mu_n) \to \psi_r(\omega),\ \omega = \omega(H) \tag{4.2}$$

with some subgroup H . Put $G_2 = G_1 H$, a (generally not normal) subgroup of G .

First consider the group G/G_1 and the measures induced by μ_n on it. By Lemma (4.1) this net can be made convergent by a suitable centering; we may assume that this net itself converges to the measure induced by ω , which in other words means

(4.3) $$\mu_n \omega_1 \to \omega \omega_1 .$$

We shall find $g_n \in G_2$ such that

(4.4) $$\delta(g_n) \mu_n \to \omega .$$

Let Γ be the dual object of G_2 , i. e. the set of (finite dimensional unitary) representations of G_2 . Each $\gamma \in \Gamma$ maps into the set of $k \times k$ matrices for some k . Now let γ' be any bounded continuous extension of γ ($\gamma'(g)$ need not be unitary for $g \notin G_2$; I do not know whether a unitary extension always exists). We define a pseudo-Fourier transform. For measures μ on G and $\gamma \in \Gamma$ we set

$$\hat{\mu}(\gamma) = \int \gamma' d\mu .$$

This does not share too much of the properties of the genuine Fourier transform; but it coincides with the Fourier transform if $\operatorname{supp}\mu \subset G_2$, and it is continuous: if $\nu_n \to \nu$, then $\hat{\nu}_n(\gamma) \to \hat{\nu}(\gamma)$. We shall find $g_n \in G_2$ so that

(4.5) $$\gamma(g_n) \hat{\mu}_n(\gamma) \to \hat{\omega}(\gamma)$$

for all $\gamma \in \Gamma$.

(4.6) LEMMA. *If* (4.5) *holds for all* $\gamma \in \Gamma$ *, then so does* (4.4) .

To the proof we recall the useful concept of a quasitight net. A net (μ_n) of distributions is *quasitight*, if for every $\varepsilon > 0$ there is a compact $K_\varepsilon \subset G$ such that

$$\liminf \nu_n(U) > 1-\varepsilon$$

for every open $U \supset K_\varepsilon$. Its main properties are:

(4.7) A quasitight net has cluster points.

(4.8) A net is convergent if and only if it is quasitight and has at most one cluster point.

(4.9) If $\lambda_n = \mu_n \nu_n$ and two of the nets (λ_n), (μ_n), (ν_n) are quasitight, then so is the third.

The concept and the above properties are due to Siebert (1976).

PROOF of Lemma (4.6). (4.3) implies that (μ_n) is quasi-tight and so is $\delta(g_n)$, since $g_n \subset G_2$ and G_2 is compact. So if (4.4) does not hold, we can find a subnet $(\mu_{f(j)})_{j\in J}$ (where J is another directed set and $f:J \to D$ is a monotonic cofinal mapping) such that with some $\mu \in \mathcal{D}(G)$ and $y \in G_2$

(4.10) $$\mu_{f(j)} \to \mu,\ g_{f(j)} \to y,\ \delta(g_{f(j)})\ \mu_{f(j)} \to \delta(y)\ \mu \neq \omega\ .$$

Now by (4.2) μ is of the form $\delta(x)\omega$ and taking into account (4.3) we obtain $x \in G_2$. (4.10) then becomes

(4.11) $$\delta(y)\ \delta(x)\ \omega \neq \omega\ .$$

Now we have

$$\gamma(g_{f(j)})\ \hat{\mu}_{f(j)}\ (\gamma) \to \gamma(y)(\delta(x)\omega)^{\wedge}(\gamma) = \gamma(y)\ \gamma(x)\ \hat{\omega}(\gamma)$$

(since $\delta(x)\omega$ is supported on G_2, its Fourier transform is what it ought to be) and (4.5) gives us

$$\gamma(y)\ \gamma(x)\ \hat{\omega}(\gamma) = \hat{\omega}(\gamma)$$

for all $\gamma \in \Gamma$, a contradiction to (4.11).

(4.12) LEMMA. *If (4.5) holds for a particular net (g_n) and γ, then for another net (h_n), $h_n \subset G_2$ and the same γ the assertions*

(4.13) $$\gamma(h_n)\ \hat{\mu}_n(\gamma) \to \hat{\omega}(\gamma)$$

and

(4.14) $$\gamma(g_n)^{-1}\hat{\omega}(\gamma) - \gamma(h_n)^{-1}\ \hat{\omega}(\gamma) \to 0$$

are equivalent.

PROOF. Since $\gamma(g_n)$ is unitary, (4.5) is clearly equivalent to

$$\hat{\mu}_n(\gamma) - \gamma(g_n)^{-1}\ \hat{\omega}(\gamma) \to 0\ .$$

We make the same transformation with (4.13) and then substract to obtai (4.14).

Now we start the search for (g_n). We shall use a version of successive approximation, finding step by step approximative solutions that satisfy (4.5) for more and more γ. More exactly, we call a pair

$$W = ((g_n)_{n\in D},\ \Delta),\quad g_n \in G_2,\quad \Delta \subset \Gamma$$

an *approximate solution*, if it satisfies (4.5) for all $\gamma \in \Delta$. We say that $W' = ((g_n'), \Delta')$ is an *extension* of W and write $W < W'$ if $\Delta \subset \Delta'$ and

$$\gamma(g_n')^{-1}\, \hat{\omega}(\gamma) = \gamma(g_n)^{-1}\, \hat{\omega}(\gamma) \tag{4.15}$$

for all $\gamma \in \Delta$ (this definition is motivated by the lemma above).

Our plan is to show that a) there is a maximal approximate solution, and b) $\Delta = \Gamma$ for it.

(4.16) LEMMA. *Every chain*

$$(W_i)_{i \in I}, \quad W_i = ((g_n^{(i)}), \Delta_i)$$

of approximate solutions (i. e. $W_i < W_j$ for $i < j$, with a linearly ordered set I of indices) has a common extension.

PROOF. We put $\Delta = \cup \Delta_i$ and we want to find g_n so that

$$\gamma(g_n)^{-1}\, \hat{\omega}(\gamma) = \gamma(g_n^{(i)})^{-1}\, \hat{\omega}(\gamma) \tag{4.17}$$

whenever $\gamma \in \Delta_i$. Denoting by K_i the set of g_n's satisfying (4.17) for a fixed i (and fixed n) and for all $\gamma \in \Delta_i$, one sees immediately that K_i is compact, not empty since $g_n^{(i)} \in K_i$ and decreasing, i. e. $K_i \subset K_j$ if $i > j$, and then for g_n one can choose any element of $\cap K_i$. To show that (g_n) satisfies (4.5) for a particular $\gamma \in \Delta$, choose an i such that $\gamma \in \Delta_i$ and then apply (4.17) and Lemma (4.12) for the nets (g_n) and $(g_n^{(i)})$.

Now we return to the proof of Theorem 2. Lemma (4.16) and Zorn's lemma yield us the existence of a maximal approximate solution W; it is sufficient to show that $\Delta = \Gamma$ for it. Suppose the contrary and choose a representation $\eta \in \Gamma \setminus \Delta$. We extend W to η.

We want to find a sequence (h_n) such that

$$\gamma(g_n)^{-1}\, \hat{\omega}(\gamma) = \gamma(h_n)^{-1}\, \hat{\omega}(\gamma) \tag{4.18}$$

for all $\gamma \in \Delta$ and moreover

$$\eta(h_n)\, \hat{\mu}_n(\eta) \to \hat{\omega}(\eta)\,; \tag{4.19}$$

the corresponding relation for the representations belonging to Δ follows from (4.18) and Lemma (4.12). The elements $h_n \in G_2$ that satisfy (4.18) for all $\gamma \in \Delta$ form a compact set K_n (not empty, since $g_n \in K_n$). Let now h_n be any of the elements of K_n minimizing

$$\|\eta(h_n)\,\hat{\mu}_n(\eta) - \hat{\omega}(\eta)\| \quad ,$$

where $\|\ldots\|$ denotes any norm inducing the usual topology (so, not the modulus of the determinant). If (4.19) does not hold, then, like in the proof of (4.6), we can find a subset converging to something else, i. e. applying the same notation we have

$$h_{f(j)} \to y, \qquad \mu_{f(j)} \to \mu, \qquad \eta(y)\,\hat{\mu}(\eta) \neq \hat{\omega}(\eta) \quad .$$

We have $\mu = \delta(x)\omega$ for some $x \in G_2$, and hence

$$\eta(y)\,\eta(x)\,\hat{\omega}(\eta) \neq \hat{\omega}(\eta) \quad .$$

On the other hand, since for $\gamma \in \Delta$

$$\gamma(h_n)\,\hat{\mu}_n(\gamma) \to \hat{\omega}(\gamma) \quad ,$$

we have

(4.20) $$\gamma(y)\,\gamma(x)\,\hat{\omega}(\gamma) = \hat{\omega}(\gamma) \qquad (\gamma \in \Delta) \quad .$$

Put now $$h_n' = x^{-1}y^{-1}h_n \quad .$$

For this net and $\gamma \in \Delta$ we have

$$\gamma(h_n')^{-1}\,\hat{\omega}(\gamma) = \gamma(h_n)^{-1}\,\gamma(y)\,\gamma(x)\,\hat{\omega}(\gamma) = \gamma(h_n)^{-1}\,\hat{\omega}(\gamma)$$

by (4.20), so it is also a solution of (4.18). Moreover

$$\eta(h'_{f(j)})\,\hat{\mu}_{f(j)}(\eta) \to \eta(x^{-1}y^{-1})\,\eta(y)\,\eta(x)\,\hat{\omega}(\eta) = \hat{\omega}(\eta) \quad ,$$

thus there is a j for which

$$\|\eta(h'_{f(j)})\,\hat{\mu}_{f(j)}(\eta)-\hat{\omega}(\eta)\| < \|\eta(h_{f(j)})\,\hat{\mu}_{f(j)}(\eta)-\hat{\omega}(\eta)\| \quad ,$$

since the left side tends to 0, the right to

$$\|\eta(y)\,\hat{\mu}(\eta)-\hat{\omega}(\eta)\| > 0 \quad .$$

This contradicts to the defining minimal property of h_n and this completes the proof of Theorem 2.

5. Proof of Theorem 1

Taking into accont Lemma (3.1), it suffices to show that (SCDH) (which was also defined in Sec. 3) holds in the class of groups described in Theorem 1.

Let $(\mu_n)_{n\in D}$ be our decreasing net. We may assume that D has a minimal element o, since otherwise appointing an arbitrary element j_o of D to this role we may change D into $D' = \{j\in D:\ j\geq j_o\}$, which does not affect the existence of limits.

Let X be a σ-compact set such that $\mu_o(X)=1$ (it exists, since we assumed our measures to be tight). We have

$$\mu_m = \nu_m^n \mu_n \qquad (m < n)$$

with suitable ν_m^n, thus

$$\mu_o(X) = 1 = \int \mu_n(y+X)\ d\nu_o^n(y) \quad ,$$

whence $\mu_n(y+X) = 1$ for a suitable $y = y_n$ (in fact, for ν_o^n--almost all values of y). Now with

$$\mu_n' = \delta(y_n^{-1})\ \mu_n$$

we have $\mu_n'(X) = 1$ for all n. X is contained in a compactly countable subgroup G_2; applying Theorem 2 for the restriction of μ_n' to G_2 we obtain Theorem 1.

References

Csiszár, I.(1966), On infinite products of random elements and infinite convolutions of probability distributions on locally compact groups, Z. Wahrscheinlichkeitstheorie verw. Geb. 5, 279-295.

Heyer, H.(1977), Probability measures on locally compact groups, Springer.

Kelley, J. L.(1955), General topology, New York, D. van Nostrand.

Kloss, B. M.(1961), Limiting distributions on compact Abelian groups (in Russian), Teor. Veroyatn. Primen. 6, 392-421.

Ruzsa, I. Z.(1982), Infinite convolution and shift-convergence of measures on topological groups, Proc. of the Conf. on Probability measures on groups, Oberwolfach 1981, Springer, LNM 928.

Siebert, E.(1976), Convergence and convolutions of probability measures on a topological group, Ann. Probab. 4, 433-443.

Tortrat, A.(1970), Convolutions dénombrables équitendues dans un groupe topologique X, Proc. of the conf. "Les probabilités sur les structures algébriques", Clermond-Ferrand 1969, Paris, CNRS.

DECOMPOSITIONS OF PROBABILITY MEASURES ON GROUPS

Imre Z. Ruzsa and Gábor J. Székely

Mathematical Institute of the Hungarian Academy of Sciences
Budapest, Hungary

Department of Probability Theory
Loránd Eötvös University
Budapest, Hungary

Abstract: Hinčin's celebrated decomposition theorem will be extended to the convolution structure of tight probability measures on first countable Abelian topological groups.

1. Introduction

Let $\mathcal{D}(R)$ denote the convolution semigroup of probability distributions on the real line. A $\mu\in\mathcal{D}(R)$ is irreducible if $\mu = \alpha\beta$ $(\alpha,\beta \in \mathcal{D}(R))$ implies that either α or β is a unit, but μ itself is not (a unit in our case means a degenerate distribution, a point mass). A $\mu \in \mathcal{D}(R)$ is antiirreducible (belongs to Hinčin's class I_0) if it has no irreducible divisor (convolution factor). Hinčin (1937) proved the following fundamental theorems.

THEOREM A. Every $\mu \in \mathcal{D}(R)$ is the convolution product of at most countable many irreducible distributions and an antiirreducible distribution.

THEOREM B. Every antiirreducible element of $\mathcal{D}(R)$ is infinitely divisible.

Parthasarathy-Rao-Varadhan (1963) extended Hinčin's theorems to $\mathcal{D}(G)$, where G is a locally compact Abelian second countable (M_2) group and $\mathcal{D}(G)$ is the convolution semigroup of probability distributions (tight probability measures) on G. Urbanik (1976) and Heinich (1975) gave further extensions of Theorem B. We shall be interested in the extension of Theorem A; we prove it for arbitrary first countable (M_1) Abelian groups. (If the group is not M_1, generally more than countably many factors are necessary, and it is not even clear what to mean by their product.)

We remark without proof that Theorem B also holds in arbitrary M_1 groups; we shall return to this in another paper.

We shall be also interested in another kind of decomposition, into an idempotent (=Haar) measure and another factor that has no idempotent divisor at all. This kind of decomposition was also introduced by

Parthasarathy-Rao-Varadhan; they first make this decomposition, then decompose the second factor and finally they have three kind of components: irreducible, antiirreducible and Haar. Though finally we shall dispose of the Haar factor, this will be applied in the course of the proof and we think it is interesting in itself.

We also give an example of a convolution semigroup without decomposition.

2. Separation of the idempotent part

It is known (see Tortrat (1965)) that the idempotent distributions are the Haar measures on compact subgroups of G. This motivates the following definition (Haar = hair).

(2.1) DEFINITION. A distribution μ is bald, if it is not divisible by any nondegenerate Haar measure.

THEOREM 1. Let G be an arbitrary commutative (not necessarily M_1) Hausdorff topological group and $\mu \in \mathcal{D}(G)$. Then μ has a decomposition

$$\mu = \omega\nu ,$$

where ω is a Haar measure and ν is bald. In general neither ω nor ν is unique. A possible choice of ω is $\omega(H)$, the Haar measure on the group H of those elements g for which $\delta(g)\mu = \mu$ (this is the "maximal" Haar divisor of μ). ($\delta(g)$ denotes the point mass at g.)

PROOF. Put $\Lambda = \{\alpha \in \mathcal{D}(G) : \omega\alpha = \mu\}$, $\omega = \omega(H)$, where H is the subgroup described in the theorem. Λ is non-empty (e.g. $\mu \in \Lambda$), convex and compact (in the weak topology), hence it can be shown by standard arguments that it has extremal points; let ν be one. We show that ν is bald.

Suppose the contrary, i.e. $\nu = \omega_1\nu$, where ω_1 is the Haar measure on a nontrivial subgroup H_1. Clearly $H_1 \subset H$ (this need not be the case if G is not commutative). Let $h \in \operatorname{supp} \nu$. Let f be any continuous function $f : G \to [0,1]$ that is not constant on hH_1. Let $\bar{f}$ be its H_1-mean, i.e.

$$\bar{f}(x) = \int f(x+y)d\omega_1(y)$$

and define the measures ν_1 and ν_2 by

$$d\nu_1 = (1+f-\bar{f})d\nu$$
$$d\nu_2 = (1-f+\bar{f})d\nu \quad .$$

Clearly they are nonnegative and $\nu = (\nu_1+\nu_2)/2$. If we show that $\nu_1,\nu_2 \in \Lambda$ and $\nu_1 \neq \nu_2$, we arrive at a contradiction to the extremal property of ν.

A routine calculation shows that $\omega_1\nu_1 = \nu$, hence by $H_1 \subset H$ $\omega\nu_1 = (\omega\omega_1)\nu_1 = \omega\nu = \mu$; $\nu_1(G) = \omega\nu_1(G) = \mu(G) = 1$, i.e. $\nu_1 \in \Lambda$ and similarly $\nu_2 \in \Lambda$. Now $\nu_1 = \nu$ would mean that $f = \bar{f}$ ν-almost everywhere, i.e. the function being continuous, $f(x) = \bar{f}(x)$ for all $x \in \text{supp}\,\nu$. But $\bar{f}$ is constant on $hH_1 \supseteq \text{supp}\,\nu$, while f, by definition, is not.

The present proof works only for the commutative case, but the result is probably more general. It would be interesting to find a proof for the noncommutative case.

3. Decomposition in abstract semigroups

We shall obtain our result on the decomposition of measures as a corollary to a general theory, which we develop in another paper (see Ruzsa-Székely (198?)). Here we quote only the main ideas and necessary results.

An attempt to such a general theory first appears in the papers of Kendall (1967), (1968) (see also Davidson (1968), (1969)). Kendall's class of semigroups, called Delphic, is required to have (besides certain quite natural ones) the following property:

> there exists a continuous homomorphism Δ on the semigroup that vanishes only at the unity.

This strong requirement renders his theory unapplicable for $\mathcal{D}(G)$ if G has nontrivial compact subsemigroups (and even if this is not the case, the applicability is not straightforward).

We are going to define a much wider class of semigroups that includes $\mathcal{D}(G)$ and for which our theory works. We consider a commutative Hausdorff topological semigroup S with unit element e.

(3.1) DEFINITION. Two elements of S are associates if each divides the other. A divisor (thus an associate) of e is a unit. If x and y are associate, we write $x \sim y$.

(3.2) DEFINITION. S is a Hungarian semigroup if

- (i) the set of associates of $x \in S$ is always a closed set, hence we can form the factor semigroup $S^* = S/\sim$;
- (ii) in S^* the set of divisors of any element is compact;
- (iii) S is first countable;
- (iv) if $x \sim y$, then $x = uy$ with some unit u.

The most generally investigated decomposition is into irreducibles, but it is by no means the only interesting one. To make our theory more widely applicable, we consider a set P of atoms, of which we assume

(3.3) (i) $e \notin P$

(ii) if $x \in P$, then all the associates of x belong to P as well.

(3.4) DEFINITION. An element of S is completely reducible (with respect to P) if it is the (finite or infinite, possibly empty) product of atoms.

An element that is not divisible by any atom clearly cannot be decomposed, but this is not the only possibility. If, say, s has atoms as divisors but in every decomposition

$$s = px, \quad p \subset P$$

x is equal to (or an associate of) s, we are also stopped. This motivates the following definition.

(3.5) DEFINITION. x is an effective divisor of y if $y = xz$, $z \neq y$.

(3.6) DEFINITION. x is an antiatom, if it is not effectively divisible by any completely reducible element. (We use the term antiirreducible to denote an antiatom if for atoms we take the irreducibles.)

(3.7) EXAMPLE. Let $S = \{0, 1, 2^{-1}, 2^{-2}, \ldots\}$ with the multiplication. 2^{-1} is the only irreducible; 0 is divisible, but not effectively divisible by it. It has, however, the infinite decomposition

$$0 = \prod_{n=1}^{\infty} 2^{-1}$$

thus it is completely reducible. If we consider a zero-extension of S, i.e. $S' = S \cup \{z\}$, where $zx = z$ for every $x \in S'$, in every (finite or infinite) decomposition of z it must itself occur as a factor, hence it is antiirreducible. Thus in the implication

x is not divisible by any atom

$\Rightarrow$ x is an antiatom

$\Rightarrow$ x is not effectively divisible by any atom

none of the arrows can be converted.

The main result of our abovementioned paper is

THEOREM H. *If S is a Hungarian semigroup and P is a set of atoms satisfying (3.3), then every element of S is the product of a completely reducible element and an antiatom.*

4. Decomposition of distributions

To be able to apply the results of the previous section, we first show

(4.1) LEMMA. *For every first countable commutative group G, $\mathcal{D}(G)$ is Hungarian.*

To the proof we need the following result of Tortrat (1965).

(4.2) LEMMA. *If $\mu\nu = \nu$ (μ,ν distributions), then there is a compact subgroup H such that $\omega(H)\nu = \nu$ and $\mu\omega(H) = \omega(H)$.*

PROOF of Lemma (4.1). From the requirements that make a semigroup Hungarian, (i) is Corollary (4.4) and (ii) is Proposition (5.6) of Ruzsa (1982). If G is M_1, so is $\mathcal{D}(G)$; it is even metrizable by a theorem of Varadarajan (1961), and this is (iii). Finally Lemma (4.2) above immediately implies that associates are translates, which is just (iv).

The idea of compactness (requirement (ii)) is an old one, but it is generally stated in the following more circumstantial way: if $\mu_1,\mu_2,\dots$ are divisors of μ, then there are elements $g_1,g_2,\dots$ such that

$$\delta(g_n)\mu_n$$

converges to some divisor of μ.

Applying Theorem H and Lemma (4.1), we obtain

THEOREM 2. *If G is commutative and first countable, then every element of $\mathcal{D}(G)$ is the convolution of (finitely or countably many) irreducible distributions and an antiirreducible one.*

Note that antiirreducibles are now defined in a rather clumsy way (Def. (3.6)). We can, however, regain what used to be the definition in the form of a theorem.

THEOREM 3. *An antiirreducible distribution has no irreducible divisors at all.*

The proof is based on the following lemmas.

(4.3) LEMMA. *If G is a compact commutative group of at least 3 elements and γ is a nonprincipal character of G, then there is an irreducible distribution μ on G such that $|\hat{\mu}(\gamma)|<1$.*

PROOF. Let $H = \ker \gamma$. If we can find an element $g \in G \setminus H$ whose order is ≥ 3, we put a mass $1/2$ to e and $1/2$ to g. Clearly

$$|\hat{\mu}(\gamma)| = \left|\frac{1+\gamma(g)}{2}\right| < 1$$

and it is easy to check that a distribution whose support consists of two elements, say x and y, is irreducible except if the order of $x-y$ is 2.

If such a g does not exist, then every element of G has order 1 or 2. Let g, h be two different elements of $G \setminus H$ (it has at least two, since G had at least 3). We define μ by putting a mass $1/3$ to e, g and h.

Suppose now $\mu = \alpha\beta$. Write $A = \operatorname{supp} \alpha$, $B = \operatorname{supp} \beta$. We know

$$A+B = \operatorname{supp} \mu = \{e,g,h\} \quad .$$

Since $e \in A+B$, there is an $x \in A$ such that $-x \in B$; by changing A into $A-x$ and B into $B+x$, we get $e \in A$, $e \in B$. Then $A,B \subset \{e,g,h\}$. Now if g is contained in one of A and B and h is contained in the other, then $g+h \in \{A+B\} = \{e,g,h\}$ which is impossible; $g+h \neq g$, $\neq h$, since non of them is e and $g+h \neq e$ since g,h are two different elements of order 2. The remaining possibilities are then $A = \{e\}$ or $B = \{e\}$ which implies that either α or β is a unit, or $A = B = \{e,g\}$ or $\{e,h\}$, in which case $A+B$ fails to contain h (resp. g).

(4.4) LEMMA. <u>The Haar measure on a compact</u> M_1 <u>group is completely reducible, if</u> $|G| \geq 3$; <u>has no irreducible divisors, if</u> $|G|=2$.

PROOF. If $|G|=2$, then $\mathcal{D}(G)$ is isomorph to the multiplicative semigroup of numbers in $[-1,1]$; a homomorphism is given by the Fourier transform at the only nonprincipal character

$$\mu \to \hat{\mu}(\gamma) = 2\mu(\{e\}) - 1 \ ,$$

which shows that there is no irreducible elements in $\mathcal{D}(G)$. If $|G| \geq 3$, let $\gamma_1, \gamma_2, \ldots$ be the nonprincipal characters of G (they are countable, since G is M_1). For each γ_j choose an irreducible μ_j such that

$$|\hat{\mu}_j(\gamma_j)| < 1.$$

Let $\nu_1, \nu_2, \ldots$ be a sequence containing each μ_j infinitely often. Then

$$\prod_{j=1}^{\infty} \hat{\mu}_j(\gamma) = 0$$

for any nonprincipal character γ, hence implying

$$\Pi \nu_j = \omega(G) .$$

PROOF OF THEOREM 3. Suppose μ is antiirreducible and it has an irreducible divisor π. It cannot be an effective divisor, thus

$$\mu = \pi\mu_1 , \quad \mu_1 \sim \mu , \quad \mu_1 = \delta(g)\mu .$$

Now replacing π by $\pi\delta(-g)$ we obtain

$$\mu = \pi\mu .$$

By Lemma (4.2) this means that

(4.5) $$\omega = \pi\omega ,$$

where $\omega = \omega(H)$ is a Haar measure on a compact subgroup. For H we can clearly choose the subgroup described in Theorem 1, and then by Theorem 1 we have a decomposition

(4.6) $$\mu = \omega\nu ,$$

where ν is bald. Now we apply Lemma (4.4); if $|H| \geq 3$ by (4.6) we find a completely reducible effective decomposition of μ, a contradiction to antiirreducibility, and if $|H| = 2$, then (4.5) is impossible.

5. A convolution semigroup without decomposability property

In this chapter we present a compact subsemigroup of the distributions on a certain compact group for which there is no decomposition.

Let $G = C \times C_2$ where C is the unit circle and C_2 is the two element group; G contains of two copies of C, which we shall denote by C^+ and C^- in the usual way. Let $\mathcal{D}^+$ and $\mathcal{D}^-$ be the sets of those distributions on G^+ resp. G^-, whose center of gravity is the origin. Our semigroup will be

$$S = \{\delta(e)\} \cup \mathcal{D}^+\mathcal{D}^+ \cup \mathcal{D}^- \quad .$$

This is clearly a compact semigroup. There are irreducibles in $\mathcal{D}^+$, but since $\delta(e) \notin \mathcal{D}^+$, they are excluded from $\mathcal{D}^+\mathcal{D}^+$. However, for any irreducible $\pi^+ \in \mathcal{D}^+$ its image π^- in $\mathcal{D}^-$ is left, and these are the irreducibles in our S. (A $\mu \in \mathcal{D}^+\mathcal{D}^+$ has by definition, the form $\alpha^+\beta^+$, $\alpha^+, \beta^+ \in \mathcal{D}^+$, hence in S it has the decomposition $\mu = \alpha^-\beta^-$.) Now in an infinite product of irreducibles

$$\pi_1^-\pi_2^-\ldots$$

its partial products oscillate between $\mathcal{D}^+$ and $\mathcal{D}^-$, thus it cannot converge (it is easy to see that not all elements have finite decompositions).

This example shows that if we omit condition (iv) in Definition (3.2), then Theorem H does not remain true.

6. Further problems

1. Characterize the commutative semigroups S having the property that $\mathcal{D}(S)$ (the convolution semigroup of tight probability measures on S endowed with the topology of weak convergence) is a Hungarian semigroup.

2. Generalize our theorems from $\mathcal{D}(G)$ to $\mathcal{D}(S)$.

3. Generalize for not necessarily commutative G and S.

4. Consider the convolution semigroup of signed measures. Here the units are not necessarily concentrated on a single point but not all signed measures are units. This problem was overlooked in Linnik-Ostrovskii (p. 446) where the authors stated that in the convolution semigroup of signed measures decomposition problems loose their meaning.

REFERENCES

Davidson, R. (1968), Arithmetic and other properties of certain Delphic semigroups: I-II, Z. für Wahrscheinlichkeitstheorie verw. Geb. 10, 120-145, 146-172.

Davidson, R. (1969), More Delphic theory and practice, Z. für Wahrscheinlichkeitstheorie verw. Geb. 13, 191-203.

Heinich, H. (1975), Sur les measure à valeurs dans des structures algébriques, Thèse de doctorat, Univ. Paris, Paris.

Hinčin, A. Ja. (1937), The arithmetic of distribution laws (in Russian), Bull. Univ. Moscow Sect. A, 1, 6-17.

Kendall, D.G. (1967), Delphic semigroups, Bull. Am. Math. Soc. 73(1), 120-121.

Kendall, D.G. (1968), Delphic semigroups, infinitely divisible regenerative phenomena, and the arithmetic of p-functions, Z. für Wahrscheinlichkeitstheorie verw. Geb. 9, 163-195.

Linnik, Ju. V. and Ostrovskii,I. V. (1972), Decompositions of random variables and vectors (in Russian), Nauka, Moscow (Translations of Mathematical Monographs, Vol. 48, American Mathematical Society, Providence, R. I., 1977).

Parthasarathy, K.R., Ranga Rao, R. and Varadhan, S.R.S. (1963), Probability distributions on locally compact Abelian groups, Illinois J. Math. 7, 337-369.

Ruzsa, I. Z. (1982), Infinite convolution and shift-convergence of measures on topological groups, Lecture Notes in Mathematics 928, Springer, Berlin Heidelberg New York, Probability Measures on Groups, Proceedings of the Sixth Conference Held at Oberwolfach, Germany, June 28 - July 4, 1981, Ed. H. Heyer), 337-353.

Ruzsa,I.Z. and Székely, G.J. (1984), Theory of decomposition in semigroups, submitted to J. Reine Angewandte Math.

Tortrat, A. (1965), Lois de probabilité sur un espace topologique complètement régulier et produits infinis à termes indépendants dans un groupe topologique, Ann. Inst. H. Poincaré 1, 217-237.

Urbanik, K. (1976), Decomposability properties of probability measures on Banach spaces, Lecture Notes in Mathematics 526, Springer, Berlin (Probability on Banach spaces, First Intern. Conf., Oberwolfach, 1975), 243-251.

Varadajan, V.S. (1961), Measures on topological spaces (in Russian), Matem. Sbornik 55, 35-100 (Transl. Amer. Math. Soc. II. Ser. 48, 161-228, 1965).

TAIL PROBABILITY OF SOME RANDOM SERIES

R. SCHOTT

The study of the density of the Brownian motion on Lie groups brings us to give precisions about the distribution of randoms series of type $S = \sum_{k=1}^{\infty} c_k \lambda_k^p$ where p is a fixed natural number, $(\lambda_k)_{k\in\mathbb{N}}$ is a sequence of gaussian random variables, the coefficients (c_k) are real and connected with the structure constants of the Lie groups.

The explicit form of the law of S is not known. The aim of this paper is to find an upper bound for the probability of the event $\{S \geq x\}$ for large x. We prove here very quickly that under some hypothesis concerning the sequence (c_k), the speed of covergence to zero is $e^{-\alpha x^{2/p}}$ where α is in $\mathbb{R}_+^*$. More precisely :

Theorem 1.-

Let $(\lambda_k)_{k\in\mathbb{N}}$ be a sequence of centered, reduced gaussian random variables. If (c_k) is a sequence of positive numbers such that $\sum_{k=1}^{\infty} c_k < +\infty$ then there exist to positive constants α and C such that :

$$\forall\, x \in \mathbb{R}^+ ,\ P\left(\sum_{k=1}^{\infty} c_k\, |\lambda_k|^p \geq x\right) \leq C\, e^{-\alpha x^{2/p}} .$$

Proof :

We use some results concerning the pseudo semi norms on gaussian fields. Let us remind first a definition and two propositions :

Definition 2.-

Let (E,B) be a mesurable vector space. A mapping N from (E,B) in $(\bar{\mathbb{R}}, B_{\bar{\mathbb{R}}})$ is called a pseudo-semi-norm on E if $N^{-1}(\mathbb{R})$ is a subvector space of E on which N induces a semi norm.

Let X be a gaussian vector with values in E. If E has a pseudo semi norm N it is proven in [1 p. 11] that :

Proposition 3.-

If $P(N(X) < +\infty)$ is strictly positive then there exists a number $\varepsilon > 0$ such that $\forall\, \alpha < \varepsilon$, $E[\exp(\alpha N^2(X))] < +\infty$.

It's easy to see that the mapping N defined by :

$$N(X) = \left(\sum_{k=1}^{\infty} c_k |\lambda_k|^p\right)^{1/p} \qquad X = (\lambda_k)_k \in \mathbb{N}^*$$

is a pseudo semi norm.

Before we give more detail about the assumption : $P(N(X) < +\infty) > 0$ we prove that the proposition 3 permits to achieve the proof of the theorem 1.

In fact : $P(N(X) \geq y) \leq P\{e^{\alpha N^2(x)} \geq e^{\alpha y^2}\}$ if $y \geq 0$

and Kolmogorov's inequality combined with the proposition 3 give :

$P\{e^{\alpha N^2(x)} \geq e^{\alpha y^2}\} \leq C\, e^{-\alpha y^2}$ where C is a strictly positive constant.

Put $y = x^{1/p}$ then

$$P\left\{\sum_{k=}^{\infty} c_k|\lambda_k|^p > x\right\} = P\left\{\left(\sum_{k=0}^{\infty} c_k|\lambda_k|^p\right)^{1/p} > x^{1/p}\right\} \leq C\, e^{-\alpha x^{2/p}}$$

and this is the desired inequality.

To finish, we have now to find assumptions under which the hypothesis $P(N(X) < +\infty)$ is fulfilled. We need here an other result about the pseudo-semi-norms.

Proposition 4.-

If N is a pseudo semi norm on E and X a gaussian vector with values in E then we have :

(i) $P\{N(X) < +\infty\} = 0$ or $P\{N(X) < +\infty\} = 1$

either

(ii) $P\{N(X) = 0\} = 0$ or $P\{N(X) = 0\} = 1$.

The proof of this result can be find in [1] .

By proposition 4, the condition $P\{N(X) < +\infty\} > 0$ is equivalent to :

$P\{N(X) < +\infty\} = 1$ and this means that the serie $[N(X)]^p = \sum_1^{\infty} c_k|\lambda_k|^p$

must be almost surely convergent.

Or by Kolmogorov's classical theorem :

$\sum_1^{\infty} c_k|\lambda_k|^p = \sum_1^{\infty} X_k$ converges almost surely if and only if the three following conditions are fulfilled for every $a > 0$:

(i) $\sum_1^{\infty} P(X_k > a) < +\infty$

(ii) $\sum_1^{\infty} \Gamma^2(X_k^a) < +\infty$ where $X_k^a = \begin{cases} X_k & \text{if } X_k \leq a \\ 0 & \text{if } X_k > a \end{cases}$

(iii) $\sum_{k=0}^{\infty} E(X_k^a) < +\infty$

or : $P\{X_k > a\} = P\{c_k|\lambda_k|^p > a\}$

$$= P\left\{|\lambda_k| > \left(\frac{a}{c_k}\right)^{1/p}\right\}$$

$$= \int_{\left(\frac{a}{c_k}\right)^{1/p}}^{+\infty} \frac{1}{\sqrt{2\pi}} e^{-\frac{x^2}{2}} dx \quad .$$

An easy calculus proves that : $P\{X_k > a\} \leqslant \frac{1}{\sqrt{2\pi}} \left(\frac{c_k}{a}\right)^{1/p} e^{-\left(\frac{a}{c_k}\right)^{2/p}}$.

The condition (ii) gives : $\sum_1^\infty c_k^2 < +\infty$ and (iii) gives : $\sum_1^\infty c_k < +\infty$.

Therefore $c_k \to 0$ where $k \to +\infty$ and we see that :

$$P\{X_k > a\} \to 0 \quad \text{if} \quad k \to +\infty \quad .$$

The three conditions are fulfilled if and only if : $\sum_1^\infty c_k < +\infty$.

Remark :

For the author the theorem 1 is only the first step in the study of the distribution of random series of the type $S = \sum_{(i_1,\dots,i_p) \in \mathbb{N}^p} c_{i_1,\dots,i_p} \lambda_{i_1} \lambda_{i_2} \dots \lambda_{i_p}$ which appear for example if we consider the Wiener "Chaos decomposition" of the Brownian motion on simply connected Lie groups.

REFERENCES :

[1] X. FERNIQUE Cours école d'été de St Flour (1974). Lecture Notes n° 480

[2] M. LOEVE Probability theory (I) : D. Van Nostrand Company

[3] R. SCHOTT Une loi du logarithme itéré pour certaines intégrales stochastiques. Note aux C.R.A.S. PARIS, t. 292, 26 Janvier 1981

[4] R. SCHOTT Une loi du logarithme itéré pour certaines intégrales stochastiques. ANNALES INSTITUT ELIE CARTAN n° 7, NANCY 1983.

R. SCHOTT
E.R.A. n° 839 du C.N.R.S.
U.E.R. Sciences Mathématiques
UNIVERSITE DE NANCY I
54506 - VANDOEUVRE LES NANCY
(France)

HOLOMORPHIC CONVOLUTION SEMIGROUPS ON TOPOLOGICAL GROUPS

Eberhard Siebert

Introduction

A convolution semigroup $(\mu_t)_{t>0}$ of probability measures on a topological group G is said to be holomorphic if roughly speaking the measures μ_t depend holomorphically on the parameter t. More precisely let us call the semigroup $(\mu_t)_{t>0}$ weakly (respectively strongly) holomorphic if for every representation π of G by isometries on a Hilbert space (respectively Banach space) E the induced operator semigroup $(\pi(\mu_t))_{t>0}$ on E extends holomorphically to an exponentially bounded operator semigroup $(T_z)_{z\in V}$ on E with an open sector V (with vertex O) of the complex plane as its parameter set.

In functional analysis holomorphic operator semigroups are well established; they have many interesting properties and applications (cf. [14,27]). In contrast holomorphic convolution semigroups have not yet found a systematic treatment; but particular classes of them appear in different places of the literature. In [15] J.Kisyński studied the holomorphy of Gaussian semigroups. A.Hulanicki [12] mentions holomorphic convolution semigroups in connection with stability and Tauberian properties for semigroups (see also [13]). T.Przebinda [19] has made some progress along these lines. L.Paquet [18] and A.M.Sinclair [26] consider strongly holomorphic convolution semigroups on the positive half line with regard to subordination. Finally in [22] we have applied weak holomorphy in studying support properties of convolution semigroups.

In the course of our subsequent investigations it has turned out that strong holomorphy is the more useful concept for convolution semigroups: First of all it has more implementations; for example with respect to absolute continuity and densities. Moreover on topological groups that are not locally compact weak holomorphy seems to be not very useful due to the lack of appropriate Hilbert space representations. Finally many of the weakly holomorphic convolution semigroups are in fact even strongly holomorphic; the symmetric convolution semigroups being a notable exception.

In the present paper we at first derive some general results on holomorphic convolution semigroups; afterwards we discuss several classes of such semigroups. In Section 1 we assemble some known facts on holomorphic operator semigroups that are at the basis of our analysis.

The definition of a (weakly or strongly) holomorphic convolution semigroup and some first examples are presented in Section 2. Some basic properties of holomorphic convolution semigroups are proved in Section 3. In particular a general result of Tauberian type is established (Theorem 1) following an idea of A.Hulanicki.

Holomorphic Gaussian semigroups are considered in Section 4. Motivated by a result of K.Yosida a forward and backward unique continuation property for these semigroups is proved (Theorem 2). In Section 5 stable convolution semigroups as defined by W.Hazod [7] are studied. With the aid of a profound result of A.Beurling [2] it is shown that a stable convolution semigroup is either strongly holomorphic or its measures are mutually singular (Theorem 3). For an absolutely continuous stable convolution semigroup this yields detailed information on its support (Theorem 4). In this context let us mention that the holomorphy of semistable convolution semigroups has been discussed in [25].

Finally it is observed in Section 6 that strongly holomorphic convolution semigroups on the positive half line give rise to strongly holomorphic convolution semigroups on topological groups by means of subordination. By this procedure one obtains plenty of strongly holomorphic convolution semigroups. Some examples of strongly holomorphic convolution semigroups on the positive half line are presented. In an Appendix the weak holomorphy of a convolution semigroup on a locally compact Abelian group is characterized with the aid of Fourier analysis.

Preliminaries

Let $\mathbb{N}$, $\mathbb{Z}$, $\mathbb{R}$, $\mathbb{C}$ be the sets of positive integers, integers, real numbers, and complex numbers respectively. By $|z|$, $\bar{z}$, Re z, Im z we denote the absolute value, the complex conjugate, the real and the imaginary part of the complex number z respectively. Let $\mathbb{Z}_+ = \{n \in \mathbb{Z} : n \geq 0\}$, $\mathbb{R}_+ = \{r \in \mathbb{R} : r \geq 0\}$, $\mathbb{R}_+^* = \{r \in \mathbb{R} : r > 0\}$, and $\mathbb{H} = \{z \in \mathbb{C} : \text{Re } z > 0\}$. Moreover the argument of $z \in \mathbb{H}$ measured between $-\pi/2$ and $\pi/2$ is denoted by arg z. Finally let V_ϑ be the open sector in $\mathbb{C}$ with vertex 0 and angle $\vartheta \in \,]0,\pi/2]$ i.e. $V_\vartheta = \{z \in \mathbb{H} : |\arg z| < \vartheta\}$.

G always denotes a topological Hausdorff group with identity e. If B is a subset of G then 1_B denotes its indicator function and $\overline{B}$ its (topological) closure. If f is a function on G and if $x \in G$ the functions ${}_xf$, f_x, f^* on G are defined by ${}_xf(y) = f(xy)$, $f_x(y) = f(yx)$, and $f^*(y) = f(y^{-1})$ (all $y \in G$).

$\mathcal{B}(G)$ denotes the σ-algebra of Borel subsets of G. Let $\mathcal{L}(G)$ be the Banach space of bounded complex valued Borel measurable functions f on G with the norm $|f|_\infty = \sup\{|f(x)| : x \in G\}$. Let $\mathcal{C}^b(G) = \{f \in \mathcal{L}(G) : f$ is continuous$\}$. Moreover let $\mathcal{C}_{lu}(G)$ (respectively $\mathcal{C}_{ru}(G)$) denote the subspace of all functions in $\mathcal{L}(G)$ that are uniformly continuous with respect to the left (respectively to the right) uniform structure on G.

$\mathcal{M}^b(G)$ denotes the linear space of all bounded τ-regular complex valued measures on G. Furnished with the norm $\|.\|$ of total variation and with the convolution product $*$ the space $\mathcal{M}^b(G)$ becomes a Banach algebra. $\mathcal{M}^1(G)$ denotes the subset of probability measures in $\mathcal{M}^b(G)$. The unit mass ε_x in $x \in G$ belongs to $\mathcal{M}^1(G)$. If $\mu \in \mathcal{M}^1(G)$ the adjoint measure $\tilde{\mu}$ is defined by $\tilde{\mu}(B) = \mu(B^{-1})$, $B \in \mathcal{B}(G)$. The support of μ is denoted by $\mathrm{supp}(\mu)$. The image $\delta(\mu)$ of μ under a continuous mapping δ of G into itself is defined by $\delta(\mu)(B) = \mu(\delta^{-1}(B))$, $B \in \mathcal{B}(G)$.

A convolution semigroup in $\mathcal{M}^1(G)$ is a family $(\mu_t)_{t>0}$ in $\mathcal{M}^1(G)$ such that $\mu_s * \mu_t = \mu_{s+t}$ for all $s,t > 0$. The semigroup is said to be <u>continuous</u> if the mapping $t \longrightarrow \int f\,d\mu_t$ of $\mathbb{R}_+^*$ into $\mathbb{C}$ is continuous for every $f \in \mathcal{C}^b(G)$. The semigroup $(\mu_t)_{t>0}$ is said to be <u>$\{e\}$-continuous</u> if $\lim_{t\downarrow 0} \int f\,d\mu_t = f(e)$ for all $f \in \mathcal{C}^b(G)$. Clearly this implies the continuity of $(\mu_t)_{t>0}$.

Now let G be a locally compact group. Put $\mathcal{C}^o(G) = \{f \in \mathcal{C}^b(G) : f$ vanishes at infinity$\}$. λ_G (or λ) always denotes a left Haar measure on G and Δ_G (or Δ) the modular function of G. If G is compact let $\lambda_G(G) = 1$. Let $L^p(G)$ be the space of (equivalence classes of) complex valued Borel measurable functions f on G such that $|f|^p$ is λ-integrable; the norm $|f|_p = (\int |f|^p\,d\lambda)^{1/p}$ turns $L^p(G)$ into a Banach space ($1 \leq p < \infty$). If $f \in L^1(G)$ then $f.\lambda$ denotes the measure with

λ-density f. Obviously one has $f.\lambda \in \mathcal{M}^b(G)$ and $\|f.\lambda\| = |f|_1$. Finally a convolution semigroup $(\mu_t)_{t>0}$ in $\mathcal{M}^1(G)$ is said to be absolutely continuous (respectively singular) if all its measures μ_t are absolutely continuous (respectively singular) with respect to λ.

1. Holomorphic operator semigroups

Let E be a complex Banach space. By $\mathcal{B}(E)$ we denote the Banach algebra of bounded linear operators on E and by E' the topological dual of E. A mapping f of an open subset D of $\mathbb{C}$ into E (respectively into $\mathcal{B}(E)$) is said to be holomorphic if for each $z \in D$ the difference quotients $\{f(z+h) - f(z)\}/h$ converge in E (respectively in $\mathcal{B}(E)$) as the complex numbers $h \neq 0$ tend to 0.

If f is a mapping of D into $\mathcal{B}(E)$ then the following assertions are known to be equivalent: (i) f is holomorphic (into $\mathcal{B}(E)$); (ii) $f(.)u$ is holomorphic (into E) for all $u \in E$; (iii) $\varphi(f(.)u)$ is holomorphic (into $\mathbb{C}$) for all $u \in E$ and $\varphi \in E'$; (iv) f admits locally a power series expansion into $\mathcal{B}(E)$ (cf. [26], Lemma 1.3; [3], 9.10).

Let $(T_t)_{t>0}$ be a semigroup of linear contractions on the Banach space E that is (weakly) measurable i.e. $t \longrightarrow \varphi(T_t u)$ is Borel measurable for all $u \in E$ and $\varphi \in E'$. $(T_t)_{t>0}$ is said to be a holomorphic operator semigroup if there exist $\vartheta \in]0, \pi/2]$ and $c,d \in \mathbb{R}_+^*$ such that the mapping $t \longrightarrow T_t$ of $\mathbb{R}_+^*$ into $\mathcal{B}(E)$ admits a holomorphic extension $z \longrightarrow T_z$ of the sector V_ϑ into $\mathcal{B}(E)$ satisfying

(BC) $\quad \|T_z\| \leq c \exp(d\,|z|) \qquad$ for all $z \in V_\vartheta$.

The family $(T_z)_{z\in V_\vartheta}$ also is a semigroup in the sense of $T_z T_w = T_{z+w}$ for all $z,w \in V_\vartheta$ (cf. [18], Proposition 1.2). Hence the boundedness condition (BC) is equivalent with

(BC') $\quad \sup\{\|T_z\| : |\arg z| < \vartheta,\ |z| < 1\} < \infty$.

The least upper bound of all $\vartheta \in]0, \pi/2]$ for which $(T_t)_{t>0}$ admits a holomorphic extension to V_ϑ in the sense above is called the angle of the holomorphic operator semigroup $(T_t)_{t>0}$.

<u>REMARK 1</u>. An operator semigroup $(T_t)_{t>0}$ admitting a holomorphic extension $(T_z)_{z\in V_\vartheta}$ not necessarily satisfying (BC) is sometimes said to be <u>pseudo-holomorphic</u> (cf. [18]). For the applications we have in mind this property would be sufficient. But for most of our examples we actually can establish holomorphy; a notable exception being the Gaussian semigroups (cf. Section 4).

Let I denote the identity operator on E. For a measurable semigroup $(T_t)_{t>0}$ of linear contractions on E let us consider the following assertions:

(i) $\lim_{t\downarrow 0} \|(T_t - I)^m\|^{1/m} < 2$ for some $m \in \mathbb{N}$;

(ii) $\lim_{t\downarrow 0, n\geq 1} \|(T_{t/n} - I)^n\|^{1/n} < 2$;

(iii) $(T_t)_{t>0}$ is a holomorphic operator semigroup.

<u>CRITERION 1</u>. (i) implies (ii) and (ii) implies (iii).

['(i)$\Longrightarrow$(ii)' There exist $\delta, \varepsilon \in \mathbb{R}_+^*$ such that $\varepsilon < 1$ and such that $\|(T_t - I)^m\|^{1/m} < 2 - \varepsilon$ if $0 < t < \delta$. Let $t \in \mathbb{R}_+^*$ and $n \in \mathbb{N}$ such that $t/n < \delta$. Then there are $k \in \mathbb{Z}_+$ and $r \in \{0,1,\ldots,m-1\}$ such that $n = km + r$. Consequently $k/n \leq 1/m$ and

$$\|(T_{t/n} - I)^n\|^{1/n} \leq 2^{r/n} \|(T_{t/n} - I)^m\|^{k/n} \leq$$
$$2^{m/n} \max\{1, \|(T_{t/n} - I)^m\|^{1/m}\} < 2^{m/n} (2 - \varepsilon).$$

Hence the assertion.

'(ii)$\Longrightarrow$(iii)' cf. [2], Theorem II.]

Now let us assume the operator semigroup $(T_t)_{t>0}$ to be <u>strongly continuous</u> i.e. $\lim_{t\downarrow 0} \|T_t u - u\| = 0$ for all $u \in E$. By $(N, \mathcal{N})$ we denote the corresponding infinitesimal generator. Then we can formulate the following assertion:

(iv) $T_t E \subset \mathcal{N}$ for all $t \in \mathbb{R}_+^*$ and
$\sup\{\|t\, NT_t\| : 0 < t < 1\} < \infty$.

CRITERION 2. The assertions (i), (ii), (iii), (iv) are all equivalent.

['(iii) $\Longrightarrow$ (iv)' cf. [27], IX.10.

'(iv) $\Longrightarrow$ (i)' There exists some $c \in \mathbb{R}_+^*$ such that $\|t\, NT_t\| < c$ for all $t \in \,]0,1[$. Choose $m \in \mathbb{N}$ such that $b := c \ln\{m/(m-1)\} < 2$.

If $0 < t < 1/m$ one has (cf. [27], p.239):

$$\|T_{mt} - T_{(m-1)t}\| = \| \int_{(m-1)t}^{mt} NT_s \, ds \| \leq \int_{(m-1)t}^{mt} \|NT_s\| \, ds <$$

$$\int_{(m-1)t}^{mt} \{c/s\} \, ds = c \, \{\ln mt - \ln(m-1)t\} = b < 2 ;$$

hence

$$\|(T_t - I)^m\| \leq (2^m - 2) + \|T_{mt} - T_{(m-1)t}\| < (2^m - 2) + b < 2^m .]$$

REMARKS. Let $(T_t)_{t>0}$ be a strongly continuous semigroup of linear contractions on the Banach space E with infinitesimal generator $(N, \mathcal{N})$.

2. If $(T_t)_{t>0}$ is a holomorphic operator semigroup then the dual semigroup $(T_t')_{t>0}$ (cf. [27], IX.13) is holomorphic too.
[For every $T \in \mathcal{B}(E)$ the dual operator T' has the same norm as T.]

3. If E is a complex Hilbert space with scalar product $<.,.>$ the following condition is sufficient for the holomorphy of the semigroup: There exist $\alpha \in \mathbb{R}_+$ and $\delta \in \,]0, \pi/2[$ such that

$$\{\alpha - <Nu,u> : u \in \mathcal{N}, \|u\| = 1\} \subset V_\delta$$

(cf. [14], p.490, Theorem 1.24).

The following two results (which are essentially known) will be useful in the course of our investigations.

LEMMA 1. Let E be a complex Banach lattice and let $(T_t)_{t>0}$ be a semigroup of positive linear contractions on E that is pseudo-holomorphic (cf. Remark 1). Moreover let $u \in E$ and $\varphi \in E'$ be positive elements. Then either $\varphi(T_t u) = 0$ for all $t > 0$ or $\varphi(T_t u) > 0$ for all $t > 0$.

PROOF. Let $\varphi(T_s u) = 0$ for some $s > 0$.
1. Assume $\lim_{t\downarrow 0} \|T_t u - u\| = 0$. Then $\varphi(T_t u) = 0$ for all $t > 0$ in view of Proposition 2 in [16].

2. Let $\varepsilon \in]0,1[$ and put $v = T_{\varepsilon s}u$. Then also v is a positive element of E and $\varphi(T_{(1-\varepsilon)s}v) = 0$. Moreover

$$\lim_{t\downarrow 0} \|T_t v - v\| = \lim_{t\downarrow 0} \|T_{\varepsilon s+t}u - T_{\varepsilon s}u\| = 0$$

in view of the pseudo-holomorphy of $(T_t)_{t>0}$. Now 1. applied to v yields $0 = \varphi(T_t v) = \varphi(T_{\varepsilon s+t}u)$ for all $t > 0$. Since ε may be chosen arbitrarily small the assertion follows. ⊥

<u>LEMMA 2</u>. Let $(\Omega, \mathfrak{A}, \mu)$ be a localizable measure space, D an open non-void subset of $\mathbb{C}$, and $F: D \times \Omega \longrightarrow \mathbb{C}$ a (product) measurable mapping with the following properties:

a) $\omega \longrightarrow F(z,\omega)$ is μ-integrable for all $z \in D$;

b) $z \longrightarrow \int |F(z,\omega)| \mu(d\omega)$ is locally bounded ;

c) $z \longrightarrow F(z,\omega)$ is holomorphic for all $\omega \in \Omega$.

Then the mapping $z \longrightarrow F(z,\,.\,)$ of D into the Banach space $L^1(\mu)$ of μ-integrable functions is holomorphic.

<u>PROOF</u>. Let $a \in D$ and choose $\delta > 0$ such that $B := \{z \in \mathbb{C} : |z - a| \leq \delta\}$ is contained in D. Moreover let $\gamma(t) = a + \delta e^{2\pi i t}$ for all $t \in [0,1]$. Then in view of property c) Cauchy's integral representation does hold for every z in the interior B_0 of B:

$$(*) \qquad \frac{\partial}{\partial z} F(z,\omega) = \frac{1}{2\pi i} \int_\gamma F(w,\omega)\ (w - z)^{-2}\, dw \quad .$$

1. There exists some $g \in L^1(\mu)$ such that $|\partial F(z,.)/\partial z| \leq g$ for all $z \in D$ with $|z - a| \leq \delta/2$.

[With the notations above let $g(\omega) = 4\delta^{-1} \int_0^1 |F(\gamma(t),\omega)|\, dt$ for all $\omega \in \Omega$. In view of b) (and of Fubini's theorem) we have $g \in L^1(\mu)$. Moreover in view of (*) we have $|\partial F(z,.)/\partial z| \leq g$ for all $z \in D$ such that $|z - a| \leq \delta/2$.]

2. Let h be a complex valued bounded measurable function defined on $(\Omega, \mathfrak{A})$. In view of a), c) and 1. the function $z \longrightarrow \int h(\omega)F(z,\omega)\mu(d\omega)$ of D into $\mathbb{C}$ is differentiable at $a \in D$ ([1], § 49, Lemma 49.1).

Since the point $a \in D$ and the function h in the dual of $L^1(\mu)$ were arbitrary the mapping $z \longrightarrow F(z,.)$ of D into $L^1(\mu)$ is weakly holomorphic. Hence the assertion (cf. the beginning of this section or [27], V.3, Theorem 1). ⌋

REMARKS. 4. Lemma 1 is essentially due to A.Kishimoto, A.Majewski and D.W.Robinson (cf. [16]).

5. Lemma 2 has been asserted by A.M.Sinclair ([26], Lemma 2.7 and Lemma 2.19).

6. Lemma 2 holds mutatis mutandis also for p-times μ-integrable functions ($p \in [1,\infty[$).

2. Definition and examples of holomorphic convolution semigroups

Let G be a topological (Hausdorff) group. If π is a strongly continuous representation of G by linear isometries on a complex Banach space E then by

$$\pi(\mu)u \quad := \quad \int \pi(x)u\,\mu(dx) \qquad\qquad (u \in E\ ;\ \mu \in \mathcal{M}^b(G))$$

there is defined a norm-contracting homomorphism $\mu \longrightarrow \pi(\mu)$ of the Banach algebra $\mathcal{M}^b(G)$ into the Banach algebra $\mathcal{B}(E)$. If $\mathcal{M}^1(G)$ is furnished with the topology of pointwise convergence on $\mathcal{C}^b(G)$ then the mapping $\mu \longrightarrow \pi(\mu)$ of $\mathcal{M}^1(G)$ into $\mathcal{B}(E)$ is also strongly continuous.

Essentially we have in mind the following two representations:

1. By $\sigma_G(x^{-1})f := {}_xf$ for all $f \in \mathcal{C}_{ru}(G)$ and $x \in G$ there is defined a strongly continuous representation σ_G of G by linear isometries on the Banach space $\mathcal{C}_{ru}(G)$. σ_G is called the left regular representation of G onto $\mathcal{C}_{ru}(G)$.

With the aid of Theorem 8.2 in [8], Vol.I, it can be easily shown that $\mathcal{C}_{ru}(G)$ is a regular algebra i.e. for every closed subset C of G and for every point $x \in G \setminus C$ there exists some $f \in \mathcal{C}_{ru}(G)$ such that $0 \leq f \leq 1 = f(x)$ and $f(y) = 0$ for all $y \in C$.

2. Let G be a locally compact group. Then by $\tau_G(x^{-1})f := {}_xf$ for all $f \in L^2(G)$ and $x \in G$ there is defined a strongly continuous representa-

tion τ_G of G by unitary operators on the Hilbert space $L^2(G)$. τ_G is called the left regular representation of G onto $L^2(G)$.

DEFINITION. Let $(\mu_t)_{t>0}$ be a convolution semigroup in $\mathcal{M}^1(G)$.

a) $(\mu_t)_{t>0}$ is said to be weakly holomorphic if for every strongly continuous representation π of G by unitary operators on a complex Hilbert space E the operator semigroup $(\pi(\mu_t))_{t>0}$ is holomorphic.

b) $(\mu_t)_{t>0}$ is said to be (strongly) holomorphic if for every strongly continuous representation π of G by linear isometries on a complex Banach space E the operator semigroup $(\pi(\mu_t))_{t>0}$ is holomorphic.

LEMMA 3. Let G be a topological group and $(\mu_t)_{t>0}$ a convolution semigroup in $\mathcal{M}^1(G)$. Then the following assertions are equivalent:

(i) $(\mu_t)_{t>0}$ is a (strongly) holomorphic convolution semigroup;

(ii) $(\sigma_G(\mu_t))_{t>0}$ is a holomorphic operator semigroup;

(iii) For some $\vartheta \in]0,\pi/2]$ the mapping $t \longrightarrow \mu_t$ admits a holomorphic extension $z \longrightarrow \mu_z$ of V_ϑ into $\mathcal{M}^b(G)$ such that $\sup\{ \|\mu_z\| : z \in V_\vartheta , |z| < 1\} < \infty$.

PROOF. '(i)$\Longrightarrow$(ii)' by definition.

'(ii)$\Longrightarrow$(iii)' Since $\mathcal{C}_{ru}(G)$ is a regular algebra (see the remark above) the homomorphism $\mu \longrightarrow \sigma_G(\mu)$ of $\mathcal{M}^b(G)$ into $\mathcal{B}(\mathcal{C}_{ru}(G))$ is an isometry; hence the assertion.

'(iii)$\Longrightarrow$(i)' Let π be a strongly continuous representation of G by linear isometries on the Banach space E. Then $\mu \longrightarrow \pi(\mu)$ is a norm-contracting homomorphism of $\mathcal{M}^b(G)$ into $\mathcal{B}(E)$; hence the assertion. ┘

DEFINITION. Let $(\mu_t)_{t>0}$ be a (strongly) holomorphic convolution semigroup in $\mathcal{M}^1(G)$. The least upper bound of all $\vartheta \in]0,\pi/2]$ for which assertion (iii) of Lemma 3 holds is called the angle of $(\mu_t)_{t>0}$.

EXAMPLES of holomorphic convolution semigroups.

a) Every Poisson semigroup (cf. [9], 6.1.1) is (strongly) holomorphic of angle $\pi/2$.

b) Every symmetric (e)-continuous convolution semigroup $(\mu_t)_{t>0}$ (i.e. $\tilde{\mu}_t = \mu_t$ for all $t > 0$) is weakly holomorphic.
[This follows by direct verification with the aid of spectral theory; cf. [22], Theorem 3.]

c) For every $t > 0$ let ν_t denote the normal distribution on $\mathbb{R}$ with mean 0 and variance t i.e. ν_t has the Lebesgue density

$$n_t(x) = (2\pi t)^{-1/2} \exp\{-x^2/2t\} \qquad (x \in \mathbb{R}).$$

Then $(\nu_t)_{t>0}$ is a convolution semigroup in $\mathcal{M}^1(\mathbb{R})$. Now let

$$n_z(x) = (2\pi z)^{-1/2} \exp\{-x^2/2z\} \qquad \text{for all } z \in \mathbb{H} \text{ and } x \in \mathbb{R}.$$

Then one has $\int |n_z(x)|\, dx = (|z|/\mathrm{Re}\ z)^{1/2}$; hence n_z is Lebesgue integrable and thus $\nu_z := n_z \cdot \lambda_{\mathbb{R}}$ is in $\mathcal{M}^b(\mathbb{R})$. Moreover $z \longrightarrow n_z(x)$ is holomorphic on $\mathbb{H}$ for every $x \in \mathbb{R}$. Consequently in view of Lemma 2 the mappings $z \longrightarrow n_z$ and $z \longrightarrow \nu_z$ of $\mathbb{H}$ into $L^1(\mathbb{R})$ respectively into $\mathcal{M}^b(\mathbb{R})$ are holomorphic. Moreover for all $z \in \mathbb{H}$ such that $\arg z = \vartheta \in]-\pi/2, \pi/2[$ one has $\|\nu_z\| = (\cos\vartheta)^{-1/2}$. Hence $(\nu_t)_{t>0}$ is a (strongly) holomorphic convolution semigroup of angle $\pi/2$.

<u>REMARKS.</u> 1. The present terminology is different from the one we have applied in [22]. In fact the convolution semigroups considered in [22] are weakly holomorphic in the sense of the definition above. Weak holomorphy is of interest mainly on a locally compact group G. There one has a powerful unitary representation available namely the left regular representation τ_G of G onto $L^2(G)$.

2. In the present paper we shall be mainly concerned with strongly holomorphic convolution semigroups. Hence we will usually omit the term 'strongly' in this context.

Of course every strongly holomorphic convolution semigroup also is weakly holomorphic.

3. One can construct symmetric convolution semigroups $(\mu_t)_{t>0}$ such

that $\| \mu_s - \mu_t \| = 2$ if $s \neq t$ (for example on the infinite-dimensional torus taking into account Kakutani's theorem; cf. [25], Remark 3.2). Hence there exist weakly holomorphic convolution semigroups (in view of Example b)) that are not strongly holomorphic (in view of Lemma 3).

3. Properties of holomorphic convolution semigroups

Let $(\mu_t)_{t>0}$ be a convolution semigroup in $\mathcal{M}^1(G)$. We begin with some basic properties following from its (weak or strong) holomorphy.

PROPERTIES. 1. Let $(\mu_t)_{t>0}$ be holomorphic. Then $(\mu_t)_{t>0}$ is also quasi-analytic (cf. [25], Section 3). Moreover the mapping $t \longrightarrow \mu_t$ of $\mathbb{R}_+^*$ into the Banach space $\mathcal{M}^b(G)$ is continuous.

2. Let $(\mu_t)_{t>0}$ be holomorphic. Then the measures μ_t, $t > 0$, are pairwise equivalent.

[Let $T_t f(x) = \int f(xy)\, \mu_t(dy)$ for all $f \in \mathcal{L}(G)$, $x \in G$ and $t \in \mathbb{R}_+^*$. In view of Lemma 3 $(T_t)_{t>0}$ is a holomorphic semigroup of positive linear contractions on the Banach lattice $\mathcal{L}(G)$. Let $\varphi(f) = f(e)$ for all $f \in \mathcal{L}(G)$. Obviously $\varphi \in \mathcal{L}(G)'$.

Now given $B \in \mathfrak{B}(G)$ let $u = 1_B$. Since $\mu_t(B) = \varphi(T_t u)$ the assertion follows from Lemma 1.]

3. Let G be locally compact and $(\mu_t)_{t>0}$ weakly holomorphic or let G be arbitrary and $(\mu_t)_{t>0}$ strongly holomorphic. Moreover let the measures μ_t, $t > 0$, be <u>tight</u>. (Of course this no restriction in the locally compact case.) Then the convolution semigroup $(\mu_t)_{t>0}$ is continuous. Moreover there exists a compact subgroup K of G such that $\lim_{t\downarrow 0} \int f\, d\mu_t = \int f\, d\lambda_K$ for all $f \in \mathcal{C}^b(G)$.

[In the first case the continuity of $(\mu_t)_{t>0}$ follows from [9],6.1.23; in the second case the continuity is immediate. The second assertion now follows from [21], Prop.5.2 .]

4. Let $(\mu_t)_{t>0}$ be holomorphic. Then there exists a closed subsemigroup S of G such that $\mathrm{supp}(\mu_t) = S$ for every $t \in \mathbb{R}_+^*$.

Moreover for tight measures μ_t, $t > 0$, one has $K \subset S$ and $KSK = S$ (where K is the compact subgroup of Property 3).

[The first assertion follows immediately from Property 2. Moreover $\lambda_K * \mu_t * \lambda_K = \mu_t$ (cf. [21], Prop. 5.2) implies $KSK = S$. Finally $\lim_{t \downarrow 0} \int f \, d\mu_t = \int f \, d\lambda_K$ for all $f \in \mathcal{C}^b(G)$ yields $K \cap S \neq \emptyset$ and hence $K \subset S$.]

5. Let G be locally compact and $(\mu_t)_{t>0}$ weakly holomorphic. Then there exists a closed subsemigroup S of G such that $\operatorname{supp}(\mu_t) = S$ for every $t \in \mathbb{R}_+^*$. Moreover $K \subset S$ and $KSK = S$ (where K is the compact subgroup of Property 3).

[Let U be an open, relatively compact and symmetric neighbourhood of $e \in G$. For some fixed $x \in G$ we put

$$\varphi(t) = \int (\mu_t * 1_{x^{-1}U}) \, 1_U \, d\lambda = \int (\tau_G(\mu_t) 1_{x^{-1}U}) \, 1_U \, d\lambda$$

for all $t > 0$. In view of Lemma 1 we have either $\varphi(t) = 0$ for all $t > 0$ or $\varphi(t) > 0$ for all $t > 0$. But $\varphi(t) > 0$ if and only if $\mu_t(U^2x) > 0$ (cf. [22], part 1 of the proof of Theorem 1). Hence the first assertion. The second assertion follows as in the proof of Property 4.]

Remark. This extends Theorem 1 in [22] where only the case $K = \{e\}$ has been considered.

6. Let G be locally compact and $(\mu_t)_{t>0}$ weakly holomorphic. Moreover let every μ_t have a λ_G-density f_t such that $f_t, f_t^* \in L^2(G)$. Since $f_{s+t} = \tau_G(\mu_s) f_t$ the mapping $t \longrightarrow f_t$ of $\mathbb{R}_+^*$ into $L^2(G)$ can be holomorphically extended to some sector V_ϑ, $0 < \vartheta \leq \pi/2$.

Furthermore we have $f_s * f_t \in \mathcal{C}^0(G)$ ([8], Vol.I, (20.16)); hence without loss of generality $f_r \in \mathcal{C}^0(G)$ and $f_s * f_t(x) = f_{s+t}(x)$ for all $x \in G$ $(r,s,t \in \mathbb{R}_+^*)$. Then the functions $t \longrightarrow f_t(x)$, $x \in G$, of $\mathbb{R}_+^*$ into $\mathbb{C}$ can be holomorphically extended to V_ϑ. Moreover the function $(t,x) \longrightarrow f_t(x)$ is simultaneously continuous on $\mathbb{R}_+^* \times G$.

[The proof is similar to that of Theorem 6 in [22].]

7. Let G be locally compact and $(\mu_t)_{t>0}$ holomorphic. Moreover let

every μ_t have a bounded λ_G-density f_t. Then $f_s * f_t \in \mathcal{L}_{ru}(G)$ ([8], Vol.I, (20.16)); hence without loss of generality $f_r \in \mathcal{L}_{ru}(G)$ and $f_s * f_t(x) = f_{s+t}(x)$ for all $x \in G$ $(r,s,t \in \mathbb{R}_+^*)$. Since $\mathfrak{S}_G(\mu_s)f_t = f_{s+t}$ the mapping $t \longrightarrow f_t$ of $\mathbb{R}_+^*$ into $\mathcal{L}_{ru}(G)$ can be holomorphically extended to some sector V_ϑ. Moreover $(t,x) \longrightarrow f_t(x)$ again is simultaneously continuous on $\mathbb{R}_+^* \times G$.

Further properties of holomorphic convolution semigroups will follow from Theorem 1 below and from the subsequent examples.

THEOREM 1. Let $\mathfrak{J}$ be a closed one-sided ideal of the Banach algebra $\mathcal{M}^b(G)$. Moreover let $(\mu_t)_{t>0}$ be a holomorphic convolution semigroup in $\mathcal{M}^1(G)$.

If $\mu_{t_0} \in \mathfrak{J}$ for some $t_0 \in \mathbb{R}_+^*$ then $\mu_t \in \mathfrak{J}$ for all $t \in \mathbb{R}_+^*$.

PROOF. Since $\mathfrak{J}$ is an ideal $\mu_{t_0} \in \mathfrak{J}$ implies

$$\mu_t = \mu_{t-t_0} * \mu_{t_0} = \mu_{t_0} * \mu_{t-t_0} \in \mathfrak{J} \qquad \text{for all } t > t_0.$$

Let ϑ be the angle of $(\mu_t)_{t>0}$. Then in view of Lemma 3 there exists a sequence $(\nu_n)_{n\geq 0}$ in $\mathcal{M}^b(G)$ such that $\mu_t = \sum_{n\geq 0} (t - t_0)^n \nu_n$ for all $t > 0$ with $|t - t_0| < t_0 \sin\vartheta$ (the series converging with respect to the norm of $\mathcal{M}^b(G)$). Obviously

$$\nu_n = (d^n/dt^n)|_{t=t_0} \mu_t \qquad \text{for all } n \in \mathbb{Z}_+.$$

Since $\mu_t \in \mathfrak{J}$ for all $t > t_0$ and since $\mathfrak{J}$ is a closed linear subspace of $\mathcal{M}^b(G)$ this yields $\nu_n \in \mathfrak{J}$ for all $n \in \mathbb{Z}_+$. Hence $\mu_t \in \mathfrak{J}$ for all $t > t_0(1 - \sin\vartheta)$.

Iterating this procedure with $t_n := t_0(1 - (\sin\vartheta)/2)^n$ instead of t_0 one finally obtains $\mu_{t_n} \in \mathfrak{J}$ for all $n \in \mathbb{N}$. Since $\lim t_n = 0$ this yields the assertion. ⌟

EXAMPLES. a) Let G be a locally compact group and $\mathcal{M}_a(G)$ the space of measures in $\mathcal{M}^b(G)$ that are absolutely continuous with respect to λ_G. As is well known $\mathcal{M}_a(G)$ is a closed two-sided ideal of $\mathcal{M}^b(G)$. Hence Theorem 1 applies: If one measure of a holomorphic convolution semigroup in $\mathcal{M}^1(G)$ is absolutely continuous (with respect to λ_G) then all of its measures are absolutely continuous.

b) Let K be a compact subgroup of G and let $\mathfrak{J}$ be the space of all measures $\mu \in \mathcal{M}^b(G)$ such that $\mu * \lambda_K = \mu$ (respectively $\lambda_K * \mu = \mu$) Then $\mathfrak{J}$ is a closed left (respectively right) ideal of $\mathcal{M}^b(G)$. Hence Theorem 1 applies to the effect that all measures of a holomorphic convolution semigroup have the same idempotent factors.

c) Let G be a non-compact locally compact group and let δ denote the point at infinity in the Alexandrov compactification of G. Fix $f \in \mathcal{L}(G)$ and put $\mathfrak{J} := \{\mu \in \mathcal{M}^b(G): \lim_{x\to\delta} \mu * f(x) = 0\}$. Obviously $\mathfrak{J}$ is a closed left ideal of $\mathcal{M}^b(G)$.

Now let $(\mu_t)_{t>0}$ be a holomorphic (e)-continuous convolution semigroup in $\mathcal{M}^1(G)$ such that $\mu_{t_0} \in \mathfrak{J}$ for some $t_0 \in \mathbb{R}_+^*$. Then in view of Theorem 1 one has $\mu_t \in \mathfrak{J}$ for all $t \in \mathbb{R}_+^*$. This implies the following results of Tauberian type (cf. [11,12]):

(i) If $f \in \mathcal{C}_{ru}(G)$ then $\lim_{t\downarrow 0} |\mu_t * f - f|_\infty = 0$ and hence $\varepsilon_e \in \mathfrak{J}$ i.e. $\mathfrak{J} = \mathcal{M}^b(G)$.

(ii) Let G be unimodular (i.e. $\Delta_G \equiv 1$) and let $g \in L^1(G)$. Then $g * \mu = (\tilde{\mu} * g^*)^*$ for all $\mu \in \mathcal{M}^b(G)$ implies

$$\|g.\lambda - (g.\lambda) * \mu_t\| = |g^* - \tilde{\mu}_t * g^*|_1 \qquad \text{for all } t > 0.$$

The (e)-continuity of $(\mu_t)_{t>0}$ yields $\lim_{t\downarrow 0} |g^* - \tilde{\mu}_t * g^*|_1 = 0$. Hence $g.\lambda \in \mathfrak{J}$ (i.e. $\mathcal{M}_a(G) \subset \mathfrak{J}$) and thus $\lim_{x\to\delta} g * f(x) = 0$. This result implies Hulanicki's 'theorem of Tauberian type for solutions of the heat equation' ([11], Theorem 5.14) since the Gaussian semigroup considered in [11] is pseudo-holomorphic in view of [15] (cf. Remark 1.1).

4. Holomorphy of Gaussian semigroups

Let G be a connected Lie group with Lie algebra $\mathfrak{g}$. Moreover let $(\mu_t)_{t>0}$ be a Gaussian semigroup in $\mathcal{M}^1(G)$ with infinitesimal generator N. Then there exist $X_0, X_1, \ldots, X_r \in \mathfrak{g}$ such that $Nf = \tilde{X}_0 f + \tilde{X}_1^2 f + \ldots + \tilde{X}_r^2 f$ for all complex valued infinitely differentiable functions f on G with compact support (cf. [9], VI and [23]). Without loss of generality we may assume that $\mathfrak{g}$ is generated (as a Lie algebra) by the system $\{X_0, X_1, \ldots, X_r\}$.

Let $\mathfrak{f}$ be the Lie subalgebra of $\mathfrak{g}$ generated by the vectors $X_1, X_2, \ldots, X_r$; $[X_{j_1}, X_{j_2}]$, $[X_{j_1}, [X_{j_2}, X_{j_3}]]$, ... for all $j_1, \ldots, j_k \in \{0,1,\ldots,r\}$ and $k \in \mathbb{N}$. Then $\mathfrak{f}$ is an ideal of $\mathfrak{g}$ such that $[\mathfrak{g}, \mathfrak{g}] \subset \mathfrak{f}$. The semigroup $(\mu_t)_{t>0}$ is absolutely continuous (respectively singular) if and only if $\mathfrak{f} = \mathfrak{g}$ (respectively $\mathfrak{f} \neq \mathfrak{g}$) ([23], Theorem 2).

REMARKS. 1. If X_0 is in the linear hull of $\{X_1, \ldots, X_r\}$ then in view of [15], Theorem 1, the semigroup $(\mu_t)_{t>0}$ is pseudo-holomorphic of angle $\pi/2$ i.e. $t \longrightarrow \mu_t$ extends to a holomorphic mapping $z \longrightarrow \mu_z$ of $\mathbb{H}$ into $\mathcal{M}^b(G)$ (cf. Remark 1.1).

2. If X_0 is in the linear hull of $\{X_i, [X_j, X_k] : 1 \leq i,j,k \leq r\}$ then the semigroup $(\mu_t)_{t>0}$ is weakly holomorphic ([22], Theorem 2).

3. A detailed inspection of the last part of the proof of Theorem 1 in [15] yields: If $(\mu_t)_{t>0}$ is weakly holomorphic then it is already pseudo-holomorphic.

4. If $(\mu_t)_{t>0}$ is weakly holomorphic then it is absolutely continuous. [In view of Remark 3 and of Lemma 3 the mapping $t \longrightarrow \mu_t$ is norm continuous on $\mathbb{R}_+^*$. On the other hand if $(\mu_t)_{t>0}$ is singular then one has $\|\mu_s - \mu_t\| = 2$ if $0 < t - s < \alpha$ for some appropriate $\alpha \in]0, \infty]$ ([23], Remark 1.3).]

5. In [24] we have shown that every absolutely continuous Gaussian semigroup $(\mu_t)_{t>0}$ on G is differentiable. But $(\mu_t)_{t>0}$ need not be

weakly holomorphic. In fact it has been shown by M.Mc Crudden and R.M. Wood [17] that there exists an absolutely continuous Gaussian semigroup $(\mu_t)_{t>0}$ on $SL(2,\mathbb{R})$ such that $supp(\mu_t)$ depends on $t \in \mathbb{R}_+^*$. Hence taking into account Property 3.5 the semigroup $(\mu_t)_{t>0}$ cannot be weakly holomorphic.

6. Let $(\mu_t)_{t>0}$ be a weakly holomorphic Gaussian semigroup. Then in view of Remarks 3 and 4 above and taking into account [24], Theorem 1, one observes that the assumptions in Properties 3.6 and 3.7 are fulfilled for $(\mu_t)_{t>0}$.

EXAMPLE. Let G be a connected nilpotent Lie group of degree 2 (for example a Heisenberg group) and let the Gaussian semigroup $(\mu_t)_{t>0}$ in $\mathcal{M}^1(G)$ be absolutely continuous. Then $(\mu_t)_{t>0}$ is weakly holomorphic and $supp(\mu_t) = G$ for all $t > 0$.

[Let $\mathfrak{m}$ denote the Lie subalgebra of $\mathfrak{g}$ generated by $\{X_1, \ldots, X_r\}$. As it has been pointed out to the author by M.Mc Crudden (and what in this particular case is easy to see) for a nilpotent Lie algebra $\mathfrak{g}$ one already has $\mathfrak{m} = \mathfrak{f}$. On the other hand by the absolute continuity of $(\mu_t)_{t>0}$ one has $\mathfrak{f} = \mathfrak{g}$. Hence $supp(\mu_t) = G$ for all $t > 0$ by [23], Theorem 4. Moreover since $\mathfrak{g}$ is nilpotent of degree 2 the weak holomorphy now follows from Remark 2 above.]

THEOREM 2. Let $(\mu_t)_{t>0}$ be a weakly holomorphic Gaussian semigroup in $\mathcal{M}^1(G)$. Let $f \in L^2(G)$ and let there exist an open non-void subset U of G and some $t_0 \in \mathbb{R}_+^*$ such that $\mu_{t_0} * f(x) = 0$ for λ-almost all x in U.

Then one has $\mu_t * f(x) = 0$ for all $x \in U$ and for all $t \in \mathbb{R}_+^*$ and moreover $f(x) = 0$ for λ-almost all $x \in U$.

PROOF. Let $T_t = \tau_G(\mu_t)$ for all $t > 0$. By assumption $(T_t)_{t>0}$ is a holomorphic operator semigroup on $L^2(G)$ (of some angle ϑ). Let $(N, \mathcal{N})$ denote its infinitesimal generator. In view of Remark 4 and of [24], Theorem 3 we have $T_{t_0} f \in \mathcal{N}$. Moreover since N is a local operator it follows $N^k T_{t_o} f(x) = 0$ for λ-almost all $x \in U$ and $k \in \mathbb{Z}_+$ (cf. [24], Theorem 2). Since

$$\mu_t * f = T_t f = \sum_{k\geq 0} [(t - t_o)^k/(k!)] N^k T_{t_o} f$$

(the series converging in $L^2(G)$ if $|t - t_o| < t_o \sin \vartheta$) this yields $\mu_t * f(x) = 0$ for λ-almost all $x \in U$ and for all $t \in \mathbb{R}_+^*$ such that $|t - t_o| < t_o \sin \vartheta$. But $x \longrightarrow \mu_t * f(x)$ is continuous ([24], Theorem 2); hence $\mu_t * f(x) = 0$ for all $x \in U$ and for all $t \in \mathbb{R}_+^*$ such that $|t - t_o| < t_o \sin \vartheta$. Iterating this procedure yields the first assertion.

Moreover in view of $\lim_{t\downarrow 0} |\mu_t * f - f|_2 = 0$ there exists a sequence $(t_n)_{n\geq 1}$ in $\mathbb{R}_+^*$ descending to 0 such that $\lim_{n\geq 1} \mu_{t_n} * f = f$ λ-almost everywhere. This yields the second assertion. ┘

REMARKS. 7. Theorem 2 extends a result of K.Yosida ([27], pp.424-425). He characterizes the first assertion of Theorem 2 as 'forward and backward unique continuation property'.

8. It follows from Remark 3 and from the results of [24] that the space $L^2(G)$ in Theorem 2 may be substituted by some space $L^p(G)$, $1 \leq p \leq \infty$.

5. Holomorphy of stable convolution semigroups

Let G be a topological group and let Aut(G) denote the group of its topological automorphisms. A continuous convolution semigroup $(\mu_t)_{t>0}$ in $\mathcal{M}^1(G)$ is said to be stable (in the strict sense) if there exists a subgroup $(\delta_t)_{t>0}$ of Aut(G) (i.e. $\delta_s \circ \delta_t = \delta_{st}$ for all $s,t > 0$) such that $\mu_t = \delta_t(\mu_1)$ for all $t > 0$. Obviously this implies $\delta_s(\mu_t) = \mu_{st}$ for all $s,t > 0$ (cf.[7]).

THEOREM 3. Let $(\mu_t)_{t>0}$ be a stable convolution semigroup in $\mathcal{M}^1(G)$. Then the following assertions are equivalent:

(i) $\|\mu_s - \mu_t\| < 2$ for some $s,t > 0$ with $s \neq t$;

(ii) $t \longrightarrow \mu_t$ is a continuous mapping of $\mathbb{R}_+^*$ into the Banach space $\mathcal{M}^b(G)$;

(iii) $(\mu_t)_{t>0}$ is a holomorphic convolution semigroup.

PROOF. '(iii) $\Longrightarrow$ (ii)' follows from Lemma 3; and '(ii) $\Longrightarrow$ (i)' is evident. It remains to prove '(i) $\Longrightarrow$ (iii)' In view of Criterion 1 it suffices to show the existence of some $c \in]0,2[$ and some $m \in \mathbb{N}$ such that $\|(\mu_p - \varepsilon_e)^m\| \leq 2^m - c$ for all $p > 0$. Now by assumption $\|\mu_s - \mu_t\| = 2 - c$ for some $c \in]0,2[$ and some $s,t \in \mathbb{R}$ with $0 < s < t$. Hence there exist some $r > 0$ and $m \in \mathbb{N}$, $m \neq 1$, such that $\|\mu_{(m-1)r} - \mu_{mr}\| \leq 2 - c$.
[Let $r := t - s$ and choose $m \in \mathbb{N}$, $m \neq 1$, such that $s < (m-1)r$. Consequently

$$\|\mu_{(m-1)r} - \mu_{mr}\| = \|\mu_{(m-1)r-s} * \mu_s - \mu_{(m-1)r-s} * \mu_t\| \leq \|\mu_s - \mu_t\| = 2 - c. \quad]$$

Let $(\delta_p)_{p>0}$ be the group of automorphisms of G associated with $(\mu_q)_{q>0}$ by the definition of stability i.e. $\delta_p(\mu_q) = \mu_{pq}$ for all $p,q > 0$. Thus

$$\|\mu_{(m-1)rp} - \mu_{mrp}\| = \|\delta_p(\mu_{(m-1)r}) - \delta_p(\mu_{mr})\| = \|\mu_{(m-1)r} - \mu_{mr}\| \leq 2 - c$$

for all $p > 0$. Consequently

$$\|(\mu_{rp} - \varepsilon_e)^m\| = \Big\|\sum_{k=0}^{m} (-1)^k \binom{m}{k} \mu_{(m-k)rp}\Big\| \leq \|\mu_{mrp} - \mu_{(m-1)rp}\| + 2^m - 2 \leq 2^m - c$$

for all $p > 0$ (where $\mu_0 := \varepsilon_e$). Hence the assertion. ⌋

REMARKS. 1. There exist similar results on the holomorphy of semistable convolution semigroups ([25], Theorems 5 and 6).

2. Let G be a locally compact group and let $(\mu_t)_{t>0}$ be a stable convolution semigroup in $\mathcal{M}^1(G)$. If one measure μ_{t_0} is absolutely continuous then all the measures μ_t are absolutely continuous and the

semigroup $(\mu_t)_{t>0}$ is holomorphic.

[Since $\mu_{st_0} = \delta_s(\mu_{t_0})$ for some $\delta_s \in \mathrm{Aut}(G)$ the measure μ_{st_0} is absolutely continuous (all $s > 0$). Hence the mapping $t \longrightarrow \mu_t$ of $\mathbb{R}_+^*$ into the Banach space $\mathcal{M}^b(G)$ is continuous. Thus the holomorphy follows by Theorem 3.]

PROPERTIES. Let $(\mu_t)_{t>0}$ be a holomorphic and stable convolution semigroup in $\mathcal{M}^1(G)$ with associated automorphism group $(\delta_t)_{t>0}$.

1. There exists a closed subsemigroup S of G such that $S = \mathrm{supp}(\mu_t)$ and $\delta_t(S) = S$ for all $t > 0$.

[The first assertion holds in view of Property 3.4. The second assertion is then evident in view of $\delta_t(\mu_1) = \mu_t$ for all $t > 0$.]

2. If the automorphism group $(\delta_t)_{t>0}$ is continuous and contracting (i.e. $\lim_{t\to s} \delta_t(x) = \delta_s(x)$ and $\lim_{t\downarrow 0} \delta_t(x) = e$ for all $x \in G$ and $s \in \mathbb{R}_+^*$) then S is connected.

[In view of Property 1 one has $[x] := \{\delta_t(x) : t > 0\}^- \subset S$ for all $x \in S$. Obviously $[x]$ is connected and $e, x \in [x]$. Hence the assertion.]

3. Let G be a locally compact group and let μ_1 admit a λ-density f_1. Then it is easy to see that $f_t := t^\beta f_1 \circ \delta_{1/t}$ is a λ-density for μ_t (where $\beta \in \mathbb{R}$ is defined by $t^\beta = \Delta(\delta_t)$ for all $t > 0$). If in addition f_1 is bounded then without loss of generality all the f_t may be assumed to be continuous. Moreover $t \longrightarrow f_t(x)$ and hence $t \longrightarrow f_1(\delta_t(x))$ are analytic on $\mathbb{R}_+^*$ for every $x \in G$. [Apply Property 3.7.]

EXAMPLE. Let $\mu \in \mathcal{M}^1(\mathbb{R}^d)$ be a full operator-stable measure (in the sense of M.Sharpe [20]). Then there exist a non-singular real $d \times d$ - matrix B and a continuous mapping b of $\mathbb{R}_+^*$ into $\mathbb{R}^d$ with $b(1) = 0$ such that the measures $\mu_t := t^B(\mu) * \varepsilon_{b(t)}$, $t > 0$, form a convolution semigroup in $\mathcal{M}^1(\mathbb{R}^d)$. If 1 is no eigenvalue of B then there exists some $a \in \mathbb{R}^d$ such that $b(t) = ta - t^B a$ ([20], Theorem 6). Consequently $(\mu_t * \varepsilon_{-ta})_{t>0}$ is a stable convolution semigroup with associated automorphism group $(t^B)_{t>0}$; in fact $(t^B)_{t>0}$ is continuous and contracting (cf. Property 2).

Without loss of generality let us assume now $a = 0$. It is well known that every μ_t admits a bounded continuous Lebesgue density f_t ([10], Theorem 1). Hence the convolution semigroup $(\mu_t)_{t>0}$ is holomorphic in view of Remark 2. By Property 3 the functions $t \longrightarrow f_t(x)$ and $t \longrightarrow f_1(t^B x)$ are thus analytic on $\mathbb{R}_+^*$ for every $x \in \mathbb{R}^d$.

Now let $d = 1$. Then B can be identified with a real number $\beta \neq 0$. In view of $f_t(x) = t^{\beta} f_1(t^{-\beta} x)$ (cf. Property 3) the analyticity of $t \longrightarrow f_t(x)$ implies the analyticity of f_1 on $\mathbb{R}^* := \mathbb{R} \setminus \{0\}$ and in turn this implies the analyticity of the mapping $(t,x) \longrightarrow f_t(x)$ on $\mathbb{R}_+^* \times \mathbb{R}^*$ (cf. [3], (9.3.2)).

For absolutely continuous stable convolution semigroups Properties 1 and 2 above can be improved. The following lemma strengthens Lemma 3 of [25].

<u>LEMMA 4</u>. Let G be a locally compact group and $(\mu_t)_{t>0}$ a convolution semigroup in $\mathcal{M}^1(G)$ such that every μ_t admits a λ-density f_t. Let S_t denote the support of μ_t. Then S_t is the closure of its interior K_t and $\mu_t(K_t) = 1$ for all $t > 0$.

<u>PROOF</u>. Let $t > 0$ and $s = t/2$ be fixed. Put $g_s = \min(f_s, 1_G)$. Then the function $f_s * g_s$ is continuous ([8], Vol.I, (20.16)).

Let $x \in G$ such that $f_t(x) > 0$ and $f_t(x) = f_s * f_s(x)$. Hence in view of $\int f_s(xy)\, f_s(y^{-1})\, \lambda(dy) > 0$ there exists some $C \in \mathfrak{B}(G)$ such that $\lambda(C) > 0$ and $f_s(xy)\, f_s(y^{-1}) > 0$ for all $y \in C$. Consequently $f_s * g_s(x) = \int f_s(xy)\, g_s(y^{-1})\, \lambda(dy) > 0$.

Conversely if $f_s * g_s(x) > 0$ and $f_t(x) = f_s * f_s(x)$ then one has $f_t(x) \geq f_s * g_s(x) > 0$.

Let $B \in \mathfrak{B}(G)$ such that $\lambda(\complement B) = 0$ and $f_t(x) = f_s * f_s(x)$ for all $x \in B$. Then $[f_t > 0] \cap B = [f_s * g_s > 0] \cap B$ as just proved. Obviously $\mu_t(B) = 1 = \mu_t([f_t > 0])$; hence $\mu_t([f_s * g_s > 0]) = 1$.

Now assume $f_s * g_s(x) > 0$ for some $x \in G \setminus S_t$. Then there exists an open neighbourhood U of x such that $U \cap S_t = \emptyset$ and $U \subset [f_s * g_s > 0$

Hence $f_t(y) = f_s * f_s(y) > 0$ for all $y \in U \cap B$ $(\neq \emptyset)$. On the other hand $\int 1_{U\cap B} f_t \, d\lambda = \int 1_U f_t \, d\lambda = \mu_t(U) = 0$. Consequently $\lambda(U \cap B) = 0$. But this contradicts $0 < \lambda(U) = \lambda(U \cap B)$.

Thus $[f_s * g_s > 0] \subset S_t$ and therefore $[f_s * g_s > 0] \subset K_t$. Hence $\mu_t(K_t) = 1$. Now the first assertion follows immediately. ┘

LEMMA 5. Let G be a topological group. Let $(\mu_t)_{t>0}$ be an {e}-continuous convolution semigroup in $\mathcal{M}^1(G)$ and T a connected closed subset of G such that $\mathrm{supp}(\mu_t) = T$ for all $t > 0$. Let K denote the interior of T and assume $\mu_t(K) = 1$ for all $t > 0$. Then K is connected too.

PROOF. Let K_n, $n \in I$, be the connected components of K. Clearly T is a subsemigroup of G with $e \in T$ and $KT = K$. Hence $K_n \subset K_n T \subset K$ and thus $K_n = K_n T$ since $K_n T$ is connected (all $n \in I$).

Let $n_0 \in I$ be fixed and put $C = K_{n_0}$, $D = \bigcup_{n \neq n_0} K_n$. Then one has $C = CT$, $D = DT$, and $C \cup D = K$. This implies for all $s,t > 0$:

$$\mu_{s+t}(C) = \mu_{s+t}(CT) \geq \mu_s(C)\mu_t(T) = \mu_s(C) ,$$

and analogously $\mu_{s+t}(D) \geq \mu_s(D)$. On the other hand one has

$$\mu_{s+t}(C) + \mu_{s+t}(D) = \mu_{s+t}(K) = 1 = \mu_s(K) = \mu_s(C) + \mu_s(D) .$$

Together this yields $\mu_{s+t}(C) = \mu_s(C)$ and hence $\mu_r(C) =: c$ for all $r > 0$.

Since $CK \subset C \subset T$ and since CK is open this implies $c > 0$. Thus $e \in \overline{C}$ since $(\mu_t)_{t>0}$ is {e}-continuous. Hence $C = CT$ implies

$$T \supset \overline{C} = \overline{CT} \supset \overline{C}T = T \qquad \text{and thus} \qquad \overline{C} = T .$$

Now $C \subset K \subset T$ and the connectedness of C imply the connectedness of K. ┘

THEOREM 4. Let G be a locally compact group. Let $(\mu_t)_{t>0}$ be a stable convolution semigroup in $\mathcal{M}^1(G)$ with respect to a continuous and contracting automorphism group $(\delta_t)_{t>0}$. Moreover let every measure μ_t be absolutely continuous.

Then there exists an open and connected subsemigroup K of G with $e \in \overline{K}$ such that $\delta_t(K) = K$, $\mu_t(K) = 1$, and $\overline{K} = \mathrm{supp}(\mu_t)$ for all $t > 0$.

PROOF. In view of Remark 2 the semigroup $(\mu_t)_{t>0}$ is holomorphic. Since the group $(\delta_t)_{t>0}$ is contracting it is easy to prove that the semigroup $(\mu_t)_{t>0}$ is {e}-continuous. Hence the assertions follow from Properties 1 and 2 in connection with Lemma 4 and Lemma 5. ┘

REMARK 3. Theorem 4 should be compared with Theorem 7 of [25] where a corresponding result has been established for semistable convolution semigroups on Euclidean spaces.

6. Holomorphy of subordinated convolution semigroups

Let $(\eta_t)_{t>0}$ be a convolution semigroup in $\mathcal{M}^1(\mathbb{R})$ such that $\mathrm{supp}(\eta_t) \subset \mathbb{R}_+$ for all $t > 0$. If $(\mu_t)_{t>0}$ is an {e}-continuous convolution semigroup in $\mathcal{M}^1(G)$ then by the weak integrals

$$\nu_t = \int \mu_s \ \eta_t(ds) \qquad (t > 0)$$

there is given a new {e}-continuous convolution semigroup $(\nu_t)_{t>0}$ in $\mathcal{M}^1(G)$. One says that $(\nu_t)_{t>0}$ is subordinated to $(\mu_t)_{t>0}$ by means of the directing semigroup $(\eta_t)_{t>0}$ (cf. [6]).

LEMMA 6. If $(\eta_t)_{t>0}$ is holomorphic then $(\nu_t)_{t>0}$ is holomorphic too.

PROOF. Let π be a strongly continuous representation of G by linear isometries on a complex Banach space E. Let $u \in E$, $\varphi \in E'$, and put $f(s) := \int \varphi(\pi(x)u)\,\mu_s(dx)$ if $s > 0$ and $f(s) := \varphi(u)$ if $s \leq 0$. Obviously $f \in \mathcal{C}_{ru}(\mathbb{R})$. Let ψ denote the continuous linear form defined on $\mathcal{C}_{ru}(\mathbb{R})$ by $\psi(g) = g(0)$. Then one has

$$\psi(G_{\mathbb{R}}(\eta_t)f) = \int f(s)\,\eta_t(ds) = \varphi(\pi(\nu_t)u) \qquad (t > 0).$$

Hence the assertion (in view of Section 1). ┘

REMARKS. 1. In fact Lemma 6 is covered by Corollary 1.5 of [18].

2. In view of Lemma 6 there exist plenty of holomorphic convolution semigroups on a topological group provided there exist enough holomorphic convolution semigroups in $\mathcal{M}^1(\mathbb{R})$ supported by $\mathbb{R}_+$. But this is true

as the following examples will reveal.

EXAMPLES. a) Let $\varkappa \in \mathcal{M}^b(\mathbb{R})$ be positive such that $\mathrm{supp}(\varkappa) \subset \mathbb{R}_+$. For every $t \in \mathbb{R}_+^*$ one denotes by $e(t\varkappa)$ the compound Poisson distribution on $\mathbb{R}$ with exponent $t\varkappa$. Of course $(e(t\varkappa))_{t>0}$ is a holomorphic convolution semigroup in $\mathcal{M}^1(\mathbb{R})$ (of angle $\pi/2$) supported by $\mathbb{R}_+$.

If $(\nu_t)_{>0}$ is subordinated to an arbitrary convolution semigroup $(\mu_t)_{t>0}$ (in $\mathcal{M}^1(G)$) by means of the directing semigroup $(e(t\varkappa))_{t>0}$ then $(\nu_t)_{t>0}$ is a Poisson semigroup ([6], Satz 4.3). In view of Example 2.a) one obtains thus no new examples in this case.

b) For $\alpha \in]0,1[$ and $t \in \mathbb{R}_+^*$ let $\sigma_t^{(\alpha)} \in \mathcal{M}^1(\mathbb{R})$ denote the unique probability measure supported by $\mathbb{R}_+$ with Laplace transform $\exp(-tr^\alpha)$, $r > 0$; i.e. $\sigma_t^{(\alpha)}$ is a (one-sided strictly) stable distribution with exponent α. Then $(\sigma_t^{(\alpha)})_{t>0}$ is a stable convolution semigroup in $\mathcal{M}^1(\mathbb{R})$ (in the sense of Section 5) supported by $\mathbb{R}_+$. Since every $\sigma_t^{(\alpha)}$ is absolutely continuous the semigroup $(\sigma_t^{(\alpha)})_{t>0}$ is holomorphic in view of Remark 5.2.

Of course this result is well known (cf. [27], IX.11 ; [18], Corollaire 2.25). In fact in [18] it is shown that the angle of $(\sigma_t^{(\alpha)})_{t>0}$ is $(1-\alpha)\pi/2$ and that the holomorphic extension is uniformly norm bounded.

c) For every $t > 0$ let γ_t denote the gamma distribution on $\mathbb{R}$ with Lebesgue density

$$g_t(x) = \Gamma(t)^{-1} x^{t-1} e^{-x} 1_{]0,\infty[}(x) \qquad (x \in \mathbb{R}).$$

Then $(\gamma_t)_{t>0}$ is a convolution semigroup in $\mathcal{M}^1(\mathbb{R})$ supported by $\mathbb{R}_+$ (cf. [4], II.2). Obviously the definition can be extended to all $z \in \mathbb{H}$:

$$g_z(x) = \Gamma(z)^{-1} x^{z-1} e^{-x} 1_{]0,\infty[}(x) \qquad (x \in \mathbb{R}).$$

Then one has $\int |g_z(x)|\, dx = \Gamma(\mathrm{Re}\, z)/|\Gamma(z)|$; hence g_z is Lebesgue integrable and thus $\gamma_z := g_z.\lambda$ is in $\mathcal{M}^b(\mathbb{R})$. Moreover $z \longrightarrow g_z(x)$ is holomorphic on $\mathbb{H}$ for every $x \in \mathbb{R}$. Consequently taking into account

Lemma 2 the mapping $z \longrightarrow g_z$ of $\mathbb{H}$ into $L^1(\mathbb{R})$ and hence also the mapping $z \longrightarrow \gamma_z$ of $\mathbb{H}$ into $\mathcal{M}^b(\mathbb{R})$ is holomorphic (see also [26], Theorem 2.6). Moreover one has for all $x \in \mathbb{R}_+^*$ and $y \in \mathbb{R}$ ([5], 8.326,2.):

$$\Gamma(x)/|\Gamma(x+iy)| = \prod_{n=0}^{\infty} |1 + i\,\frac{y}{n+x}| \quad .$$

This implies for x,y such that $x^2 + y^2 \leq 1$ and $|y| \leq cx$ $(c \in \mathbb{R}_+^*)$

$$|\Gamma(x)/\Gamma(x+iy)|^2 \leq e^2\,(1 + c^2)$$

and thus for $|z| \leq 1$ and $0 < |\arg z| \leq \operatorname{arctg} c$ $(z \in \mathbb{H})$

$$\|\gamma_z\| \leq e\,(1 + c^2)^{1/2} \quad .$$

Hence $(\gamma_t)_{t>0}$ is a holomorphic convolution semigroup of angle $\pi/2$.

d) For every $z \in \mathbb{H}$ let I_z denote the following Bessel function:

$$I_z(x) = \sum_{k=0}^{\infty} (1/(\Gamma(k+z+1)\,k!))\,(x/2)^{2k+z} \qquad (x \in \mathbb{R}_+^*)$$

([5], 8.445). Moreover let

$$v_z(x) = e^{-x}\,(z/x)\,I_z(x)\,1_{]0,\infty[}(x) \qquad (x \in \mathbb{R}).$$

If $t \in \mathbb{R}_+^*$ then v_t is the Lebesgue density of a probability measure ν_t on $\mathbb{R}$. In addition we have $\nu_s * \nu_t = \nu_{s+t}$ for all $s,t > 0$ i.e. $(\nu_t)_{t>0}$ is a convolution semigroup in $\mathcal{M}^1(\mathbb{R})$ supported by $\mathbb{R}_+$ ([4], II.7 and XIII.3).

From the series expansion of I_z and from the functional equation of Γ it follows easily that

$$|I_z(x)| \leq (\Gamma(\operatorname{Re} z)/|\Gamma(z)|)\,I_{\operatorname{Re} z}(x) \qquad (x \in \mathbb{R}_+^*).$$

Hence every v_z is Lebesgue integrable. Thus $\nu_z := v_z.\lambda$ is in $\mathcal{M}^b(\mathbb{R})$ such that

$$\|\nu_z\| = |v_z|_1 \leq (\Gamma(\operatorname{Re} z)/\operatorname{Re} z)/|\Gamma(z)/z| \qquad (z \in \mathbb{H}).$$

Since in addition $z \longrightarrow I_z(x)$ is holomorphic on $\mathbb{H}$ (for every $x \in \mathbb{H}$) the mappings $z \longrightarrow v_z$ and $z \longrightarrow \nu_z$ of $\mathbb{H}$ into $L^1(\mathbb{R})$ respectively into $\mathcal{M}^b(\mathbb{R})$ are holomorphic taking into account Lemma 2. Moreover in view of the estimation established in c) one has $\|\nu_z\| \leq e\,(1 + c^2)$

if $|z| \leq 1$ and $0 < |\arg z| \leq \operatorname{arctg} c$ $(z \in \mathbb{H})$ for every $c \in \mathbb{R}_+^*$. Hence $(\nu_t)_{t>0}$ is a holomorphic convolution semigroup of angle $\pi/2$.

e) If $(\eta_t)_{t>0}$ and $(\eta_t')_{t>0}$ are two holomorphic convolution semigroups in $\mathcal{M}^1(\mathbb{R})$ supported by $\mathbb{R}_+$ then it is easy to see that the semigroup $(\eta_t * \eta_t')_{t>0}$ is of the same type.

Appendix: Characterization of weakly holomorphic convolution semigroups on a locally compact Abelian group

Let G be a locally compact Abelian group with character group $\hat{G}$. For the results from harmonic analysis needed in the sequel we refer to [8]. Let $(\mu_t)_{t>0}$ be an {e}-continuous convolution semigroup in $\mathcal{M}^1(G)$ with corresponding continuous negative-definite function $\psi : \hat{G} \longrightarrow \mathbb{C}$ i.e. $\hat{\mu}_t = \exp(-t\psi)$ for all $t > 0$ (cf. [9], 1.5.18). We put

$$\hat{\mu}^z := \exp(-z\psi) \qquad (z \in \mathbb{C}).$$

Obviously we have $\hat{\mu}_t = \hat{\mu}^t$.

<u>THEOREM.</u> The following assertions are equivalent:

(i) $(\mu_t)_{t>0}$ is a weakly holomorphic convolution semigroup;

(ii) $(\tau_G(\mu_t))_{t>0}$ is a holomorphic operator semigroup on $L^2(G)$;

(iii) there exists some $\vartheta_0 \in]0, \pi/2]$ such that $|\hat{\mu}^z|_\infty := \sup\{|\hat{\mu}^z(\chi)| : \chi \in \hat{G}\}$ is finite for all $z \in V_{\vartheta_0}$.

<u>PROOF.</u> '(i)$\Longrightarrow$(ii)' is obvious.

'(ii)$\Longrightarrow$(iii)' There exist some $\vartheta_0 \in]0, \pi/2]$ and a holomorphic mapping $z \longrightarrow S_z$ of V_{ϑ_0} into $\mathcal{B}(L^2(G))$ such that $S_t f = \tau_G(\mu_t)f = \mu_t * f$ for all $f \in L^2(G)$ and $t > 0$.

Let $f \in L^2(G)$ such that the support of $\hat{f}$ is compact and define $\varphi_1(z) = (S_z f)^\wedge$ as well as $\varphi_2(z) = \hat{\mu}^z \hat{f}$ for all $z \in V_{\vartheta_0}$. We consider φ_1 and φ_2 as mappings of V_{ϑ_0} into the Banach space $L^2(\hat{G})$. Then φ_1 is holomorphic by assumption; and φ_2 is holomorphic since the functions $z \longrightarrow \hat{\mu}^z(\chi)$, $\chi \in \hat{G}$, are holomorphic on V_{ϑ_0} and since the support of $\hat{f}$ is compact (cf. Lemma 2 and Remark 1.6). But the identity $\varphi_1(t) = (\mu_t * f)^\wedge = \varphi_2(t)$ for all $t > 0$ then yields $\varphi_1(z) = \varphi_2(z)$ for all $z \in V_{\vartheta_0}$.

Now fix $\chi_o \in \hat{G}$. For every compact neighbourhood U of χ_o we put $g_U = (\lambda_{\hat{G}}(U)^{-1/2}).1_U$. This yields for all $z \in V_{\vartheta_0}$:

$$\|S_z\|^2 \geq |S_z \check{g}_U|_2^2 = |(S_z \check{g}_U)^\wedge|_2^2 = |\hat{\mu}^z g_U|_2^2 =$$
$$= \lambda_{\hat{G}}(U)^{-1} \int 1_U |\hat{\mu}^z|^2 d\lambda_{\hat{G}} .$$

Letting U shrink to χ_o we get $\|S_z\| \geq |\hat{\mu}^z(\chi_o)|$; thus $\|S_z\| \geq |\hat{\mu}^z|_\infty$

Hence (iii).

'(iii)$\Longrightarrow$(i)' Let $\vartheta \in]0,\vartheta_o[$ and let π be a strongly continuous representation of G by unitary operators on a complex Hilbert space E.

1. $\sup\{|\hat{\mu}^z|_\infty : |\arg z| < \vartheta, |z| < r\} < \infty$ for all $r \in \mathbb{R}_+^*$.

[Let $z_o = x_o + iy_o \in \mathbb{H}$ such that $\arg z_o = \vartheta$. Since $\operatorname{Re} \psi(\chi) \geq 0$ and $\psi(\overline{\chi}) = \overline{\psi(\chi)}$ for all $\chi \in \hat{G}$ one has $|\hat{\mu}^{x_o+iy}|_\infty \leq |\hat{\mu}^{z_o}|_\infty$ for all $y \in \mathbb{R}$ such that $|y| \leq y_o$. Moreover

$$|\hat{\mu}^{r(x_o+iy)}|_\infty = |\hat{\mu}^{x_o+iy}|_\infty^r \leq |\hat{\mu}^{z_o}|_\infty^r \qquad \text{for all } r \in \mathbb{R}_+^* .$$

Hence the assertion.]

2. By Bochner's theorem there exists for all $u,v \in E$ some $\alpha_{u,v} \in \mathcal{M}^b(\hat{G})$ such that $\langle\pi(x)u,v\rangle = \int \chi(x)\alpha_{u,v}(d\chi)$ for all $x \in G$ and such that $\|\alpha_{u,v}\| \leq \|u\|\,\|v\|$. Hence

$$\langle\pi(\mu_t)u,v\rangle = \int\chi(\mu_t)\alpha_{u,v}(d\chi) = \int \hat{\mu}^t(\chi)\,\alpha_{u,v}(d\chi)$$

for all $t > 0$. By assumption $f_{u,v}(z) = \int \hat{\mu}^z(\chi)\,\alpha_{u,v}(d\chi)$ is defined for all $z \in V_\vartheta$.

3. $f_{u,v}$ is holomorphic on V_ϑ.

[Let $F(z,\chi) := \hat{\mu}^z(\chi)$ for all $z \in V_\vartheta$ and $\chi \in \hat{G}$. It is easy to check (taking into account 1.) that F satisfies the assumptions of Lemma 2 (with $D = V_\vartheta$ and $(\Omega,\mathfrak{A},\mu) = (\hat{G}, \mathcal{B}(\hat{G}), \alpha_{u,v})$). Hence $z \longrightarrow F(z,.)$ is a holomorphic mapping of V_ϑ into $L^1(\alpha_{u,v})$. Since the function $1_{\hat{G}}$ is in the dual of $L^1(\alpha_{u,v})$ the function

$$z \longrightarrow \int 1_{\hat{G}}(\chi) F(z,\chi) \alpha_{u,v}(d\chi) \;=\; f_{u,v}(z)$$

is holomorphic too.]

4. Since $|f_{u,v}(z)| \leq |\hat{\mu}^z|_\infty \|\alpha_{u,v}\| \leq |\hat{\mu}^z|_\infty \|u\|\,\|v\|$ $\quad (u,v \in E)$

there exist bounded linear operators S_z on E such that

$$\langle S_z u, v\rangle = f_{u,v}(z) \quad \text{and} \quad \|S_z\| \leq |\hat{\mu}^z|_\infty \qquad (u,v \in E\ ;\ z \in V_\vartheta).$$

Obviously $S_t = \pi(\mu_t)$ for all $t > 0$. In view of 1. and 3. then $(S_z)_{z\in V_\vartheta}$ is a holomorphic operator semigroup on E that extends $(\pi(\mu_t))_{t>0}$. Since $\vartheta \in]0,\vartheta_0[$ was arbitrary assertion (i) follows. ⌟

<u>REMARKS</u>. 1. The proof of the theorem yields $S_z f = (\hat{\mu}^z \hat{f})^\vee$, $f \in L^2(G)$, and $\|S_z\| = |\hat{\mu}^z|_\infty$ for every holomorphic extension $(S_z)_{z\in V_\vartheta}$ of $(\tau_G(\mu_t))_{t>0}$. Moreover the angle ϑ_0 of holomorphy of $(\tau_G(\mu_t))_{t>0}$ is the least upper bound of all $\vartheta \in]0,\pi/2]$ such that $|\hat{\mu}^z|_\infty$ is finite for all $z \in V_\vartheta$.

2. If the function $\mathrm{Im}\,\psi$ is bounded on $\hat{G}$ we have $\vartheta_0 = \pi/2$. Otherwise let $c(\psi)$ and $d(\psi)$ denote the limes inferior and the limes superior respectively of $\mathrm{Re}\,\psi(\chi)/|\mathrm{Im}\,\psi(\chi)|$ as χ tends to the point at infinity of the Alexandrov compactification of $\hat{G}$. In this case we have $c(\psi) \leq \operatorname{arctg} \vartheta_0 \leq d(\psi)$.

[We have $|\hat{\mu}^{x+iy}| = \exp(-x\,\mathrm{Re}\,\psi + y\,\mathrm{Im}\,\psi)$ if $x+iy \in \mathbb{H}$. Now let $x+iy \in V_{\vartheta_0}$ i.e. $|\hat{\mu}^{x+iy}|_\infty$ is finite. Then there exists some $c > 0$ such that $|y\,\mathrm{Im}\,\psi| \leq c + x\,\mathrm{Re}\,\psi$ (observe $\mathrm{Im}\,\psi(\bar{\chi}) = -\mathrm{Im}\,\psi(\chi)$). Hence $|y/x| \leq c/(x\,|\mathrm{Im}\,\psi|) + \mathrm{Re}\,\psi/|\mathrm{Im}\,\psi|$. Since $\mathrm{Im}\,\psi$ is unbounded this yields $|y/x| \leq d(\psi)$. Thus $\operatorname{arctg} \vartheta_0 \leq d(\psi)$.

On the other hand let $x+iy \in \mathbb{H}$ such that $|y/x| < c(\psi)$. Then there exists some compact subset C of $\hat{G}$ such that

$$|y/x| < \inf\{\mathrm{Re}\,\psi(\chi)/|\mathrm{Im}\,\psi(\chi)| : \chi \in \hat{G}\setminus C\}.$$

Hence

$$|y\,\mathrm{Im}\,\psi(\chi)| \leq x\,\mathrm{Re}\,\psi(\chi) + |y| \sup\{|\mathrm{Im}\,\psi(\chi')| : \chi' \in C\}$$

for all $\chi \in \hat{G}$. Thus $|\hat{\mu}^{x+iy}|_\infty < \infty$ i.e. $x+iy \in V_{\vartheta_0}$. Consequently

$c(\psi) \leq \operatorname{arctg} \vartheta_o$. Hence the assertion.]

3\. Let $(\mathcal{G}_t^{(\alpha)})_{t>o}$ be the stable convolution semigroup of Example 6.b). The corresponding negative-definite function is

$$\psi_\alpha(r) = |r|^\alpha \{\cos(\alpha\pi/2) + i \sin(\alpha\pi/2)\} \qquad (r \in \mathbb{R}).$$

Obviously we have $\lim_{|r|\to\infty} |\operatorname{Im} \psi_\alpha(r)| = \infty$ and

$$\operatorname{Re} \psi_\alpha(r)/|\operatorname{Im} \psi_\alpha(r)| = \cos(\alpha\pi/2)/\sin(\alpha\pi/2) = \operatorname{tg}((1-\alpha)\pi/2).$$

Hence $\vartheta_o = (1 - \alpha)\pi/2$ in view of Remark 2. Moreover $\|S_z\| = 1$ for all $z \in V_{\vartheta_o}$ where $(S_z)_{z\in V_{\vartheta_o}}$ is the holomorphic extension of $(\tau_G(\mu_t))'_t$ (in view of Remark 1).

This should be compared with the corresponding results of [18] (see also Example 6.b)).

References

1. Bauer, H.: Wahrscheinlichkeitstheorie und Grundzüge der Masstheorie 3.Aufl. Berlin-New York: Walter de Gruyter 1978
2. Beurling, A.: On analytic extension of semigroups of operators. J.Funct.Anal.6,387-400 (1970)
3. Dieudonné, J.: Foundations of Modern Analysis. New York and London: Academic Press 1960
4. Feller, W.: An Introduction to Probability Theory and its Applications, Vol.II, 2nd ed. New York-London-Sydney-Toronto: Wiley 1971
5. Gradshteyn, I.S., Ryzhik, I.M.: Table of integrals, series and products. Corrected and enlarged edition. New York-London-Toronto-Sydney-San Francisco: Academic Press 1980
6. Hazod, W.: Subordination von Faltungs- und Operatorhalbgruppen. In: Probability Measures on Groups. Proceedings, Oberwolfach 1978, pp.144-202. Lecture Notes in Math. Vol.706. Berlin-Heidelberg-New York: Springer 1979
7. Hazod, W.: Stable probabilities on locally compact groups. In: Probability Measures on Groups. Proceedings, Oberwolfach 1981, pp.183-208. Lecture Notes in Math. Vol.928. Berlin-Heidelberg-New York: Springer 1982
8. Hewitt, E., Ross, K.A.: Abstract Harmonic Analysis I, II. Berlin-Göttingen-Heidelberg-New York: Springer 1963/70
9. Heyer, H.: Probability Measures on Locally Compact Groups. Berlin-Heidelberg-New York: Springer 1977
10. Hudson, W.N.: Operator-stable distributions and stable marginals. J.Multivar.Anal.10,26-37 (1980)
11. Hulanicki, A.: Subalgebra of $L^1(G)$ associated with Laplacian on a Lie group. Coll.Math.31,259-287 (1974)

12. Hulanicki,A.: A class of convolution semi-groups of measures on a Lie group. In: Probability Theory on Vector Spaces II. Proceedings, Błazejewko 1979, pp.82-101. Lecture Notes in Math.828. Berlin-Heidelberg-New York: Springer 1980

13. Hulanicki,A.: A functional calculus for Rockland operators on nilpotent Lie groups. To appear in 'Studia Math.'

14. Kato,T.: Perturbation Theory for Linear Operators. Berlin-Heidelberg-New York: Springer 1966

15. Kisyński,J.: Holomorphicity of semigroups of operators generated by sublaplacians on Lie groups. In: Function Theoretic Methods for Partial Differential Equations. Proceedings, Darmstadt 1976, pp. 283-297. Lecture Notes in Math. Vol.561. Berlin-Heidelberg-New York: Springer 1976

16. Majewski,A., Robinson,D.W.: Strictly positive and strongly positive semigroups. J.Austral.Math.Soc.34,36-48 (1983)

17. Mc Crudden,M., Wood,R.M.: On the support of absolutely continuous Gauss measures on SL(2,$\mathbb{R}$). These Proceedings

18. Paquet,L.: Semi-groupes holomorphes en norm du sup. Sém. de Théorie du Potentiel, Paris, No.4, pp.194-242. Lecture Notes in Math. Vol.713. Berlin-Heidelberg-New York: Springer 1979

19. Przebinda,T.: Holomorphicity of a class of semigroups of measures operating on $L^p(G/H)$. Proc.Amer.Math.Soc.87,637-643 (1983)

20. Sharpe,M.: Operator-stable probability distributions on vector groups. Trans.Amer.Math.Soc.136,51-65 (1969)

21. Siebert,E.: Convergence and convolutions of probability measures on a topological group. Ann.Probab.4,433-443 (1976)

22. Siebert,E.: Supports of holomorphic convolution semigroups and densities of symmetric convolution semigroups on a locally compact group. Arch. der Math.36,423-433 (1981)

23. Siebert,E.: Absolute continuity, singularity, and supports of Gauss semigroups on a Lie group. Monatsh.Math.93,239-253 (1982)

24. Siebert,E.: Densities and differentiability properties of Gauss semigroups on a Lie group. To appear in 'Proc.Amer.Math.Soc.'

25. Siebert,E.: Semistable convolution semigroups on measurable and topological groups. To appear in 'Ann.Inst.H.Poincaré'

26. Sinclair,A.M.: Continuous Semigroups in Banach Algebras. London Math.Soc.Lect.Note Ser.63. Cambridge University Press 1982

27. Yosida,K.: Functional Analysis, 3rd ed. Berlin-Heidelberg-New York: Springer 1971

Eberhard Siebert
Mathematisches Institut
der Universität
Auf der Morgenstelle 10

D-7400 Tübingen 1

Bundesrepublik Deutschland

Positive and conditionally positive sesquilinear forms on anticommutative coalgebras

Wilhelm von Waldenfels

Abstract. For sesquilinear forms K, L on complex coalgebras E a convolution $K,L \longmapsto K*L$ can be defined. If E is anticocommutative and K and L are even and positive, then $K*L$ is positive. This generalizes Schur's theorem that the coefficientwise product of positive-definite matrices is positive definite. If E is anticocommutative and K an even sesquilinear form, the exponential with respect to convolution, $\exp_* Kt$, is positive iff K is Hermitian and conditionally positive. This generalizes a theorem of Parthasarathy and Schmidt [5] on conditionally positive kernels. If E is a $*$-bialgebra, $(x,y) \longmapsto \varphi(x^* y)$ is a sesquilinear form on E for any linear functional $\varphi : E \longmapsto \mathbb{C}$ and the results above can be applied to linear functionals. A free algebra F is an anticocommutative $*$-bialgebra with respect to the coproduct defined by $\xi \longmapsto \xi \otimes 1 + 1 \otimes \xi$ for the generating indeterminates ξ. These considerations yield a simple proof of the fact that a Gaussian functional on F is positive iff its covariance matrix is positive definite.

1. Introduction

This paper originated from the question why Gaussian functionals ([4], [7]) on free algebras are positive if their covariance matrix is positive. A crucial observation is that a Gaussian functional is the exponential with respect to convolution of a conditionally positive functional. Therefore, the connection between positive and conditionally positive functionals on free algebras had to be investigated. It is better, however, to generalize the problem and to deal with positive and conditionally positive sesquilinear forms on free algebras. Then it is possible to use the main ideas of Parthasarathy and Schmidt's proof that a matrix of the form $(\exp tK_{ij})_{i,j}$ is positive definite iff K is conditionally positive definite. We obtained in this way a theorem for free algebras, analogous to that of Parthasarathy and Schmidt. Now the question arises whether there exists a theory covering both theorems. This is indeed the case. $\mathbb{C}^N$ and the free algebra F are both coalgebras and there is a theorem on anticocommutative coalgebras which contains both theorems mentioned.

For any $*$-bialgebra it is possible to specialize from sesquilinear forms to linear forms by setting $K(x,y) = \varrho(x^*y)$ for any linear form ϱ and x,y in the $*$-bialgebra. Thus the results for sesquilinear forms yield the results for positive and conditionally positive linear forms and give in particular an easy proof of the positivity of Gaussian functionals.

The results for free algebras in the Bose (commutative) case, i.e. all indeterminates are of degree 2 in our notation, were found by the author some years ago and are written down in the Diplomarbeit of J. Canisius [3]. This paper also covers the Fermi (anticommutative) case where all the indeterminates are of degree 1.

2. Preliminaries

In the following we shall use the notions and results of Bourbaki [1],[2],§11. We consider a graded complex vector space $E = \bigoplus_{n=0}^{\infty} E_n$. A coproduct c is a linear mapping $c: E \longrightarrow E \otimes E$ which we assume to be homogeneous of degree 0. The space E together with c is a graded coalgebra or a cogebra in Bourbaki's notation. The coproduct c is called coassociative if the diagram

$$\begin{array}{ccc} E & \xrightarrow{\quad c \quad} & E \otimes E \\ {\scriptstyle c}\downarrow & & \downarrow{\scriptstyle 1_E \otimes c} \\ E \otimes E & \xrightarrow{\; c \otimes 1_E \;} & E \otimes E \otimes E \end{array}$$

is commutative. A counit γ of a coalgebra E is a linear mapping $\gamma : E \longrightarrow \mathbb{C}$ such that the following diagramms are commutative

$$\begin{array}{ccc} E & \xrightarrow{\quad c \quad} & E \otimes E \\ & {\scriptstyle x \mapsto x \otimes 1} \searrow \quad \swarrow {\scriptstyle 1_E \otimes \gamma} & \\ & E \otimes \mathbb{C} & \end{array} \qquad \begin{array}{ccc} E & \xrightarrow{\quad c \quad} & E \otimes E \\ & {\scriptstyle x \mapsto 1 \otimes x} \searrow \quad \swarrow {\scriptstyle \gamma \otimes 1_E} & \\ & \mathbb{C} \otimes E & \end{array}$$

We shall assume that all considered coalgebras are coassociative and possess a counit.

Let $\sigma g : E \otimes E \longrightarrow E \otimes E$ be the linear mapping defined by

$$\sigma g(x \otimes y) = (-1)^{\deg(x)\deg(y)} y \otimes x$$

for homogeneous x and y. The coalgebra is anticocommutative if the diagram

$$\begin{array}{ccc} & E & \\ {}^{c}\swarrow & & \searrow^{c} \\ E \otimes E & \xrightarrow{\ \sigma g\ } & E \otimes E \end{array}$$

is commutative.

Let $J : \mathbb{C} \longrightarrow \mathbb{C}$ be the mapping $z \longrightarrow \bar{z}$. If E is any $\mathbb{C}$-vector space E^J denotes the complex conjugate vector space of E, i.e., E and E^J coincide as $\mathbb{R}$-vector spaces, but the multiplication by i is different (cf. [2], p. 11). Denote by $\bar{x} = x^J$ the element of E corresponding to $x \in E$. Then $\overline{ix} = -i\bar{x}$, more generally, $\overline{cx} = \bar{c}\bar{x}$ for $c \in \mathbb{C}$. Denote by J the mapping $x \in E \longrightarrow \bar{x} \in E^J$ and the mapping $\bar{x} \in E^J \longrightarrow x \in E$ as $(E^J)^J = E$. Let E and F be two $\mathbb{C}$-vector spaces and $f : E \rightarrow F$ a $\mathbb{C}$-linear mapping. By f^J we denote the $\mathbb{C}$-linear mapping $f^J : E^J \longrightarrow F^J$, $f^J = JfJ$. There is a canonical isomorphism between $(E \otimes F)^J$ and $E^J \otimes F^J$ identifying $(x \otimes y)^J$ with $x^J \otimes y^J$ ([2], p. 27).

Let (E, c, γ) be a coalgebra. Define in E^J a coproduct $c = c_{E^J}$ by the following procedure

$$c_{E^J} : E^J \xrightarrow{J} E \xrightarrow{c} E \otimes E \xrightarrow{J} E^J \otimes E^J \xrightarrow{\sigma} E^J \otimes E^J$$

where σ is defined by $\sigma(\bar{x} \otimes \bar{y}) = \bar{y} \otimes \bar{x}$.

So if $c(x) = \sum x_i' \otimes x_i''$ then $c(\bar{x}) = \sum \bar{x}_i'' \otimes \bar{x}_i'$. The coalgebra (E^J, c_{E^J}) is coassociative and possesses the counit $\gamma^J : \gamma^J(\bar{x}) = \overline{\gamma(x)}$.

If (E, c, γ) is anticocommunicative then (E^J, c, γ^J) is anticocommunicative.

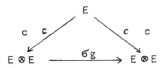

A sesquilinear form on a $\mathbb{C}$-vector space E is an $\mathbb{R}$-bilinear mapping $K: E \times E \longrightarrow \mathbb{C}$ such that $K(-ix,y) = K(x,iy) = iK(x,y)$ or $K(\bar{c}x,y) = K(x,cy) = cK(x,y)$ for $c \in \mathbb{C}$, $x,y \in E$. It can be identified with a $\mathbb{C}$-linear mapping $E^J \otimes E \longrightarrow \mathbb{C}$ (cf. [2], p.14) and shall be denoted by the same letter K. So $K(x,y) = K(\bar{x} \otimes y)$. A sesquilinear form is called Hermitian if $K(x,y) = K(y,x)$, or if $K(\bar{x} \otimes y) = \overline{K(\bar{y} \otimes x)}$. A sesquilinear form on E is called positive if $K(x,x) = K(\bar{x} \otimes x) \geqslant 0$ for all $x \in E$. If (E, c, γ) is a coalgebra, a sesquilinear form K on E is called conditionally positive if $K(x,x) = K(\bar{x} \otimes x) \geqslant 0$ for all $x \in E$ such that $\gamma(x) = 0$.

Let E be a graded $\mathbb{C}$-vector space $E = \bigoplus_{n=0}^{\infty} E_n$. We denote by $\hat{E}$ the completion of E consisting of all sequences $x = (x_0, x_1, \ldots)$, $x_n \in E_n$. If E is a graded algebra, $E_k E_\ell \subset E_{k+\ell}$, then $\hat{E}$ is an algebra where the product z of x and y is defined $z_n = \sum_{k+\ell=n} x_k y_\ell$. Assume that E is commutative and that the subalgebra $E_0 \subset E$ is finite-dimensional. Then for $x_0 \subset E_0$ the power series defining $\exp x_0$ converges. Let $x' \in E$ be an element whose E_0-component vanishes, $x' = (0, x_1, x_2, \ldots)$, then the power series $\exp x'$ has only finitely many terms belonging to E_n for fixed n. So $\exp x'$ is defined as well. Define for $x \in \hat{E}$ the element

$$\exp x = \exp x_0 \exp x'$$

with $x = (x_0, x_1, \ldots)$, $x' = (0, x_1, x_2, \ldots)$. One has $\exp(x+y) = \exp x \exp y$ for $x, y \in \hat{E}$.

Let E be a graded vector space $E = \bigoplus_{n=0}^{\infty} E_n$. A linear mapping $\varphi: E \longrightarrow \mathbb{C}$ is called homogeneous of degree n if $\varphi(E_k) = 0$ for $k \neq n$. Call E_n^* the vector space of all homogeneous linear mappings of degree n. The direct sum $\bigoplus_{n=0}^{\infty} E_n^*$ is the graded dual of E and is denoted by E^{*grad}. The dual E^* of E, i.e., the set of all linear mappings $E \longrightarrow \mathbb{C}$ is the completion $E^* = (E^{*grad})^\wedge$.

Let (E, c, γ) be a graded coassociative coalgebra and recall the complex conjugate coalgebra (E^J, c, γ^J) defined above. We make $E^J \otimes E$ into a coalgebra by defining the coproduct $c = c_{E^J \otimes E}$ by

$$c : E^J \otimes E \xrightarrow{c_{E^J} \otimes c_E} E^J \otimes E^J \otimes E \otimes E$$

$$\xrightarrow{1_{E^J} \otimes \sigma g \otimes 1_{E^J}} E^J \otimes E \otimes E^J \otimes E$$

where σg is the mapping

$$\sigma g : E^J \otimes E \longrightarrow E \otimes E^J$$

$$\bar{x} \otimes y \longrightarrow (-1)^{\deg(x)\deg(y)} y \otimes \bar{x}$$

for homogeneous elements x, y. The counit is $\gamma^J \otimes \gamma$. If E is anticocommutative, then $E^J \otimes E$ is anticocommutative.
The graduation of $E^J \otimes E$ is given by

$$(E^J \otimes E)_n = \bigoplus_{k+\ell=n} E^J_k \otimes E_\ell \ .$$

A sesquilinear form K on E is called homogeneous of degree n if the associated linear mapping $E^J \otimes E \longrightarrow \mathbb{C}$ is homogeneous of degree n. Call $(E^J \otimes E)^*_n$ the space of all homogeneous sesquilinear forms of degree n and by $(E^J \otimes E)^{*grad}$ the direct sum as above. The vector space of all sesquilinear forms $(E^J \otimes E)^*$ is the completion $((E^J \otimes E)^{*grad})^\wedge$.

If (E, c, γ) is a coalgebra we introduce a convolution in E^* by $\rho * \sigma : E \xrightarrow{c} E \otimes E \xrightarrow{\rho \otimes \sigma} \mathbb{C}$, if $\rho, \sigma \in E^*$. Similarly, if K and L are two sesquilinear forms we define a convolution $K * L$ by $K * L : E^J \otimes E \xrightarrow{c} (E^J \otimes E) \otimes (E^J \otimes E) \xrightarrow{K \otimes L} \mathbb{C}$. So E^* and $(E^J \otimes E)^*$ become associative $\mathbb{C}$-algebras with units γ resp. $\gamma^J \otimes \gamma$. The algebra E^* and $(E^J \otimes E)^*$ are the completions of the subalgebras E^{*grad} resp. $(E^J \otimes E)^{*grad}$. If E is anticocommutative, then E^* and $(E^J \otimes E)^*$ are anticommutative. A sesquilinear form K is called <u>even</u> if K is of the form $K = (K_0, 0, K_2, 0, K_4, 0, \ldots)$, i.e., if

$K(x, y) = 0$ for x, y homogeneous elements of E such that $\deg(x) + \deg(y)$ is odd. If E is anticocommutative the even sesquilinear forms on E form a commutative algebra with respect to convolution. Similar definitions apply to linear functionals $E \longrightarrow \mathbb{C}$.

3. Results for coalgebras

In this chapter E is an anticocommutative graded coalgebra.

Lemma 1. If K and L are two even sesquilinear forms, then

$$K \otimes L(\bar{x} \otimes y) = \sum K(\bar{x}_i' \otimes y_i')L(\bar{x}_i'' \otimes y_j'')$$

if $c(x) = \sum x_i' \otimes x_i''$ and $c(y) = \sum y_j' \otimes y_j''$

and the x_i', x_i'', y_j', y_j'' are homogeneous.

Proof. We have

$$c(\bar{x}) = \sum \bar{x}_i'' \otimes \bar{x}_i' = \sum (-1)^{\deg(x_i')\deg(x_i'')}\bar{x}_i' \otimes \bar{x}_i''$$

as E^J is anticocommutative and

$$c(\bar{x} \otimes y) = \sum (-1)^{\deg(x_i')\deg(x_i'')+\deg(x_i'')\deg(y_j')} \bar{x}_i' \otimes y_j' \otimes \bar{x}_i'' \otimes y_j'' .$$

As K and L are even, only terms with

$$\deg(x_i) + \deg(y_j') \equiv 0 \mod 2$$

$$\deg(x_i'') + \deg(y_j'') \equiv 0 \mod 2$$

are of importance. From there one concludes that the exponent of (-1) in the last formula is even. This proves the lemma.

Theorem 1. Let K and L be two positive even sesquilinear forms. Then $K * L$ is positive.

Proof. By lemma 1

$$(K * L)(\bar{x} \otimes x) = \sum K(\bar{x}_i' \otimes x_j')L(\bar{x}_i'' \otimes x_j'')$$

if $c(x) = \sum_{i=1} x_i' \otimes x_i''$. The $n \times n$ matrices $K(\bar{x}_i' \otimes x_j')$ and $L(\bar{x}_i'' \otimes x_j'')$, $i,j = 1, \ldots, n$, are, however, positive definite. Hence the sum is $\geqslant 0$.

Lemma 2. Let $\varkappa_1, \varkappa_2, \lambda_1, \lambda_2$ be linear functionals on E. Define two sesquilinear forms by $K = \varkappa_1^J \otimes \varkappa_2$, $L = \lambda_1^J \otimes \lambda_2$. If K and L are even then

$$K * L = (\varkappa_1 * \lambda_1)^J \otimes (\varkappa_2 * \lambda_2).$$

Proof. By lemma 1

$$(K*L)(\bar{x} \otimes y) = \sum \varkappa_1^J(\bar{x}_i') \varkappa_2(y_j') \lambda_1^J(\bar{x}_i'') \lambda_2(y_j'')$$

$$= \sum \bar{\varkappa}_1(x_i') \varkappa_2(y_j') \bar{\lambda}_1(x_i'') \lambda_2(y_j'')$$

$$= \overline{\varkappa_1 * \lambda_1(x)} \, (\varkappa_2 * \lambda_2)(y).$$

Theorem 2. Let the dimension of E_0 be finite. Let K be an even sesquilinear form. Then

$$\exp_* tK$$

is positive for all $t \geqslant 0$ if and only if K is Hermitian and conditionally positive.

Proof. If $\exp_* tK$ is positive for all $t \geqslant 0$ and if $\gamma(x) = 0$, then $(\gamma^J \otimes \gamma)(\bar{x} \otimes x) = 0$

$$\frac{1}{t} (\exp_* tK)(\bar{x} \otimes x) = \frac{1}{t} (\exp_* tK - \gamma^J \otimes \gamma)(\bar{x} \otimes x) \geqslant 0$$

for all $t > 0$. But for $t \downarrow 0$ the last expression converges to $K(\bar{x} \otimes x)$ hence $K(\bar{x} \otimes x) \geqslant 0$. As positivity implies harmiticity and as $\gamma^J \otimes \gamma$ is Hermitian, we get that K is Hermitian.

Assume that K is conditionally positive and Hermitian. The linear functional γ is of degree 0 and $\gamma \neq 0$ if E is not trivial. There exists an element $e \in E_0$ such that $\gamma(e) = 1$. Call P the mapping $E \longrightarrow E$ defined by $Px = e\,\gamma(x)$ and call P' the mapping

$P' = 1_E - P$. Then

$$K = K_1 + K_2 + K_3 + K_4 \quad \text{with}$$

$$K_1(x,y) = K(Px, Py)$$

$$K_2(x,y) = K(Px, P'y)$$

$$K_3(x,y) = K(P'x, Py)$$

$$K_4(x,y) = K(P'x, P'y).$$

As $P'(x) = x - e\,\gamma(x)$ one has $\gamma(P'(x)) = 0$ and K_4 is positive and as K is conditionally positive, $\exp_* tK_4$ is positive for all $t \geqslant 0$ by theorem 1.

Now $K_1(x,y) = \overline{\gamma(x)}\,\gamma(y)K(e, e)$. So $K_1 = K(e, e)\,\gamma^J \otimes \gamma$. Hence $\exp_* tK_1 = e^{tK(e,e)}\,\gamma^J \otimes \gamma$ and positive as $\gamma^J \otimes \gamma$ is positive and $K(e, e)$ is real as K is Hermitian.

As $K_2(x,y) = \overline{\gamma(x)}\,K(e, P'y) = \overline{\gamma(x)}\,\alpha(y)$ we have $K_2 = \gamma^J \otimes \alpha$ and by lemma 2 $\exp_* K_2 t = \gamma^J \otimes \exp_* \alpha t$.

We have $K_3(x,y) = \gamma(y)\,K(P'x, e) = \overline{\alpha(x)}\,\gamma(y)$ as K is Hermitian and hence $K_3 = \alpha^J \otimes \gamma$ and $\exp_* tK_3 = (\exp_* t\alpha)^J \otimes \gamma$.

By lemma 2

$$(\exp_* tK_2) * (\exp_* tK_3) = (\exp_* t\alpha)^J \otimes \exp_* t\alpha$$

which is clearly positive. So finally

$$\exp_* tK = (\exp_* tK_1) * (\exp_* tK_4) * ((\exp_* tK_2) * (\exp_* tK_3))$$

is positive as it is a product of three positive factors by theorem 1.

4. $*$-bialgebras

A (skew) bialgebra E (cf. [1], p. 148) is a $\mathbb{C}$-vector space in which the structures of a graded $\mathbb{C}$-algebra and a graded $\mathbb{C}$-co-algebra are defined sucht that i) the $\mathbb{C}$-algebra E is associative and has a unit e, ii) the $\mathbb{C}$-coalgebra E is coassociative and has the counit γ ; iii) the coproduct $c : E \longrightarrow E \otimes E$ is a homomorphism of E into the $\mathbb{C}$-algebra $E \otimes E$ with skew multiplication, i.e.

$$(x \otimes y)(z \otimes w) = (-1)^{\deg(y)\deg(z)} xz \otimes yw$$

for homogeneous x, y, z, $w \in E$; iv) the counit γ is a homogeneous homomorphism of degree 0 from $E \longrightarrow \mathbb{C}$ such that $\gamma(e) = 1$. Furthermore we assume that E is a $*$-algebra, i.e., there exists an involution $I : x \longrightarrow x^*$ from $E \longrightarrow E$ which is $\mathbb{R}$-linear and of degree 0 such that a) $x^{**} = x$, b) $e^* = e$, c) $(cx)^* = \bar{c}x^*$, d) $(xy)^* = y^*x^*$ for $x, y \in E$, $c \in \mathbb{C}$.

Define an involution I in $E \otimes E$ by $(x \otimes y)^* = y^* \otimes x^*$. Check e.g. (d):

$$((x \otimes y)(r \otimes w))^* = (-1)^{\deg(y)\deg(z)}(xz \otimes yw)^*$$

$$= (-1)^{\deg(y)\deg(z)}(w^*y^* \otimes z^*x^*)$$

$$= (w^* \otimes z^*)(y^* \otimes x^*) = (z \otimes w)^*(x \otimes y)^* .$$

Let us assume that c and γ are $*$-homomorphisms from E into $E \otimes E$ resp. $\mathbb{C}$. This means

(v) $c(x^*) = c(x)^*$

(vi) $\gamma(x^*) = \overline{\gamma(x)}$.

If E is a bialgebra which is a $*$-algebra such that (v) and (vi) hold, E is called a <u>$*$-bialgebra</u>.

<u>Lemma 3.</u> Let E be a $*$-bialgebra. The mapping $\eta : E^J \otimes E \longrightarrow E$, $\bar{x} \otimes y \longrightarrow x^* y$ is a homomorphism of degree 0 from the coalgebra $E^J \otimes E$ into the coalgebra E, i.e., η is $\mathbb{C}$-linear and the two following diagrams are commutative:

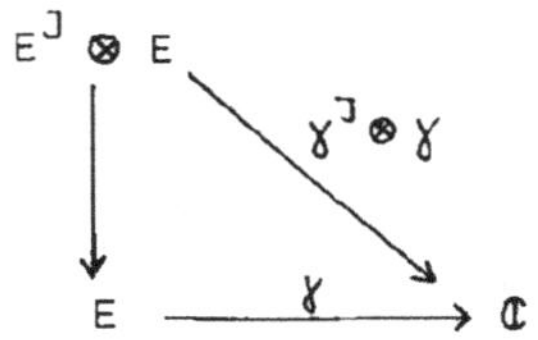

$$\begin{array}{ccc} E^J \otimes E & \xrightarrow{\ c\ } & (E^J \otimes E) \otimes (E^J \otimes E) \\ \downarrow \eta & & \downarrow \eta \otimes \eta \\ E & \xrightarrow{\ c\ } & E \otimes E \end{array}$$

Proof. Check the first diagram.

$$\gamma(x^* y) = \gamma(x^*)\gamma(y) = \overline{\gamma(x)}\gamma(y) = (\gamma^J \otimes \gamma)(\bar{x} \otimes y).$$

Let $x, y \in E$ and $c(x) = \sum x_i' \otimes x_i''$ and $c(y) = \sum y_j' \otimes y_j''$ where the x_i', x_i'', y_j', y_j'' are homogeneous elements. Then

$$c(\bar{x} \otimes y) = \sum (-1)^{\deg x'_i \deg y'_j} \bar{x}_i'' \otimes y_j' \otimes \bar{x}_i' \otimes y_j''$$

and

$$\eta(c(\bar{x} \otimes y)) = \sum (-1)^{\deg x_i' \deg y_j'} x_i''^* y_j' \otimes x_i'^* y_j'' .$$

On the other hand

$$c(\eta(\bar{x} \otimes y)) = c(x^* y) = c(x^*)c(y) = c(x)^* c(y)$$

$$= \sum (x_i''^* \otimes x_i'^*)(y_j' \otimes y_j'')$$

$$= \sum (-1)^{\deg x_i' \deg y_j'} x_i''^* y_j' \otimes x_i'^* y_j'' .$$

This proves the second diagram.

Let E be a $*$-bialgebra and let $\varrho \in E^*$ be a linear functional on E. Consider the sesquilinear form K_ϱ on E given by $K_\varrho(x,y) = \varrho(x^* y)$. The last lemma shows that the mapping $\varrho \longrightarrow K_\varrho$ is a homomorphism with respect to convolution, i.e., $K_\gamma = \gamma^J \otimes \gamma$ and $K_{\varrho * \sigma} = K_\varrho * K_\sigma$ for $\varrho, \sigma \in E^*$.

We call a functional ϱ on E Hermitian if $\varrho(x^*) = \overline{\varrho(x)}$ for all $x \in E$, we call it positive if $\varrho(x^* x) \geqslant 0$ for all $x \in E$, we call it conditionally positive if $\varrho(x^* x) \geqslant 0$ for all $x \in E$ with $\gamma(x) = 0$.

Corollary of theorem 1. Let E be an anticocommutative $*$-bialgebra and let ϱ and σ be two positive, even functionals on E. Then $\varrho * \sigma$ is an even positive functional on E.

Corollary of theorem 2. Let E be an anticocommutative $*$-bialgebra and let ϱ be an even functional on E. Then $\exp_* t\varrho$ is positive for all $t \geq 0$ if and only if ϱ is Hermitian and conditionally positive.

5. Applications

1) Let $E = \mathbb{C}^n$ and define the coproduct $c(e_i) = e_i \otimes e_i$ if $e_1, \ldots, e_n$ is the standard basis of $\mathbb{C}^n$. The counit is given by $\gamma(e_i) = 1$ for all $i = 1, \ldots, n$. We define the graduation by $E_0 = E$, $E_1 = E_2 = \ldots = 0$. As only the degree 0 occurs, all powers of (-1) occurring in the formulae, get 1 and anticocommutative becomes equivalent to cocommutative. With c and γ the vector space E forms a cocommutative coalgebra. A linear functional $\varrho \in E^*$ is given by $\varrho = (\varrho_1, \ldots, \varrho_n)$, $\varrho_i = \varrho(e_i)$. Then $\varrho * \sigma = (\varrho_1\sigma_1, \ldots, \varrho_n\sigma_n)$. A sesquilinear form on E is given by a $n \times n$-matrix $K_{ij} = K(e_i, e_j) = K(\bar{e}_i \otimes e_j)$. By lemma 1 the convolution $K * L$ is given by

$$(K*L)(\bar{e}_i \otimes e_j) = K(\bar{e}_i \otimes e_j)L(\bar{e}_i \otimes e_j).$$

So $(K*L)_{i,j} = K_{ij}L_{ij}$. So the convolution is nothing else but Schur's product of coefficientwise multiplication and theorem 1 gets Schur's theorem: The Schur product of two positive definite matrices is positive definite. Theorem 2 becomes the theorem of Schmidt and Parthasarathy: If $K = (K_{ij})_{i,j}$ is a matrix, the matrix $(\exp tK_{ij})_{i,j}$ is positive definite iff K is Hermitian and conditionally positive, i.e.

$$\sum K_{ij}\bar{\xi}_i\xi_j \geq 0 \quad \text{for} \quad \xi = (\xi_1, \ldots, \xi_n) \in \mathbb{C}^n, \ \gamma(\xi) = \sum \xi_i = 0.$$

2) Let Ξ be a finite set of non-commutative indeterminates and $F = \mathbb{C}\langle \Xi \rangle$ the free complex algebra generated by $\xi \in \Xi$. It consists of the $\mathbb{C}$-linear combinations of monomials $\xi_1 \cdots \xi_n$ $\xi_i \in \Xi$. We split Ξ into two parts Ξ_1 and Ξ_2, such that $\Xi \cap \Xi_2 = \emptyset$ and $\Xi_1 \cup \Xi_2 = \Xi$. We define a graduation in $\mathbb{C}(\Xi) = F$ by putting $\deg \xi = 1$ for $\xi \in \Xi_1$ and $\deg \xi = 2$ for $\xi \in \Xi_2$ and by setting $\deg(\xi_1 \cdots \xi_n) = \deg \xi_1 + \ldots + \deg \xi_n$ for monomials.

Consider the tensor product $F \otimes F$ with skew multiplication and the homomorphism $c : F \longrightarrow F \otimes F$ given by

$$\xi \in \Xi \longrightarrow \xi \otimes 1 + 1 \otimes \xi \in F \otimes F.$$

Let $w = \xi_1 \cdots \xi_n$ be a monomial and define for $I' = \{i_1 < \ldots < i_k\} \subset I = \{1, \ldots, n\}$, $\xi_{I'} = \xi_{i_1} \cdots \xi_{i_k}$.

Then

$$c(\xi_1 \cdots \xi_n) = c(\xi_I) = \sum_{I_1 \dot\cup I_2 = I} \varepsilon(I_1, I_2)\, \xi_{I_1} \otimes \xi_{I_2}$$

where $I_1 \dot\cup I_2 = I$ means that I is the disjoint union of I_1 and I_2. The number $\varepsilon(I_1, I_2) = \pm 1$ more precisely

$$\varepsilon(I_1, I_2) = \prod_{(i,j) \in S(I_1, I_2)} (-1)^{\deg(\xi_i)\deg(\xi_j)}$$

where

$$S(I_1, I_2) = \{(i,j) \in I \times I \mid i < j,\ i \in I_2,\ j \in I_1\}.$$

One has

$$(c \otimes 1_F)\, c(\xi_I) = \sum_{I_1 \dot\cup I_2 \dot\cup I_3 = I} \varepsilon(I_1, I_2, I_3)\, \xi_{I_1} \otimes \xi_{I_2} \otimes \xi_{I_3}$$

where

$$\varepsilon(I_1, I_2, I_3) = \prod_{(i,j) \in S(I_1, I_2, I_3)} (-1)^{\deg \xi_i \deg \xi_j}$$

and

$$S(I_1, I_2, I_3) = \{(i,j) \in I \times I \mid i < j,\ i \in S_\alpha,\ j \in S_\beta,\ \alpha > \beta\}.$$

This proves the coassociativity of the coproduct c. The linear functional $\delta_e : F \longrightarrow \mathbb{C}$ defined by

$$\delta_e(w) = \begin{cases} 0 & \text{for } w \neq e \\ 1 & \text{for } w = e \end{cases}$$

where w is any monomial and e is the void monomial forming the unit element of F. So F together with c and δ_e forms a bialgebra.

For $I = I_1 \cup I_2$ one has

$S(I_1, I_2) \cap S(I_2, I_1) = \emptyset$ and

$$\varepsilon(I_1, I_2)\,\varepsilon(I_2, I_1) = \prod_{(i,j)\in I_1\times I_2} (-1)^{\deg(\xi_i)\deg(\xi_j)} =$$

$$= (-1)^{\deg \xi_{I_1} \deg \xi_{I_2}} .$$

Hence

$$c(\xi_I) = \sigma_g\, c(\xi_I)$$

with $\sigma_g(\xi_{I_1} \otimes \xi_{I_2}) = (-1)^{\deg \xi_{I_1} \deg \xi_{I_2}} (\xi_{I_2} \otimes \xi_{I_1})$ as defined above. So the bialgebra F is anticocommutative.

Define an involution $x \longmapsto x^*$ in F by putting for monomials $(\xi_1 \dots \xi_n)^* = \xi_n \dots \xi_1$. Of course $\delta_e(x^*) = \overline{\delta_e(x)}$ for all polynomials $x \in F$. We prove that

$$c(x^*) = c(x)^*.$$

It is sufficient to do this for monomials.
Consider the mapping $\sigma : \{1, \dots, n\} \longrightarrow \{1, \dots, n\}$, $\sigma(1) = n$, $\sigma(2) = n-1, \dots, \sigma(n) = 1$. Define $\xi_k = \eta_{\sigma(k)}$. Then

$$c(\xi_I^*) = \sigma(\eta_I) = \sum_{I_1 \cup I_2 = I} \varepsilon(I_2, I_1)\, \eta_{I_2} \otimes \eta_{I_1} =$$

$$= \sum_{I_1 \dot{\cup} I_2 = I} \varepsilon(I_2, I_1)\, \xi^*_{\sigma(I_2)} \otimes \xi^*_{\sigma(I_1)}$$

$$= \sum_{I_1 \dot{\cup} I_2 = I} \varepsilon(\sigma(I_2), \sigma(I_1))\, \xi^*_{I_2} \otimes \xi^*_{I_1} .$$

Now

$$\varepsilon(\sigma(I_2), \sigma(I_1)) = \varepsilon(I_1, I_2).$$

This proves the assertion.

So finally F together with c and δ_e is an anticocommutative $*$-bialgebra and the theory applies.

The special case $\Xi = \Xi_2$ which implies that $F = \sum_{n=0} F_{2n}$ where F is cocommutative was treated by me some years ago and written down in the Diplomarbeit of J. Canisius [3].

3) We treat a special class of linear functionals on F which are called Gaussian functionals. Let Q be a $\Xi \times \Xi$-matrix, $Q = (Q(\xi, \eta))_{\xi, \eta \in \Xi}$. We define a linear functional δ_Q on F by putting for monomial

$$\delta_Q(w) = \begin{cases} Q(\xi, \eta) & \text{for } w = \xi\eta \\ 0 & \text{if } w \text{ is not of this form.} \end{cases}$$

A Gaussian functional is a functional of the form

$$\gamma_Q = \exp_* \delta_Q .$$

One sees easily that

$$\gamma_Q(e) = 1$$

$$\gamma_Q(\xi\eta) = Q(\xi, \eta)$$

$$\gamma_Q(\xi_1 \ldots \xi_{2m+1}) = 0 \quad \text{for } m = 0, 1, 2, \ldots$$

$$\gamma_Q(\xi_1, \ldots, \xi_{2m}) = \frac{1}{m!} \sum_{I_1 \dot{\cup} \ldots \dot{\cup} I_m = I} \varepsilon(I_1, \ldots, I_m) Q(\xi_{I_1}) \ldots Q(\xi_{I_m})$$

where $I = \{1, \ldots, 2m\}$ and the sum runs over all m-tuples of subsets $I_1, \ldots, I_m$ of I which have exactly two elements and where I is the disjoint union of $I_1, \ldots, I_m$. If $I_k = \{i < j\}$ is such a subset $Q(\xi_{I_K})$ is defined by $Q(\xi_i, \xi_j)$. The number

$$\varepsilon(I_1,\ldots,I_m) = \prod_{(i,j)\in S(I_1,\ldots,I_m)} (-1)^{\deg \xi_i \deg \xi_j}$$

where

$$S(I_1,\ldots,I_m) = \{i < j \mid i \in I_\alpha,\ j \in I_\beta,\ \beta < \alpha\}$$

If $\Xi = \Xi_2$, $\Xi_1 = \emptyset$ we get the familiar expression for Bose quasi-free states in quantum mechanics

$$\gamma_Q(\xi_1 \ldots \xi_{2m}) = \sum_{\{I_1,\ldots,I_m\}} Q(\xi_{I_1}) \ldots Q(\xi_{I_m})$$

where $\{I_1, \ldots, I_m\}$ runs through all partitions of I into pairs.

If $\Xi = \Xi_1$, $\Xi_2 = \emptyset$ we get the expression for Fermi quasi-free states

$$\gamma_Q(\xi_1 \ldots \xi_{2m}) = \sum_{\{I_1,\ldots,I_m\}} \operatorname{sgn}(I_1,\ldots,I_m) Q(\xi_{I_1}) \ldots Q(\xi_{I_m})$$

where

$$\operatorname{sgn}(I_1,\ldots,I_m) = (-1)^{\# S(I_1,\ldots,I_m)}$$

and is equal to the signature of the permutation

$$\begin{pmatrix} 1 & 2 & 3 & 4 & \ldots & 2m-1 & 2m \\ i_1 & j_1 & i_2 & j_2 & & i_m & j_m \end{pmatrix}$$

if $I_k = \{i_k < j_k\}$.

We prove in the general case a theorem which was already stated in [3] for the Bose case.

<u>Theorem 3.</u> A Gaussian functional γ_Q on E is positive if and only if Q is positive definite.

<u>Proof.</u> Let $x = \sum_{\xi \in \Xi} \alpha(\xi)\xi$, $\alpha(\xi) \in \mathbb{C}$.

Then $\gamma_Q(x^* x) = \sum_{\xi,\eta \in \Xi} \bar{\alpha}(\xi)\alpha(\eta) Q(\xi,\eta) \geqslant 0$ for arbitrary $\alpha(\xi)$.

Hence Q is positive definite.

In order to prove the converse we show that δ_Q is Hermitian and conditionally positive and apply the corollary of theorem 2. That δ_Q is Hermitian means that $\delta_Q(\xi\eta) = \delta_Q((\xi\eta)^*)^- = \delta_Q(\eta\xi)^-$. But this means that $Q(\xi,\eta) = Q(\eta,\xi)^-$ which follows from the positivity of Q.

Let $x \in F$, $\delta_e(x) = 0$. Then x is of the form

$$x = \sum \alpha(\xi)\xi + x'$$

where x' is a linear combination of monomials of the type $\xi_1 \cdots$ $\cdots \xi_n$, $n \geqslant 2$. So $x^*x = \sum \alpha(\xi)\alpha(\eta)\xi\eta$ + higher terms and

$$\delta_Q(x^*x) = \sum \bar{\alpha}(\xi)\alpha(\eta)Q(\xi,\eta).$$

This proves theorem 3.

Literature

[1] Bourbaki, N., Elements de Mathématique. Algèbre, Chap. III, Hermann, Paris, 1970.

[2] Bourbaki, N., Elements de Mathématique. Algèbre, Chap. IX, Hermann, Paris, 1959.

[3] Canisius, J., Algebraische Grenzwertsätze und unbegrenzt teilbare Funktionale. Diplomarbeit. Fakultät für Mathematik. Universität Heidelberg. 1979.

[4] Giri, N., von Waldenfels, W., An algebraic version of the central limit theorem. Z. Wahrscheinlichkeitstheorie verw. Gebiete 42, 129-134 (1978).

[5] Parthasarathy, K.R., Schmidt, K., Positive kernels, continuous tensor products and central limit theorems of probability theory. Lecture Notes in Mathem. 272, Springer Verlag, Berlin, Heidelberg, New York, 1972.

[6] Sweedler, M.E., Hopf algebras. BENJAMIN, New York, 1969.

[7] von Waldenfels, W., An algebraic central limit theorem in the anti-commuting case. Z. Wahrscheinlichkeitstheorie verw. Gebiete 42, 135-140 (1978).

Wilhelm von Waldenfels
Institut für Angewandte Mathematik
Universität Heidelberg
Im Neuenheimer Feld 294
6900 Heidelberg/BRD

A RANDOM WALK ON FREE PRODUCTS OF FINITE GROUPS

Wolfgang WOESS

1. Introduction

Let $G = \prod_{j=1}^{N}{}^{*} G_j$ be the free product of finite groups G_j of order s_j, $j = 1,\dots,N$. We exclude $N = 2$ and $s_1 = s_2 = 2$ in which case G is amenable. On G, we consider a probability measure μ which is a convex combination of the uniform distributions on each of the factors.
The asymptotic behaviour of $(\mu^{(n)}(e))$ is determined, where e is the identity of G and $\mu^{(n)}$ the n'th convolution power of μ.

2. Result

By assumption there are positive numbers α_j, $j = 1,\dots,N$, with $\sum_{j=1}^{N} \alpha_j = 1$ such that for $x \in G$

$$\mu(x) = \frac{\alpha_j}{s_j} \quad \text{if} \quad x \in G_j - \{e\}, \quad \mu(e) = \sum_{j=1}^{N} \frac{\alpha_j}{s_j} \quad \text{and} \quad \mu(x) = 0 \quad \text{otherwise.}$$

μ defines a random walk on G which is irreducible and aperiodic. (See [6] for terminology.)
The generating function

$$E(z) = \sum_{n=0}^{\infty} \mu^{(n)}(e) z^n \qquad (z \in \mathbb{C})$$

may be described by the equation

$$E(z) = P(zE(z)) \quad , \text{ where}$$

$$P(z) = 1 + \frac{1}{2} \sum_{j=1}^{N} \left(\sqrt{(1-\alpha_j z)^2 + 4\frac{\alpha_j}{s_j} z} - (1-\alpha_j z) \right) \qquad (z \in \mathbb{C}) .$$

The radius of convergence r of $E(z)$ is given by

$$r = 1/P'(u) \quad ,$$

where u is the unique positive real solution of the equation

$$P(z) = zP'(z) \; .$$

Therefore [4]

$$\lim_{n\to\infty} \frac{\mu^{(n+1)}(x)}{\mu^{(n)}(x)} = \frac{1}{r} \qquad \forall\, x \in G \quad ,$$

and as μ is symmetric, by [3] $\rho = 1/r$ is also the norm of the convolution operator on $L^2(G)$ defined by μ:

$$\rho = \|\mu\|_2 = \operatorname{Min} \{ P(x)/x \mid x > 0 \} = P(u)/u \; .$$

There is a close similarity with the formula given by *Akemann* and *Ostrand* [1].

As $n\to\infty$, we have

$$\mu^{(n)}(e) \sim \frac{c}{\sqrt{2\pi\rho}}\cdot n^{-3/2}\cdot\rho^{n} \quad , \text{ where } \quad c = \left(\frac{P(u)P'(u)}{u^{2}P''(u)}\right)^{1/2}$$

This result extends §5 of [6] and is similar to the result of [7].

3. Proof

With each finite sequence of integers $j_1\dots j_m$, $j_k \in \{1,\dots,N\}$ and $j_{k+1} \neq j_k$, we associate the set $C(j_1\dots j_m)$ of all elements of G whi
have the form

$$x = x_1\cdots x_m \text{ , where } x_k \in G_{j_k}-\{e\} \text{ for } k=1,\dots,m .$$

$\{e\}$ is associated with the "void sequence".
Thus we obtain a partition of G , and by definition of μ the transition probabilities $p_{x,y} = \mu(x^{-1}y)$ depend only on the classes of x and y in the following sense: If we start in an element of $C(j_1\dots j_m$ then after one step we arrive at an element of $C(j_1\dots j_{m-1})$ with proba bility $\frac{\alpha_{j_m}}{s_{j_m}}$ (by multiplying with $x_{j_m}^{-1}$), we remain in $C(j_1\dots j_m)$ with probability $\sum_{j=1}^{N}\frac{\alpha_j}{s_j} + \frac{s_{j_m}-2}{s_{j_m}}\alpha_{j_m}$ (by multiplying with e or any element o $G_{j_m}-\{e,x_{j_m}^{-1}\}$) and we arrive at an element of $C(j_1\dots j_m j_{m+1})$, $j_{m+1} \neq j_m$ with probability $\frac{s_{j_{m+1}}-1}{s_{j_{m+1}}}\alpha_{j_{m+1}}$ (by multiplying with any element of $G_{j_{m+1}}-\{e\}$). Thus we can map the random walk onto the Markov chain illustrated by figure 1 on the homogeneous tree with vertices $j_1\dots j_m$ and edges $[j_1\dots j_m , j_1\dots j_m j_{m+1}]$, admitting all sequences as defined above. Having started at e (void sequence) on this tree, the probability to be back at e at the n'th step is still $\mu^{(n)}(e)$ .....

$$p_j = \frac{s_j-1}{s_j}\alpha_j \;,\; q_j = \frac{\alpha_j}{s_j} \;,\; r_j = \frac{s_j-2}{s_j}\alpha_j \quad (\, j=1,\dots,N \,)$$

$$r_o = \sum_{j=1}^{N}\frac{\alpha_j}{s_j}$$

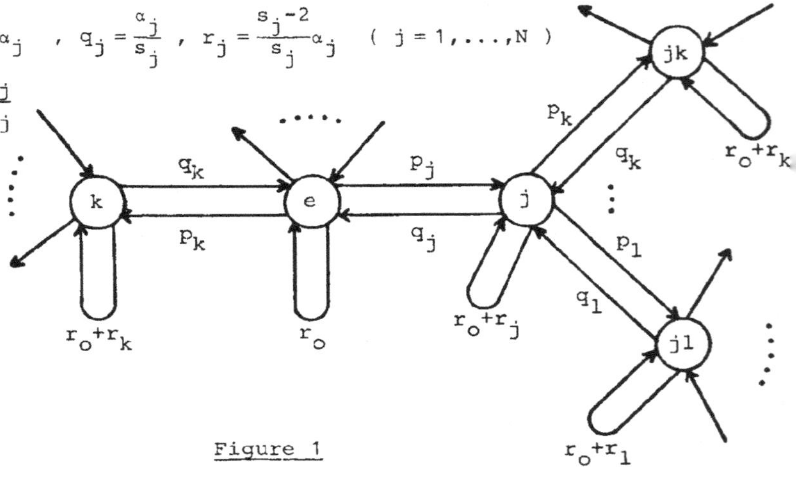

Figure 1

For $j \in \{1,\dots,N\}$ let $a_j^{(n)}$ be the probability to return to e for the first time at the n'th step after first step from e to j ($a_j^{(0)} = 0$). We write $A_j(z) = \sum_{n=0}^{\infty} a_j^{(n)} z^n$ for the corresponding generating function. If $a^{(n)}$ denotes the probability of first return to e at the n'th step without specifying the first step then

$\sum_{n=0}^{\infty} a^{(n)} z^n = \sum_{j=1}^{N} (A_j(z) + \frac{\alpha_j}{s_j} z)$, and from the relation

$\mu^{(n)}(e) = \sum_{k=0}^{n} a^{(k)} \mu^{(n-k)}(e)$ we get

$$\text{(1)} \quad E(z) = \frac{1}{1 - \sum_{j=1}^{N} (A_j(z) + \frac{\alpha_j}{s_j} z)}$$

To obtain $a_j^{(n)}$ we have to make the first step from e to j with probability $\frac{s_j - 1}{s_j}\alpha_j$. Then we can "walk" in any direction except back to e and we have to be back at vertex j after $n-1$ steps. Finally we make the last step from j to e with probability $\frac{\alpha_j}{s_j}$. This can be expressed by

$$\text{(2)} \quad A_j(z) = \frac{\frac{s_j - 1}{s_j^2} \alpha_j^2 z^2}{1 - \frac{s_j - 1}{s_j}\alpha_j z - \sum_{k \neq j} (A_k(z) + \frac{\alpha_k}{s_k} z)}$$

Combinig (1) and (2) we obtain

$$A_j(z) = \frac{1}{2}\left(\sqrt{\left(\frac{1}{E(z)} - \frac{s_j - 2}{s_j}\alpha_j z\right)^2 + 4\frac{s_j - 1}{s_j^2}\alpha_j^2 z^2} - \left(\frac{1}{E(z)} - \frac{s_j - 2}{s_j}\alpha_j z\right)\right)$$

and

$$\text{(3)} \quad E(z) = P(zE(z)) \quad ,$$

where $P(z)$ is the function defined in §2 . We have to take the branches of the roots for which $E(0) = 1$ and $A_j(0) = 0$.

To determine the radius of convergence of $E(z)$ we proceed like in [7] :

As $E(z)$ is a power series with nonnegative coefficients, r has to be the least positive singularity of $E(z)$ by a well known theorem of *Pringsheim*.

For real x , $P(x)$ is a sum of branches of hyperbolas, strictly increasing and convex, approaching the asymptotes $y = x + 1 - \sum_{j=1}^{N} \frac{s_j - 1}{s_j}$ for $x \to \infty$ and $y = 1 - \sum_{j=1}^{N} \frac{1}{s_j}$ for $x \to -\infty$. If $z > 0$ then $E(z)$ is the second coordinate of the point of intersection of the line $y = \frac{1}{z}x$ with $y = P(x)$ (if there are two such points we have to take the left one). Increasing z we can find a positive real solution $E(z)$ until we reach the tangent to $y = P(x)$ through the origin. This line is $y = P'(u)x$, where u satisfies $uP'(u) = P(u)$. Because of the shape of $P(x)$ for real x it is clear that u is the only real solution of this equation.

If we put

$$F(z,w) = P(zw) - w$$

then F is regular in every real point. For $0 < z < 1/P'(u)$, $F(z,E(z)) = 0$ and $F_w(z,E(z)) \neq 0$. By the theorem on implicit functions E is regular in z . For $z > P'(u)$ we can find no real solution $E(z)$ For $z = 1/P'(u)$ we have $F(z,E(z)) = 0$ and $F_w(z,E(z)) = 0$. Therefore the least positive singularity of E is $r = 1/P'(u)$.

Observe that equation (3) may be transformed into an algebraic equation for $E(z)$. Therefore $E(z)$ has finitely many singularities on the circl of convergence $|z| = r$, which are all algebraic and no poles, as $|E(z)| \leq E(r) = P(u) < \infty$. Under these conditions

$$\lim_{n\to\infty} \frac{\mu^{(n+1)}(e)}{\mu^{(n)}(e)} = \frac{1}{r}$$ yields that r is the only singularity of $E(z)$ on the circle of convergence (see [2]).

Now we can apply a special case of a theorem of *Darboux* (see [2],[5]) t determine the asymptotic behaviour of $(\mu^{(n)}(e))$:

$$F_z(r,E(r)) = E(r)P'(u) \neq 0 \quad \text{and} \quad F_{ww}(r,E(r)) = u^2P''(u) \neq 0$$

which implies

$$\mu^{(n)}(e) \sim \left(\frac{rE(r)P'(u)}{2\pi u^2P''(u)}\right)^{1/2} \cdot n^{-3/2} \cdot r^{-n} \quad \text{as} \quad n\to\infty .$$

References

[1] AKEMANN, CH.A., and PH.A.OSTRAND : Computing norms in group C^*-algebras. Am.Journ.of Math. 98, 1015-1047 (1976).

[2] BENDER, E.A.: Asymptotic methods in enumeration. Siam Review 16, 485-515 (1974).

[3] BERG, CH., and J.P.R.CHRISTENSEN : Sur la norme des opérateurs de convolution. Invent.Math. 23, 173-178 (1974).

[4] GERL, P.: Wahrscheinlichkeitsmaße auf diskreten Gruppen. Archiv Math. 31, 611-619 (1978).

[5] PLOTKIN, J.M. and J.ROSENTHAL : Some asymptotic methods in combinatorics. J.Austral.Math.Soc. Ser.A 28, 452-460 (1979).

[6] WOESS, W.: A local limit theorem for random walks on certain discrete groups. In "Probability Measures on Groups", Springer Lecture Notes in Math. 928, 467-477 (1982).

[7] WOESS, W.: Puissances de convolution sur les groupes libres ayant nombre quelconque de générateurs. Publ.Inst.Elie Cartan 7 (Nancy), 180-190 (1983).

W.Woess, Institut für Mathematik u.Angew.Geometrie
Montanuniversität, A-8700 Leoben, Austria

COMPLEX LEVY MEASURES

Hansmartin Zeuner
Mathematisches Institut
der Universität Tübingen
auf der Morgenstelle 10
D-7400 Tübingen

Introduction

The connection between a convolution semigroup $(\mu_t : t>0)$ on a locally compact group G and its infinitesimal generator is an important subject of probability theory. The complete treatment including a detailed description of the historical development of this problem is given in Heyer [4]. For convolution semigroups of probability measures on a Lie group this problem has been solved by Hunt [5]. Hazod [2] and Siebert [8] generalized Hunt's solution to arbitrary locally compact groups.

With a new method Duflo [1] generalizes this result into two directions: on the one hand the measures μ_t will not longer be supposed to be probability measures, the restriction to contraction measures (i.e. complex measures of total mass less or equal to 1) is sufficient. On the other hand the idempotent measure μ_o (i.e. $\mu_o * \mu_o = \mu_o$) will not be restricted to be the Dirac measure at the unit e of G. By a result of Hazod [3] each idempotent contraction measure is of the form $\mu_o = \chi\omega_K$ with a compact subgroup K of G and a continuous homomorphism χ from K into the group $\mathbb{T}$ of all complex numbers of norm 1.

In the case of the trivial idempotent $\mu_o = \varepsilon_e$, Siebert [8] proves a Lévy-Khintchine formula, that means a decomposition of the infinitesimal generator into a local part and a non-local part determined by the Lévy measure of the infinitesimal generator.

For two reasons the generalization of Siebert's result into the direction of Duflo's approach [1] is difficult: on the one hand, in the case of a non-trivial idempotent $\mu_o \neq \varepsilon_e$, the existence of a Lévy mapping (the most important technical tool for the Lévy-Khintchine formula) has not been proved up to now (compare [3], p. 136). On the other hand the correct generalization of the notion of a Lévy measures in the case of a convolution semigroup of contraction measures has yet to be found.

Both problems are solved by the author in his dissertation. This article takes care of the second problem. For the sake of simplicity, we restrict ourselves to the case of the trivial idempotent $\mu_o = \varepsilon_e$. This permits a direct and independent approach to the problem. The most important technical tool used in this article is Roth's theorem ([7], Corollaire II.2.4.) whereas it becomes possible to reduce to the case of probability measures and apply [8].

In the first part of this article, an abstract definition and some properties of Lévy measures are given. In (1.9) an analogue of Roth's theorem is proved which provides a connection with the usual definition of a positive Lévy measure. In the second part we motivate our definitio of a complex Lévy measure by showing that a complex Radon measure on $G^* := G \setminus \{e\}$ is a Lévy measure if and only if it is the Lévy measure of some dissipative distribution on G. The third part is devoted to the Lévy-Khintchine formula.

Throughout this article we keep the notations of [8], p. 314 with the usual typographical simplifications. If $\mathcal{V}$ is a linear space of function or measures on G we denote by $\mathcal{V}_{\mathbb{C}}$ its complexification. For every linear functional on $\mathcal{V}$ we denote by the same symbol its canonical $\mathbb{C}$-linear extension to $\mathcal{V}_{\mathbb{C}}$.

The author would like to express his gratitude to Professor H. Heyer and Professor E. Siebert for many encouraging discussions.

1. Lévy measures: definition and properties

(1.1) Definition: A complex Radon measure η on G^* is called a Lévy measure for G if the following conditions are satisfied:

(LM1) $\int_{G^*} f \, d|\eta| < \infty$ for every $f \in D(G)$ with $0 = f(e) \leq f$,

(LM2) $|\eta|(\complement U) < \infty$ for every open neighbourhood U of e,

(LM3) $|\eta| - \mathrm{Re}(\eta) \in M^b(G^*)$.

At first we discuss some elementary properties of Lévy measures.

(1.2) The set of all Lévy measures for G is a convex cone.

(1.3) For every Lévy measure η the negative real part $(\mathrm{Re}\ \eta)^-$ is a bounded measure on G^*.

This is a consequence of (LM3) and the inequality

$$2(\mathrm{Re}\ \eta)^- \leq |\eta| - \mathrm{Re}(\eta).$$

(1.4) Let η be a Lévy measure for G. Then up to an $\mathrm{Im}(\eta)$-null set, the support of $\mathrm{Im}(\eta)$ is σ-compact, and $\mathrm{Im}(\eta)$ is a σ-finite measure on G^*.

Proof: It is easily seen that $\mathrm{Im}(\eta)$ is dominated by the measure $|\eta| - \mathrm{Re}(\eta)$. Therefore the support of $\mathrm{Im}(\eta)$ is contained in the support of $|\eta| - \mathrm{Re}(\eta)$ which by (LM3) is the union of a σ-compact set M and a null set N with respect to $|\eta| - \mathrm{Re}(\eta)$. Since $\mathrm{Im}(\eta)$ is dominated by

$|\eta| - Re(\eta)$, N is a null set with respect to $Im(\eta)$ too. Thus $Im(\eta)$ is concentrated on the σ-compact set M and therefore σ-finite.

(1.5) The real part of a Lévy measure is not necessarily σ-finite. See for example [6], Corollary 5.

(1.6) Let $f \in D(G)$ satisfy $f(e) = 0$. Then the restriction of f to G^* is $Im(\eta)$-integrable.

(1.7) Let η be a positive Radon measure on G^*. Then η is a Lévy measure for G if and only if it satisfies (LM1) and (LM2).

(1.8) An analogue of Roth's theorem II.2.4. is valid for Lévy measures. Prior to the formulation of this analogue some notations have to be introduced. For every $f \in K(G^*)$ let the function $Zf \in K((G\times\mathbb{T})^*)$ be defined by

$$Zf(g,z) := \begin{cases} zf(g) & \text{if } g \neq e, \\ 0 & \text{if } g = e \text{ and } z \neq 1. \end{cases}$$

The adjoint of the linear mapping $Z : K(G^*) \to K((G\times\mathbb{T})^*)$ assigns to each Radon measure η' on $(G\times\mathbb{T})^*$ the Radon measure $Z\eta' : f \to \eta'(Zf)$ $(f \in K(G^*))$ on G^*.

(1.9) Theorem: Let η be a complex Radon measure on G^*. Then the following assertions are equivalent:

(i) η is a Lévy measure for G,

(ii) there is a positive Lévy measure η' for $G\times\mathbb{T}$ satisfying $\eta = Z\eta'$.

Proof: (i)$\Rightarrow$(ii) : By (LM3), (1.3) and (1.4) there is a σ-compact set $S \subset G^*$ such that $1_S\eta$ is a σ-finite measure and $1_{\complement S}\eta$ is a positive measure on G^*. Let ψ_o be a density of $1_S\eta$ with respect to $|1_S\eta|$. Then $\psi := \psi_o 1_S + 1_{\complement S}$ is a density of η with respect to $|\eta|$. After a modification on a $|\eta|$-null set, the range of ψ is contained in $\mathbb{T}$. Now we define the positive measure η' on $(G\times\mathbb{T})^*$ by

$$\eta'(f') := \int_{G^*} f'(g,\psi(g))\, d|\eta|(g) \qquad \text{for all } f' \in K((G\times\mathbb{T})^*).$$

At first we show the existence of $\eta'(f')$ and the continuity of the linear form η' on $K((G\times\mathbb{T})^*)$: Let M be a compact subset of $(G\times\mathbb{T})^*$. Then there exist a neighbourhood U of e and an $\varepsilon>0$ such that M and $U\times B_\varepsilon(1)$ are disjoint. Thus for every $f' \in K((G\times\mathbb{T})^*)$ supported by M,

$$\int_{G^*} |f'(g,\psi(g))|\, d|\eta|(g)$$
$$= \int_{G^*} |f'(g,\psi(g))|\, 1_{\complement(U\times B_\varepsilon(1))}(g,\psi(g))\, d|\eta|(g)$$
$$\leq \int_{G\setminus U} |f'(g,\psi(g))|\, d|\eta|(g) + \int_{G^*} |f'(g,\psi(g))|\, 1_{\complement B_\varepsilon(1)}(\psi(g))\, d|\eta|(g)$$

$$\leq \|f'\|_\infty (|\eta|(\complement U) + |\eta|(\{g \in G : |\psi(g)-1| \geq \varepsilon\}))$$
$$\leq \|f'\|_\infty (|\eta|(\complement U) + \frac{2}{\varepsilon^2} \| |\eta| - \mathrm{Re}(\eta) \|)$$

The mapping $g \to f'(g,\psi(g))$ is measurable and hence $|\eta|$-integrable. Since the constant following $\|f'\|_\infty$ at the right hand of the above inequality does not depend on f' but only on M, η' is continuous and therefore a Radon measure on $(G\times\mathbb{T})^*$.

Our next step is to prove that η' satisfies (LM1). Let f' be a function in $D(G\times\mathbb{T})$ with $0 = f'(e,1) \leq f'$. By Taylor's formula there are a positive constant a and a function $f \in D(G)$ such that $0 = f(e) \leq f$ and

$$f'(g,z) \leq f(g) + a\cdot \mathrm{Re}(1-z) \qquad \text{for every } g\in G,\ z \in \mathbb{T}.$$

Therefore

$$\begin{aligned}\int_{(G\times\mathbb{T})^*} f'\, d\eta' &\leq \int_{G^*} f(g)\, d|\eta|(g) + a \cdot \int_{(G\times\mathbb{T})^*} \mathrm{Re}(1-z)\, d\eta'(g,z)\\ &= \int_{G^*} f\, d|\eta| + a\, \| |\eta| - \mathrm{Re}(\eta) \| \\ &< \infty .\end{aligned}$$

In order to prove (LM2) suppose U' to be a neighbourhood of $(e,1)$ in $G\times\mathbb{T}$. Then there exist an $\varepsilon>0$ and a neighbourhood U of e in G such that $U' \supset U\times B_\varepsilon(1)$. It follows

$$\begin{aligned}\eta'(\complement U') &\leq \eta'(\complement U\times B_\varepsilon(1))\\ &\leq |\eta|(\complement U) + |\eta|(\{g\in G : |\psi(g)-1| \geq \varepsilon\})\\ &\leq |\eta|(\complement U) + 2\| |\eta| - \mathrm{Re}(\eta)\|/\varepsilon^2 \\ &< \infty.\end{aligned}$$

By (1.7) η' is a positive Lévy measure. For every $f\in K(G^*)$ we have

$$\begin{aligned}\eta(f) &= \int_{G^*} f(g)\psi(g)\, d|\eta|(g)\\ &= \int_{G^*} Zf(g,\psi(g))\, d|\eta|(g)\\ &= \eta'(Zf)\end{aligned}$$

which implies

$$\eta = Z\eta'.$$

(ii)$\Rightarrow$(i) : As an analogue to the definition of $Z\eta'$ we denote by $B\eta'$ and $R\eta'$ the Radon measures

$$f \to \int f(g)\, d\eta'(g,z)$$

and

$$f \to \int f(g)\, \mathrm{Re}(z)\, d\eta'(g,z)$$

on G^*. Since η' is a positive measure we have $|\eta| \leq B\eta'$ and $\mathrm{Re}(\eta) = R\eta'$, and therefore

$$0 \le |\eta| - \mathrm{Re}(\eta) \le B\eta' - R\eta'.$$

Since η' satisfies (LM1),

$$\begin{aligned} \| |\eta| - \mathrm{Re}(\eta) \| &= (|\eta| - \mathrm{Re}(\eta))(G^*) \\ &\le (B\eta' - R\eta')(G^*) \\ &= \int (1-\mathrm{Re}(z))\, d\eta'(g,z) \\ &< \infty . \end{aligned}$$

This implies (LM3).

In order to show that η satisfies (LM1) we consider a function $f \in D(G)$ with $0 = f(e) \le f$ and define the function $f' \in D(G \times \mathbb{T})$ by

$$f'(g,z) := f(g) \qquad \text{for every } g \in G \text{ and } z \in \mathbb{T}.$$

We have $0 = f'(e,1) \le f'$ and by (LM1)

$$\int_{G^*} f\, d|\eta| \le \int_{G^*} f\, dB\eta' = \int_{(G\times\mathbb{T})^*} f'\, d\eta' < \infty .$$

Hence η satisfies (LM1).

It is easy to see that the validity of (LM2) for η' implies that η satisfies (LM2). Therefore η is a Lévy measure for G.

(1.10) <u>Corollary</u>: Let η be a complex Radon measure on G^*. Then the following conditions are equivalent:

(i) η is a Lévy measure for G,

(ii) η satisfies (LM1), (LM2), η possesses a density ϕ with respect to $|\eta|$ and $1_G - \phi \in L^2(\eta)$.

The <u>proof</u> is a consequence of $|\eta| - \mathrm{Re}(\eta) = (1 - \mathrm{Re}(\phi)) \cdot |\eta|$ and the equation $1 - \mathrm{Re}(z) = \frac{1}{2}|1-z|^2$ for all $z \in \mathbb{T}$.

2. Lévy measures and dissipative distributions

(2.1) <u>Definition</u>: A distribution $T \in D'_{\mathbb{C}}(G)$ on G is called <u>dissipative</u> if $\mathrm{Re}(T(f)) \le 0$ for every $f \in D_{\mathbb{C}}(G)$ with $f(e) = \max_{x\in G} |f(x)|$.
A distribution $T \in D'(G)$ is called a <u>generalized laplacian</u> if $T(f) \le 0$ for every $f \in D(G)$ with $f(e) = \max_{x \in G} f(x) \ge 0$.
A linear form T on D(G) is called <u>primitive</u> respectively <u>quadratic form</u> if it satisfies
$T(f_o f_1^*) = T(f_o) f_1(e) - T(f_1) f_o(e)$ for all $f_o, f_1 \in D(G)$
respectively if it is a generalized laplacian and satisfies
$T(f_o f_1) + T(f_o f_1^*) = 2T(f_o) f_1(e) + 2T(f_1) f_o(e)$ for all $f_o, f_1 \in D(G)$.

(2.2) <u>Remark</u>: Every primitive or quadratic form is a generalized laplacian. Every generalized laplacian is a dissipative distribution.

(2.3) As in (1.8) we define for every $f \in D_{\mathbb{C}}(G)$ the function $Zf \in D_{\mathbb{C}}(G \times \mathbb{T})$

by $Zf(g,z) := zf(g)$ for every $g \in G$, $z \in \mathbb{T}$. The adjoint of the linear mapping $Z : \mathcal{D}_{\mathbb{C}}(G) \to \mathcal{D}_{\mathbb{C}}(G\times\mathbb{T})$ is also denoted by $Z : \mathcal{D}'_{\mathbb{C}}(G\times\mathbb{T}) \to \mathcal{D}'_{\mathbb{C}}(G)$. It is easily seen that for every dissipative distribution T' on $G\times\mathbb{T}$, the distribution ZT' on G is dissipative too.

(2.4) Theorem: Let T be a dissipative distribution on G. Then the linear form which maps each $f \in \mathcal{D}_{\mathbb{C}}(G)$ vanishing on a neighbourhood of e onto the complex number $T(f)$ extends to a Radon measure η on G^*. η is a Lévy measure.

(2.5) Definition: η is called the Lévy measure of the dissipative distribution T.

Proof of (2.4): The first statement is a consequence of [1], Lemme 1. From [7], Corollaire II.2.4. we get a generalized laplacian T' on $G\times\mathbb{T}$ such that $ZT' = T$. Let η' be the Lévy measure of T'. For every $f \in \mathcal{D}_{\mathbb{C}}(G)$ vanishing on a neighbourhood of e, the function $Zf \in \mathcal{D}_{\mathbb{C}}(G\times\mathbb{T})$ vanishes on a neighbourhood of $(e,1)$ and

$$\eta(f|G^*) = T(f) = T'(Zf) = \eta'(Zf|(G\times\mathbb{T})^*) = \eta'(Z(f|G^*)).$$

Therefore we have

$$\eta = Z\eta'.$$

In order to prove that η is a Lévy measure, by (1.9) and (1.6) it is suffcient to show that η' is a positive measure on $(G\times\mathbb{T})^*$ which satisfies (LM1) and (LM2).

Let $f' \in D(G\times\mathbb{T})$ be a non negative function vanishing on a neighbourhood of $(e,1)$. Then

$$-f'(e,1) = \max_{g,z}(-f'(g,z)) = 0$$

and, because of T being a generalized laplacian, we have

$$-\eta'(f') = T'(-f') \leq 0.$$

Hence η' is a positive measure.

In order to prove (LM1) consider a function $f' \in D(G\times\mathbb{T})$ with $0 = f'(e,1) \leq f'$. For every $\phi \in D(G\times\mathbb{T})$ vanishing on a neighbourhood of $(e,1)$ and with $0 \leq \phi \leq 1$ we have

$$(f'\phi - f')(e,1) = \max_{g,z}(f'\phi - f')(g,z) = 0,$$

whence

$$T'((f'\varphi - f') \leq 0$$

and

$$T'(f'\phi) \leq T'(f').$$

Therefore

$$\int_{(G\times\mathbb{T})^*} f' \, d\eta' = \sup_{\phi} T'(f'\phi) \leq T'(f') < \infty.$$

Finally from [1], Lemme 1, we get condition (LM2).

(2.6) In the following let Γ be a Lévy mapping for G in the sense of Siebert [8], that means an endomorphism of D(G) satisfying

$P \circ \Gamma = P$ for every primitive form P on G,

$\Gamma(f)^* = -\Gamma(f)$ for every $f \in D(G)$, and such that the mapping

$f \to \Gamma(f)(g)$ is a primitive form for every $g \in G$.

(2.7) Theorem: Let η be a Lévy measure for G. Then there exist a primitive form P_η and a real number a such that the mapping

$$L : D_{\mathbb{C}}(G) \to \mathbb{C},$$

$$f \to L(f) := A_o(f) + iP_\eta(f) + af(e) \qquad (f \in D_{\mathbb{C}}(G))$$

(where $A_o(f) := \int_{G^*} (f - f(e) - \Gamma(f))\, d\eta \qquad (f \in D_{\mathbb{C}}(G))$)

is a dissipative distribution on G.

P_η is uniquely determined by η.

The Lévy measure of the dissipative distribution L is η.

Proof: Let $\phi \in D(G)$ be a function with values in $[0,1]$ such that $\phi^* = \phi$ and ϕ takes the value 1 on a neighbourhood of e. For every function f' on $G \times \mathbb{T}$ define the function f'_G on G by

$$f'_G(g) := f'(g,1) \qquad \text{for } g \in G.$$

It is an easy calculation that the mapping $\Gamma' : D(G\times\mathbb{T}) \to D(G\times\mathbb{T})$,

$$\Gamma'(f')(g,z) := \mathrm{Re}(z)\cdot\Gamma(f'_G)(g) + \mathrm{Im}(z)\cdot\phi(g)\cdot\frac{d}{dt}f'(e,\exp(it))|_{t=o}$$

(for $f' \in D(G\times\mathbb{T})$, $g \in G$ and $z \in \mathbb{T}$) is a Lévy mapping for $G\times\mathbb{T}$. For every $f \in D_{\mathbb{C}}(G)$ we get

$$\Gamma'(Zf)(g,z) = Z(\Gamma(f) + f(e)\cdot 1_G)(g,z) - i\,\mathrm{Im}(z)\Gamma(f)(g)$$
$$-f(e)(z - i\,\mathrm{Im}(z)\,\phi(g)) \qquad (g \in G,\ z \in \mathbb{T}) \quad (1).$$

By (1.9) there exists a positive Lévy measure η' on $G\times\mathbb{T}$ with $\eta = Z\eta'$. By a slight generalization of [8], Lemma 6 (i) the linear form A_o on $D_{\mathbb{C}}(G)$ is well defined as well as the linear form A_1 on $D_{\mathbb{C}}(G\times\mathbb{T})$ determined by

$$A_1(f') := \int (f' - f'(e,1) - \Gamma'(f'))\, d\eta' \qquad \text{for all } f' \in D_{\mathbb{C}}(G\times\mathbb{T}).$$

From (1) we deduce that for every $f \in D_{\mathbb{C}}(G)$

$$A_1(Zf) - \int Z(f-f(e)-\Gamma(f))\, d\eta' + i\int \mathrm{Im}(z)\,\Gamma(f)(g)\, d\eta'(g,z)$$
$$+ f(e)\cdot\int (z-1-i\,\mathrm{Im}(z)\,\phi(g))\, d\eta'(g,z).$$

In correspondence to the proof of [8], Lemma 6 (ii), one sees that the linear functional P_η on D(G)

$$P_\eta(f) := \int \mathrm{Im}(z)\,\Gamma(f)(g)\, d\eta'(g,z) \qquad \text{for all } f \in D(G)$$

is a primitive form. If we define

$$b := \int (z-1-i\,\mathrm{Im}(z)\,\phi(g))\, d\eta'(g,z) \quad \in \quad \mathbb{C}$$

we get

$$ZA_1 = A_o + iP_\eta + b\varepsilon_e \tag{2}$$

Since by [8], Korollar on page 332, A_1 is a distribution, the same is true for A_o. By [8], Satz 3 (ii), A_1 is a generalized laplacian on $G\times\mathbb{T}$. From (2.2) and (2.3) we conclude that the distribution

$$L := A_o + iP_\eta + a\varepsilon_e = ZA_1 - i\,\mathrm{Im}(b)\varepsilon_e$$

(where $a := \mathrm{Re}(b)$) is dissipative.

For every $f\in D(G)$ vanishing on a neighbourhood of e we have $\Gamma(f)=0$ and $P_\eta(f)=0$, whence $L(f) = \int_{G^*} f\,d\eta$. Therefore η is the Lévy measure of L.

In the sequel we shall show that the dissipativity of the distribution $L = A_o + iP_\eta + a\varepsilon_e$ implies

$$P_\eta(f) = \int_{G^*} \Gamma(f)\,d\,\mathrm{Im}(\eta) \qquad \text{for all } f\in D(G) \text{ with } f(e)=0 \tag{3}$$

which proves the unicity of P_η.

Consider $f\in D(G)$ with $f(e)=0$ and for each $n\geq 1$ let $\psi_n\in D(G)$ be a function with values in $[0,1]$, equal to 1 on a neighbourhood of e and such that $|f^2\psi_n| < 1/2n^2$, $\int f^2\psi_n\,d\,|\mathrm{Re}(\eta)| < 1/n^2$ and $\int |f\psi_n|\,d\,|\mathrm{Im}(\eta)| < 1/n$. The first inequality is obtained from $f(e)=0$, the second from $0=f^2(e)\leq f^2$ and (LM1) and the third from (1.6).

For the functions

$$f_n^{\pm} := 1_G - n^2\psi_n f^2 \pm in\psi_n f \in D_{\mathbb{C}}(G) \qquad (n\geq 1)$$

we have by construction

$$1 = f_n^{\pm}(e) = \max_{x\in G} |f_n^{\pm}(x)|,$$

and hence

$$\begin{aligned} 0 &\geq \mathrm{Re}(L(f_n^{\pm})) \\ &= -n^2\int \psi_n f^2\,d\,\mathrm{Re}(\eta) + a \mp \left(n\int(\psi_n f - \Gamma(f))\,d\,\mathrm{Im}(\eta) + nP_\eta(f)\right). \end{aligned}$$

This implies

$$\left|\int \psi_n f\,d\,\mathrm{Im}(\eta) - \int\Gamma(f)\,d\,\mathrm{Im}(\eta) + P_\eta(f)\right| \leq -\frac{a}{n} + n\int\psi_n f^2\,d\,\mathrm{Re}(\eta).$$

For $n\to\infty$ we get (3).

3. The Lévy-Khintchine formula

(3.1) Definition: A dissipative distribution T on G is normed if for no $a > 0$ the distribution $T + a\varepsilon_e$ is dissipative.

(3.2) If T is a dissipative distribution on G, there exists a unique $a\geq 0$ such that $T + a\varepsilon_e$ is normed.

Therefore in (2.7) there is a unique $a =: a_\eta$ such that $L_\eta := A_o + iP_\eta + a_\eta\varepsilon_e$ is a normed dissipative distribution.

(3.3) Theorem: Let $T \in \mathcal{D}'_{\mathbb{C}}(G)$ be a dissipative distribution with Lévy measure η. Then there exist primitive forms P_o and P_1 on G, a quadratic form Q on G, $a_o, a_1 \leq 0$, $a_2 \in \mathbb{R}$, all uniquely determined, such that

$$T = a_o \varepsilon_e + i a_2 \varepsilon_e + P_o + (a_1 \varepsilon_e + i P_1 + Q) + L_\eta$$

and such that $a_1 \varepsilon_e + i P_1 + Q$ is a normed dissipative distribution.

Proof: At first define a_o to be the unique non positive number such that $T_o := T - a_o \varepsilon_e$ is normed. By [7], Corollaire II.2.4. there is a generalized laplacian T' on $G \times \mathbb{T}$ with $T_o = ZT'$. Let η' be the Lévy measure of T'. By [8], Satz 3 (i), there are a primitive form P' on $G \times \mathbb{T}$, a quadratic form Q' on $G \times \mathbb{T}$ and a real number $b' \leq 0$ such that

$$T' = b' \varepsilon_{(e,1)} + P' + Q' + A_1 \tag{1}$$

where A_1 and the Lévy mapping Γ' for $G \times \mathbb{T}$ are defined as in the proof of (2.7). From (3.2) and formula (2) of the proof of (2.7) we get

$$ZA_1 = A_o + iP_\eta + b\varepsilon_e = L_\eta + b''\varepsilon_e \tag{2}$$

with some $b'' \leq 0$.

Let us now consider the functions ψ_o, ψ_1, ψ_2 on $G \times \mathbb{T}$ defined by $\psi_o(g,z) := z$, $\psi_1(g,z) := \mathrm{Re}(z)$ and $\psi_2(g,z) := \mathrm{Im}(z)$ for $(g,z) \in G \times \mathbb{T}$. For every function $f \in \mathcal{D}_{\mathbb{C}}(G)$ define $f' \in \mathcal{D}_{\mathbb{C}}(G \times \mathbb{T})$ by $f'(g,z) := f(g)$ for every $(g,z) \in G \times \mathbb{T}$. Putting $P_o(f) := P'(f')$ for $f \in D(G)$ it is easily seen that P_o is a primitive form on G and (with $a_2 := -iP'(\psi_o)$)

$$\begin{aligned}(ZP')(f) &= P'(\psi_o f') = P'(f')\psi_o(e,1) + P'(\psi_o)f'(e,1)\\ &= P_o(f) + i a_2 f(e) \qquad \text{for } f \in \mathcal{D}_{\mathbb{C}}(G)\end{aligned} \tag{3}$$

Furthermore the distribution Q on G defined by $Q(f) := Q'(f')$ is easily seen to be a quadratic form. We have

$$(ZQ')(f) = Q'(\psi_1 f') + iQ'(\psi_2 f') \qquad \text{for } f \in \mathcal{D}_{\mathbb{C}}(G) \tag{4}$$

Since $\psi_1^* = \psi_1$ and Q' is a quadratic form, we get

$$\begin{aligned}Q'(\psi_1 f') &= \tfrac{1}{2}Q'(f'\psi_1) + \tfrac{1}{2}Q'(f'\psi_1^*) = Q'(f')\psi_1(e,1) + Q'(\psi_1)f'(e,1)\\ &= Q(f) + a_1 f(e) \qquad \text{for } f \in \mathcal{D}_{\mathbb{C}}(G)\end{aligned} \tag{5}$$

where $a_1 := Q'(\psi_1) \leq 0$.

The distribution P_1 defined by $P_1(f) := Q'(\psi_2 f')$ for $f \in D(G)$ is a primitive form since for every functions $f_o, f_1 \in D(G)$

$$\begin{aligned}P_1(f_o f_1^*) &= Q'(\psi_2 f_o' f_1'^*)\\ &= \tfrac{1}{2}(Q'(\psi_2 f_o' f_1') + Q'(\psi_2 f_o' f_1'^*)) - \tfrac{1}{2}(Q'(\psi_2 f_1' f_o') + Q'(\psi_2 f_1' f_o'^*))\\ &= Q'(\psi_2 f_o')f_1'(e,1) - Q'(\psi_2 f_1')f_o'(e,1)\\ &= P_1(f_o)f_1(e) - P_1(f_1)f_o(e).\end{aligned}$$

By (4) and (5), $a_1\varepsilon_e + iP_1 + Q = ZQ'$ is a dissipative distribution. Combining (1), (2), (3), (4) and (5) we get

$$T_o = ZT' = b'\varepsilon_e + P_o + ia_2\varepsilon_e + a_1\varepsilon_e + iP_1 + Q + L_\eta + b''\varepsilon_e.$$

Since T_o is normed, $(a_1\varepsilon_e + iP_1 + Q)$ is normed too and $b' = b'' = 0$. Hence

$$T = T_o + a_o\varepsilon_e = a_o\varepsilon_e + ia_2\varepsilon_e + P_o + (a_1\varepsilon_e + iP_1 + Q) + L_\eta.$$

As in [3], I. Satz 5.3, Q is uniquely determined by T, and consequently so are a_1, a_2, P_o and P_1.

Bibliography

[1] Duflo, M.: Représentations de semi-groupes de mesures sur un groupe localement compact. Ann. Inst. Fourier (Grenoble) 28/3 (1978), 225-249.

[2] Hazod, W.: Über die Lévy-Hinčin-Formel auf lokalkompakten topologischen Gruppen. Z. Wahrscheinlichkeitstheorie und verw. Gebiete 25 (1973), 301-322.

[3] Hazod, W.: Stetige Faltungshalbgruppen von Wahrscheinlichkeitsmaßen und erzeugende Distributionen. Lecture Notes in Mathematics Vol. 595, Springer Berlin-Heidelberg-New York, 1977.

[4] Heyer, H.: Probability measures on Locally Compact Groups. Ergebnisse der Mathematik und ihrer Grenzgebiete 94, Springer, Berlin-Heidelberg-New York, 1977.

[5] Hunt, G.A.: Semigroups of measures on a Lie group. Transactions of the AMS 81 (1956), 264-294.

[6] Janssen, A.: Continuous convolution semigroups with unbounded Lévy measures on locally compact groups. Arch. Math. 38 (1982), 565-576.

[7] Roth, J.P.: Opérateurs dissipatifs et semi-groupes dans les espaces de fonctions continues. Ann. Inst. Fourier (Grenoble) 26/4 1-97.

[8] Siebert, E.: Über die Erzeugung von Faltungshalbgruppen auf beliebigen lokalkompakten Gruppen. Math. Z. 131 (1973), 313-333.

Probability theory on hypergroups: A survey

by Herbert Heyer

Introduction

Probability theory on hypergroups has been promoted by attempts to study the central limit theorem for measures on such structures as homogeneous spaces, double coset spaces, Sturm-Liouville systems, generalized translation spaces, spherical systems and others. It was S. Bochner who in his early papers [9] and [10] started a systematic analysis of spherically symmetric probability measures on spheres and Euclidean spaces and related his results to spherical functions. In the framework of Sturm-Liouville equations Bochner considers eigenfunctions which are ultraspherical polynomials or Bessel functions. These types of spherical functions constitute what one calls the duals of the Gelfand pairs of compact or Euclidean type respectively. It appears that these duals $\mathbb{Z}_+$ and $\mathbb{R}_+$ carry the structure of a hypergroup which is given in terms of convolution of measures determined by the corresponding spherical functions. In this context one might recall that the general theory of hypergroups was initiated by the search for an appropriate structure in the dual of a compact group. Both aspects of the theory, the probabilistic as well as the harmonic analytic one, have contributed to remarkable new ideas and results. The notion of a hypergroup has proved to serve as an efficient tool to study infinitely divisible probability measures, convolution semigroups and Markov chains on double coset and orbit spaces, in particular on double coset spaces arising from Euclidean spheres and spaces, hyperbolic spaces, homogeneous trees, and their duals. While the general theory of hypergroups seems to be developed in order to obtain more information on the representation theory of certain classical groups, the more specific aspects exposed in this article are motivated by probabilistic demands: In recent papers of L. Gallardo [14], [15] hypergroup analysis has been successfully applied in studying the central limit theorem, strong laws of large numbers, the iterated logarithm and the rate of decay for certain classes of classical Markov chains.

I Preparations from harmonic analysis

§ 1 Hypergroups: Definition and Examples

Let K be a locally compact space. By $\mathfrak{M}^1(K)$ we denote the space of probability measures on K furnished with the weak topology $\mathfrak{T}_w$. The Dirac measure in a point $x \in K$ will be abbreviated by ε_x. We will further need the space $\mathfrak{K}(K)$ of all nonempty compact subsets of K furnished with the Michael topology.

1.1 Definition. K is said to be a *weak hypergroup* if the following axioms are fulfilled:

(HG 1) There exists a mapping $(x,y) \to \varepsilon_x * \varepsilon_y$ from $K \times K$ into $\mathfrak{M}^1(K)$ which is continuous and satisfies the equality

$$\varepsilon_x * (\varepsilon_y * \varepsilon_z) = (\varepsilon_x * \varepsilon_y) * \varepsilon_z$$

for all $x, y, z \in K$.

(HG 2) For any $x, y \in K$ the set $\operatorname{supp}(\varepsilon_x * \varepsilon_y)$ is compact.

(HG 3) There exists a homeomorphism $x \to x^-$ from K into K such that for all $x, y \in K$ the equalities

$$x = x^{--}$$

and

$$(\varepsilon_x * \varepsilon_y)^- = \varepsilon_{y^-} * \varepsilon_{x^-}$$

hold.

(HG 4) There exists an element $e \in K$ such that

$$\varepsilon_e * \varepsilon_x = \varepsilon_x * \varepsilon_e = \varepsilon_x$$

holds for all $x \in K$.

The mapping * introduced in (HG 1) is called the *convolution* in K; it can be extended to a convolution in the set $\mathcal{M}^b(K)$ of all bounded measures on K. The homeomorphism $^-$ introduced in (HG 3) is called the *involution* of K; it can also be extended to $\mathcal{M}^b(K)$. Naturally the element e defined in (HG 4) is said to be the *unit element* of K.

A weak hypergroup K is called a *hypergroup* if the following additional axioms are satisfied:

(HG 5) For any $x,y \in K$ we have $e \in \operatorname{supp}(\varepsilon_x * \varepsilon_{\bar{y}})$ iff $x=y$.

(HG 6) The mapping $(x,y) \to \operatorname{supp}(\varepsilon_x * \varepsilon_y)$ from $K \times K$ into $\mathfrak{K}(K)$ is continuous.

Hypergroups in the sense of Definition 1.1 have been introduced for the first time by Jewett [26]. See also Spector [37].

A weak hypergroup K is said to be *commutative* if

$$\varepsilon_x * \varepsilon_y = \varepsilon_y * \varepsilon_x$$

holds for all $x,y \in K$.

Commutative hypergroups have been investigated thoroughly by Jewett [26], Dunkl [11] and Spector [37].

Let $^b(K)$ denote the space of bounded continuous functions on K. Given $f \in \mathcal{C}^b(K)$ and $x,y \in K$ we write

$$f(x*y) = \int f\,d(\varepsilon_x * \varepsilon_y) = f_x(y) = f^y(x),$$

where $x*y$ is considered as the set $\operatorname{supp}(\varepsilon_x * \varepsilon_y)$.

For any subhypergroup H of a hypergroup K one defines a Radon measure on K to be *left* H-*invariant* if

$$\varepsilon_x * \mu = \mu$$

for every $x \in H$ in the sense of

$$\mu({}_x f) = \mu(f)$$

for every $f \in \mathcal{C}^b(K)$ and $x \in H$.

In the case where μ is left K-invariant and positive, we refer to it as a *left Haar measure* on K and write ω_K or ω in place of μ. Left Haar measures exist if K is discrete, compact or commutative (See Jewett [26] and Spector [38]). For general hypergroups the problem of existence of a left Haar measure is still open, whereas the problem of uniqueness has been solved in the affirmative sense.

A measure on K that is both left and right H-invariant will be referred to as *H-invariant*. Clearly a left Haar measure on a compact hypergroup K is also right K-invariant and thus K-invariant.

For the entire exposition that follows we shall always assume that the underlying hypergroup admits a left Haar measure.

1.2. First interesting examples of hypergroups are provided by the coset spaces G//H of a locally compact group G by a compact subgroup H of G. They are known as *double coset hypergroups*. Among those one can recover the *orbit hypergroups* which are usually introduced as the orbit spaces $G_B = \{\overline{B}x : x \in G\}$ of a locally compact group G with respect to a relatively compact subgroup B of the automorphism group Aut(G) of G.

1.3. In the following we are going to introduce a class of commutati hypergroups which turns out to be of particular importance for proba bility theory. This class has been studied in great detail by Lasser in [29].

Let $(a_n)_{n\geq 1}$, $(b_n)_{n\geq 1}$ and $(c_n)_{n\geq 1}$ be sequences in $\mathbb{R}$ with $a_n > o$, $c_n > o$, $b_n \geq o$ and $a_n + b_n + c_n = 1$ for all $n\geq 1$. We fix $a_o, b_o \in \mathbb{R}_+$, $a_o > o$ such that $a_o + b_o = 1$. Then we define a sequence $(P_n)_{n\geq o}$ of orthogonal polynomials over $\mathbb{R}$ by

$$P_o(x) := 1$$
$$P_1(x) := \frac{1}{a_o} x - \frac{b_o}{a_o}, \quad \text{and}$$
$$P_{n+1}(x) := \frac{1}{a_n} P_1(x) P_n(x) - \frac{b_n}{a_n} P_n(x) - \frac{c_n}{a_n} P_{n-1}(x)$$

for all $n\geq 1$.

Linearizing the products $P_m(x)P_n(x)$ for $m,n \in \mathbb{Z}_+$ in the form

$$P_m(x)P_n(x) = \sum_{k=o}^{2(m\wedge n)} g(m,n,n+m-k) P_{n+m-k}(x)$$

we observe that the coefficients $g(m,n,n+m-k)$ are uniquely determined by the sequences $(a_n)_{n\geq 1}$, $(b_n)_{n\geq 1}$ and $(c_n)_{n\geq 1}$.

Introducing in $\mathbb{Z}_+$ the convolution

$$\varepsilon_m * \varepsilon_n := \sum_{k=o}^{2(m\wedge n)} g(m,n,n+m-k)\varepsilon_{n+m-k}$$

for all $m,n \in \mathbb{Z}_+$ and assuming that all coefficients $g(m,n,n+m-k)$ are $\geq o$ one can show that $\mathbb{Z}_+$ becomes a hypergroup with the identity mapping as involution and with o as unit element. It is called the *polynomial hypergroup defined by the sequences* $(a_n)_{n\geq 1}$, $(b_n)_{n\geq 1}$ and $(c_n)_{n\geq 1}$ or defined by the sequence $(P_n(x))_{n\geq o}$.

The Haar measure ω of $\mathbb{Z}_+$ turns out to be the measure $\sum_{n\in\mathbb{Z}_+} h(n)\varepsilon_n$ with

$$h(n) := \frac{1}{\varepsilon_n * \varepsilon_n(\{o\})} = \frac{1}{g(n,n,o)}$$

for all $n \in \mathbb{Z}_+$. In terms of the defining sequences $(a_n)_{n\geq 1}$, $(b_n)_{n\geq 1}$ and $(c_n)_{n\geq 1}$ one has

$$h(o) = 1,$$

$$h(1) = \frac{1}{c_1} \quad \text{and}$$

$$h(n) = \frac{\prod_{k=1}^{n-1} a_k}{\prod_{k=1}^{n} c_k}$$

for all $n\geq 2$.

Now we give a list of <u>subclasses of the class of polynomial hypergroups</u>.

1.3.1. *Jacobi hypergroups* are defined by the sequence $(P_n(z))_{n\geq 1}$ of Jacobi polynomials

$$P_n(z) := P_n^{(\alpha,\beta)}(z)$$

for $\alpha,\beta \in \mathbb{R}$ such that $\alpha\geq\beta > -1$, $\alpha+\beta+1\geq o$, and $n\geq 1$.

Special cases are

1.3.2 the *ultraspherical (Gegenbauer) hypergroups* defined by the additional condition $\alpha=\beta$. Here the ultraspherical polynomials $P_n^{\alpha}(z)$ occur as the defining polynomials.

Ultraspherical hypergroups $\mathbb{Z}_+$ appear as duals $(G,K)^\wedge$ of Gelfand pairs of compact type, where G is a compact, connected Lie group and K a compact subgroup arising from an involutive automorphism of G. See Heyer [23].

Further specialization yields

1.3.3 the *first kind Tschebychev hypergroups* in the case $\alpha=\beta=-\frac{1}{2}$,

1.3.4 the *Legendre hypergroups* in the case $\alpha=\beta=0$, and

1.3.5 the *second kind Tschebychev hypergroups* in the case $\alpha=\beta=\frac{1}{2}$.

We note that the hypergroup $\mathbb{Z}_+$ of type 1.3.2 for $\alpha=\frac{d}{2}-1$ appears as the dual $({}^G/_K)^\wedge$ of a sphere ${}^G/_K$ with $G := SO(d+1)$ and $K := SO(d)$ for $d\geq 2$, that of type 1.3.5. as the dual of $SU(2)$.

Besides the Jacobi hypergroups we also have

1.3.6 the *Arnaud hypergroups* defined by the sequence $(P_n(z))_{n\geq 1}$ of Arnaud polynomials

$$P_n(z) := P_n(z|a)$$

for some $a \in \mathbb{R}$, $a\geq 2$.

Arnaud hypergroups are connected with the study of probability measures on a homogeneous tree. See the work of Letac and his school, f.e. [32] and [33].

§ 2 Representations and Fourier transforms

We recall that the underlying hypergroup K is assumed to admit a left Haar measure $\omega = \omega_K$.

For any given Hilbert space $\mathcal{H}$ we denote by $B(\mathcal{H})$ the Banach$^\sim$-algebra of bounded linear operators of $\mathcal{H}$, with identity I.

2.1 Definition. We refer to D as a *representation of* K *in* $\mathcal{H}$ if the following axioms are fulfilled :

(R 1) D is a $^\sim$-representation of the Banach$^\sim$-algebra $\mathfrak{M}^b(K)$ in $B(\mathcal{H})$.

(R 2) $D(\varepsilon_e) = I$.

(R 3) For all $u,v \in \mathcal{H}$ the mapping $\mu \to \langle D(\mu)u,v \rangle$ is continuous on $\mathfrak{M}^b_+(K)$ with respect to the weak topology.

By Rep (K) we denote the set of (unitary) equivalence classes and by Rep_f (K) and $K^\wedge$ the subsets consisting of equivalence classes of finite dimensional and irreducible representations respectively. $K^\wedge$ is called the *dual* of K, although in general it is far from being a hypergroup.

For $D \in \text{Rep}(K)$ and $x \in K$ we set

$$D(x) := D(\varepsilon_x)$$

2.2 Some results due to Vrem [39] .

2.2.1. There are sufficiently many irreducible representations of K to separate its points.

2.2.2. If K is compact, then the separation of 2.2.1 can be achieved with finite dimensional representations, since

2.2.3 in the case if a compact hypergroup K all irreducible representations of K are finite dimensional. Indeed, for a compact hypergroup K, $K^\wedge$ consists of unitary operators iff K is already a group.

2.2.4 In general one cannot assume that irreducible representations of a hypergroup K are unitary.

2.2.5. The class of hypergroups K satisfying $K^\wedge \subset \mathrm{Rep}_f(K)$ includes both the compact and the commutative hypergroups. One calls these *Moore hypergroups*.

For any subset A of Rep(K) let $\mathfrak{L}(A)$ denote the $^\sim$-algebra

$$\Pi\{B(\mathcal{H}(D)) : D \in A\}.$$

2.3 <u>Definition.</u> Given $\mu \in \mathcal{M}^b(K)$ we define $\hat{\mu} \in \mathfrak{L}(\mathrm{Rep}(K))$ by

$$\hat{\mu}(D) := \overline{D}(\mu)$$

for all $D \in \mathrm{Rep}(K)$. $\hat{\mu}$ is called the *Fourier transform* of μ.

2.4 <u>Properties.</u>

2.4.1 $\hat{\mu}(1) = I$

2.4.2 $\widehat{\mu^\sim}(D) = \hat{\mu}(D)$.

2.4.3 $(\alpha\mu + \beta\nu)^\wedge(D) = \alpha\hat{\mu}(D) + \beta\hat{\nu}(D) \quad (\alpha, \beta \in \mathbb{C})$.

2.4.4 $(\mu * \nu)^\wedge(D) = \hat{\mu}(D)\hat{\nu}(D)$.

2.4.5 $\|\hat{\mu}(D)\| \leq \|\mu\| \quad (D \in \mathrm{Rep}(K))$.

2.4.6 The mapping $\mu \to \hat{\mu}$ from $\mathcal{M}^b(K)$ into $\mathfrak{L}(K^\wedge)$ is one-to-one.

For the proof of property 2.4.6. see Bloom, Heyer [7].

2.5 <u>An application.</u> Let us recall that a subhypergroup H of K is called *normal* if

$$\varepsilon_x * \omega_H = \omega_H * \varepsilon_x$$

holds for every $x \in K$. Here ω_H denotes the *normalized Haar measure* of H. For any compact normal subhypergroup K we have

$$\hat{\omega}_H = 1_{A(K^\wedge, H)} ,$$

where $A(K^\wedge, H)$ is the *annihilator*

$$\{U \in K^\wedge : U(x) = I \quad \text{for all} \quad x \in H\}$$

of H *in* K .

If, in particular, K is compact, then

$$\hat{\omega}(D) = \begin{cases} I & \text{if} \quad D = 1 \\ 0 & \text{if} \quad D \neq 1 . \end{cases}$$

For a converse of this statement we quote

2.6 <u>Theorem.</u> Let K be a hypergroup, $\mu \in \mathfrak{M}^1(K)$ with $\mu^- = \mu$ and $\operatorname{supp}(\mu) \subset H$ for some subhypergroup H of K. Let $\hat{\mu}(D) = 0$ for all $D \notin A(K^\wedge, H)$. Then $\mu = \omega_H$ and H is compact.

2.7 <u>Corollary.</u> Let K be a Moore hypergroup, $\mu \in \mathfrak{M}^1(K)$ with $K = [\operatorname{supp}(\mu)]$ $e \in \operatorname{supp}(\mu)$ and $\mu^- = \mu$. If for all $D \in K^\wedge$ either $\hat{\mu}(D) = 0$ or $\|\hat{\mu}(D)\| = 1$ holds, then K is compact and $\mu = \omega_K$.

<u>Proofs</u> of the preceding and the following two theorems are contained in Bloom, Heyer [7] .

2.8 <u>Theorem</u> (Continuity of the Fourier transform). Let K be a hypergroup, let $(\mu_n)_{n \geq 1}$ be a sequence in $\mathfrak{M}^b_+(K)$ and let $\mu \in \mathfrak{M}^b(K)$. The following statements are equivalent:

(i) $\mathcal{T}_w\text{-}\lim_{n\to\infty} \mu_n = \mu$.

(ii) $\lim_{n\to\infty} \langle \hat{\mu}_n(D)u,v \rangle = \langle \hat{\mu}(D)u,v \rangle$ for all $D \in K^\wedge$, $u,v \in \mathcal{H}(D)$.

2.9 Theorem (Continuity in the compact case). Let K be a compact hypergroup, $(\mu_n)_{n\geq 1}$ a sequence in $\mathcal{M}_+^b(K)$, and let there exist a $\psi \in \mathcal{E}(K^\wedge)$ satisfying

$$\lim_{n\to\infty} \langle \hat{\mu}_n(D)u,v \rangle = \langle \psi(D)u,v \rangle$$

for all $D \in K^\wedge, u,v \in \mathcal{H}(D)$. Then there exists a measure $\mu \in \mathcal{M}_+^b(K)$ such that $\hat{\mu} = \psi$ and $\mathcal{T}_w\text{-}\lim_{n\to\infty} \mu_n = \mu$.

§ 3 Strong hypergroups

In order to present stronger versions of the continuity theorem for the Fourier transform and also some deeper harmonic analysis we assume for this section that G is commutative.

3.1. Let us introduce the set

$$\mathfrak{X}(K) := \{\chi \in \mathcal{C}^b(K) : \chi \neq o, \chi(x*y) = \chi(x)\chi(y) \quad \text{for all} \quad x,y \in K\}.$$

Then the dual of K appears as the set

$$K^\wedge = \{\chi \in \mathfrak{X}(K) : \chi(x^-) = \overline{\chi(x)} \quad \text{for all} \quad x \in K\}.$$

Furnished with the compact open topology $\mathfrak{X}(K)$ and $K^\wedge$ become locally compact spaces.

The *Fourier transform* $\hat{\mu}$ of a measure $\mu \in \mathcal{M}^b(K)$ has the form

$$\hat{\mu}(\chi) = \int_K \overline{\chi(x)}\mu(dx)$$

for all $\chi \in K^\wedge$, and we have the notion of the *inverse Fourier transform* $\check{\nu}$ of a measure $\nu \in \mathcal{M}^b(K^\wedge)$ given by

$$\check{\nu}(x) := \int_{K^\wedge} \chi(x)\nu(d\chi)$$

for all $x \in K$.

It is an early result of Levitan reorganized in Jewett [26] that there exists a unique nonnegative measure π on $K^\wedge$ satisfying

$$\int_K |f|^2 d\omega = \int_{K^\wedge} |\hat{f}|^2 d\pi$$

for all $f \in L^1 \cap L^2(K)$. π is called the *Plancherel* measure of $K^\wedge$, and the equality above is known as the *Plancherel identity*.

3.2 Theorem (Continuity of the Fourier transform in the commutative case). Let K be a commutative hypergroup such that supp $(\pi) = K^\wedge$, and let $(\mu_n)_{n\geq 1}$ be a sequence in $\mathcal{M}^b_+(K)$. We assume that there exists a function $\psi \in \mathcal{C}(K^\wedge)$ such that

$$\lim_{n\to\infty} \hat{\mu}_n(\chi) = \psi(\chi)$$

for π-almost all $\chi \in K^\wedge$ and $\chi = 1$.

Then there exists a measure $\mu \in \mathcal{M}^b_+(K)$ such that $\hat{\mu} = \psi$ and $\mathcal{T}_w\text{-}\lim_{n\to\infty} \mu_n = \mu$.

For the proof of this result see Bloom, Heyer [7].

For many applications in probability theory the following version of the continuity theorem appears to be the proper tool.

3.3 Theorem. Let K be a commutative hypergroup having a countable basis of its topology. It is assumed that supp (π) contains the unit character $\chi = 1$. Given any sequence $(\mu_n)_{n\geq 1}$ of measures in $\mathcal{M}^1(K)$ we suppose that there exists a function ψ which is continuous at 1 satisfying

$$\lim_{n\to\infty} \hat{\mu}_n(\chi) = \psi(\chi)$$

for π-almost all $\chi \in K^\wedge$.

Then there exists a probability measure $\mu \in \mathcal{M}^1(K)$ such that $\hat{\mu} = \psi$ π - a.e. and $\mathcal{T}_v - \lim_{n\to\infty} \mu_n = \mu$.

A proof of this result is contained in Gallardo, Gebuhrer [16].

3.4 Definition. The dual $K^\wedge$ of a commutative hypergroup K is said to be a *hypergroup with respect to pointwise multiplication* if for all $\chi, \rho \in K^\wedge$ there exists a measure $\varepsilon_\chi * \varepsilon_\rho \in \mathcal{M}^1(K^\wedge)$ such that

$$\chi(x)\rho(x) = \int_{K^\wedge} \tau(x)\, \varepsilon_\chi * \varepsilon_\rho(d\tau)$$

holds whenever $x, y \in K$, and if $K^\wedge$ is a hypergroup with $*$ as its convolution, complex conjugation as its involution, and 1 as unit.

3.5 Remark. We shall see later that for general hypergroups K the dual $K^\wedge$ is not necessarily a hypergroup with respect to pointwise multiplication.

But $K^\wedge$ is always a *weak* hypergroup if the defining relationship of Definition 3.3 holds with $\varepsilon_\chi * \varepsilon_\rho$ having a compact support.

3.6 Definition. Let K be a commutative hypergroup and let $K^\wedge$ be a hypergroup with respect to pointwise multiplication. K is called a *strong* (or Pontryagin) *hypergroup* if $K^{\wedge\wedge} = K$.

3.7 Remark. It will be indicated later that strong hypergroups are fairly rare.

But every discrete commutative hypergroup K such that $K^\wedge$ is a hypergroup with respect to pointwise multiplication is indeed a strong hypergroup.

3.8. We return to the class of polynomial hypergroups introduced in 1.3. Their underlying space is $\mathbb{Z}_+$, and their convolution is defined via sequences $(a_n)_{n\geq 1}$, $(b_n)_{n\geq 1}$ and $(c_n)_{n\geq 1}$ of real numbers, or via a sequence $(P_n)_{n\geq 0}$ of orthogonal polynomials over $\mathbb{R}$ depending on those sequences.

For a polynomial hypergroup $\mathbb{Z}_+$ the spaces $\mathfrak{X}(\mathbb{Z}_+)$ and $\mathbb{Z}_+^\wedge$ are compact with respect to the Gelfand topology, and they admit representations

$$\mathfrak{X}(\mathbb{Z}_+) = \{\chi_z : z \in D\} \quad \text{and}$$
$$\mathbb{Z}_+^\wedge = \{\chi_x : x \in D_S\} ,$$

where $D := \{z \in \mathbb{C} : (P_n(z))_{n\geq o}$ is bounded$\}$,

$D_S := D \cap \mathbb{R}$, and

$\chi_z : \mathbb{Z}_+ \to \mathbb{C}$ is defined by

$$\chi_z(n) := P_n(z)$$

for all $n \in \mathbb{Z}_+$.

The mappings $z \to \chi_z$ from D into $\mathfrak{X}(\mathbb{Z}_+)$ and $x \to \chi_x$ from D_S into $\mathbb{Z}_+^\wedge$ are in fact homeomorphisms, whence D and D_S are compact spaces, and moreover $D_S \subset [1-2a_o, 1]$. These facts and a detailed discussion of the following example are contained in Lasser's work [30].

We note that the Plancherel measure π on D_S is given by the identity

$$\sum_{n\geq o} f(n)\overline{g(n)}h(n) = \int_{D_S} \hat{f}(x)\overline{\hat{g}(x)}\pi(dx)$$

for all $f, g \in \ell^1(\mathbb{Z}_+)$, where the Fourier transform $\hat{f}$ of a function $f \in \ell^1(\mathbb{Z}_+)$ is defined by

$$\hat{f}(x) := \sum_{n\geq o} P_n(x)f(n)h(n)$$

for all $x \in D_S$.

In particular one obtains the *orthogonality relations*

$$\int_{D_S} P_n \cdot P_m d\pi = \begin{cases} \frac{1}{h(m)} & \text{if } n = m \\ o & \text{if } n \neq m \end{cases}$$

which justify the terminology of *orthogonal* polynomials for the defining $P_n(x)$.

For a polynomial hypergroup $\mathbb{Z}_+$ the dual $\mathbb{Z}_+^\wedge \cong D_S$ is a hypergroup with respect to pointwise multiplication if for all $x,y \in D_S$ there exists a measure $\varepsilon_x * \varepsilon_y \in \mathcal{M}^1(D_S)$ such that

$$P_n(x)P_n(y) = \int_{D_S} P_n(w)\,\varepsilon_x*\varepsilon_y(dw)$$

holds whenever $n \in \mathbb{Z}_+$, and if D_S is a hypergroup with $*$ as its convolution, the identity as involution and $1 \in D_S$ as unit.

3.9 Examples.

3.9.1. Jacobi hypergroups $\mathbb{Z}_+$ and their duals $\mathbb{Z}_+^\wedge \cong D_S$ are strong hypergroups.

Moreover one has $\pi = (1-x)^\alpha(1+x)^\beta\,dx$ and $D_S = D = [-1,1]$ with $dx := \lambda_{[-1,1]}$.

In particular one obtains supp $(\pi) = D_S$.

3.9.2. Orbit hypergroups of the form G_B for a locally compact group G and a relatively compact subgroup B of Aut (G) and their duals $G_B^\wedge$ are strong hypergroups. See Hartmann, Henrichs, Lasser [21] and Hartmann [20] .

In particualar

3.9.3. *Conjugacy hypergroups* of the form G_B with a $G \in [FC]^- \cap SIN$ and $B := I(G)$ are strong hypergroups, and

3.9.4 the orbit hypergroup G_B with $G := \mathbb{R}^p$ and $B := SO(p)$ $(p \geq 2)$ is a strong hypergroup.

For the terminology in Example 3.9.3 see Hartmann [20] .

3.9.5. Arnaud hypergroups $\mathbb{Z}_+$ defined by the sequence $(P_n(z))_{n\geq 1}$ of Arnaud polynomials $P_n(z) = P_n(z|a)$ are *not* strong hypergroups if $a > 2$.

In fact, $\mathbb{Z}_+^\wedge \cong D_S = [-1,1]$ is *not* a hypergroup with respect to pointwise multiplication. Moreover,

$$\pi = \frac{a\sqrt{\frac{4}{a^2}(a-1) - x^2}}{1 - x^2}\, dx \qquad ,$$

and

$$\operatorname{supp}(\pi) = \left[-\frac{2\sqrt{a-1}}{a} , \frac{2\sqrt{a-1}}{a} \right]$$

$$\subsetneq D_S \subsetneq D$$

unless $a = 2$.

II Convergence of convolution products

§ 4 Convolution equations, idempotents

We are going to discuss *convolution equations* of the form

(CE) $\quad \nu * \mu = \mu$

for measures $\nu, \mu \in \mathcal{M}_+^b(K)$, where K is an arbitrary hypergroup admitting a left Haar measure.

For any $\mu \in \mathcal{M}^b(K)$ we introduce the set

$$L(\mu) := \{x \in K : \varepsilon_x * \mu = \mu\}.$$

If $\mu \in \mathcal{M}_+^b(K)$, then $L(\mu)$ is seen to be a compact *set*.

4.1 Theorem. Suppose that $\mu \in \mathcal{M}_+^b(K)$ and let $\nu = \nu^- \in \mathcal{M}^1(K)$ satisfy (CE). Then

$$\varepsilon_x * \mu = \mu$$

for all $x \in [\mathrm{supp}(\nu)]$.

As a consequence we obtain the following important

4.2 Theorem. Suppose that $\mu \in \mathcal{M}_+^b(K)$ is a nontrivial idempotent in the sense that

$$\mu * \mu = \mu \neq 0 .$$

Then μ is the normalized Haar measure of a compact subhypergroup H of K.

The proofs of the two preceding results make use of Fourier transform theory. See Bloom, Heyer [7] . Theorem 4.2 can be established without harmonic analysis, as has been shown by Jewett [26].

4.3 Corollary. Let H be a nonempty compact subset of K satisfying $H*H \subset H$. Then $H^- = H$ and H is a subhypergroup of K.

Applying the corollary we get the desired generalization of Theorem 4.1.

4.4 Theorem. Suppose that $\mu \in \mathcal{M}_+^b(K)$ and let $\nu \in \mathcal{M}^1(K)$ satisfy (CE). Then

$$\varepsilon_x * \mu = \mu$$

for all $x \in [\mathrm{supp}(\nu)]$.

4.5 Corollary. For each $\mu \in \mathcal{M}_+^b(K)$, $L(\mu)$ is a compact *subhypergroup* of K.

Given a subhypergroup H of K we introduce the set

$$\mathcal{M}_H^1(K) := \{\mu \in \mathcal{M}^1(K) : \omega_H * \mu = \mu * \omega_H = \mu\}$$

of H-*invariant probability measures* on K.

4.6 Theorem. Let H be a compact subhypergroup of K with normalized Haar measure ω_H. Suppose the measures $\mu \in \mathcal{M}_H^1(K)$ and $\nu \in \mathcal{M}^1(K)$ satisfy

$$\mu * \nu = \omega_H.$$

Then, for all $x \in \mathrm{supp}(\nu)$ we have

(i) $\mu = \omega_H * \varepsilon_{x^-}$ and

(ii) $\{x^-\} * H * \{x\} \subset H$.

If, moreover, $\nu \in \mathcal{M}_H^1(K)$ then

(iii) $\nu = \varepsilon_x * \omega_H$ and

(iv) $\{x^-\} * H * \{x\} = H$.

A proof of this theorem is contained in Bloom, Heyer [7].

§ 5 Convolution powers

Let K be a hypergroup admitting a left Haar measure ω, which in the case of a compact hypergroup will be assumed to be normalized.

5.1 Preparations. Let $\mu \in \mathcal{M}^1(K)$. We introduce the sets

$Q(\mu) := \{\mu^n : n \in \mathbb{N}\}^-$ and

$A(\mu) :=$ set of weak accumulation points of $Q(\mu)$.

If K is compact then

5.1.1. $Q(\mu)$ is a compact Abelian subsemigroup of $\mathcal{M}^1(K)$.

5.1.2. $A(\mu)$ is the maximal group and the kernel of $Q(\mu)$.

5.1.3. If λ denotes the unit element of $A(\mu)$, then

$$Q(\mu) * \lambda = \lambda * Q(\mu) = A(\mu) = \{\nu^n : \nu = \mu * \lambda = \lambda * \mu,\ n \in \mathbb{N}\}^- .$$

5.1.4. $(\mu^n)_{n \geq 1}$ converges weakly iff $A(\mu) = \{\lambda\}$.

The proofs of these facts do not involve the special structure of a hypergroup.

In what follows, however, the assumption that K be a hypergroup becomes crucial, as the following definition shows.

The results in the remainder of this section are due to Bloom, Heyer [8].

5.2 Definition. A subhypergroup H of K is said to be *supernormal* in K if

$$\{x^-\} * H * \{x\} \subset H$$

for all $x \in K$.

5.3 Remark. If K is a group, then "normal" equals "supernormal", but for general hypergroups K this is no longer the case. Indeed, $\{e\}$ is clearly normal, but generally not supernormal.

5.4 Theorem. Let K be a hypergroup, G(K) a group of measures in $\mathcal{M}^1(K)$ with unit element λ, and write $H := \text{supp}(\lambda)$ and $K_o := (\text{supp}\, G(K))^-$.

Then

(i) K_o is a subhypergroup of K.

(ii) H is a compact supernormal subhypergroup of K_o.

(iii) For all $\nu \in G(K)$ and $x \in \text{supp}(\nu)$ we have

$$\nu = \lambda * \varepsilon_x = \varepsilon_x * \lambda .$$

5.5 Corollary. Let K be a compact hypergroup, $\mu \in \mathcal{M}^1(K)$ and let λ denote the unit element of $A(\mu)$ with $H := \text{supp}(\lambda)$. Then $\text{supp}(\mu)$ is contained in an H-hypercoset in $\text{supp}(A(\mu))^-$ and hence is an H-hypercoset in K.

5.6 Theorem. Let K be a compact hypergroup, $\mu \in \mathcal{M}^1(K)$.

Then

$$\text{supp}(A(\mu))^- = \overline{\lim}\, \text{supp}(\mu^n) = \text{supp}(Q(\mu)).$$
$$= [\text{supp}(\mu)]^- .$$

5.7 Theorem (Itô, Kawada). Let K be a compact hypergroup, $\mu \in \mathcal{M}^1(K)$ and let λ denote the unit element of $A(\mu)$ with $H := \text{supp}(\lambda)$. The following conditions are equivalent:

(i) $\mathcal{T}_w\text{-}\lim \mu^n$ exists.

(ii) $\lim \text{supp}(\mu^n)$ exists.

(iii) $[\mathrm{supp}(\mu)] = [\bigcup_{n\geq 1} \mathrm{supp}(\mu * \mu^-)^n]$.

(iv) $\mathrm{supp}(\mu)$ is not contained in a hypercoset of any proper supernormal subhypergroup of $[\mathrm{supp}(\mu)]$.

(v) $\mathrm{supp}(\mu)$ is not contained in any proper hypercoset of H in $[\mathrm{supp}(\mu)]$.

(vi) $H = [\mathrm{supp}(\mu)]$.

5.8 Corollary. Let K be a compact hypergroup and let $\mu \in \mathfrak{M}^1(K)$.

The following statements are equivalent:

(i) $\mathcal{T}_w\text{-}\lim \mu^n = \omega$.

(ii) $\underline{\lim}\, \mathrm{supp}\,(\mu^n) = K$.

5.9 Theorem. Let K be a compact connected hypergroup and let $\mu \in \mathfrak{M}^1(K)$ satisfy the conditions $\mu^- = \mu$ and $\mu \ll \omega$. Then
$\mathcal{T}_w\text{-}\lim_{n\to\infty} \mu^n = \omega$.

5.10 Theorem (Convergence in the noncompact case). Let K be a hypergroup and $\mu \in \mathfrak{M}^1(K)$ with $e \in \mathrm{supp}(\mu)$, $\mu^- = \mu$ and $[\mathrm{supp}(\mu)]$ a Moore hypergroup. Suppose that $(\mu^n)_{n\geq 1}$ admits a weakly convergent subnet in $\mathfrak{M}^1(K)$.

Then $[\mathrm{supp}(\mu)]$ is a compact subhypergroup of K and the limit of the subnet is its normalized Haar measure.

§ 6 General convolution products

We keep the assumption that K is a hypergroup admitting a left Haar measure. If one replaces the sequence $(\mu^n)_{n\geq 1}$ studied in the preceding section (for some $\mu \in \mathcal{M}^1(K)$) by the sequence $(\nu_n)_{n\geq 1}$ of n-fold convolution products

$$\nu_n := \mu_1 * \ldots * \mu_n$$

(for a given sequence $(\mu_j)_{j\geq 1}$ in $\mathcal{M}^1(K)$), not much is known concerning the limiting behavior whenever $n\to\infty$. We shall present a limiting result at least for normal sequences in the sense of the following

6.1 Definition. A sequence $(\mu_j)_{j\geq 1}$ in $\mathcal{M}^1(K)$ is said to be *normal* if for every $n\geq 1$ there exists a strictly increasing sequence $(j_\ell)_{\ell\geq 1}$ in $\mathbb{N}$ such that $\mu_s = \mu_{j_\ell + s}$ for all $s = 1,\ldots,n$ and $\ell \in \mathbb{N}$.

Clearly, the sequence $(\mu_j)_{j\geq 1}$ with $\mu_j := \mu \in \mathcal{M}^1(K)$ for all $j\geq 1$ is normal.

6.2 Theorem. Let K be a compact hypergroup, $(\mu_j)_{j\geq 1}$ a normal sequence in $\mathcal{M}^1(K)$, and let $(\nu_n)_{n\geq 1}$ be the corresponding sequence of n-fold products. Assume that $\mathcal{T}_w\text{-}\lim\limits_{n\to\infty} \nu_n = \nu$.

Then $\nu = \omega_H$ for $H := [\bigcup\limits_{j\geq 1} \operatorname{supp}(\mu_j)]$.

6.3 Theorem. Let K be a compact hypergroup, $(\mu_j)_{j\geq 1}$ a normal sequence in $\mathcal{M}^1(K)$ and $\nu_n := \mu_1 * \ldots * \mu_n$ for all $n\geq 1$.

The following statements are equivalent:

(i) $\mathcal{T}_w\text{-}\lim\limits_{n\to\infty} \nu_n$ exists.

(ii) $[\bigcup\limits_{j\geq 1} \operatorname{supp}(\mu_j)] = [\bigcup\limits_{n\geq 1} \operatorname{supp}(\mu_1 * \mu_2 * \ldots * \mu_n * \mu_n^- * \ldots * \mu_2^- * \mu_1^-)]$.

In the proofs of both results the method of Fourier transforms is applied. See Bloom, Heyer [8].

If for every given sequence $(\mu_j)_{j\geq 1}$ in $\mathfrak{M}^1(K)$ the corresponding sequence $(\nu_n)_{n\geq 1}$ of n-fold convolution products

$$\nu_n := \mu_1 * \ldots * \mu_n$$

converges at least after suitable shifting, then the underlying hypergroup will be of very special kind. We are going to justify this statement by quoting two results.

6.4 Definition. A hypergroup K is said to have the *normed convergence property* if for every sequence $(\mu_j)_{j\geq 1}$ in $\mathfrak{M}^1(K)$ there exists a sequence $(x_j)_{j\geq 1}$ in K such that the sequence $(\nu_n)_{n\geq 1}$ defined by

$$\nu_n := \mu_1 * \ldots * \mu_n * \varepsilon_{x_n}$$

converges weakly in $\mathfrak{M}^1(K)$.

6.5 Theorem. Let K be a compact hypergroup, $(\mu_j)_{j\geq 1}$ a sequence in $\mathfrak{M}^1(K)$ and write

$$\nu_n := \mu_1 * \ldots * \mu_n$$

for each $n \in \mathbb{N}$. If ν', ν'' are two weak accumulation points of $(\nu_n)_{n\geq 1}$ then there exists an $x \in K$ such that $\nu' = \nu'' * \varepsilon_x$.

6.6 Theorem (Normed Convergence Characterization). Let K be a Moore hypergroup with a countable basis for its topology.

The following statements are equivalent:

(i) K has the normed convergence property.

(ii) K is compact.

Proofs of these two theorems are given in Bloom, Heyer [8].

III Convolution semigroups

§ 7 Embedding of infinitely divisible measures.

The problem of embedding infinitely divisible probability measures on a hypergroup K at least in rational convolution semigroups on K has been studied by Bloom in [6]. The approach is similar to that for locally compact groups.

Again we assume the given hypergroup K to admit a left Haar measure.

7.1 Definition. Let $n \in \mathbb{N}$. A hypergroup K is called *n-root compact* (written $K \in \mathfrak{R}_n$) if the following condition holds: For every compact $C \subset K$ there exists compact $C_n \subset K$ such that all finite sets $\{x_1, x_2, \ldots, x_n\}$ in K with $x_n = e$ satisfying

$$\{x_i\} * C * \{x_j\} * C \cap \{x_{i+j}\} * C = \emptyset$$

for $i+j \leq n$ are contained in C_n.

Write $\mathfrak{R} := \bigcap_{n \geq 1} \mathfrak{R}_n$ for the class of all root compact hypergroups.

For each subset $\mathfrak{N}$ of $\mathfrak{M}^1(K)$ and every $C \subset K$ we introduce the set

$$R(n, \mathfrak{N}) := \{\mu \in \mathfrak{M}^1(K) : \mu^n \in \mathfrak{N}\}.$$

Moreover we put

$$C^{\frac{1}{n}} := \{x \in K : \{x\}^n \subset C\}.$$

7.2 Theorem. Let K be a hypergroup and consider the following conditions:

(i) $K \in \mathfrak{R}_n$.

(ii) $R(n, \mathfrak{N})$ is relatively compact for each relatively compact

$\mathcal{N} \subset \mathcal{M}^1(K)$.

(iii) $C^{\frac{1}{n}}$ is compact for every $C \subset K$.

Then (i) => (ii) => (iii).

In order to reverse the implications of the preceding theorem we need to restrict the class of hypergroups involved.

For $x, y \in K$ we introduce the equivalence relation

$$x \sim y :\Longleftrightarrow \text{There exist } z_1, \ldots, z_n \in K \text{ such that } y \in Z_n * \{x\} * Z_n^- , \text{ where } Z_n := \{z_1\} * \ldots * \{z_n\} .$$

For every $x \in K$ the set

$$C_x := \{y \in K : y \sim x\}$$

is said to be the *conjugacy class of* K *containing* x.

A set $F \subset K$ is called *H-invariant* for a set $H \subset K$ if $\{x\} * F * \{x^-\} \subset F$ for all $x \in H$.

It is easily seen that every class C_y is in fact K-invariant

With this notation *class compact hypergroups* can be introduced as hypergroups whose conjugacy classes are relatively compact.

One notes that in general not even commutative hypergroups are class compact.

7.3 Corollary. If K is assumed to be class compact and has a compact K-invariant neighborhood, then the statements (i) to (iii) of the theorem are equivalent.

7.4 <u>Definition.</u> A measure $\mu \in \mathfrak{M}^1(K)$ is called *infinitely divisible* if for every $n \in \mathbb{N}$ there exists $\mu_n \in \mathfrak{M}^1(K)$ such that $\mu_n^n = \mu$.

The set of infinitely divisible measures in $\mathfrak{M}^1(K)$ will be denoted by $\mathfrak{I}(K)$.

7.5 <u>Theorem.</u> If $K \in \mathfrak{R}$ then $\mathfrak{I}(K)$ is weakly closed in $\mathfrak{M}^1(K)$.

Let $\mathbb{K}$ denote either of the cones $\mathbb{Q}_+$ and $\mathbb{R}_+$.

7.6 <u>Definition.</u> A family $(\mu_t)_{t \in \mathbb{K}}$ of measures in $\mathfrak{M}^1(K)$ is called a (continuous) *convolution semigroup* on K if

(SG 1) $\mu_t * \mu_s = \mu_{t+s}$ for all $t, s \in \mathbb{K}$.

(SG 2) $\lim\limits_{t \to o} \mu_t = \varepsilon_e$ in the sense of the vague topology $\mathfrak{T}_v$.

In the case $\mathbb{K} := \mathbb{Q}_+$ we talk about *rational* convolution semigroups, in the case $\mathbb{K} := \mathbb{R}_+$ about *real* convolution semigroups. For $(\mu_t)_{t \in \mathbb{R}_+}$ we usually write $(\mu_t)_{t \geq o}$.

7.7 <u>Definition.</u> A measure $\mu \in \mathfrak{M}^1(K)$ is said to be *rationally embeddable* if there exists a rational convolution semigroup $(\mu_t)_{t \in \mathbb{Q}_+}$ on K such that $\mu_1 = \mu$.

7.8 <u>Theorem.</u> If $K \in \mathfrak{R}$ then every $\mu \in \mathfrak{I}(K)$ is rationally embeddable.

7.9 <u>Application.</u> The double coset hypergroup $K := SO(d+1)//SO(d)$ is compact. Therefore $K \in \mathfrak{R}$ and the assertions of Theorems 7.5 and 7.8 are established in this case.

7.10 <u>Remark.</u> We note, however, that the double coset hypergroup $SL(2,\mathbb{R})//SO(2,\mathbb{R})$ is not in $\mathfrak{R}_n$ for any $n \geq 2$. Indeed, if it were for some n then we would deduce that $SL(2,\mathbb{R}) \in \mathfrak{R}_n$ which is not the case. Here one uses the fact that for a compact subhypergroup H of K, $K//H \in \mathfrak{R}_n$ implies $K \in \mathfrak{R}_n$.

Typical examples of infinitely divisible measures in $\mathcal{M}^1(K)$ which are continuously embeddable are the (compound) Poisson measures introduced as follows

7.11 Definition. A measure $\mu \in \mathcal{M}^1(K)$ is said to be a *Poisson measure* if there exists a measure $\nu \in \mathcal{M}_+^b(K)$ such that

$$\mu = \exp(-\|\nu\|)\exp(\nu)$$

where $\exp(\nu)$ is given in $\mathcal{M}^b(K)$ as a norm convergent series starting with ε_e .

The class of all Poisson measures on K will be denoted by $\mathcal{P}ois(K)$.

As mentioned above we have $\mathcal{P}ois(K) \subset \mathcal{I}(K)$. For the reverse inclusion we restrict ourselves to commutative hypergroups K. In this case any $\mu \in \mathcal{P}ois(K)$ of the form

$$\mu = \exp(-\|\nu\|)\exp(\nu)$$

for $\nu \in \mathcal{M}_+^b(K)$ has a Fourier transform of the form

$$\hat{\mu} = \exp\left[-(\|\nu\|-\hat{\nu})\right].$$

The following result and its consequences are due to Gallardo and Gebuhrer [16].

7.12 Theorem. Let K be a countably discrete commutative hypergroup with connected (compact) dual $X^\wedge$ and such that $\mathrm{supp}(\pi)$ contains the unit character $\chi = 1$.

Then $\mathcal{I}(K) = \mathcal{P}ois(K)$.

More precisely for every $\mu \in \mathcal{I}(K)$ there exists a unique measure $\nu \in \mathcal{M}_+^b(K)$ with $\nu(\{e\}) = 0$ such that

$$\hat{\mu}(\chi) = \exp\left[- \sum_{x \in K} (1 - \overline{\chi(x)})\nu(\{x\}) \right]$$

for all $\chi \in K$.

7.13 <u>Application.</u> Let $(\mu_{nj})_{j=1,\dots,j_n;\; n\geq 1}$ be a triangular system of measures in $\mathfrak{M}^1(K)$ which is

(a) *infinitesimal* in the sense of

$$\lim_{n\to\infty} \sup_{1\leq j\leq j_n} \mu_{nj}(K\setminus\{e\}) = o ,$$

and

(b) *convergent* in the sense of

$$\mathfrak{T}_v\text{-}\lim_{n\to\infty} \mathop{\star}_{j=1}^{j_n} \mu_{nj} = \mu .$$

Under the assumption that either the measures μ_{nj} $(j=1,\dots,j_n; n\geq 1)$ are symmetric or that they satisfy the condition

$$\sup_{n\geq 1} \sum_{j=1}^{j_n} (1 - \mu_{nj}(\{e\}) < \infty$$

we obtain $\mu \in \mathfrak{I}(K)$.

7.14 <u>Remark.</u> For the ultraspherical hypergroup $\mathbb{Z}_+$ of Example 1.3.2 Theorem 7.12 follows from the discussion in Kennedy [27].

§ 8 Positive and negative definite functions.

Let K be an arbitrary hypergroup.

8.1 Definition. A continuous complex-valued function ϕ on K is called *positive definite* if for each choice of complex numbers $c_1,\dots,c_n$ and points $x_1,\dots,x_n \in K$ one has

$$\sum_{i=1}^{n}\sum_{j=1}^{n} c_i\bar{c}_j\,\phi(x_i * x_j^-) \geq 0 .$$

By $\mathcal{P}(K)$ we shall denote the set of all positive definite functions on K.

8.2 Properties of $\phi \in \mathcal{P}(K)$.

8.2.1 $\phi(e) \geq 0$.

8.2.2 $\phi(x*x^-) \geq 0$ for all $x \in K$.

8.2.3 $\phi(x^-) = \overline{\phi(x)}$ for all $x \in K$.

8.2.4 $\phi(e) = \|\phi\|$ whenever ϕ is bounded.

It has been shown in Jewett [26], that bounded functions in $\mathcal{P}(K)$ correspond to representations in Rep(K). If K is a commutative hypergroup, then the obvious generalization of Bochner's theorem holds, i.e. there exists a one-to-one correspondence between bounded functions $\phi \in \mathcal{P}(K)$ and measures $\mu \in \mathcal{M}_+^b(K^\wedge)$ given by

$$\check{\mu} = \phi .$$

For the proof of this and related results see Jewett [26] .

8.3 Definition. A continuous complex-valued function ϕ on K is said to be *negative definite* if for each choice of complex numbers $c_1,\dots,c_n$ and points $x_1,\dots,x_n \in K$ one has

$$\sum_{i=1}^{n}\sum_{j=1}^{n} c_i\bar{c}_j(\psi(x_i) + \overline{\psi(x_j)} - \psi(x_i * x_j^-)) \geq 0.$$

Let us abbreviate the set of all negative definite functions on K by $\mathcal{N}(K)$.

8.4 <u>Properties</u> of $\psi \in \mathcal{N}(K)$.

8.4.1 $\psi(e) \geq 0$.

8.4.2 $\psi(x * x^-)$ is real.

8.4.3 $\psi(x^-) = \overline{\psi(x)}$ for all $x \in K$.

8.4.4 $\psi(x) + \psi(x^-) \geq \psi(x * x^-)$.

The following characterization of negative definite functions is proved as in the case of a locally compact group.

8.5 <u>Theorem.</u> For any complex-valued function ψ on a hypergroup K the following statements are equivalent:

(i) $\psi \in \mathcal{N}(K)$.

(ii) (a) ψ is continuous.

(b) $\psi(e) \geq 0$.

(c) $\psi(x^-) = \overline{\psi(x)}$ for all $x \in K$.

(d) For each choice of $c_1, \ldots, c_n \in \mathbb{C}$ with $\sum_{i=1}^{n} c_i = 0$ and each choice of $x_1, \ldots, x_n \in K$ we have

$$\sum_{i=1}^{n}\sum_{j=1}^{n} c_i\bar{c}_j\psi(x_i * x_j^-) \leq 0 .$$

8.6 <u>Consequences.</u>

8.6.1 If $\psi \in \mathcal{N}(K)$, then $\psi - \psi(e) \in \mathcal{N}(K)$.

8.6.2 If $\psi \in \mathcal{P}(K)$, then $\psi(e) - \psi \in \mathcal{N}(K)$.

8.6.3 Let ψ be a continuous complex-valued function on K with $\psi(e)\geq o$ and such that for each $t \in \mathbb{R}_+^*$ the function $\exp(-\psi)$ belongs to $\mathcal{P}(K)$. Then $\psi \in \mathcal{N}(K)$.

The following list of examples of negative definite functions is due to Lasser [30].

8.7 Types of negative definite functions.

8.7.1 $\alpha 1 \in \mathcal{N}(K)$ for all $\alpha \geq o$.

8.7.2. Let $\ell: K \to \mathbb{R}$ be a *homomorphism* in the sense that

(a) $\ell(x^-) = -\ell(x)$ and

(b) $\ell(x*y) = \ell(x) + \ell(y)$ for all $x,y \in K$.

Then $q := i\ell \in \mathcal{N}(K)$.

8.7.3. Let $q \in \mathcal{C}_+(K)$ be a *quadratic form* in the sense that

$$q(x*y) + q(x*y^-) = 2\,[q(x) + q(y)]$$

for all $x,y \in K$. Then $q \in \mathcal{N}(K)$.

8.8 Example. Let $K := \mathbb{Z}_+$ denote the polynomial hypergroup defined in 1.3 by the sequences $(a_n)_{n\geq 1}$, $(b_n)_{n\geq 1}$ and $(c_n)_{n\geq 1}$:

8.8.1. The only homomorphism on K is the function identically zero.

8.8.2. The defining property of quadratic forms is equivalent to

$$q(m*n) = q(m) + q(n)$$

whenever $m,n \in K$.

8.8.3. Any nonnegative quadratic form q on K is of the form

$$q(n) = s_n \cdot a$$

for all $n \in K$ and some $a \in \mathbb{R}_+$, where

$$s_o := o,$$

$$s_1 := 1, \quad \text{and}$$

$$s_{n+1} := \frac{1}{a_n}\left(1+(1-b_n)s_n - c_n s_{n-1}\right).$$

for all $n \geq 1$.

Conversely, any function q of the form $q(n) := s_n \cdot a$ for all $n \in K$ with $a \in \mathbb{R}_+$ is a nonnegative quadratic form on K.

8.9 <u>Subexample.</u> Let $K := \mathbb{Z}_+$ be an ultraspherical hypergroup in the sense of 1.3.2, given by

$$a_n := \frac{n + 1 + 2\alpha}{2n + 1 + 2\alpha},$$

$$b_n := o \quad \text{for all} \quad n \in \mathbb{Z}_+, \text{ and}$$

$$c_n := \frac{n}{2n + 1 + 2\alpha},$$

where $\alpha \geq -\frac{1}{2}$. Then any nonnegative quadratic form q on K is of the form

$$q(n) = s_n \cdot a$$

with $a \in \mathbb{R}_+$, where

$$s_n := \frac{n(n+1+2\alpha)}{2 + 2\alpha}$$

whatever $n \in K$.

§ 9 Schoenberg's correspondence theorem

From now on we assume that K is a commutative hypergroup such that also its dual $K^\wedge$ is a hypergroup with respect to pointwise multiplication.

The following results have been established for measures in $\mathcal{M}^{(1)}_+(K)$ by Lasser [30]. For their specialization to measures in $\mathcal{M}^1(K)$ we need the notion of a *normed* negative definite function ψ on $K^\wedge$ given by the additional property $\psi(1) = o$.

9.1 Theorem (Schoenberg's theorem part I). For every convolution semigroup $(\mu_t)_{t\geq o}$ there exists exactly one normed negative definite function $\psi \in \mathcal{N}(K^\wedge)$ satisfying

(i) $\operatorname{Re} \psi \geq o$.

(ii) $\hat{\mu}_t = \exp(-t\psi)$ for all $t \geq o$.

Clearly the proofs of this and the following results are performed in the spirit of the case that K is an Abelian group.

9.2 Convention. The function ψ on $K^\wedge$ introduced in the theorem will be called the *negative definite function corresponding* to the convolution semigroup $(\mu_t)_{t\geq o}$.

9.3 Examples.

9.3.1 Let $\nu \in \mathcal{M}^b_+(K)$. For every $t \geq o$ we define the measure

$$\mu_t := \exp(-t\|\nu\|)\exp(t\nu)$$

where $\exp(t\nu)$ is given in $\mathcal{M}^b(K)$ as a norm convergent series (with $\mu_o = \varepsilon_e$).

$(\mu_t)_{t\geq o}$ is called the *Poisson semigroup on* K *with defining measure* ν .

We note that

$$\hat{\mu}_t = \exp[-t(\|\nu\|-\hat{\nu})]$$

for all $t\geq o$, whence that $\|\nu\|-\hat{\nu}$ is the negative definite function corresponding to $(\mu_t)_{t\geq o}$.

9.3.2. If we specialize ν of 9.3.1 to be the measure ε_{x_o} for some $x_o \in K$ we obtain the (elementary) *Poisson semigroup* $(\mu_t)_{t\geq o}$ *with parameter* x_o, whose corresponding negative definite function ψ is given by

$$\psi(\chi) = 1 - \chi(x_o)$$

for all $\chi \in K^\wedge$.

A partial converse of the preceding theorem is

9.4 Theorem (Schoenberg's theorem part II). Let K be a strong hypergroup. For every normed negative definite function $\psi \in \mathcal{N}(K^\wedge)$ such that the conditions

(i) Re $\psi\geq o$, and

(ii) $\phi_t := \exp(-t\psi) \in \mathcal{P}(K^\wedge)$ for all $t\geq o$

are satisfied, there exists exactly one convolution semigroup $(\mu_t)_{t\geq o}$ on K with the property

$$\hat{\mu}_t = \phi_t$$

for all $t\geq o$.

9.5 Convention. The family $(\mu_t)_{t\geq o}$ of measures in $\mathcal{M}^1(K)$ introduced in the theorem will be called the *convolution semigroup corresponding to* the negative definite function ψ .

9.6 Remark. In the sense of Schoenberg's theorem parts I and II and under their joint assumptions we have a one-to-one correspondence

$$(\mu_t)_{t\geq o} \longleftrightarrow \psi$$

between convolution semigroups $(\mu_t)_{t\geq o}$ on K and certain normed negative definite functions ψ on $K^\wedge$ given by

$$\hat{\mu}_t = \exp(-t\psi)$$

for all $t\geq o$.

9.7 Applications. Let us consider the ultraspherical hypergroup $\mathbb{Z}_+$ of Example 1.3.2 and its dual hypergroup $\mathbb{Z}_+^\wedge \cong D_S = D = [-1,1]$ discussed in Example 3.9.1.

9.7.1. Let $(\mu_t)_{t\geq o}$ be a Poisson convolution semigroup on $\mathbb{Z}_+$ of the form

$$\mu_t = \exp(-t\|\nu\|)\exp(t\nu)$$

for some $\nu = \sum_{n\geq o} r_n\varepsilon_n \in \mathcal{M}_+^b(\mathbb{Z}_+)$ with $r_n \in \mathbb{R}_+$ and $\sum_{n\geq o} r_n < \infty$ $(t\geq o)$.

Then by Example 9.3.1 we obtain

$$\hat{\mu}_t(\chi) = \exp\left[-t\sum_{n\geq o} r_n(1-\hat{\varepsilon}_n(\chi))\right]$$

for all $\chi \in \mathbb{Z}_+^\wedge$ or in the case of Example 1.3.2

$$\hat{\mu}_t(x) = \exp\left[-t\sum_{n\geq o} r_n(1-P_n^\alpha(x))\right]$$

for all $x \in [-1,1]$, where $(P_n^\alpha)_{n\geq o}$ denotes the sequence of ultraspherical polynomials (for $\alpha \geq -\frac{1}{2}$) .

In the special case of an elementary Poisson convolution semigroup $(\mu_t)_{t\geq o}$ on $\mathbb{Z}_+$ with parameter x_o we get

$$\hat{\mu}_t(x) = \exp\left[-t(1-P^{\alpha}_{x_o}(x))\right]$$

for all $x \in [-1,1]$.

For a generalization of these examples from Fourier transforms to more general positive definite functions one might consult Kennedy [27] and Bingham [5].

9.7.2. Let $(\mu_t)_{t\geq o}$ be the convolution semigroup on $\mathbb{Z}_+^\wedge$ corresponding to the normed negative definite function

$$n \to q(n) := s_n \cdot a$$

on $\mathbb{Z}_+$, where $a \in \mathbb{R}_+$ and

$$s_n := \frac{n(n+1+2\alpha)}{2+2\alpha}$$

as in Subexample 8.9. We realize that q is a nonnegative quadratic form on $\mathbb{Z}_+$.

$(\mu_t)_{t\geq o}$ is called the *Brownian semigroup on* $\mathbb{Z}_+^\wedge$. In order to establish its existence by Theorem 9.4 we have to show that the functions

$$n \to \phi_t(n) := \exp(-tq(n))$$

on $\mathbb{Z}_+$ $(t\geq o)$ are positive definite. Clearly the functions

$$n \to \exp\left[-t(1-P^{\alpha}_n(x))\right]$$

are positive definite for all $x \in [-1,1]$ $(t\geq o)$. Let $x \in [-1,1]$. Replacing t by $t(1-x)^{-1}$ we obtain that the functions

$$n \to \exp\left(-t\,\frac{1-P^{\alpha}_n(x)}{1-x}\right)$$

are positive definite $(t\geq o)$. But

$$\lim_{x\to 1} \frac{1 - P_n^\alpha(x)}{1 - x} = \frac{n(n + 2\alpha + 1)}{2 + 2\alpha} = s_n .$$

yields the assertion.

For the origin of this reasoning see Bochner [9], [10] and Lasser [30].

In Letac [33] we find a similar discussion for the dual $[-1,1]$ of the <u>Arnaud hypergroup</u> $\mathbb{Z}_+$, which as we know from Example 3.8.5 is *not* a hypergroup with respect to pointwise multiplication. Nevertheless the convolution in $[-1,1]$ can be computed as there exists for each pair $z_1, z_2 \in [-1,1]$ a measure $\varepsilon_{z_1} * \varepsilon_{z_2} \in \mathfrak{M}^1([-1,1])$ such that for all $n \geq 1$

$$P_n(z_1|q)P_n(z_2|q) = \int_{[-1,1]} P_n(z|q)\varepsilon_{z_1} * \varepsilon_{z_2}(dz)$$

holds. Here $(P_n(\cdot|q))_{n\geq 1}$ denotes the sequence of Arnaud polynomials (for $q=a-1>1$). The corresponding computation and the following theorem are due to Letac [33].

9.8 <u>Theorem.</u> For every $n \geq 1$ let

$$Q_n(x) := \begin{cases} \dfrac{1 - P_n(x|q)}{1 - x} & \text{if } x \neq 1 \\[2ex] \dfrac{q+1}{q-1}\left[n - \dfrac{2q}{(q-1)^2}(1 - q^{-n})\right] & \text{if } x=1 . \end{cases}$$

Then, given a convolution semigroup $(\mu_t)_{t\geq o}$ in $\mathfrak{M}^1([-1,1])$ there exists a unique measure $\eta \in \mathfrak{M}_+^b([-1,1])$ such that

$$\int_{[-1,1]} P_n(x|q)\mu_t(dx) = \exp(-t \int_{[-1,1]} Q_n(x)\eta(dx))$$

for all $t \geq o$.

Now one can introduce the *Brownian semigroup* on $[-1,1]$ as a convolution semigroup $(\mu_t)_{t\geq o}$ in $\mathcal{M}^1([-1,1])$ given by

$$\hat{\mu}_t(n) := \exp\left[-t\left(n - \frac{2q}{(q-1)^2}(1-q^{-n})\right)\right]$$

for all $n \in \mathbb{Z}_+$.

Obviously the Brownian semigroup $(\mu_t)_{t\geq o}$ on $[-1,1]$ corresponds to the representing measure $\eta := \frac{q-1}{q+1}\varepsilon_1$ on $[-1,1]$.

9.9 Remarks.

9.9.1 One has a statement of *central limit type*

$$\mathcal{T}_w\text{-}\lim_{n\to\infty} \varepsilon^n_{1-\frac{h}{n}} = \mu_{h\frac{q-1}{q+1}}$$

whatever h>o, which clearly contrasts the

9.9.1 *ergodicity result* that for any measure $\mu \in \mathcal{M}^1([-1,1])$ which is not concentrated on $\{-1,+1\}$

$$\mathcal{T}_w\text{-}\lim_{n\to\infty} \mu^n = \pi,$$

where π denotes the Plancherel measure of the Arnaud hypergroup.

§ 10 The Lévy-Khintchine representation

Let K be a commutative hypergroup such that its dual $K^\wedge$ is a hypergroup with respect to pointwise multiplication and with Haar measure π. We introduce the set

$$\mathcal{S} := \{\mu \in \mathcal{M}^1(K^\wedge) : \mu = \mu^- , \ \mathrm{supp}(\mu) \text{ is compact}\}.$$

We want to apply this set in the discussion of a Lévy-Khintchine representation for convolution semigroups $(\mu_t)_{t\geq o}$ on K with corresponding normed negative definite functions ψ on $K^\wedge$. The following approach due to Lasser [30] has been inspired by the analoguous treatment in the case of an Abelian group K.

10.1 Construction of the Lêvy measure.

Let $\sigma \in \mathcal{S}$. For all $\chi \in K$ one obtains

$$\begin{aligned}\frac{1}{t}\,[(1-\check{\sigma})\cdot\mu_t\,]^\wedge(\chi) &= \frac{1}{t}(\hat{\mu}_t(\chi) - \int_{K^\wedge}\int_{K^\wedge} \hat{\mu}_t(\xi)\,\varepsilon_\chi * \varepsilon_{\zeta^-}(d\xi)\,\sigma(d\zeta)) \\ &= \frac{1}{t}(\hat{\mu}_t(\chi) - \hat{\mu}_t * \sigma(\chi)) \\ &= \frac{1}{t}(1 - \exp(-t\psi)) * (\sigma - \varepsilon_1)(\chi).\end{aligned}$$

Moreover one has

$$\lim_{t\to o} \frac{1}{t}(1 - \exp(-t\psi)) = \psi.$$

uniformly on compact subsets of $K^\wedge$.

Consequently

$$\lim_{t\to o} \frac{1}{t}\,[(1-\check{\sigma})\cdot\mu_t\,]^\wedge = \psi * \sigma - \psi$$

uniformly on compact subsets of $K^\wedge$, and $\psi * \sigma - \psi$ is a bounded function in $\mathcal{P}(K^\wedge)$. Therefore

$$\mathcal{T}_v\text{-}\lim_{t\to o}\frac{1}{t}\left[(1-\check{\sigma})\cdot\mu_t\right]^\wedge\cdot\pi = (\psi*\sigma-\psi)\cdot\pi .$$

From Theorem 3.2 follows that there exists a measure $\mu_\sigma \in \mathcal{M}_+(K)$ such that

$$\hat{\mu}_\sigma = \psi*\sigma - \psi$$

and

$$\mathcal{T}_w\text{-}\lim_{t\to o}\frac{1}{t}(1-\check{\sigma})\cdot\mu_t = \mu_\sigma .$$

Next one shows that there exists a measure $\eta \in \mathcal{M}_+(K\setminus\{e\})$ such that

$$\eta = \mathcal{T}_v\text{-}\lim_{t\to o}\frac{1}{t}\mu_t\Big|_{K\setminus\{e\}}$$

and

$$(1-\check{\sigma})\cdot\eta = \mu_\sigma\Big|_{K\setminus\{e\}}$$

for all $\sigma \in \mathcal{S}$. The measure η turns out to be uniquely determined by the convolution semigroup $(\mu_t)_{t\geq o}$; it is called the *Lévy measure* of $(\mu_t)_{t\geq o}$.

10.2 Integration with respect to the Lévy measure.

First of all we note that the Lévy measure η corresponding to the given convolution semigroup $(\mu_t)_{t\geq o}$ on K satisfies the following conditions

10.2.1. $\int_{K\setminus\{e\}}(1-\mathrm{Re}\chi)d\eta < \infty$ for all $\chi \in K^\wedge$.

10.2.2. For any compact neighborhood V of e in K

$$\eta\Big|_{K\setminus V} \in \mathcal{M}_+^b(K\setminus V).$$

Next we describe homomorphisms and nonnegative quadratic forms in $\mathcal{N}(K^\wedge)$ in terms of the set $\mathcal{S}$.

10.2.3. A function $\ell \in \mathcal{L}(K^\wedge, \mathbb{R})$ satisfying $\ell(1) = o$ is a homomorphism iff

$$\ell * \sigma - \ell = o$$

for all $\sigma \in \mathcal{S}$.

10.2.4. A function $q \in \mathcal{L}(K^\wedge, \mathbb{R})$ satisfying

(a) $q(\chi) = q(\chi^-)$ for all $\chi \in K^\wedge$, and

(b) $q(1) = o$

is quadratic iff

$$q * \sigma - q \quad \text{is constant}$$

for all $\sigma \in \mathcal{S}$.

In the affirmative case $q \gtrsim o$ iff

$$q * \sigma - q \geqq o$$

for all $\sigma \in \mathcal{S}$.

The following discussion will be limited to convolution semigroups $(\mu_t)_{t \gtrsim o}$ on K having a *symmetric* Lévy measure.

10.2.5. If η is symmetric, then $\mathrm{Im}\psi$ is a homomorphism. In particular, $i\,\mathrm{Im}\psi$ is a homomorphism.

10.2.6. If η is symmetric, then η is the Lévy measure of the *symmetrized convolution semigroup* $(\nu_t)_{t \geqq o}$ on K defined by

$$\nu_t := \mu_{\frac{t}{2}} * \mu^-_{\frac{t}{2}}$$

for all $t \geqq o$.

10.3 Auxiliary Property (F).

Again we learn from the special case of a locally compact group K what estimates have to be provided in order to obtain the principle results.

10.3.1 Definition. The underlying hypergroup K is said to have *Property* (F) if for any compact subset C of $K^\wedge$ there exist a constant $M_C \geq 0$, a neighborhood U_C of e in K and a finite subset N_C of C such that for each $x \in U_C$ we have

$$\sup\{1 - \mathrm{Re}\chi(x) : \chi \in C\} \leq M_C \cdot \sup\{1 - \mathrm{Re}\chi(x) : \chi \in N_C\}.$$

10.3.2 Examples of hypergroups having Property (F) are

10.3.2.1 the compact hypergroups

10.3.2.2 the discrete hypergroups

10.3.2.3. the hypergroups G_B of Example 1.2 and their duals.

In the proof of the latter statement some structure theory of $[FC]^-$-groups is applied.

10.4 Main Lemma. Let K have Property (F). We are given a symmetric positive Radon measure μ on $K \setminus \{e\}$ satisfying the conditions 10.2.1. and 10.2.2. Then the function $\psi_\mu : K^\wedge \to \mathbb{R}$ defined by

$$\psi_\mu(\chi) := \int_{K\setminus\{e\}} (1 - \mathrm{Re}\chi(x))\mu(dx)$$

for all $\chi \in K^\wedge$ belongs to $\mathcal{N}(K^\wedge)$.

10.5 Theorem (Lévy-Khintchine representation in the symmetric case). Let K be a commutative hypergroup such that $K^\wedge$ is a hypergroup with respect to pointwise multiplication. We assume that K has Property (F).

Let $(\mu_t)_{t\geq 0}$ be a convolution semigroup on K with corresponding normed negative definite function ψ on $K^\wedge$ and Lévy measure η on $K\setminus\{e\}$ We assume that η is symmetric. Then

(i) ψ has the form

$$\psi(\chi) = i\ell(\chi) + q(\chi) + \int_{K\setminus\{e\}} (1 - \mathrm{Re}\,\chi(x))\eta(dx)$$

for all $\chi \in K^\wedge$, where ℓ is a homomorphism and q a nonnegative quadratic form on $K^\wedge$,

and

(ii) $(\mu_t)_{t\geq o}$ determines ℓ and q in (i) uniquely, i.e.

$\ell = \mathrm{Im}\,\psi$, and

$$q(\chi) = \lim_{n\to\infty} \left[\frac{\varepsilon_\chi^n(\psi)}{n^2} + \frac{\varepsilon_\chi^n * \varepsilon_{\bar{\chi}}^n(\psi)}{2n}\right]$$

whenever $\chi \in K^\wedge$.

10.6. <u>Examples</u> for the application of the theorem are

10.6.1 all Jacobian hypergroups of Example 1.3.1, and

10.6.2 all hypergroups of type G_B introduced in Example 1.2 .

Explicit computations of the corresponding negative definite forms have been performed in Lasser [31].

IV Markov chains on polynomial hypergroups

§ 11 Generalities on random walks

Let P be a Markov kernel on a measurable space $(K,\mathfrak{K})$. We introduce the set $\Omega := K^{\mathbb{Z}_+}$, the sequence $(X_n)_{n\in\mathbb{Z}_+}$ of n-th coordinate mappings $X_n : \Omega \to K$, the sequence $(\mathfrak{A}_n)_{n\in\mathbb{Z}_+}$ of σ-algebras $\mathfrak{A}_n$ generated by the sets of the form $X_o^{-1}(B_o) \cap \ldots \cap X_n^{-1}(B_n)$ with $B_o,\ldots,B_n \in \mathfrak{K}$, and the σ-Algebra $\mathfrak{A} := \mathfrak{A}(\bigcup_{n\in\mathbb{Z}_+} \mathfrak{A}_n)$.

The subsequent existence result is wellknown in the theory of Markov chains

11.1 <u>Theorem.</u> For every measure $\nu \in \mathfrak{M}^1(K)$ there exists a probability measure $\underline{\underline{P}}^\nu$ on $(\Omega,\mathfrak{A})$ having the following properties:

(i) $\underline{\underline{P}}^\nu[X_o \in B] = \nu(B)$ for all $B \in \mathfrak{K}$.

(ii) The quadruple $(\Omega,\mathfrak{A},\underline{\underline{P}}^\nu,(X_n)_{n\in\mathbb{Z}_+})$ is a *Markov chain with transition kernel P and starting measure* ν ; i.e.

$$E_{\underline{\underline{P}}^\nu}(f\circ X_{n+1}|\mathfrak{A}_n) = Pf\circ X_n \quad \underline{\underline{P}}^\nu\text{ - a.s.}$$

for every bounded $\mathfrak{K}$-measurable function $f \geq o$ on K and every $n \in \mathbb{Z}_+$.

11.2 <u>Notation.</u> The family $((\Omega,\mathfrak{A},\underline{\underline{P}}^\nu,(X_n)_{n\in\mathbb{Z}_+}))_{\nu\in\mathfrak{M}^1(K)}$ of Markov chains $(\Omega,\mathfrak{A},\underline{\underline{P}}^\nu,(X_n)_{n\in\mathbb{Z}_+})$ with transition kernel P and starting measure ν will be called the *canonical Markov chain with transition kernel* P.

In most of the applications it suffices to consider canonical chains of the type $((\Omega,\mathfrak{A},\underline{\underline{P}}^x,(X_n)_{n\in\mathbb{Z}_+}))_{x\in K}$ where the measures $\underline{\underline{P}}^x$

are defined as P^{ε_x} for starting measures ε_x $(x \in K)$.

11.3 Now let $(K, \mathfrak{B}(K))$ denote the Borel measurable space arising from a hypergroup K. For every measure $\mu \in \mathfrak{M}^1(K)$ we consider the Markov kernel $P = P_\mu$ on $(K, \mathfrak{B}(K))$ defined by

$$P(x,B) := \varepsilon_x * \mu(B)$$

for all $x \in K$, $B \in \mathfrak{B}(K)$, where $*$ denotes the convolution of $\mathfrak{M}^1(K)$. By Theorem 11.1 there exists the canonical Markov chain $(\Omega, \mathfrak{A}, \underline{\underline{P}}^x, (X_n)_{n \in \mathbb{Z}_+})_{x \in K}$ with transition kernel P. It is called the (generalized) *random walk on* K *with transition kernel* P and is abbreviated by $\mathfrak{X}(P)$.

11.4. For any random walk of the form $\mathfrak{X}(P)$ on a hypergroup K the *transition probabilities* $P_k = P_{n,n+k}$ $(n \in \mathbb{Z}_+,\ k \in \mathbb{N})$ are given by

$$P_k(x,B) = \varepsilon_x * \mu^k(B)$$

for all $x \in K$, $B \in \mathfrak{B}(K)$.

Applying the notion of *right translation* $T_y f$ of a function $f \in \mathfrak{L}^b(K)$ by an element $y \in K$, defined through

$$T_y f(x) := \int f(z)(\varepsilon_x * \varepsilon_y)(dz)$$

for all $x \in K$ we arrive at the following

11.5 <u>Property.</u> Let $(Y_j)_{j \geq 1}$ be a sequence of independent random variables on a probability space $(\Omega, \mathfrak{A}, \underline{\underline{P}})$ taking values in K, which have the same distribution $\mu \in \mathfrak{M}^1(K)$. Then we obtain for every function $f \in \mathfrak{L}^b(K)$, every $x \in K$ and all $k \in \mathbb{N}$ the formula

$$E_{\underline{\underline{P}}}(T_{Y_k} T_{Y_{k-1}} \cdot \ldots \cdot T_{Y_1} f(x)) = \int f(z) P_k(x,dz).$$

11.6 From now on we only consider discrete hypergroups $(K, \mathfrak{b}(K))$. In this case any measure $\mu \in \mathcal{M}^1(K)$ is of the form

$$\mu = \sum_{y \in \mathbb{Z}_+} a_y \varepsilon_y$$

with coefficients $a_y \geq 0$ such that $\sum_{y \in \mathbb{Z}_+} a_y = 1$. The corresponding Markov kernels $P = P_\mu$ on $(K, \mathfrak{b}(K))$ are determined by the transition probabilities

$$P(x,y) = \varepsilon_x * \mu(\{y\})$$

which are the coefficients of ε_y in the representation of the measure $\varepsilon_x * \mu$. Analoguously we denote by $P_k(x,y)$ for every $k \in \mathbb{Z}_+$ the probability that the random walk $\mathfrak{X}(P)$ on K arrives in y at time k after having started in x at time o $(x,y \in K)$. Clearly $P_k(x,y)$ is the coefficient of ε_y in the representation of the measure $\varepsilon_x * \mu^k$.

As in the classical set up the *potential kernel* of the random walk $\mathfrak{X}(P)$ is defined by

$$U(x,y) := \sum_{k \geq o} P_k(x,y)$$

for all $x,y \in K$. $U(x,y)$ can be interpreted as the expected value of the number of visits of $\mathfrak{X}(P)$ in the point y after having started in x.

11.7 Definition. $\mathfrak{X}(P)$ is said to be *transient* if for all $x,y \in K$

$$\underline{\underline{P}}^x [X_n = y \text{ for at most finitely many } n \geq 1] = 1 .$$

$\mathfrak{X}(P)$ is called *recurrent* if it is not transient.

11.8 Fact. For any random walk $\mathfrak{X}(P)$ on a discrete hypergroup K the following statements are equivalent:

(i) $\mathfrak{X}(P)$ is transient.

(ii) $U(x,y) < \infty$ for all $x,y \in K$.

11.9 Definition. A subset B of K is said to be *recurrent* with respect to the random walk $\mathfrak{X}(P)$ if for every fixed $x \in K$ one has

$$\underline{\underline{P}}^x(\overline{\lim_{n\to\infty}}\,[X_n \in B]) = 1 .$$

With this definition we restate the recurrence of $\mathfrak{X}(P)$ by requiring that for all $x,y \in K$

$$\underline{\underline{P}}^x\,[X_n = y \text{ for infinitely many } n \geq 1] = 1$$

holds.

§ 12 Transience of random walks on ultraspherical hypergroups

In what follows we are going to study random walks on the ultraspherical hypergroup $\mathbb{Z}_+$ defined by the sequence $(P_n^\alpha)_{n\geq o}$ of ultraspherical polynomials P_n^α with $\alpha \geq -\frac{1}{2}$. In this case the random walk under discussion will be of the form $\mathfrak{X}_\alpha := \mathfrak{X}(p_\alpha)$ where $p_\alpha = (p_\alpha(n,m))_{n,m\in\mathbb{Z}_+}$ is a Markov matrix intimately connected with the sequence $(P_n^\alpha)_{n\geq o}$. The probabilistic results quoted in the sequel are due to C. George [18] ; they have been reproduced in Guivarc'h, Keane, Roynette [19] and Roynette [35].

For any $f \in \ell^1(\mathbb{Z}_+)$ we define the function $f_\alpha \in \mathcal{C}([-1,1])$ by

$$f_\alpha(x) := \sum_{n\geq o} f(n) P_n^\alpha(x)$$

for all $x \in [-1,1]$, the series involved being normally convergent. We note that the set

$$\mathcal{A}_\alpha := \{f_\alpha : f \in \ell^1(\mathbb{Z}_+)\}$$

is a subalgebra of $\mathcal{C}([-1,1])$.

12.1 Proposition. Let $f \in \mathcal{M}^1(\mathbb{Z}_+)$, i.e. f denotes a function $\mathbb{N} \to \mathbb{R}_+$ satisfying $\sum_{n\geq o} f(n) = 1$. If

$$f_\alpha \cdot P_n^\alpha = \sum_{m\geq o} p_\alpha(n,m) P_m^\alpha ,$$

for all $n\geq o$, then $p_\alpha = (p_\alpha(n,m))_{n,m\in\mathbb{Z}_+}$ is a Markov matrix, and for every $k\geq o$ the k-th power p_α^k of p_α is given by the transition probabilities

$$p_\alpha^k(n,m) = h(m) \int_{[-1,1]} f_\alpha^k P_n^\alpha P_m^\alpha d\pi$$

for all $n,m \in \mathbb{Z}_+$, where h and π denote Haar and Plancherel measure on $\mathbb{Z}_+$ and $[-1,1] = \mathbb{Z}_+^\wedge$ respectively.

12.2 <u>Special case.</u> Let $f : \mathbb{Z}_+ \to \mathbb{R}_+$ be given by

$$f(n) := \begin{cases} 1 & \text{for } n=1 \\ o & \text{for } n \neq 1. \end{cases}$$

Then $f \leftrightarrow p_\alpha$ with

$$\begin{cases} p_\alpha(o,1) = 1 & \\ p_\alpha(n,n-1) = \dfrac{n}{2n+2\alpha+1} & \text{for } n \geq 1 \\ p_\alpha(n,n+1) = \dfrac{n+2\alpha+1}{2n+2\alpha+1} & \text{for } n \geq 1 \end{cases}$$

12.3 <u>Remark.</u> Proposition 12.1 supplies us for every $\alpha \geq -\frac{1}{2}$ with a canonical Markov chain $\mathfrak{X}_\alpha = \mathfrak{X}(p_\alpha)$ on $\mathbb{Z}_+$ having the Markov matrix p_α as its transition matrix. Thus, for every $\alpha \geq -\frac{1}{2}$ we have a family of Markov chains with state space $\mathbb{Z}_+$ indexed by functions $f \in \mathfrak{M}^1(\mathbb{Z}_+)$.

We note that in the case $\alpha = o$ we obtain random walks on the Legendre hypergroups $\mathbb{Z}_+ \cong ({}^G/_K)^\wedge$ and in the case $\alpha = \frac{1}{2}$ random walks on the 2nd kind Tschebychev hypergroups $\mathbb{Z}_+ \cong SU(2)^\wedge$. The latter case will be studied in greater detail later.

12.4 <u>Fact.</u> For the canonical random walks $\mathfrak{X}_\alpha$ with transition matrix $p_\alpha \leftrightarrow f \in \mathfrak{M}^1(\mathbb{Z}_+)$ the following statements are equivalent:

(i) $\mathfrak{X}_\alpha$ is *irreducible* in the sense that for all $n,m \in \mathbb{Z}_+$ there exists an $N \geq 1$ such that

$$p_\alpha^N(n,m) > o .$$

(ii) f_α is not an even function.

(iii) f charges an odd integer.

12.5 General assumption. For the remainder of this section $\mathfrak{X}_\alpha$ will be assumed to be irreducible.

12.6 Theorem. The potential kernel U_α of the random walk $\mathfrak{X}_\alpha$ with transition matrix p_α admits the representation

$$U_\alpha(n,m) = h(m) \int_{[-1,1]} \frac{P_n^\alpha P_m^\alpha}{1-f_\alpha}\, d\pi$$

with both sides of the equality eventually equal to $+\infty$.

12.7 Corollary. For every $\alpha > o$ the random walk $\mathfrak{X}_\alpha$ with transition matrix p_α is transient.

In the special case $\alpha = \frac{1}{2}$ this result is due to Eymard, Roynette [12].

12.8 Corollary. Let f possess a *2nd order moment* in the sense that $\sum_{n \geq o} n^2 f(n) < \infty$. Then the following statements are equivalent:

(i) $\mathfrak{X}_\alpha$ is transient.

(ii) $\alpha > o$.

12.9 Example. Let $\mathfrak{X}_\alpha$ be a random walk on the ultraspherical hypergroup $\mathbb{Z}_+$, with transition matrix $p_{\alpha,\lambda}$ for $\lambda > -\frac{1}{2}$ defined by

$$\begin{cases} p_{\alpha,\lambda}(o,1) := 1 \\ p_{\alpha,\lambda}(n,n-1) := \frac{1}{2}\left(1 - \frac{\lambda}{n+\lambda}\right) \\ p_{\alpha,\lambda}(n,n+1) := \frac{1}{2}\left(1 + \frac{\lambda}{n+\lambda}\right) \quad \text{for all } n \geq 1 . \end{cases}$$

Then $\mathfrak{X}_\alpha$ is recurrent iff $\lambda \in \left]-\frac{1}{2}, \frac{1}{2}\right]$.

12.10 Example. Let $\mathfrak{X}_\alpha$ be a random walk on the ultraspherical hypergroup $\mathbb{Z}_+$, with transition matrix p_α defined by

$$\begin{cases} p_\alpha(n,n-1) := a_n \\ p_\alpha(n,n+1) := b_n \\ p_\alpha(n,n) := c_n \quad \text{with} \quad a_n+b_n+c_n=1,\ a_n, b_n > o, \quad n \geqq 1, \\ p_\alpha(o,1) := b_o \\ p_\alpha(o,o) := c_o \quad \text{with} \quad b_o+c_o=1,\ b_o > o. \end{cases}$$

Then

$$\underline{\lim}_n\, n \frac{b_n - a_n}{b_n + a_n} > \frac{1}{2} \quad \text{or} \quad < \frac{1}{2}$$

iff $\mathfrak{X}_\alpha$ is transient or recurrent respectively.

In the case $c_n = o$ for all $n \geqq o$ we get

$$\underline{\lim}_n\, n(b_n - a_n) > \frac{1}{2} \quad \text{or} \quad < \frac{1}{2}$$

iff $\mathfrak{X}_\alpha$ is transient or recurrent respectively.

12.11 Theorem (Central limit theorem).

Let $\mathfrak{X}_\alpha := (\Omega, \mathfrak{A}, \underline{\underline{P}}^x, (X_n)_{n \in \mathbb{Z}_+})_{x \in \mathbb{Z}_+}$ be a random walk on the ultraspherical hypergroup $\mathbb{Z}_+$ with transition matrix p_α arising from a function $f \in \mathcal{M}^1(\mathbb{Z}_+)$ such that

$$C := \frac{1}{4(\alpha+1)} \sum_{n \geq o} n(n+2\alpha+1) f(n) < \infty$$

$(\alpha \geqq -\frac{1}{2})$. We form the sequence $(Y_n)_{n \geqq o}$ of random variables

$$Y_n := \frac{1}{\sqrt{2\,Cn}} X_n$$

(taking values in $\mathbb{R}_+$).

Then for every $x \in \mathbb{Z}_+$

$$\mathfrak{T}_v\text{-}\lim_{n\to\infty} (\underline{\underline{P}}^x)_{Y_n} = \nu \in \mathcal{M}^1(\mathbb{R}_+) .$$

with $\nu := n \cdot \lambda_{\mathbb{R}_+}$, where

$$n(x) := \frac{x^{2\alpha+1}}{2^{\alpha}\Gamma(\alpha+1)} e^{-\frac{x^2}{2}}$$

for all $x \in \mathbb{R}_+$.

For a <u>proof</u> of this theorem see George [18] or Eymard, Roynette [12], where the special case $\alpha = \frac{1}{2}$ has been treated.

12.12 <u>Theorem</u> (Asymptotic behavior of the potential kernel).
Let $\mathfrak{X}_\alpha$, P_α and f be as in the preceding theorem such that in particular f possesses a 2nd order moment. Assume that $\alpha > o$.
Then for every $m \in \mathbb{Z}_+$ one has

$$\lim_{n\to\infty} \frac{1}{n} U_\alpha(m,n) = \lim_{n\to\infty} \frac{1}{n} U_\alpha(n,n) = \frac{1}{2\alpha C} ,$$

where C is the constant of the preceding theorem.

The <u>proof</u> of this and related results has been given by George in [18]. See also Guivarc'h, Keane, Roynette [19], and, of course, Eymard, Roynette [12] for the special case $\alpha = \frac{1}{2}$.

12.13 <u>An application.</u> For every subset B of $\mathbb{Z}_+$ we consider the *return time* to B of the random walk $\mathfrak{X}_\alpha$, defined by

$$S_B := \inf\{n>o: X_n \in B\}.$$

It is well known that S_B being the first hitting time of B is a stopping time.

We note that under the assumptions of Theorem 12.12 we obtain

$$\lim_{n\to\infty} \underline{\underline{P}}^x [S_{\{n\}} < \infty] = 1$$

valid for all $x \in \mathbb{Z}_+$. In fact the Markov property of $\mathfrak{X}_\alpha$ implies

$$\underline{\underline{P}}^x [S_{\{n\}} < \infty] = \frac{U_\alpha(x,n)}{U_\alpha(n,n)}$$

for all $x, n \in \mathbb{Z}_+$, and by Theorem 12.12 the quotient tends to 1 as $n \to \infty$, for all $x \in \mathbb{Z}_+$.

Again under the assumptions of Theorem 12.12 we have the following <u>fact:</u> A subset B of $\mathbb{Z}_+$ is recurrent with respect to $\mathfrak{X}_\alpha$ iff B is infinite.

In fact, since $\mathfrak{X}_\alpha$ is assumed to be transient, the recurrent sets are necessarily infinite. The converse follows from the above relation.

Very recently laws of the iterated logarithm and tests in the spirit of Dvoretzky and Erdös on the rate of decay for random walks on an ultraspherical hypergroup have been obtained by Gallardo [15], [14]. We shall reproduce two sample results.

12.14 <u>Theorem</u> (Law of the iterated logarithm).
Let $\mathfrak{X}_\alpha := (\Omega, \mathfrak{A}, \underline{\underline{P}}^x, (X_n)_{n\in\mathbb{Z}_+})_{x\in\mathbb{Z}_+}$ be a random walk on the ultraspherical hypergroup $\mathbb{Z}_+$ with transition matrix $p_\alpha \leftrightarrow f \in \mathfrak{M}^1(\mathbb{Z}_+)$ admitting a 4th moment in the sense of $\sum_{n\geq 0} n^4 f(n) < \infty$ $(\alpha \geq -\frac{1}{2})$. Then for every $x \in \mathbb{Z}_+$

$$\overline{\lim_n} \frac{X_n}{\sqrt{4Cn \log\log n}} \leq 1 \qquad \underline{\underline{P}}^x\text{ - a.s.}$$

In the special case 12.2 which appears as Example 12.9 for

$\lambda = \alpha + \frac{1}{2}$ the above inequality is in fact an equality with $C = \frac{1}{2}$.

12.15 Theorem (Dvoretzky-Erdös test).

Let $\mathfrak{X}_\alpha := (\Omega, \mathfrak{A}, \underline{\underline{P}}^x, (X_n)_{n \in \mathbb{Z}_+})_{x \in \mathbb{Z}_+}$ be a random walk on the ultraspherical hypergroup $\mathbb{Z}_+$ with transition matrix $p_{\alpha,\lambda}$ given in Example 12.9 with $\lambda = \alpha + \frac{1}{2}$ $(\alpha > o)$. Then for any sequence $(g(n))_{n \geq o}$ in $\mathbb{R}_+^*$ which is monotone for sufficiently large n one has

$$\underline{\lim_n} \frac{X_n}{\sqrt{n}\, g(n)} = o \quad \text{or} \quad +\infty \qquad \underline{\underline{P}}^o\text{ - a.e.}$$

according to whether the integral

$$\int \frac{g(t)^{2\lambda-1}}{t}\, dt \quad \text{diverges or converges.}$$

Here g is the function on $\mathbb{R}_+$ which satisfies $g(t) = g(n)$ for all $t \in [n, n+1[$ $(n \in \mathbb{Z}_+)$.

12.6 Remark. Both the preceding theorems can be generalized to random walks $\mathfrak{X}_\alpha$ on $\mathbb{Z}_+$ having transition matrices p_α of the form

$$\begin{cases} p_{\alpha,\lambda}(o,1) = 1 \\ p_{\alpha,\lambda}(n,n+1) = \frac{1}{2}\left(1 + \frac{\lambda}{n} + o\left(\frac{1}{n}\right)\right) \\ p_{\alpha,\lambda}(n,n-1) = 1 - p_{\alpha,\lambda}(n,n+1) \end{cases}$$

where $\lambda = \alpha + \frac{1}{2}$ $(\alpha > o)$.

§ 13 Random walks on the dual of SU(2)

In this section we intend to illustrate some of the more general theory of the last section in the special case of the dual hypergroup $\mathbb{Z}_+ \cong SU(2)^\wedge$ or equivalently of the 2nd kind Tschebychev hypergroup $\mathbb{Z}_+$. Moreover we quote within this restricted framework versions of the central limit theorem and of the strong law of large numbers for irreducible random walks. In our presentation we follow the sources Gallardo [13] and Gallardo, Ries [17].

Let $\mathbb{Z}_+$ denote the 2nd kind Tschebychev hypergroup defined by the sequence $(P_n^\alpha)_{n\geq o}$ of 2nd kind Tschebychev polynomials P_n^α with $\alpha = \frac{1}{2}$. For any given measure $\mu := \sum_{x\in\mathbb{Z}_+} a_x \varepsilon_x \in \mathfrak{M}^1(\mathbb{Z}_+)$ we consider the random walk $\mathfrak{X}_\alpha := (\Omega, \mathfrak{A}, \underline{\underline{P}}^x, (X_n)_{n\geq o})_{x\in\mathbb{Z}_+}$ on $\mathbb{Z}_+$ with transition kernel $P = P_\mu$ defined by

$$P(x,y) := \varepsilon_x * \mu(\{y\})$$

for all $x,y \in \mathbb{Z}_+$. We assume that $\mathfrak{X}_\alpha$ is

(a) *aperiodic* in the sense that there exist at least one $r \in \mathbb{Z}_+$ odd and one $s \in \mathbb{Z}_+$ even such that $a_r \neq o$ and $a_s \neq o$, and admits

(b) *moments of order* $1+\delta$ with $\delta \geq o$ in the sense that

$$\sum_{r\geq o} a_r r^{1+\delta} < \infty \quad .$$

13.1 Theorem (Central limit theorem). Let $\mathfrak{X}_\alpha$ admit a moment of order 2 which implies that for any fixed $x \in \mathbb{Z}_+$

$$C := \frac{1}{6} E_{\underline{\underline{P}}^x} (X_1^2 + 2X_1) < \infty \quad .$$

Then the sequence $(Y_n)_{n\geq o}$ of random variables

$$Y_n := \frac{X_n}{\sqrt{2Cn}}$$

(taking values in $\mathbb{R}_+$) converges in distribution to a random variable Z (with values in $\mathbb{R}_+$) whose distribution $(\underline{\underline{P}}^x)_Z$ (under the measure $\underline{\underline{P}}^x$) is of the form $n \cdot \lambda_{\mathbb{R}_+}$ with

$$n(x) := \sqrt{\frac{2}{\pi}}\, x^2 e^{-\frac{x^2}{2}}$$

for every $x \in \mathbb{R}_+$.

13.2 Remark. We note that Theorem 13.1 is the special case of Theorem 12.11 arising for $\alpha = \frac{1}{2}$.

13.3 Theorem (Strong law of large numbers). Let $\mathfrak{X}_\alpha$ admit a moment of order $1+\delta$ with $\delta \in [o,1[$.

Then

$$\frac{X_n}{n^{1-\frac{\delta}{2}}} \to o \quad \underline{\underline{P}}^x\text{-a.s.}$$

as $n \to \infty$ for every $x \in \mathbb{Z}_+$.

13.4 Discussion. In Gallardo, Ries [17] the main idea of the proofs of Theorems 3.1 and 3.3 is to represent every X_n as the sum $Y_1 \oplus \ldots \oplus Y_n$ of independent identically distributed random variables Y_j $(j=1,\ldots,n,\ n \geq 1)$ with values in $\mathbb{Z}_+$ having μ as their distribution, where $\oplus$ is defined with respect to the operation in $\mathbb{Z}_+$ arising from the convolution in $\mathbb{Z}_+$. The operation $\oplus$ is constructed in such a way that given two independent random variables Y_1 and Y_2 with values in $\mathbb{Z}_+$ having distributions μ_1 and μ_2 in $\mathfrak{M}^1(\mathbb{Z}_+)$ respectively, the distribution of the sum $Y_1 \oplus Y_2$ is just $\mu_1 * \mu_2$ where * denotes the convolution defining the 2nd kind Tschebychev hypergroup $\mathbb{Z}_+$.

Some open problems

For the general theory of hypergroups and its *harmonic analysis* a large number of important questions can be posed. We only pick one the affirmative answer to which would have useful consequences in the study of probability measures.

1. Does any Moore hypergroup admit a (left) Haar measure? This existence problem will doubtlessly be connected with the structure of hypergroups about which very little is known. Recent results of Hartmann [20] on orbit hypergroups and of Hauenschild, Kaniuth and Kumar on central hypergroups might be of some help in taking up the problem.

Concerning *probability* and *potential theory* on hypergroups challenging open problems emerge from the corresponding results achieved for Gelfand pairs. See Letac's survey article "Problèmes classiques de probabilité sur un couple de Gelfand" quoted in [32], [33] and also Heyer [23]. We can only present a short selection.

2. More information is needed about the *limiting behavior of convolution sequences* of probability measures in the case of a noncompact hypergroup. One might think of the generalizations of contributions of Csiszár, Mukherjea, Rusza and others from groups or semigroups to hypergroups. See the articles of Mukherjea and Rusza in this volume.

In order to settle the *central limit problem* for probability measures on hypergroups

3. detailed studies are required on the set of *infinitely divisible probability measures* on a hypergroup. In particular one wants

more knowledge about the roots of an infinitely divisible law and a description of infinitely divisible measures as limits of infinitesimal triangular systems. In a special case such work has been done by Gallardo, Gebuhrer [16] (included in this volume).

Moreover

4. the problem of *embedding* an infinitely divisible probability measure on a hypergroup in a continuous convolution semigroup is still open. The work of Bloom [6] should be extended in some fashion.

In connection with Lasser's work on *positive* and *negative definite functions* on commutative hypergroups

5. a proper version of *Schoenberg's correspondence theorem* is still missing. See § 9 of this paper or the original article [30] , in which the problem appears for the first time. An axiomatization of the Schoenberg correspondence based on an idea of C. Berg's has been attempted by Bloom and Heyer in a forthcoming paper.

6. There is some hope in obtaining a *canonical representation* à la Lévy-Khintchine *for all infinitely divisible probability measures or convolution semigroups* at least on a commutative hypergroup. The ideas of Harzallah and Lasser [30] have certainly to be modified.

What will be most important for a further development of the theory is a more sophisticated analysis of particular *convolution semigroups*.

7. Studies on the *decomposability* (or factorization) of probability measures and positive definite functions along the lines of Khintchine's classical contributions are in order. See the papers of Bingham [5], Lamperti [28], Schwartz [36] and Rusza, Szekély (this volume).

8. *Poisson semigroups* can be studied in the framework of a general Banach algebra. Here is a method providing interesting properties of such semigroups on hypergroups. In the case of locally compact groups the corresponding results have been obtained by Hazod and Schmetterer. See Heyer [22], 3.2,6.1.
   In this connection de Finetti's theorem on the Poisson approximation of infinitely divisible probabilities could be taken up.

9. There are various ways of introducing *Gaussian* (Brownian) *semigroups* on (nondiscrete) hypergroups: via Fourier transforms, via Bernstein's property and via locality. In the commutative case the first two approaches should be attempted along the classical lines. For the group case see Heyer [22], 5.2, 5.3. Local semigroups of measures on a hypergroup must be studied if possible in connection with the canonical representation. Details in the group case are given in Berg, Forst [3], § 18.

Some theory is already available for *stochastic processes on hypergroups*. Particular interest has been given to Markov chains, especially to random walks.

10. An important progress would be reached if *dichotomy theorems* could be proved for random walks on at least commutative hypergroups. This would lead to studies on recurrence and transience.

For pioneering results see the work of Eymard, Roynette [12] and George [18]. Gallardo and Gebuhrer intend to proceed in this direction.

More generally

11. It will be desirable to develop a *potential theory of convolution semigroups* on a hypergroup which would serve as an efficient tool in the analysis of continuous time stationary Markov processes. Again, the notion of transience will be of special interest. Some work has been done by Bloom and Heyer, some more can be done along the lines of Berg, Forst [3], §§ 8, 13.

In the context of stochastic processes with a hypergroup as their state space only a few problems can be stated yet in precise terms. Referring to the Brownian motion parametrized by a hypergroup we pose the question

12. Is there a notion of *Brownian function* for certain hypergroups and if so, what is the relation between its existence and the structure of the underlying hypergroup? For Gelfand pairs these questions have been studied by Gangolli, Faraut and Askey, Bingham [1].

Bibliography

[1] R. Askey, N. H. Bingham: Gaussian processes on compact symmetric spaces. Z. Wahrscheinlichkeitstheorie verw. Gebiete 37 (1976), 127-143

[2] C. Berg: Suites définies négatives et espaces de Dirichlet sur la sphère. In: Séminaire Brelot-Choquet-Deny (Théorie du potentiel), 13e année (1969/70), 12-01 to 12-18

[3] C. Berg, G. Forst: Potential theory on locally compact Abelian groups. Springer-Verlag Berlin-Heidelberg-New York 1975

[4] N. H. Bingham: Random walks on spheres. Z. Wahrscheinlichkeitstheorie verw. Gebiete 22 (1972), 169-192

[5] N. H. Bingham: Positive definite functions on spheres. Proc. Camb. Philos. Soc. 73 (1973), 145-156

[6] W. R. Bloom: Infinitely divisible measures on hypergroups. In: Proceedings of the 6th Conference on Probability Measures on Groups. Lecture Notes in Math. Vol. 928, pp. 1-15. Springer 1982

[7] W. R. Bloom, H. Heyer: The Fourier transform for probability measures on hypergroups. Rendiconti di Matematica (2) 1982 Vol. 2, Serie VII, 315-334

[8] W. R. Bloom, H. Heyer: Convergence of convolution products of probability measures on hypergroups. Rendiconti di Matematica (3) 1982 Vol. 2, Serie VII, 547-563

[9] S. Bochner: Positive zonal functions on spheres. Proc. Nat. Acad. Sci. (USA) 40 (1954), 1141-1147

[10] S. Bochner: Sturm-Liouville and heat equations whose eigenfunctions are ultraspherical polynomials or associated Bessel functions. In: Proceedings of the Conference on Differential Equations, pp. 23-48. University of Maryland 1955

[11] C. F. Dunkl: The measure algebra of a locally compact hypergroup. Trans. Amer. Math. Soc. 179 (1973), 331-348

[12] P. Eymard, B. Roynette: Marches aléatoires sur le dual de SU(2). In: Analyse Harmonique sur les Groupes de Lie. Lecture Notes in Math. Vol. 497, pp. 108-152 Springer 1975

[13] L. Gallardo: Théorème limites pour les marches aléatoires sur le dual de SU(2). Ann. Scient. de l'Univ. de Clermont N° 67 (1979), 63-68

[14] L. Gallardo: Le test de Dvoretzky et Erdös pour les marches aléatoires sur $\mathbb{N}$. To appear.

[15] L. Gallardo: Une loi du logarithme itéré pour les chaînes de Markov sur IN associées à des polynômes orthogonaux.
To appear.

[16] L. Gallardo, O. Gebuhrer: Lois infiniment divisibles sur les hypergroupes de Spector commutatifs, discrets, dénombrables et à dual connexe.
To appear in the Proceedings of the 7th Conference on Probability Measures on Groups (1983)

[17] L. Gallardo, V. Ries: La loi des grands nombres pour les marches aléatoires sur le dual de SU(2).
Studia Mathematica LXVI (1979), 93-105

[18] C. George: Les chaînes de Markov associées à des polynômes orthogonaux.
Thèse de Doctorat d'Etat ès-Sciences Mathématique, Université de Nancy 1975

[19] Y. Guivarc'h, M. Keane, B. Roynette: Marches aléatoires sur les groupes de Lie.
Lecture Notes in Mathematics Vol. 624, Springer 1977

[20] K. Hartman: $[FIA]_B^-$-Gruppen und Hypergruppen.
Mh. Math. 89 (1980), 9-17

[21] K. Hartman, R. W. Henrichs, R. Lasser: Duals of orbit spaces in groups with relatively compact inner automorphism groups are hypergroups.
Mh. Math. 88 (1979), 229-238

[22] H. Heyer: Probability measures on locally ocmpact groups. Springer-Verlag Berlin-Heidelberg-New York 1977

[23] H. Heyer: Convolution semigroups of probability measures on Gelfand pairs. Exp. Math. 1 (1983), 3-45

[24] I. I. Hirschmann jr.: Harmonic analysis and ultraspherical polynomials. Symposium on Harmonic Analysis and Related Integral Transforms. Cornell University 1956

[25] I. I. Hirschmann jr.: Sur les polynômes ultrasphériques. C. R. Acad. Sci. Paris Sér. A-B 242, 2212-2214 (1956)

[26] R. I. Jewett: Spaces with an abstract convolution of measures. Advances in Math. 18 (1975), 1-101

[27] M. Kennedy: A stochastic process associated with ultraspherical polynomials. Proc. Roy. Irish Acad. 61 (1961), 89-100

[28] J. Lamperti: The arithmetic of certain semigroups of positive operators. Proc. Camb. Phil. Soc. 64 (1968), 161-166

[29] R. Lasser: Orthogonal polynomials and hypergroups. Rendiconti di Mat. (VII) 3, 2 (1983), 185-209.

[30] R. Lasser: Orthogonal polynomials and hypergroups: Contributions to Analysis and Probability Theory. Technische Universität München, Institut für Mathematik 1981

[31] R. Lasser: On the Lévy-Hinčin formula for commutative hypergroups.
To appear in the Proceedings of the 7th Conference on Probability Measures on Groups (1983)

[32] G. Letac: Les fonctions sphériques d'un couple de Gelfand symétrique et les chaînes de Markov.
Adv. Appl. Prob. 14 (1982), 272-294

[33] G. Letac: Dual random walks and special functions on homogeneous trees.
In: Marches aléatoires et processus stochastiques sur les groupes de Lie. Equipe associée d'Analyse Globale n° 839, Institut Elie Cartan, Université de Nancy (1983), 96-142

[34] P. H. Roberts, H. D. Ursell: Random walks on a sphere and on a Riemannian manifold.
Phil. Trans. Roy. Soc. London, Ser. A, 252 (1960), 317-356

[35] B. Roynette: Marches aléatoires sur les groupes de Lie.
In: Ecole d'Eté de Probabilité de Saint-Flour VII-1977, pp. 237-379.
Lecture Notes in Math. Vol. 678
Springer 1978

[36] A. Schwartz: Generalized convolutions and positive definite functions associated with general orthogonal series.
Pac. J. Math. 55 (2) (1974), 565-582

[37] R. Spector: Aperçu de la théorie des hypergroupes. In: Analyse Harmonique sur les Groupes de Lie. Lecture Notes in Math. Vol. 497, pp. 643-673. Springer 1975

[38] R. Spector: Mesures invariantes sur les hypergroupes. Trans. Amer. Math. Soc. 239 (1978), 147-165

[39] R. C. Vrem: Harmonic analysis on compact hypergroups. Pacific J. Math. 85 (1979), 239-251

Mathematisches Institut
der Universität Tübingen
Auf der Morgenstelle 10

7400 Tübingen
West-Germany

A survey about zero - one laws for probability measures on linear spaces and locally compact groups

Arnold Janssen

It is the purpose of this paper to summarize the development of the last thirty years of zero-one laws for infinitely divisible probability measures μ and measures P which are induced by certain stochastic processes on suitable function spaces. We are interested in zero-one laws in the sense that a given measurable subgroup of the underlying space has either probability zero or one. The simplest laws can be proved by Kolmogorov's zero-one law for tail events. But in general special topological and algebraic arguments are used.
The first zero-one laws appear in the context of Gaussian processes dealing with path properties. The history of these laws is presented in part 1. The second part is devoted to stable measures which are defined on linear spaces. The third section contains the generalization of zero-one laws for infinitely divisible probability measures on measurable or topological groups. The results mainly rely on algebraic properties of the measures. In part 4 zero-one laws for continuous convolution semigroups on real locally convex vector spaces and locally compact groups are studied. It is pointed out that various applications can be deduced from the main theorem which allows a description of zero-one laws in terms of the Lêvy-Khintchine formula. This formula is an important analytic tool. In section 5 we want to stimulate the study of extensions which may be possible in future. There is a relationship between zero-one laws and other dichotomy theorems and purity laws. Therefore it should be useful to continue the work in this direction. For example it is an open problem wether two symmetric stable laws on $\mathbb{R}^{\infty}$ are either equivalent or mutually singular (Chatterji and Ramaswamy's conjecture).

Finally the author wants to thank Professor H. Heyer for the invitation to give the survey talk in Oberwolfach. I am also grateful to Professor A. Tortrat for various helpful comments.

1. Gaussian processes

In 1951 Cameron and Graves proved the following result for the Wiener process:

Suppose that P is the Wiener measure on $C([a,b])$ and let G be a measurable module over the rationals of $C([a,b])$ then $P(G)$ is either zero or one.

Recall that a subset M of a vector space X is called a module over the rationals (or an r-module) if for every x_1 and x_2 in M and rational numbers r_1 and r_2 the elements $r_1x_1 + r_2x_2$ ly in M.

This result was the beginning of the examination of zero-one laws for Gaussian processes. Zero-one laws are very useful in view of applications for path properties of Gaussian stochastic processes. In many cases one wants to know wether the paths of a process ly in a certain set.

In the subsequent text we shall consider a group X, a probability measure μ or P on X and a subgroup G which is assumed to be P-measurable with respect to the completion of the underlying σ-field. We say that P satisfies the zero-one law (0 - 1 law) for G if $P(G)$ is either zero or one.

In order to distinguish measurable sets and sets which are measurable with respect to the P-completion we shall speak about measurable and P-measurable sets.

In 1970 Jamison and Orey proved the zero-one law for the infinite product of the normalized Gaussian measure μ on $\mathbb{R}^\infty$ and μ-measurable groups. Moreover the same assertion is proved for a Gaussian process $(\Omega, \mathfrak{F}, P, \{X_t, 0 \leq t \leq 1\})$ with continuous sample paths and mean zero.

At the same time Kallianpur (1970) published his fundamental paper. He considered Gaussian measures P (which arise from Gaussian processes with mean 0) on a linear subspace $X \subset \mathbb{R}^T$ such that

(i) T is a complete separable metric space,

(ii) the covariance R is continuous on T x T,

(iii) the reproducing kernel Hilbert space H(R) is included in X and

(iv) X carries the Borel structure induced by the projections $x \longrightarrow x(t)$.

Under these assumptions it is shown that the zero-one law is satisfied for P-measurable r-modules and measurable groups $G \subset X$. The results have been extended by Jain in 1971. He showed that Kallianpur's result also holds for P-measurable groups. Moreover Baker remarked in 1973 that the result carries over for Gaussian processes with arbitrary mean function. Until now the proofs mainly utilized the structure of the reproducing kernel Hilbert space H(R) and the covariance R.
Let us sketch some applications, Kallianpur 1970: Suppose that x(t), $a \leq t \leq b$, is a sample continuous Gaussian stochastic process with zero mean function. Let f be a real measurable function on [a,b] and let p be a positive number. Then either $f(t)|x(t)|^p \in L^1([a,b], \lambda)$ almost everywhere or $f(t)|x(t)|^p \notin L^1([a,b], \lambda)$ almost everywhere (λ denotes the Lebesgue measure). This theorem extends results of Shepp (1966)[1].
The ideas of Kallianpur and Jain have inspired many other authors. In 1973 Baker extended the zero-one law for Gaussian measures on real separable Banach spaces (with respect to P-measurable subgroups). Rajput (1972) pointed out that the theorem holds for Gaussian measures on Fréchet spaces. Moreover Cambanis and Rajput (1973) remarked that the zero-one law is true for Gaussian processes with arbitrary parameter space T (not necessaryly complete, metric, separable). We note that the papers of Baker (1973) and Cambanis and Rajput (1973) contain a lot of important examples of measurable subgroups of the sample path space of the underlying process.

---

1) see also Varberg (1967).

Consider for example a real separable Gaussian process $(X(t,\omega), t \in T)$ on an interval T of the real line and $(\Omega, \mathfrak{A}, P)$. Then with probability zero or one the paths of $X(t,\omega)$ are for example bounded on T, of bounded variation on every compact subinterval of T, continuous on T, differentiable on T, absolutely continuous on every compact subinterval of T, compare with Cambanis and Rajput (1973).
Moreover in 1974 Zinn proved 0 - 1 laws for non-Gaussian measures on $\mathbb{R}^{\infty}$.

2. Stable measure

Another extension of the zero-one law has been done for stable measures. Suppose that $(X, \mathfrak{B})$ is a real measurable vector space which means that the addition and multiplication is pointly measurable on $X \times X$ and $X \times \mathbb{R}$ respectively. Define $T_a x = ax$ for $a \in \mathbb{R}$, $x \in X$. Then a probability measure μ on $(X, \mathfrak{B})$ is said to be stable if for all positive real numbers a and b there exist a positive real number c and $x \in X$ such that $T_a \mu * T_b \mu = T_c \mu * \varepsilon_x$. (Note that $T_a \mu$ denotes the image measure of μ with respect to T_a and ε_x is the point measure of x).
In 1974 Dudley and Kanter proved the zero-one law for stable measures and measurable linear subspaces. In the same year Fernique (1974) gave a short proof of this result.
A few years later Krakowiak (1978) proved the 0 - 1 law for stable measures μ on separable real Banach spaces X and measurable subgroups G. The proof contains a new idea. It is used that μ can be embedded in a continuous convolution semigroup (c.c.s) $(\mu_t)_{t>0}$ such that $\mu_t = T_{t^{1/\alpha}} \mu * \varepsilon_{x_t}$ for some $x_t \in X$. Recall that μ_t is a c.c.s. if $t \longrightarrow \mu_t$ is weakly continuous and $\mu_t * \mu_s = \mu_{t+s}$ holds for all $t,s > 0$. Finally Krakowiak remarks that the proof caries over for tight stable measures on locally convex spaces.
Let us remark that it is sufficient to consider σ-compact subgroups G if X is a topological Hausdorff group and μ is tight. If G is a subgroup (perhaps non-measurable) then there exists a σ - compact sub-

group $H \subset G$ such that $\mu(H)$ and the inner measure $\mu_*(G)$ coincide. Choose an increasing sequence of symmetric compact sets $K_n \subset G$ such that $\mu(K_n) \longrightarrow \mu_*(G)$. Then $H = \bigcup_n (K_n)^n$ fulfils the assertion. Hence in this case all zero-one laws are valid for $\mu_*(G)$.

3. Recent development

The development of this area in the last years can be devided in two parts. First the Polish school around Byczkowski tried to study Gaussian measures on Abelian groups and secondly extensions for semistable measures were considered by Tortrat, Rajput and further authors.
At first a definition for Gaussian measures on Abelian groups $(X,+)$ is needed. A tight probability measure μ on a Hausdorff group is said to be a Gauss measure in the sense of Bernstein if the formula $\psi(\mu \otimes \mu) = (\mu * \mu) \otimes (\mu * \tilde{\mu})$ is valid where ψ is defined by $\psi(x,y,\ x-y)$ and $\tilde{\mu}(A) = \mu(-A)$. The same definition is used for measurable groups which admit a product measurable map $(x,y) \longrightarrow x-y$ of $X \times X$ in X. We note that there are further concepts of Gaussian measures on groups, compare with Heyer (1977), Hazod (1977).
In these books the connection between the different definitions is discussed and the generating functions of Gauss semigroups are studied in view of Bernstein's result.
On the real line the definition above is just Bernstein's theorem which shows that for two independent random variables X_1, X_2 with common distribution μ the variables X_1+X_2 and X_1-X_2 are independent iff μ is a Gauss measure.
We note that on Banach spaces the definition above coincides with the usual definition of Gauss measures which demands that for each continuous real functional f the image measure $f(\mu)$ is Gaussian.
In (1977) a) Byczkowski proved the 0 - 1 law for Gaussian measures μ and μ-measurable r-modules on complete separable metric linear spaces. The same result received Inglot (1979) for measurable vector spaces.

There is a series of papers of Byczkowski and Byczkowska dealing with 0 - 1 laws for measurable subgroups and Gaussian measures in the sense of Bernstein on metrizable Abelian groups, see Byczkowski (1977) b), (1978), (1980) and Byczkowska, Byczkowski (1981). We will only present one result which is proved for topological groups by Byczkowska, Byczkowski (1981). Similar theorems hold for measurable groups. Suppose that X is a metric Abelian group such that X is a standard Borel space with respect to the Borel σ-field and X contains no elements of order two. Then the 0 - 1 law is satisfied for Gaussian measures without idempotent factors and measurable subgroups. The proofs are based on algebraic arguments and the algebraic definition of Gaussian measures. In view of the earlier results of Kallianpur and Jain this approach is completely different. The space D [0,1] of all left-continuous real functions defined on the unit interval, without discontinuities of second kind is a typical example of a measurable group which is not a topological group. We suppose that D [0,1] carries the Skorohod topology and the corresponding σ-field. Since this space is important in view of applications for stochastic processes this is one reason why many authors study measurable groups.

A probability measure μ on a measurable vector space X is called r-semistable for $r \in (0,1)$ if there exist a convolution semigroup $(\mu_t)_t$, a sequence x_m in X, and a positive real number $c \neq 1$ such that $\mu_1 = \mu$ and

(1) $\mu_{r^m} = T_{c^m} \mu * \varepsilon_{x_m}$ for all $m \in \mathbb{N}$.

This definition extends the classical concept of r-semistable distributions which is for example well-known for separable Banach spaces. Let Q(c) denote the smallest field containing the rational numbers and c. Then the zero-one law holds for r-semistable measures μ and measurable subspaces G over the field Q(c). This result is due to Louie, Rajput and Tortrat (1980).

In 1981 Tortrat extends the definition of stable laws on measurable groups by using algebraic equalities. Moreover a 0 - 1 law for measurable subgroups is established.

At the same time the authors mentioned above made much effort to weaken the definition (1) of semistable laws and to sharpen the 0 - 1 laws. The equality (1) was substituted by weaker algebraic divisibility relations being valid for μ. For example probability measures μ on a measurable group satisfying $n\cdot\mu = \mu^{*k} * \varepsilon_b$ for some integers $n,k \geq 2$ were considered. (Let $n\cdot\mu$ denote the image of μ with respect to $x \rightarrow x^n$). The assumptions are sometimes very complicated. Therefore we don't go into details. The papers of Rajput and Tortrat (1980), Rajput, Louie and Tortrat (1981), Tortrat (1980) and Tortrat (1982) contain further 0 - 1 laws in this direction. The proofs of the 0 - 1 laws are purely of algebraic nature.

4. Zero-one laws for continuous convolution semigroups

There is another recent development in the theory of 0 - 1 laws which mainly bases on analytic and topological arguments. In this section we consider a continuous convolution semigroup (c.c.s.) $(\mu_t)_{t>0}$ of Radon-probability measures on a locally compact group such that μ_t tends weakly to the point measure of the natural element e if $t \longrightarrow 0$. Moreover infinitely divisible tight probability measures μ on locally convex real vector spaces X are studied.

Note that μ is continuously embeddable in a c.c.s. if X is quasi-complete.

Perhaps it is a surprise that 0 - 1 laws can be described by the Lévy measure η. The Lêvy measure appears in the Lêvy-Khintchine representation of the c.c.s.$(\mu_t)_t$. If X is locally compact then $\frac{1}{t}\mu_t|X - \{e\}$ converges vaguely on $X - \{e\}$ to η for $t \longrightarrow 0$. If X is Abelian then each member μ of a c.c.s. without Gaussian part is the weak limit of a family of compound Poisson measures $e(\nu_i) * \varepsilon_{x_i}$ such that $\nu_i \uparrow \eta$

(let $e(\nu)$ denote the compound Poisson measure generated by a finite measure ν). The same result is valid for infinitely divisible tight probability measures on locally convex spaces.

There is another meaning of the Lêvy measure. Let X have a countable basis of it's topology. Then each c.c.s. determines a stochastically continuous process $((X_t)_{t\geq o}, X_o = e)$ with stationary and independent left increments in X whose paths are right continuous and have left-hand limits such that X_t has the distribution on μ_t. The Lêvy measure counts the jumps of the process: For each Borel set B with $e \notin \bar{B}$ the value $\eta(B)$ is the expected number of jumps of the process $(X_t)_{0\leq t\leq 1}$ that fall into B.

A c.c.s. is said to be Gaussian if the Lêvy measure vanishes.

In 1982 Siebert proved that every symmetric Gauss semigroup on a connected Lie group arises from a semigroup of absolutely continuous measures. More precise: There is an absolutely continuous Gauss semigroup $(\nu_t)_t$ on a connected Lie group H, a monomorphism m of H into X such that $m(\nu_t) = \mu_t$. Hence the 0 - 1 law is satisfied for all μ_t and all measurable subgroups. The proof is based on Hörmander's theorem for hypoelliptic differential operators.

A c.c.s. is said to be normal if μ_t and $\tilde{\mu}_t$ comute for each $t > o$ ($\tilde{\mu}(A) = \mu(A^{-1})$). The next result has been proved by Janssen (1982) a). Suppose that $(\mu_t)_{t>o}$ is a normal c.c.s. on a locally compact group X. Suppose that G is a normal measurable subgroup of X.

a) If the Lêvy measure η is unbounded on the complement G^c of G, i.e. $\eta(G^c) = \infty$, then $\mu_t(xG) = o$ is valid for all $x \in X$.

b) If $\eta(G^c) = o$ then μ_t satisfies the 0 - 1 law for G.

We remark that this result contains a 0 - 1 law for normal Gauss semigroups.

The results a) and b) carry over if μ is an infinitely divisible tight probability measure without Gaussian part on locally convex spaces X, see Janssen (1982) a), section 3. These measures can be received as weak limit points of translations of compound Poisson measures $e(\nu_i) * \varepsilon_{x_i}$, $\nu_i \uparrow \eta$, mentioned above. Moreover our paper contains a new proof of Baker's 0 - 1 law for Gaussian measures on separable Banach spaces.

We note that the proofs of a) and b) agree for Abelian locally compact groups and locally convex spaces. But for non-Abelian locally compact groups different arguments are needed. We remark that the theorem above seems to be new for $\mathbb{R}^n$.

We shall give two applications for infinitely divisible μ on locally convex spaces, Janssen (1982) b), (1982) a).

1. Choose $G = \{0\}$. Then $\mu(\{x\}) > 0$ for some x iff μ has a finite Lēvy measure ν and it has the form $\mu = e(\nu) * \varepsilon_y$.

2. (Krakowiak's result). Let μ be a stable non-Gaussian measure on a locally convex space X. Then $\mu(G+x) = 0$ or 1. The proof is an application of our main result. First we note that it is sufficient to deal with symmetric μ. We note that the Lēvy measure fulfils the well-known equality $t\,\eta = T_t{}^{1/\alpha}\,\eta$ for all $t > 0$ and some $\alpha \in (0,2)$, $T_a(x) = ax$. Choose $t = 2^{-\alpha} < 1$. Then

$$2^{-\alpha}\,\eta(G^c) = \eta((2^{-\alpha})^{-1/\alpha}\,G^c) = \eta(2G^c) \geq \eta(G^c) \quad \text{since} \quad 2G^c \supset G^c.$$

Hence $\eta(G^c) = 0$ or ∞.

There is a very nice application of the main result due to Tortrat (1982) for separable Banach spaces. Suppose that $T_c\mu$ devides the infinitely divisible measure μ for some c, $0 < c < 1$, such that the cofactor is infinitely divisible. Then μ satisfies the 0 - 1 law for measurable groups G if $c\,G \supset G$. (Note that $T_c\,\eta \leq \eta$).

Tortrat's paper contains further 0 - 1 laws relying on divisibility assumptions.

The result above can be applied to semistable distributions fulfilling $\mu_r = T_{r^\beta}\mu * \varepsilon_b$, $r < 1$, $c = r^\beta$, $\beta > 1/2$. Moreover the 0 - 1 law is satisfied for measurable subgroups and self-decomposable measures μ. Recall that μ is self-decomposable if for each $a \in (0,1)$ the measure $T_a\mu$ is a factor of μ. The same 0 - 1 laws appear in the author's paper, Janssen (1982) b), which is mainly devoted to measures on arbitrary locally convex spaces. Krakowiak's paper of 1978 also includes 0 - 1 laws for semistable distributions.

5. Possible extension in future

The proof of the main result of the preceding section mainly depends on a general purity law for infinite convolution products, Janssen (1980). Zero-one laws often seem to be a special form of this purity law . Let us mention some other results of this direction. For example infinite convolution products of discrecte measures on a locally compact group are either discrete or singular and diffuse or absolutely continuous.

It is my fealing that there is a deeper relationship between zero-one laws and other dichotomy results. For example Hájek-Feldman's dichotomy theorem states that two Gaussian measures on a suitable function space are either equivalent or mutually singular. Good references concerning this problem are included in the survey article of Chatterji and Mandrekar (1978).

There is only little known about non-Gaussian measures in view of dichotomy results and absolute continuity. For example Chatterji and Ramaswamy (1982) posed the question wether two symmetric stable distributions on $\mathbb{R}^\infty$ always are equivalent or mutually singular. To my best knowledge this question is in general open until now. In view of applications it seems to be useful to study the equivalence for infinitely divisible measures in fucture. This is one possible extension of zero-one and purity laws. It is also interesting to study the problem how to find conditions which ensure that one of the possible dichotomy cases is fulfilled. For example for Hájek-Feldman's result such conditions are known, see Chatterji and Mandrekar (1978). But in general this problem is very hard to attack. For instance it is in general not exactely clear what's the behaviour of an infinite convolution product of discrete probability measures in view of the purity law mentioned above. Of course special results are known for example for the coin tossing problem. Good references are included in the book of Graham and McGehee (1979).

Finally let us summarize some comments of colleges visiting Oberwolfach. Professor Urbanik announced a result of a vietnamese fellow who solved Chatterji's and Ramaswamy's conjecture. The answer is positive if the [2)] stable laws have discrete Lévy-measure. There is another recent paper of Byczkowski and Hulanicki (1983) who studied Gaussian semigroups $(\mu_t)_{t>o}$ on a separable, metric and complete group X in the sense of part 4. They showed that $\mu_t(H) = 1$ for a normal measurable subgroup H if $\mu_t(H) > o$ for each $t > o$.

Moreover T. Byczkowski refered about zero-one laws for continuous convolution semigroups on the arbitrary product of separable locally compact groups.

REFERENCES

Baker, C.R. (1973). Zero-one laws for Gaussian measures on Banach spaces. Trans. Amer. Math. Soc. 186, 291-308. MR 49 # 1569.

Byczkowska, H., Byczkowski, T., (1981). A generalization of the zero-one law for Gaussian measures on metric Abelian groups. Bull. l'Acad. Polon. Sc., Sér. Math. 29, 187-191.

Byczkowski, T. (1977) a). Gaussian measures on L_p spaces, $0 \leq p < \infty$. Studia Math. 59, 249-261. MR 56 # 6777.

Byczkowski, T.(1977)b).Some results concerning Gaussian measures on metric spaces. Lect. Notes Math. 656, 1-16, Springer-Verlag. MR 80j : 60021.

Byczkowski, T. (1978). Gaussian measures on metric Abelian groups. In Probability measures on groups. Proc. Oberwolfach. Lect. Notes Math. 706, 41-53. Springer. MR 80k : 60013.

Byczkowski, T. (1980). Zero-one laws for Gaussian measures on metric Abelian groups. Studia Math. 69,159-189.

2) The general problem seems to be open until now.

Byczkowski, T., Hulanicki, A. (1983). Gaussian measure of normal subgroups. Annals of Probability (in print).

Cambanis, S. Rajput, B. S. (1973). Some zero-one laws for Gaussian processes. Ann. Prob. 1, 304-312. MR 52 # 6851.

Cameron, R.H., Graves, R.E. (1951). Additive functionals on a space of continuous functions. I. Trans. Amer. Math. Soc. 70, 160-176. MR 12 # 718.

Chatterji,S.D., Mandrekar, V. (1978). Equivalence and singularity of Gaussian measures and applications. In: Probabilistic analysis and related topics, 169-195. Edited by A.T. Bharucha-Reid. Academic press, London.

Chatterji,S.D., Ramaswamy, S. (1982). Mesures Gaussiannes et mesures produit. Lect. Notes Math. 920, 570-580. Springer.

Dudley, R.M., Kanter. M. (1974). Zero-one laws for stable measures. Proc. Amer. Math. Soc. 45, 245-252. MR 51 # 6901.

Fernique, X.M. (1974). Une démonstration simple du théorème de R.M. Dudley et M. Kanter sur les loi 0-1 pour les mesures stables. Séminaire de Prob. VII. Lect. Notes Math. 381, 78-79. Springer. MR 53 # 6664.

Graham, C.C., McGehee, O. C. (1979). Essays in commutative harmonic analysis. Grundlehr. d. math. Wissensch. 238, Springer Verlag.

Hazod, W. (1977). Stetige Faltungshalbgruppen und erzeugende Distributionen. Lect. Notes Math. 595. Springer.

Heyer, H. (1977). Probability measures on locally compact groups. Ergebnisse d. Math. u. ihrer Grenzgebiete. Springer.

Inglot, T. (1979). An elementary approach of the zero-one laws for Gaussian measures. Colloqu. Math. 40, 319-325. MR 80j : 60012.

Jain, N.C. (1971). A zero-one law for Gaussian processes. Proc. Amer. Math. Soc. 29,585-587. MR 43 # 4099.

Jamison, B. Orey, S. (1970). Subgroups of sequences and paths. Proc. Amer. Math. Soc. 24, MR 40 # 8121.

Janssen, A. (1980). A general purity law for convolution semigroups with discrete Lévy measure. Semigroup Forum 20, 55-71.

Janssen, A. (1982) a). Zero-one laws for infinitely divisible probability measures on groups. Z. Wahrscheinlichkeitstheorie verw. Geb. 60, 119-138.

Janssen, A. (1982) b). Some zero-one laws for semistable and selfdecomposable measures on locally convex spaces. In: Probability measures on groups. Proc. Oberwolfach. Lect. Notes Math. 928, 236-246. Springer.

Kallianpur, G. (1970). Zero-one laws for Gaussian processes. Trans. Amer. Math. Soc. 149, 199-221. MR 42 # 1200.

Krakowiak, W. (1978). Zero-one laws for stable and semi-stable measures on Banach spaces. Bull. d. l'acad. Polon., ser. sci. math. 27, 1045-1049. MR 80i : 60005.

Louie, D., Rajput, B. Tortrat, A. (1980). A zero-one dichotomy theorem for r-semistable laws on infinite dimensional linear spaces. Sankhyā, Ser. A. 42, 9-18.

Rajput, B. (1972). Gaussian measures on L_p spaces, $1 \leq p \leq \infty$. J. Mult. Anal. 2, 382-403. MR 49 # 9896.

Rajput, B., Louie, D., Tortrat, A, (1981). Une loi zéro-un pour une class de mesures sur les groupes. Ann. Inst. Henri Poincaré 17, 331-335.

Rajput, B., Tortrat, A. (1980). Un théorème de probabilité zéro ou un dans un groupe mesurable ou topologique quelconque. C.R. Acad. Sc. Paris, t. 290, Sér. A., 251-254.

Shepp, L.A. (1966). Radon-Nikodym derivatives of Gaussian measures. Ann. Math. Statist. 37, 321-354. MR 32 # 8408.

Siebert, E. (1982). Absolute continuity, singularity, and supports of Gauss semigroups on a Lie group. Mh. Math. 93, 239-253.

Tortrat, A. (1980). Lois de zéro-un pour des probabilités semi-stables ou plus générales, dans un espace vectoriel ou un groupe (abêlien ou non). Publication du C.N.R.S., Colloque de Saint-Flour.

Tortrat, A. (1981). Loi stables dans un groupe. Ann. Inst. Henri Poincaré 17, 51-61. MR 82m : 60012.

Tortrat, A. (1982). Loi de zéro-un et lois semi-stables dans un group. In: Probability measures on groups. Proc. Oberwolfach. Lect. Notes Math. 928, 452-465.

Varberg, D.E. (1967). Equivalent Gaussian measures with a particularly simple Radon-Nikodym derivative. Ann. Math. Statist. 38, 1027-1030. MR 35 # 4981.

Zinn, J. (1974). Zero-one laws for non-Gaussian measures. Proc. Amer. Math. Soc. 44, 179-185. MR 49 # 9897.

Universität Dortmund
Abteilung Mathematik
Postfach 50 05 00
D-46 Dortmund West Germany

RANDOM WALKS ON HOMOGENEOUS SPACES

R. SCHOTT

The aim of this paper is to report the main results obtained during the last years about random walks on homogeneous spaces.
It's first part is devoted to the dichotomy theorem. In the second part a partial classification of the homogeneous spaces which have an invariant measure into recurrent and transient is given using the notions of growth and amenability. For several reasons that are explicited belove, the study of random processes on homogeneous spaces is more difficult than on groups and some open problems still exist in this area.

I. PRELIMINAIRES.

Let G be a locally compact group with countable basis (we write L.C.B.), (X_n) a sequence of independent random variables with the same law μ. We suppose that μ is adapted (i.e. the closed subgroup generated by the support of μ is equal to G).
$Y_n^g = g \cdot X_1 \dots X_n$ is the right random walk starting on g at time zero. A state $x \in G$ is said to be recurrent if for each neighbourhood V of x we have :

$$P\{\overline{\lim_{n\to+\infty}} (Y_n^g \in V)\} = 1 \quad .$$

A state which is not recurrent is said transient.

Loynes dichotomy theorem asserts that if G is L.C.B. and μ adapted then all the states of Y_n^g are recurrent or all the states are transient.

A group G is said to be recurrent if there exists a recurrent random walk of it.
Let m be a right Haar measure on G, a random walk is recurrent in the sense of Harris (H-recurrent) if $m(A) > 0$ implies $P\{\sum_{n=1}^{\infty} 1_A(Y_n^x) = +\infty\} = 1$ for all x.
Remember that a H-recurrent random walk is recurrent and that a recurrent random walk is H-recurrent if and only if μ is spread out (i.e. $\exists\, p_0 \in \mathbb{N}$ such that μ^{*p} is not singular with respect to a right Haar measure on G)

Growth of a group :

Let G be a L.C.B. group, compactly generated, V a compact neighbourhood of e which generates G, m a Haar measure.
If there exists $k \in \mathbb{N}$ such that : $0 < \lim_{n\to+\infty} \frac{m(V^n)}{n^k} < +\infty$, we can prove that k is unic, independent of m and V. We say that G has polynomial growth of degree k.
If $\lim_{n\to+\infty} [m(V^n)]^{1/n} > 1$ (resp. $= 1$) we can prove also that this result is independent of V and m and we say that G has exponential growth (resp. non exponential growth).

The following result proves partially a conjecture of H. Kesten :

Theorem I.1.- If G is a connected Lie group then G is recurrent if and only if G has a polynomial growth of degree at most two.

See [1] and [8] for the proof.
A partial result can also be given if G is a finitely generated group.

Theorem I.2.- Let G be a finitely generated solvable group then G is recurrent if an only if G containing a subgroup $G^* \subset G$ of finite index $[G\ ;\ G^*] < +\infty$ and such that $G^* \simeq \mathbb{Z}$ or $\mathbb{Z}^2$.

See [19] for the proof.

II. HOMOGENEOUS SPACES

Let G be a L.C.B. group, compactly generated, H a closed subgroup of G and π the canonical mapping $G \to M = {}_H\backslash G$, M is the right homogeneous space. $Z_n^{\pi(g)} = \pi(g \cdot X_1 \ldots X_n) = \pi(g)\, X_1 \ldots X_n$ is the right random walk on M associated to μ (or the induced random walk on M). The transition probability of this Markov Chain is : $P(x, A) = \varepsilon_x * \mu(A)$, $x \in M$, A is a Borel set of M . We know that Loynes dichotomy theorem is false for the induced random walks on M .

Conter examples can be given on the homogeneous spaces of the affine group :

$G = \mathbb{R}_+^* \ltimes \mathbb{R}$, $(a,b)\cdot(a',b') = (a+a', b+ab')$ $\forall (a,b) \in G$, $\forall (a',b') \in G$.

G operates on the left on the homogeneous space $M\ (\simeq \mathbb{R})$ by : $g \cdot x = ax + b$, $x \in M$ and $g = (a,b) \in G$.

It's easy to see that Loynes theorem is false for some measures on G , example : μ spread out, $\text{supp}\,\mu \in\ ]0, \frac{1}{2}] \times [-1,1]$

for $x = 0$ we have $U1_{[-2,2]}(0) = +\infty$ (U = potential)

if K is a compact such that $K \cap [-2,2] = \emptyset$: $U1_K(0) = 0$

because $Y_n^x = X_n X_{n-1} \ldots X_1 \cdot x + X_n X_{n-1} \ldots X_2 Z_1 + X_n X_{n-1} \ldots X_2 Z_2 + \ldots + X_n Z_{n-1} + Z_n$

the hypothesis on μ implies that : $|Y_n^x| \leqslant |x| \cdot 2^{-n} + 2^{-(n+1)} + \ldots + 2^{-2} + 1$ a.s.

$$|Y_n^x| \leqslant 2 + \frac{|x|}{2^n} \quad \text{a.s.} \quad \forall n \ .$$

Hovewer dichotomy holds if some other hypothesis are made on (G,H,μ) .

II.1.- Dichotomy theorem of H. Hennion and B. Roynette

Theorem II.1.1.- Let G , H as above, μ a measure which is adapted and spread out

If there exists a measure λ *on* M *which is relatively invariant under the* G-*action (i.e. there exists a continu character* χ *of* G *such that :* $\varepsilon_g * \lambda = \chi(g)\lambda$) *and if* λ *is an excessive measure for the induced random walk of law* μ *(i.e. :* $\mu * \lambda \leqslant \lambda$) *then.*

- *either all the states of* M *are transient and the potential of all compact is bounded*
- *either all the states of* M *are recurrent and* $Z_n^{\pi(g)}$ *is* H-*recurrent with respect to the measure* λ *(i.e. :* $\lambda(A) > 0 \Rightarrow P_\chi[\overline{\lim}\ \{Z_n^{\pi(g)} \in A\}] = 1$ *).*

A more shorter proof of this result was given by D. Revuz (see [14]) . He uses some results of the ergodic theory, in particular the Hopf decomposition.

<u>A sketch of Revuz' proof</u> :

Let $\mathbb{P}_{(x,\cdot)}$ be the transition probability of the random walk $Z_n^{\pi(g)}$. We know that $P(x,\cdot) = \varepsilon_x * \mu \quad (x \in M)$.

Let T be the positive contraction induced by P on $L^1(M, \lambda)$
$\forall f \in L^1(M, \lambda)$, $Tf = \frac{d}{d\lambda}((f\lambda)P)$ is the Radon-Nýkodim derivative of $f \lambda P$ with respect to λ where $\lambda P(A) = \int_M \lambda(dx)\, P(x, A)$.

<u>The Hopf decomposition</u> [14]

There exists a set C of M , unic up to equivalence such that :

$$\forall f \in L^1_+(M, \lambda) \quad : \quad \sum_0^\infty T^n f = 0 \text{ or } +\infty \text{ on } C$$

$$\sum_0^\infty T^n f < +\infty \text{ on } D = C^c$$

C is called the conservative part and D the dissipative part.

D may be viewed as a "transient" part while C may be viewed as a recurrent part.

Let T^* be the adjoint operator of T : T^* is the positive contraction on $L^\infty(M,\lambda)$ defined if $f \in L^\infty(M, \lambda)$ by the equivalence class of Pf in $L^\infty(M, \lambda)$.

We know that : 1) $T^* 1_D = 0 \quad \lambda$ a.s. on C
then $P\, 1_C = 1 \quad \lambda$ a.s. on C

2) the conservative part C of T (resp. D) is the same as for the contraction $f \to Pf$ on $L^1(M, \lambda)$ because λ is excessive.

<u>D. Revuz proves now that the contraction</u> $f \to Pf$ <u>is either conservative either dissipative and that if the contraction is conservative all the points of</u> M <u>are recurrent</u>.

Extensions of the dichotomy theorem were given by L. Elie [3] . She studies the decomposition of M into ergodic classes and transient sets and proves in particular :

Theorem II.1.2.- Let T_μ be the closed semi group generated by the support of μ.
If μ is adapted, spread out on G.
If $T_\mu^{-1} T_\mu$ operates transitively on M then :

- *all the states of M are transient*
- *or there exists an absorbing set F such that the random walk on M is recurrent in the sense of Harris with invariant measure m and such that $\overline{F}^c$ is transient. In addition, the invariant measure m is the restriction to F of a quasi-invariant measure (under the G-action) : m is the only measure on M such that : $\mu * m \leq m$ (up to a multiplicative constant).*

Remark II.1.3.- 1) the hypothesis on T_μ is verified if the bounded harmonic functions are constant because in this case : $T_\mu^{-1} T_\mu = G$

2) if $T_\mu^{-1} T_\mu$ operates transitively on M, P is ergodic.

L. Elie proves also that the dichotomy theorem II.1.1. is true under the following hypotheses :

i) μ is adapted and spread out

ii) there exists on M a quasi-invariant measure λ which is excessive for the induced random walk with law μ (i.e. : $\mu * \lambda \leq \lambda$) .

II.2.- A partial classification of the homogeneous spaces

Our goal is to prove a result similar to theorem I.1.

For this reason we have to introduce the notion of growth of a homogeneous space and to find which homogeneous spaces have exponential growth and which ones have polynomial growth and to calculate exactly the degree.

Definition II.2.1.- Let G be a L.C.B. group compactly generated, H a closed subgroup of G such that the homogeneous space $M = {}_H\backslash G$ has an invariant measure λ (i.e. : $\lambda(A \cdot g) = \lambda(A)$, $\forall g \in G$ and A Borel set of M).

Let $x \in M$ and V a compact neighbourhood of e (the neutral element of G) which generates G.

If there exists $k \in \mathbb{N}$ such that : $0 \leq \underline{\lim} \frac{\lambda(x \cdot V^n)}{n^k} \leq \overline{\lim} \frac{\lambda(x \cdot V^n)}{n^k} < +\infty$ then k is unic, independent of x, λ and V. We say that the homogeneous space M has polynomial growth of degree k.

If : $\overline{\lim}_{n\to+\infty} [\lambda(x \cdot V^n)]^{1/n} > 1$ (resp. $= 1$) wa can prove that this result is also independent of x, λ and V, we say that M has an exponential growth (resp. exponential).

Properties II.2.2. [7]

i) $G/_H$ and $_H\backslash G$ have the same growth.

ii) If H and H' are closed subgroup of G such that H is uniform in H' (i.e. : $H'/_H$ is compact) then $G/_H$ and $G/_{H'}$ have the same growth.

Remark II.2.3.- The following result do to Y. Guivarc'h [9] allow us to say that the homogeneous spaces of a group which has a polynomial growth have polynomial growth but the calculus of the degree is not possible.

Proposition II.2.4.- $\forall\ (p,q) \in \mathbb{N} \times \mathbb{N} : m(V^{p+q}) \geq \lambda(V^p \cdot O) \cdot m(W^q)$ *where* $W = V \cap H$, m *is a Haar measure on* G *and* λ *is an invariant measure on* M, V *is a compact neighbourhood of* e *in* V.

If we fixe q so : $m(W^q)$ is constant and $\lambda(V^p \cdot O) \leq C'\, m(V^{p+q})$.

Definition II.2.3.- (amenability). $G/_H$ *is G-amenable (we shall say amenable) if there exists a G-invariant mean on* $\mathcal{UCB}(G/_H)$ *the space of uniformly continuous and bounded functions on* $G/_H$ *(i.e. a mean* m *such that :* $\forall\ s \in G$, $m({}_s f) = m(f)$ *where* ${}_s f(\dot{x}) = f(s^{-1} \cdot \dot{x})$, $\forall\ \dot{x} \in G/_H$.

Further we shall use often the following properties (see [4] for the proof).

Properties II.2.4.-

1) A homogeneous space of an amenable group is amenable
2) If H is an amenable subgroup of G then $G/_H$ is amenable if and only if G is amenable.

Growth and amenability :

Consider the two examples.

Example 1 : $G = H_1$ the first Heisenberg group (i.e. : $H_1 = (\mathbb{R}^3, \cdot)$,

$(x,y,z) \cdot (x',y',z') = x + x',\ y + y',\ z + z' + \frac{1}{2}\,(xy' - yx'))$

and $H = \mathbb{R}$ a subgroup of dimension 1 which is not normal in G .

G and H are amenable then $H_1/_R$ is amenable.

An explicit calculus which we not reproduce here proves that $H_1/_{\mathbb{R}}$ has a polynomial growth of d° 3 .

Example 2 : $G = SL(2, \mathbb{R}) = \left\{\begin{pmatrix} a & b \\ c & d \end{pmatrix} ;\ ad - bc = 1\ ,\ (a,b,c,d) \in \mathbb{R}^4\right\}$.

Consider the Iwasawa decomposition : $SL(2, \mathbb{R}) = K.\ A.\ N.$

with $K = \left\{\begin{pmatrix} \cos\theta & \sin\theta \\ -\sin\theta & \cos\theta \end{pmatrix},\ \theta \in \mathbb{R}\right\}$ compact, $A = \left\{\begin{pmatrix} e^t & 0 \\ 0 & e^{-t} \end{pmatrix}\right\}$, $t \in \mathbb{R}$ abelian and

$N = \left\{\begin{pmatrix} 1 & u \\ 0 & 1 \end{pmatrix},\ u \in \mathbb{R}\right\}$ nilpotent.

Property II.2.4 (2) implies that $SL(2,\mathbb{R})/_N$ is not amenable.

We can prove that this homogeneous space has an exponential growth. These two examples suggest us a relation between growth and amenability. Before we give the result which we have proved we remember the following definition.

Definition II.2.5.- Let G be a simply connected Lie group. We say that G is of rigid type if all the proper values of the adjoint mapp ad_G are of absolute value 1.

Example : $G = \mathbb{R} \ltimes C$ with the multiplication $(t,z)\cdot(t',z') = (t+t',z+z'e^{2\pi it})$ is of rigid type.

Theorem II.2.6.-

i) If $G/_H$ is not amenable then $G/_H$ has an exponential growth.

ii) If $G/_H$ is amenable :

if G is simply connected, solvable and of rigid type then $G/_H$ has a polynomial growth and we can calculate explicitly the degree k .

See [6], [17], [18] for the proof.

A sketch of the proof :

i) We prove that if $G/_H$ has not exponential growth then $G/_H$ is amenable.

Let K be a compact of G , there exists a neighbourhood V of e , symmetric, relatively compact such that : $G = \bigcup_n V^n$ and $K \subset V$.

We proof now the following lemmas.

Lemma II.2.7.- If $G/_H$ has an exponential growth, there exists a sequence (n_i) in $\mathbb{N}$ such that : $\lim\limits_{n_i \to +\infty} \dfrac{\lambda(V^{n_i+1}\cdot 0)}{\lambda(V^{n_i}\cdot 0)} = 1$.

Lemma II.2.8.-

$$\forall\ x \in K \quad \frac{\lambda(x(V^{n_i}\cdot 0)\,\Delta\, V^{n_i}\cdot 0)}{\lambda(V^{n_i}\cdot 0)} \leqslant 2\left[\frac{\lambda(V^{n_i+1}\cdot 0) - \lambda(V^{n_i}\cdot 0)}{\lambda(V^{n_i}\cdot 0)}\right] \underset{n_i\to+\infty}{\longrightarrow} 0$$

and we use now the following caracterization of amenability.

Theorem II.2.9.- Let G be a L.C.D, H a closed subgroup, such that G/H has an invariant measure λ , G/H is amenable $\Longleftrightarrow \forall\ \varepsilon > 0$, ∀ K compact of G, ∃ U a Borel set *of* $G/_H$ *such that* : $\frac{\lambda(xU\Delta U)}{\lambda(U)} < \varepsilon \quad \forall x \in K$.

Remark II.2.10.- Let $\varepsilon > 0$, there exists $i \in \mathbb{N}$ such that $U_i = V^{n_i}.0$ is convenient.

So $G/_H$ is amenable.

We give now the main steps of the proof of (ii).

Step 1 : $G = N$ a simply connected nilpotent group.

$$g . 0 = (P_1(x_1,\ldots,x_n), P_2(x_1,\ldots,x_n), \ldots, P_{n-k}(x_1,\ldots,x_n)), \ (n,k) \in \mathbb{N}^2_{k<n}$$

(P_i) are polynomial.

We prove that $N/_H$ has polynomial growth of degree, $d = \sum_{i=1}^{n-k} \partial^o P_i$.

Step 2 : Let $G = K \ltimes N$ be a compact extension of a simply connected nilpotent group.

a) If $H \subset N$ is a closed subgroup of N we prove that

$K \ltimes N/_H$ and $N/_H$ has the same growth

b) If $H \subset K \ltimes N$ is a closed subgroup of $K \ltimes N$

H is uniform in a closed connected subgroup $\widetilde{H}$ of $K \ltimes N$ and $\widetilde{H} = K_o \ltimes N_o$ where K_o is compact , N_o is a closed subgroup of $\widetilde{H}$, $\widetilde{H}$ is the algebraic hull of H (i.e. the smallest algebraic group containing H) .

We prove that : $K \ltimes N/_H$, $K \ltimes N/_{\widetilde{H}}$, $K \ltimes N/_{N_o}$, $N/_{N_o}$ have the same growth.

Step 3 : If G is simply connected, solvable of rigide type.

We use the method of semi-simple splitting :

$\exists$ K compact, $\exists!$ N nilpotent group, simply connected such that:

$K \ltimes G = K \ltimes N$, N is the nilshadow of G

if $H \subset G$ closed in G :

$K \ltimes G/_H$, $G/_H$ have the same growth.

We apply now steps ① and ② .

Now we can give the classification for the random walks.

Result II.2.11.- μ is adapted, spread out, λ an invariant measure on $M = G/_H$

i) if M is not amenable then Γ is transient

ii) if M is amenable :

The results are partial but then allows us to conjecture the following

statement :
If G is simply connected of rigid type then $G/_H$ is recurrent if and only if $G/_H$ has a polynomial growth of degree at most 2.

The result i) was announced in [2], a more explicit proof can be found in [16] . Let now look at the current situation of this conjecture.

Case I : $G = N$ nilpotent simply connected.
We have completely solved the conjecture in this cas in [13] .

Let me remind here the main steps of the proof.

Step 1 : We prove that we can assume N nilpotent of class $r = 2$. For this we make a clever application of the following lemma :

Lemma II.2.12.- Let G be a Lie group, V a normal abelian subgroup of G and $G' = \frac{G}{V} \ltimes V$ (semi-direct product) where the action of $G/_V$ on V is given by : $\alpha_{\bar{g}}(v) = g \cdot vg^{-1}$ (this depends only of the class $\bar{g}$ of g) .
If H is a closed subgroup of G let $H' = \frac{HV}{H} \ltimes H \cap V$
If $G'/_{H'}$ is transient then $G/_H$ is transient.

Step 2 : If $\mathcal{H}$ is the Lie algebra of H and $\mathcal{L}$ such that $\mathcal{L} \oplus \mathcal{G}_2 = \mathcal{G}$ where $\mathcal{G}_2 = [\mathcal{G}, \mathcal{G}]$, we can suppose $\mathcal{H} \subset \mathcal{L}$.

Step 3 : It's sufficient to prove the result if $\dim \mathcal{G}_2 = 1$.

Step 4 : Let B be the antisymmetric bilinear form associated with $[\cdot,\cdot]$ in $\mathcal{G}_2$. If $\mathcal{C}$ is the kernel of B , we can suppose that $\mathcal{H}$ is included in $\mathcal{C}'$ where : $\mathcal{C}' \oplus \mathcal{C} = \mathcal{G}_2$.

Step 5 : For achieving the proof we have to examine the two cases :

1) $\dim \mathcal{L} = 2p > 3$
2) $\dim \mathcal{L} = 2p < 3$.

Case II : $G = K \ltimes N$ (compact extension of a simply connected nilpotent group).
We have also completely solved the conjecture in this case in [6].
The proof is very difficult :
in the first step we prove that we can assume : $H \subset N$ (we use again the results concerning the closed subgroups of $G = K \ltimes N$ and the fact that if H' is uniform in H : $G/_H$ and $G/_{H'}$ have the same kind.

Step 2 : If $H \subset N$ then

a) $N/_H$ is recurrent $\Rightarrow$ $K \ltimes N/_H$ is recurrent

b) $K \ltimes N/_H$ recurrent $\Rightarrow$ $\dim N - \dim H \leqslant 2$.

Step 3 : If $\dim N - \dim H = 2$ then $K \ltimes N/_H$ is transient.
To prove this we use the method of the barrier function.

Lemma II.2.13.- *Let G be a Lie group, H a closed subgroup of G, $M = G/_H$ and μ as before. Suppose that for some compact C, there exists a function $f : M \to \mathbb{R}$ such that :*

i) $0 \leqslant f \leqslant 1$, $\lim_{x \to +\infty} f(x) = 1$, $\sup_{x \in C} f(x) = d < 1$

ii) $E[f(x \cdot X)] \geqslant f(x)$ *for* $x \notin C$ *(where X is a random variable with law μ .*

Then the random walk of law μ is transient.

In the last part of this paper, we shall examine an example where H is a discrete subgroup of G .

II.3.- Homogeneous spaces of $SL(d, \mathbb{R})$

Let H be a discrete subgroup of $SL(d, \mathbb{R})$, the following result gives only an answer if $d > 3$ but the problem still open if $d = 2$.

Theorem II.3.1.- *Let μ be a probability measure on $SL(d, \mathbb{R})$ which is adapted and spread out. If $d > 3$ and if H is a discrete subgroup of $SL(d, \mathbb{R})$ then $SL(d, \mathbb{R})_{/H}$ is recurrent $\Leftrightarrow$ $SL(d, \mathbb{R})_{/H}$ bas an invariant measure of finite mass.*

In the proof of this theorem, we use the fact that $SL(d, \mathbb{R})$ has the KAZHDAN property (see [4]) :

If G is a locally compact group, let $\widetilde{G}$ be the set of classes of unitary, continuous representations of G in Hilbert spaces $\mathcal{H}_\pi$ and $\hat{G}$ the set of classes of representations which are irreducible.

If $\pi \in \widetilde{G}$ and if $\nu \in M_1(G)$ is a bounded Radon measure on G let
$\pi(\nu) = \int_G \pi(x)\, d\nu(x)$, $< \pi(\nu)\, \xi \mid \eta > \; = \int_G < \pi(x)\, \xi \mid \eta > d\nu(x)$ for $\xi, \eta \in \mathcal{H}_\pi$.

The functions of positive type associated to π are defined by :
$x \overset{h}{\longmapsto} < \pi(x)\, \xi, \xi >$, $\xi \in \mathcal{H}_\pi$ (i.e. h continuous and for all $x_1, x_2, \dots, x_n \in G$ and $c_1, c_2, \dots, c_n \in \mathbb{C}$: $\sum_{i,j} c_i \overline{c_j}\, h(x_i x_j^{-1}) > 0)$.

Definition II.3.2.- Let $w \in \tilde{G}$ *and* $\mathcal{S} \subset \tilde{G}$ *, we say that* w *is weakly contained in* $\mathcal{S}$ *if all function of positive type associated to* w *is the uniform limit on all compact of* G *of finite summs of functions of positive type associated to representations belonging to* $\mathcal{S}$ *.*

Definition II.3.3.- Let $\mathcal{S} \subset \hat{G}$ *, the closure of* $\mathcal{S}$ *is* $\overline{\mathcal{S}} = \{w \in \hat{G}$ *,* w *weakly contained in* $\mathcal{S}\}$ *. We say that* $\mathcal{S}$ *is closed in* $\hat{G}$ *if and only if* $\mathcal{S} = \overline{\mathcal{S}}$ *.*

This notion of closure defines a topology on $\hat{G}$ called the Fell topology.

Definition II.3.4.- Let G *be a locally compact separable group,* G *has the KAZHDAN property (K) if* G *has the following (equivalent) properties :*

i) $\{i_G\}$ *is an open set in* $\hat{G}$ *, (*i_G *is the trivial unitary representation of dimension 1 of* G*)*

ii) if G *is weakly contained in a unitary representation* π *of* G *, then* i_G *is strongly contained in* π *(i.e.* $\exists\, \xi \in \mathcal{H}_\pi$ *,* $\xi \neq 0$ *such that* $\forall\, x \in G$ *,* $\pi(x)\xi = \xi$*) .*

Examples II.3.5.- $SL(d, \mathbb{R})$ has the property (K) if $d \geqslant 3$

$SL(2, \mathbb{R})$ has not this property .

All compact group has the property (K) .

For groups which have the property (K) we can prove the following result :

Theorem II.3.6.- Let G *be a group which has the property (K) and* H *a closed unimodular subgroup of* G *. Suppose that the homogeneous space* G/H *is amenable. Then* G/H *is of finite volume for the invariant measure.*

(see [4] p. 59).

Now we can achieve the proof of the theorem II.3.1.
First we can remark that under the hypotheses which we have made $SL(d, \mathbb{R})_{/H}$ is dichotomic for the induced random walks (theorem II.1.1.) .

If $SL(d, \mathbb{R})_{/H}$ has an invariant measure of finite mass, it's easy ot prove that $SL(d, \mathbb{R})_{/H}$ is recurrent.

If $SL(d, \mathbb{R})_{/H}$ is recurrent we know by the result II.2.11 that this space is amenable. Or $SL(d, \mathbb{R})$ has the property (K) for $d \geqslant 3$, the theorem II.3.6. permits now to conclude.

II.4.- Some open problems

The study of random processes on homogeneous spaces has future because a lot of problems still open.

Examples : 1) if G is a solvable group and H a closed subgroup of G , what happen for the random walks on G/H ?

2) if $\mathbb{R}$ or $GL(d)$ operates transitively on $\mathbb{R}^d$, what can we say about the homogeneous space $\mathbb{R}^d$?

BIBLIOGRAPHY

[1] P. BALDI, N. LOHOUE, J. PEYRIERE. Sur la classification des groupes récurrents (Note aux C.R.A.S. t. 285, Série A, p. 1103, 1977)

[2] Y. DERRIENNIC et Y. GUIVARC'H. Théorème de renouvellement pour les groupes non moyennables, (Note aux C.R.A.S. Paris t. 281 (1975), p. 985-988)

[3] L. ELIE. Sur le théorème de dichotomie pour les marches aléatoires sur les espaces homogènes (L. N. n° 928 Proceeding OBERWOLFACH 1981)

[4] P. EYMARD. Moyennes invariantes et représentations unitaires (L. N. n° 300)

[5] L. GALLARDO et V. RIES. Marches aléatoires sur les espaces homogènes du groupe des déplacements de $\mathbb{R}^d$ (ASTERISQUE n° 74, 1980, p. 123-138)

[6] L. GALLARDO et R. SCHOTT. Marches aléatoires sur les espaces homogènes de certains groupes de type rigide (ASTERISQUE n° 74, 1980, p. 149-170)

[7] L. GALLARDO et R. SCHOTT. Un théorème de structure pour les sous-groupes fermés connexes des groupes extensions compactes de groupes nilpotents simplement connexes (L. N. n° 739, Analyse harmonique sur les groupes de Lie, p. 283-293)

[8] Y. GUIVARC'H, M. KEANE, B. ROYNETTE. Marches aléatoires sur les groupes de Lie (L. N. n° 624, 1977)

[9] Y. GUIVARC'H. Croissance polynomiale et période des fonctions harmoniques (Bull. Soc. Math. de France, 101, 1973, p. 333-379)

[10] H. HENNION. Marches aléatoires sur les espaces homogènes des groupes nilpotents à génération finie (Zeitschrift für Warsch 34, 1976, p. 245-267)

[11] H. HENNION et B. ROYNETTE. Un théorème de dichotomie pour les marches aléatoires sur les espaces homogènes (ASTERISQUE n° 74, 1980, p. 99-122)

[12] A. HUARD. Récurrence des marches aléatoires sur les espaces homogènes récurrents du groupe des déplacements de $\mathbb{R}^d$ (ASTERISQUE n° 74, 1980, p. 139-148)

[13] D. PREVOT et R. SCHOTT. Marches aléatoires sur les espaces homogènes des groupes nilpotents simplement connexes (L. N. n° 739, p. 404-427)

[14] D. REVUZ. Sur le théorème de dichotomie de HENNION-ROYNETTE (Annales de l'INSTITUT E. CARTAN n° 7, 1983, p. 143-146)

[15] R. SCHOTT. Marches aléatoires sur les groupes de Lie nilpotents à génération compacte (ANNALES SCIENTIFIQUES, Université de Clermont-Ferrand 2, n° 69, 1980, p. 237-242)

[16] R. SCHOTT. Irrfahrten auf nicht mittelbaren homogenen Räumen (Publ. Mathematisches INSTITUT SALZBURG, Arbeitsbericht 1-2, 1981, Seite 63-76)

[17] R. SCHOTT. Croissance et moyennabilité des espaces homogènes (Prépublication INSTITUT E. CARTAN, NANCY, 1982)

[18] R. SCHOTT. Quelques remarques à propos des marches aléatoires sur les espaces homogènes du groupe de MAUTNER (Prépublication INSTITUT E. CARTAN, NANCY, 1982)

[19] N. Th. VAROPOULOS. Random walks on soluble groups (preprint).

R. SCHOTT
E.R.A. n° 839 du C.N.R.S.
U.E.R. Sciences Mathématiques
Université de NANCY I
B.P. 239
54506 - VANDOEUVRE-les-NANCY (France)

The Lévy-Khinchin Formula and Order Structure

by Martin E. Walter

Introduction. In this article we discuss a natural generalization of the Lévy-Khinchin formula on $\mathbb{R}^n$, n-dimensional Euclidean space, to the "dual" of a (not necessarily commutative) locally compact group. We do this by considering the continuous unitary representations of the group, or equivalently, the closely related object P(G), the semigroup of the continuous positive definite functions on G, which is in the case of abelian G the inverse Fourier transform of the convolution semigroup of positive, finite, regular Borel measure on the dual group.

Let $P(G)_1$ be the elements of P(G) that are 1 at the identity of G. We study "derivatives" at the identity $\mathbb{1}$ of $P(G)_1$, namely $\mathbb{1}(g) = 1$ for all g in G. Geometrically a "derivative" at $\mathbb{1}$ will be called a <u>semitangent vector</u> at $\mathbb{1}$; the collection of all such vectors is denoted by $N_0(G)$. Algebraically these semitangents have a characterization as the class of normalized, i.e. 0 at the identity of G, continuous functions of negative type on G. The following five examples on $\mathbb{R}^n$ which we give below all generalize to the context of a locally compact group G. Also, even in this general context the five examples give essentially equivalent definitions of what we are calling semitangents at $\mathbb{1}$ in $P(G)_1$.

(1) $D = d^2/dx^2$ on $\mathbb{R}^1$; $D(f\bar{f}) \geqq (D\bar{f})f + \bar{f}(Df)$, pointwise inequality of functions. This is a simple example of a competely dissipative operator from physics. The operator D is a semitangent vector concretely realized by what we shall call a <u>semiderivation.</u>

(2) The Lévy-Khinchin formula on $\mathbb{R}^n$:

$$-\psi(y) = c + i\ell(y) + q(y)$$

$$+ \int_{\mathbb{R}^n - (0)} \left[1 - \exp(-i(x|y)) - \frac{i(x|y)}{1+\|x\|^2}\right]\left[\frac{1+\|x\|^2}{\|x\|^2}\right] d\mu(x)$$

where $x,y \in \mathbb{R}^n$, $c \geqq 0$, ℓ is a continuous linear form, q is a continuous, nonnegative quadratic form and μ is a non-negative bounded measure on $\mathbb{R}^n - \{0\}$ such that the above integral converges.

(3) A function ψ on $\mathbb{R}^1$ that satisfies $\psi(e) \leqq 0$, $\overline{\psi(-x)} = \psi(x)$, and

$$\int_{\mathbb{R}^1} \psi(x) \left(\frac{d\phi}{dx}\right)^{\#} * \frac{d\phi}{dx}(x)dx \geqq 0 \quad \text{for all} \quad \phi \in C_c^\infty(\mathbb{R}^1) \quad ,$$

where $\phi^{\#}(x) = \overline{\phi(-x)}$, the bar denoting the complex conjugate, $*$ denoting convolution, and ϕ an infinitely differentiable function with compact support.

(4) $H^1(G,H(\pi)) = Z^1(G,H(\pi))/B^1(G,H(\pi)))$, the first cohomology group of continuous, unitary representation π of G .

(5) The "screw functions" of J. von Neumann and I.J. Schoenberg.

Though the rigorous proofs and a complete understanding of the "theory of semiderivatives" seems to require an acquaintance with C^*-algebras, we will do our best in this article to explain the "theory of semiderivatives" in a way that is intelligible to mathematicians not well acquainted with C^*-algebras. Complete proofs can be found in [11] .

1. A dual for locally compact group G . Recall that if G is a locally compact topological group that a function $p : G \to \mathbb{C}$, where $\mathbb{C}$ is

the set of complex numbers, is positive definite or of positive type if $\sum_{i,j=1}^{n} \lambda_i \bar{\lambda}_j p(x_j^{-1} x_i) \geqq 0$ for any choice of $x_1, \ldots, x_n$ in G, any $\lambda_1, \lambda_2, \ldots, \lambda_n$ in $\mathbb{C}$ and any natural number n. Let $P(G)_1$ be the set of continuous positive definite functions on G which are 1 at the identity e of G. There are "sufficiently many" functions in $P(G)_1$, cf. 13.66 [5].

Recall that $P(G)_1$ is a semigroup under pointwise multiplication since the pointwise product of two positive definite functions is positive definite, cf. p. 683 [7], vol. II. Though $P(G)_1$ needs some additional structure, e.g. geometric structure, to be a complete dual for G in the sense we usually use the term dual, cf. [10] we will nevertheless refer to $P(G)_1$ as the dual of G in this paper.

By Bochner's theorem, cf. [9], if G is abelian $P(G)_1$ is none other than the set of inverse Fourier transforms of the probability measures on $\hat{G}$, the dual group of G. Thus the Lévy-Khinchin formula and related concepts can be developed in an almost classical fashion as it is done on $\mathbb{R}^n$, see [2] for this theory for abelian G.

In the general case of nonabelian G we will look at $P(G)_1$ directly and try to find infinitesimal generators of one-parameter semigroups $\{p_t\}_{t \geq 0}$ in $P(G)_1$ which satisfy $p_0 = \mathbb{1}$. These infinitesimal generators or "semitangents" are defined in the classical spirit of elementary calculus, to wit:

Definition. A semitangent vector at $\mathbb{1}$ to $P(G)_1$ is any continuous, complex-valued function ψ on G satisfying $\psi(g) = \lim_{n \to \infty} n(p_n(g) - 1)$ for each G

and some $\{p_n\} \subset P(G)_1$, $\{n\}$ the natural numbers.

Remark. To see that this is really a classical limit of a difference quotient, we write $n(p_n(g)-1) = \dfrac{p_n(g)-1}{\frac{1}{n} - 0}$.

2. Semitangents and functions of negative type. Now we will need the following.

Definition. A function $\phi : G \to \mathbb{C}$ is said to be <u>negative definite</u> or <u>negative type</u> if

(i) $\phi(g^{-1}) = \overline{\phi(g)}$ for all $g \in G$, the over bar denoting complex conjugation; and

(ii) $\sum_{i,j=1}^{n} \{\phi(g_j^{-1}g_i) - \phi(g_j^{-1}) - \phi(g_i)\}\lambda_i\overline{\lambda}_j \geq 0$ for any $g_1, g_2, \ldots g_n$ in G , any $\lambda_1, \lambda_2, \ldots, \lambda_n$ in $\mathbb{C}$ and any natural number n . We denote the collection of <u>continuous</u> functions of negative type on G which are 0 at the identity of G by $N_0(G)$.

We have the following.

Theorem. Let ϕ be a continuous, complex-valued function on G . Then ϕ is a semitangent at $\mathbb{1}$ to $P(G)_1$ if and only if $\phi \in N_0(G)$, i.e., ϕ is of <u>negative type</u> and 0 at the identity of G .

Idea of the proof. If we denote by T_p the operator of "pointwise multiplication by p " then $T_p f = pf$ if $f \in L^1(G)$, the convolution algebra of functions absolutely integrable with respect to left Haar measure, $d\lambda$, on G . Also if $\sum_{i=1}^{n} \lambda_i \delta g_i$ is a linear combination of point-masses where $g_i \in G$, $\lambda_i \in \mathbb{C}$ $i = 1,2,\ldots,n$, then

$T_p \sum_{i=1}^{n} \lambda_i \delta g_i = \sum_{i=1}^{n} \lambda_i p(g_i) \delta g_i$. Both of the above are examples of finite regular Borel measures on G , the set of all such measures is denoted by $M^1(G)$. Note that $f \in L^1(G)$ when thought of as an element of $M^1(G)$ is really $f(\bullet)d\lambda(\bullet)$.

There is a natural involution, $*$, on $M^1(G)$ which in some sense "extends" the inverse operation on G , namely, $\langle\mu^*, f\rangle = \int f(x)d\mu^*(x) = \overline{\int \overline{f(x)}d\mu(x^{-1})} = \overline{\int \overline{f^b(x)d\mu(x)}} = \overline{\langle f^b(x), \mu\rangle}$ where $f^b(x) = \overline{f(x^{-1})}$ for any bounded continuous function f on G .

There is a notion of positivity that accompanies this involution, namely, a measure of the form $\mu^* * \mu$ is said to be (globally) positive as opposed to the notion of pointwise positivity, i.e. "local positivity." Recall that if $\mu, \nu \in M^1(G)$ then the convolution $\mu * \nu$ is defined by the formula

$$\langle\mu*\nu, f\rangle = \int_G \int_G f(xy)d\mu(x)d\nu(y)$$

for f bounded and continuous. Note that $\mu \in M^1(G)$ is locally positive if and only if $\langle\mu, \overline{f}f\rangle = \int |f(x)|^2 d\mu(x) \geq 0$ for every bounded, continuous function f on G . A $\mu \in M^1(G)$ is globally positive if and only if $\langle\mu, f^b * f\rangle = 0$ for all compactly supported continuous f on G . Recall $f*h(x) = \int f(y)h(y^{-1}x)d\lambda(y)$ for functions f,h , if the integral makes sense. We thus see that local positivity of μ is "dual" in some sense to the involution $\overline{}$, complex-conjugation, and pointwise multiplication of functions. Global positivity is "dual" in some sense to the involution ^{b} and convolution. Now it is a fact, proved in the theory of operator algebras, cf. p. 503 [11] , that T_p preserves global positivity of measures and in fact that an inequality holds:

Kadison-Cauchy-Schwarz inequalty

$$T_p(\mu^* * \mu) \geq (T_p\mu)^* * (T_p\mu) \quad ,$$

where the $\geq$ here means that $T_p(\mu^* * \mu) - (T_p\mu)^* * (T_p\mu)$ is globally positive.

Using the above inequality we can easily prove one direction of the theorem, viz., if $\psi(g) = \lim_{n\to\infty} n(p_n(g)-1)$ for all $g \in G$, then considering the global positivity of $\lim_{n\to\infty} n\{Tp_n(\mu^* * \mu) - (Tp_n\mu)^* * (Tp_n\mu)\}$, we get that $\sum_{i,j=1}^{n} \lambda_i \bar{\lambda}_j \{\psi(g_j^{-1} g_i) - \psi(g_j^{-1}) - \psi(g_i)\} \geq 0$ for $g_1, g_2, \ldots, g_n \in G$. Note: take μ above to be $\sum_{i=1}^{n} \lambda_i g_i$, cf. p. 508 [11]. It is trivial to verify that $\psi(e) = 0$, e = identity of G; and $\psi(g^{-1}) = \overline{\psi(g)}$ for all $g \in G$. Conversely, if $\psi \in N_0(G)$, then $\{e^{t\psi}\}_{t\geq 0}$ is a one-parameter semigroup of continuous functions of positive type. The proof here is essentially the same as that of Schoenberg's theorem in the abelian case, cf., [2], so we omit it here. This ends our discussion of the proof of the theorem.

We would like to point out that the above characterization of semitangents as negative definite functions is based on the use of the order structures, in this case, the order induced by global positivity. Though the use of such arguments using order are common in operator algebras they have not been as common in commutative harmonic analysis or probability theory. In particular the notion of complete positivity, until recently, was not explicitly used at all in these "non-operator algebraic" areas, cf. [11] .

3. Semiderivations. We will now note that if $\psi \in N_0(G)$ then ψ defines a <u>semiderivation</u> ∂_ψ on $M^1(G)$ in the sense that for $\mu \in M^1(G)$

$$\partial_\psi(\mu^* * \mu) \geqq (\partial_\psi \mu^*) * \mu + \mu^* * (\partial_\psi \mu)$$

where ∂_ψ is to be interpreted as pointwise multliplication by ψ and the $\geqq$ means that the left hand side minus the right hand side of the inequality is globally positive. For a fuller discussion of this see [11] .

Just as all functions are not differentiable all $\mu \in M^1(G)$ are not necessarily in the domain of ∂_ψ .

4. The Lévy-Khinchin formula and the orderd cone $N_0(G)$. We now comment all too briefly on the Lévy-Khinchin formula. The Lévy-Khinchin formula on $\mathbb{R}^n$ gives a formula for an arbitrary $\psi \in N_0(\mathbb{R}^n)$. This formula may be interpreted in terms of global order. In particular, $N_0(G)$ is a convex cone. In this convex cone of semitangents is a linear subspace of "tangents", i.e. $\psi \in N_0(G)$ is a tangent if $\sum_{i,j=1}^{n} \lambda_i \bar{\lambda}_j \{\psi(g_j^{-1} g_i) - \psi(g_j^{-1}) - \psi(g_i)\} = 0$ for all $\lambda_1, \ldots, \lambda_n \in \mathbb{C}$, $g_1, \ldots, g_n \in G$, n a natural number, i.e., $\psi : G \to C$ is a homomorphism. In fact $\psi : G \to i\mathbb{R} \subset \mathbb{C}$ and $\psi(gh) = \psi(g) + \psi(h)$ for all $g,h \in G$. We can look at the cone $N_0(G)$ modulo the subspace of tangents. This subspace of tangents is thus identified with the "zero" element of $N_0(G)$.

If we then apply the Choquet theory to this cone we can expect to write an arbitrary element in $N_0(G)$ as an integral over extreme rays in this cone. This is precisely what the Lévy-Khinchin formula for $\mathbb{R}^n$ does, since for each $x \in \mathbb{R}^n$ the function $e^{-i(x|y)} - 1$ is an extreme negative definite function. More troublesome is the "correction term" $-\frac{i(x|y)}{1+\|x\|^2}$ which is a tangent for each $x \in \mathbb{R}^n$, and thus it is "essentially zero" from this order theoretic point of view. For the <u>explicit</u> formula, though, this correction term is quite necessary in order that the integral converge. This "zero correction" term is well motivated from the probabilistic point of view. Some alternative motivation must be found for a particular explicitly given group G in order to find the explicit formula for this "zero correction" term. See [6] , [8] for some further discussion of this point for a few explicit groups. There is much work remaining to be done regarding the <u>explicit</u> determination of the Lévy-Khinchin formula of a given locally compact group.

5. Cohomology. Though we did not go into any detail regarding the role unitary representations play in this theory, let us remind the reader that $P(G)$ is just the set of "diagonal coefficients" of the continuous unitary representations of G, i.e. $P(G) = \{g \in G \mapsto (\pi(g)\xi|\xi) : \pi$ is a continuous unitary representation of G on a Hilbert space H_π and $\xi \in H_\pi\}$. Conversely, given $P(G)$ the collection of continuous unitary representations of G can be constructed by a process which involves something known as the G.N.S. (Gelfand-Naimark-Segal) construction.

It turns out that an analogous relationship holds for $N_0(G)$ and continuous unitary representations of G with a cocycle.

Let us first see the natural unitary representation with cocycle that appears, given $\psi \in N_0(G)$. Let $\mu, \nu \in M_c^1(G)$, where $M_c^1(G)$ are the elements of $M^1(G)$ with compact support. We can then define a sesquilinear form on $M^1(G)$ as follows:

$$(\mu|\nu)_\psi = \int[\psi(h^{-1}g)-\overline{\psi(h)}-\psi(g)]d\mu(g)d\overline{\nu}(h)$$

Let $K_\psi = \{\mu \in M^1(G) : \|\mu\|_\psi^2 \equiv (\mu|\mu)_\psi = 0\}$, and define $\Pi_\psi(g)\mu = \delta_g * \mu - \mu(G)\delta_g$, δ_g the unit point mass at $g \in G$ and $\mu(G) = \int_G d\mu(g)$. It is routine to check that $(\Pi_\psi(g)\mu|\Pi_\psi(g)\nu)_\psi = (\mu|\nu)_\psi$ for all $\mu,\nu \in M^1(G)$. Also K_ψ is a subspace invariant under Π_ψ. If H_ψ is defined to be the completion of $M_c^1(G)/K_\psi$ with respect to the inner product $(\cdot|\cdot)_\psi$, then a continuous unitary representation is defined by $\pi_\psi(g)(\mu+K_\psi) = \delta_g * \mu - \mu(G)\delta_g + K_\psi$. The map $c_\psi : g \in G \mapsto c_\psi(g) = \delta_g + K_\psi \in H_\psi$ (definition of c_ψ) has the cocycle property

$$c_\psi(gh) = c_\psi(g) + \pi_\psi(g)c_\psi(h) \quad .$$

We also note that $\|c_\psi(g)-c_\psi(h)\|_\psi^2 = -2\mathrm{Re}\psi(h^{-1}g)$. Thus we have the following.

Proposition. Given $\psi \in N_0(G)$, there exists a pair (π_ψ, c_ψ), where c_ψ is a continuous Hilbert-space valued cocycle for continuous unitary representation π_ψ of G, i.e.

$$c_\psi(gh) \equiv c_\psi(g) + \pi_\psi(g)c_\psi(h) \quad ,$$

for $g,h \in G$. Conversely we have the following.

Proposition. Given a continuous complex Hilbert space valued 1-cocycle c for continuous unitary representation π of group G on Hilbert space H_π, there exists a continuous function ψ of negative type, 0 at the identity, such that $c = c_\phi$ and $\pi = \pi_\phi$, where c_ϕ and π_ϕ are defined as above in this section. The function ϕ is defined either on G or $\tilde{G}$, an extension of G.

Remark. The group $\tilde{G}$ may always be taken to be a central extension of G by $\mathbb{R}$, the additive reals, with respect to the trivial action of G on $\mathbb{R}$. In fact $\tilde{G}$ may be taken to be the "multiplier", i.e., factor set or 2-cocycle, $\mathrm{Im}(c(h), c c g^{-1}))$, for $g,h \in G$, cf. p. 528 [11].

6. The "screw functions" of von Neumann and Schoenberg. Having established the explicit relationship between $N_0(G)$ and cocycles for continuous unitary representations of G it is easy to establish the connection between $N_0(G)$ and the so-called screw functions.

Von Neumann and Schoenberg wanted to know all semimetrics ϕ on $\mathbb{R}$ the real line, such that the (semi) metric space $(\mathbb{R}, \rho)$ could be imbedded isometrically in a (real) Hilbert space. In particular, they wanted to know all screw functions of F. A screw function F is defined by the property that $F(x-y) = \phi(x,y)$ and (R,ϕ) is embeddable in a (real) Hilbert space. The key to their eventual complete understanding of the problem was the following calculation. Note first that $x \in \mathbb{R} \mapsto -\sin^2(\omega x) = \frac{\cos\omega 2x - 1}{2}$ is in $N_0(\mathbb{R})$ since $x \mapsto \cos(\omega 2x) = \mathrm{Re}\, e^{i\omega 2x}$ is in $P(\mathbb{R})_1$. Then $-(\sin^2 \omega x)^{1/2} = -|\sin\omega x|$ is in $N_0(\mathbb{R})$ since if $\psi \in N_0(G)$, then $-(-\psi)^\alpha \in N_0(G)$ for $\alpha \in (0,1)$. This followed from the beautiful formula established by von Neumann and Schoenberg

$$x^\alpha = \frac{\alpha}{\Gamma(1-\alpha)} \int_0^\infty (1-e^{-xs}) s^{-\alpha-1}\, ds$$

for $x > 0$, $\alpha \in (0,1)$, Γ = the gamma-function. Von Neumann and

Schoenberg went on to establish a general formula for screw functions whi[ch] was essentially the formula of Lévy and Khinchin. The reason for this is now clear by the following.

Proposition. A necessary and sufficient condition for F to be a screw function on locally compact group G is that $-F^2 \in N_0(G)$.

Remark. The embedding that is associated with F is just the cocycle associated with $-F^2$, cf. p. 531 [11] .

6. Concluding remarks and applications. The characterization of $N_0(G)$ using differential operators, cf. condition (3) in the introduction, is discussed in [11] . We now go to our last topic.

A most useful characterization of property (T) groups of Kazhdan follows from our work. Recall that a locally compact group has property (T) if the function $\mathbb{1}$ is an isolated point in the space of continuous unitary representations of G , cf. [11] , [1] . We have the following

Theorem. Let G be a locally compact, G-compact group. The following are equivalent:

(1) G has property (T) ;

(2) Every $\phi \in N_0(G)$ is bounded as a function on G ;

(3) Every semiderivation ∂_ϕ , induced on $C^*(G)$ by a $\phi \in N_0(G)$ is bounded as an operator on $C^*(G)$;

(4) $H^1(G,H(\pi)) = 0$ for all continuous unitary representations π of [G]

(For a discussion of $H^1(G,H(\pi))$ see §3 [11] .)

Remark. Since $\{\mathbb{1}\}$ is an isolated point in the "dual of unitary representations" of G if G has property (T) it seems intuitively clear that the only type of differentiation possible at $\mathbb{1}$ in $P(G)_1$ should be of a trivial type. In the sense that bounded (semi) derivations are trivial and bounded semitangent functions are also trivial, we can expect that the above theorem is true. In fact just these intuitive considerations led us to conjecture the above result and prove it. For a complete simplified proof see [1] and [11] .

Applications of these ideas which are of a recent nature include an affirmative solution of the Delorme conjecture by E. Larsen, [8] . Joachim Cuntz relates K-theory to these ideas, cf. [4] . Watatani used these ideas in an application to arithmetic groups to obtain a result of Margolis and Tits, cf. [12] ; and Choda, [3] , has found an application to von Neumann algebra theory.

References

1. C.A. Akemann and M.E. Walter, Unbounded negative definite functions, Canadian Journal of Mathematics, 33 (1981), 862-871.

2. C. Berg and G. Forst, Potential Theory on Locally Compact Abelian Groups, Springer Verlag, New York 1975.

3. M. Choda, Group factors of the Haagerup type, preprint.

4. J. Cuntz, K-theoretic amenability for discrete groups, preprint.

5. J. Dixmier, Les C^*-algebres et leures représentations, Gauthier-Villars, Paris, 1965.

6. J. Erven and B. Falkowski, Low order cohomology and applications, Lecture Notes in Math., vol. 877, Springer-Verlag, 1981.

7. E. Hewitt and J. Ross, Abstract Harmonic Analysis, volumes I and II, Springer-Verlag, New York, 1970.

8. E. Larsen, Extreme Negative definite functions, Ph.D. thesis, University of Colorado, 1982.

9. W. Rudin, Fourier Analysis in Groups, Interscience, New York, 1962.

10. M.E. Walter, Duality theory for nonabelian locally compact groups, Symposia Mathematica, XXII, 1977, 47-59.

11. M.E. Walter, Differentiation on the dual of a group: an introduction, Rocky Mountain Journal of Mathematics, vol. 12, No. 3, Summer 1982.

12. Y. Watatani, Property T of Kazhdan implies property FA of Serre, Math. Japonica, No. 1, (1982), 97-103.

Martin E. Walter
Department of Mathematics
University of Colorado
Boulder Co 80309
USA